GW00692288

OXFORD REVIEWS OF REPRODUCTIVE BIOLOGY

Volume 9
1987

OXFORD REVIEWS OF REPRODUCTIVE BIOLOGY
EDITORIAL BOARD

Dr A. McLaren (chairman), MRC Mammalian Development Unit, University College, London NW1 2HE, UK.

Dr J. R. Clarke (editor), Department of Zoology, University of Oxford, Oxford OX1 3PS, UK.

Professor D. T. Baird, Department of Obstetrics and Gynaecology, University of Edinburgh, Centre for Reproductive Biology, 37 Chalmers Street, Edinburgh EH3 9EW, UK.

Professor A. R. Bellvé, Department of Anatomy and Cell Biology, and Center of Cellular and Molecular Urology, Columbia Presbyterian Medical Center, New York, NY 10032, USA.

Professor I. D. Cooke, Department of Obstetrics and Gynaecology, University of Sheffield S3 7RE, UK.

Professor C. A. Finn, Department of Veterinary Physiology and Pharmacology, University of Liverpool, Liverpool L69 3BX, UK.

Professor W. Hansel, Department of Physiology, New York State College of Veterinary Medicine, Cornell University, 816 Veterinary Research Tower, Ithaca, N.Y. 14853–6401, USA.

Dr R. B. Heap, AFRC Institute of Animal Physiology, Babraham, Cambridge CB2 4AT, UK.

Miss Anne M. Jequier, Department of Obstetrics and Gynaecology, University Hospital, Queen's Medical Centre, Nottingham NG7 2UH, UK.

Professor G. E. Lamming, Department of Physiology and Environmental Science, University of Nottingham LE12 5RD, UK.

Dr S. R. Milligan, Department of Physiology, King's College London, Strand, London WC2R 2LS, UK.

Professor R. V. Short, Department of Physiology, Monash University, Clayton, Victoria 3168, Australia.

OXFORD REVIEWS OF REPRODUCTIVE BIOLOGY

EDITED BY
J. R. CLARKE

Volume 9
1987

CLARENDON PRESS · OXFORD
1987

Oxford University Press, Walton Street, Oxford OX2 6DP
Oxford New York Toronto
Delhi Bombay Calcutta Madras Karachi
Petaling Jaya Singapore Hong Kong Tokyo
Nairobi Dar es Salaam Cape Town
Melbourne Auckland

and associated companies in
Beirut Berlin Ibadan Nicosia

Oxford is a trade mark of Oxford University Press

Published in the United States
by Oxford University Press, New York

© Oxford University Press, 1987

ISBN 0 19 857645-5

All rights reserved. No part of this publication may be reproduced,
stored in a retrieval system, or transmitted, in any form or by any means,
electronic, mechanical, photocopying, recording, or otherwise, without
the prior permission of Oxford University Press

This book is sold subject to the condition that it shall not, by way
of trade or otherwise, be lent, re-sold, hired out or otherwise circulated
without the publisher's prior consent in any form of binding or cover
other than that in which it is published and without a similar condition
including this condition being imposed on the subsequent purchaser

British Library Cataloguing in Publication Data
Oxford reviews of reproductive biology. ——
Vol. 9 (1987)
1. Reproduction——Periodicals
591.1'6'05   QP251
ISBN 0-19-857645-5

Set by Activity Limited, Salisbury

Printed in Great Britain
at the University Printing House, Oxford
by David Stanford
Printer to the University

Contents

Contributors

F. H. de Jong: Department of Biochemistry, (Division of Chemical Endocrinology), Erasmus University Rotterdam, P. O. Box 1738, 3000 DR, Rotterdam, The Netherlands.

Iain J. Clarke: Medical Research Centre, Prince Henry's Hospital, St. Kilda Road, Melbourne, Australia 3004.

F. J. Cunningham: Department of Physiology & Biochemistry, University of Reading, Whiteknights, P. O. Box 228, Reading, RG6 2AJ, UK.

Charles H. Rodeck: Institute of Obstetrics and Gynaecology, University of London, Queen Charlotte's Maternity Hospital, Goldhawk Road, London, W6 0XG, UK.

R. J. Levin: Department of Physiology, University of Sheffield, Sheffield, S10 2TN, UK.

K. Griffiths, P. Davies, C. L. Eaton and M. E. Harper: Tenovus Institute for Cancer Research, University of Wales College of Medicine, The Heath, Cardiff, CF4 4XX, UK.

W. B. Peeling: South Gwent Urological Centre, St. Woolos Hospital, Newport, NP9 4SZ, UK.

A. O. Turkes, A. Turkes, D. W. Wilson and C. G. Pierrepoint: Tenovus Institute for Cancer Research, University of Wales College of Medicine, The Heath, Cardiff, CF4 4XX, UK.

Trevor A. Howlett: Department of Endocrinology, St. Bartholomew's Hospital, London EC1A 7BE, UK.

Lesley H. Rees: Department of Chemical Endocrinology, St. Bartholomew's Hospital, London EC1A 7BE, UK.

Susanne Fournier-Delpech: Station de Physiologie de la Reproduction, Institut National de la Recherche Agronomique, 37380 Nouzilly, France.

Michel Courot: Station de Physiologie de la Reproduction, Institut National de la Recherche Agronomique, 37380 Nouzilly, France.

Gerald Schatten: Integrated Microscopy Facility for Biomedical Research, Zoology Research Building, University of Wisconsin, 1117 West Johnson Street, Madison, Wisconsin, Wi 53706, USA.

Heide Schatten: Integrated Microscopy Facility for Biomedical Research, Zoology Research Building, University of Wisconsin, 1117 West Johnson Street, Madison, Wisconsin, Wi 53706, USA.

H. M. Charlton: Department of Human Anatomy, Oxford University, South Parks Road, Oxford OX1 3QX, UK.

D. J. Hill: Department of Paediatrics, University of Sheffield, Children's Hospital, Sheffield S10 2TH, UK.

A. J. Strain: Department of Paediatrics, University of Sheffield, Children's Hospital, Sheffield S10 2TH, UK.

R. D. G. Milner: Department of Paediatrics, University of Sheffield, Children's Hospital, Sheffield S10 2TH, UK.

Gary P. Moberg: Department of Animal Science, University of California, Davis, California 95616, USA.

1 Inhibin – Its nature, site of production and function

F. H. DE JONG

1 Introduction

The pituitary gland secretes two gonadotrophic hormones which stimulate the separate compartments of the testes in male and of the ovaries in female animals: lutropin (luteinizing hormone, LH) and follitropin (follicle-stimulating hormone, FSH). The secretion of both gonadotrophins is stimulated by the hypothalamic decapeptide LH-releasing hormone (LHRH), and roughly parallel changes in the secretion of LH and FSH can be observed for instance at the beginning of puberty, during the midcycle gonadotrophin peak in female animals and after orchidectomy. The latter observation had led to the concept of negative feedback regulation of gonadotrophin secretion by gonadal steroid hormones (Moore and Price 1932). However, when the changes in gonadotrophin levels are scrutinized in more detail, it becomes clear that discrepancies between the changes in the peripheral concentration of gonadotrophic hormones may occur: during puberty FSH levels in girls increase earlier than LH levels (Jenner, Kelch, Kaplan, and Grumbach 1972), the midcycle peak of FSH in some species shows a secondary peak which is not accompanied by increased concentrations of LH (Gay, Midgley, and Niswender 1970) and the increases of FSH after gonadectomy may be much faster than those of LH (Hermans, van Leeuwen, Debets, and de Jong 1980). Finally, the administration of antagonists to LHRH to orchidectomized animals leads to a suppression of LH, which is far larger than that of FSH concentrations (Charlesworth, Grady, Shin, Vale, Rivier, Rivier, and Schwartz 1984), indicating that the pituitary secretion of FSH contains a component which is not dependent on stimulation with LHRH.

These discrepancies between the secretion of LH and FSH have been explained in a number of ways. Firstly, the existence of a hypothalamic FSH-releasing hormone (FSHRH) has been postulated, which might specifically or preferentially stimulate FSH release, instead of the release of both LH and FSH, as LHRH does (Currie, Warberg, Folkers, Bowers, Paulsen, Mulder, and Paulsen 1977; Mizunuma, Samson, Lumpkin, and McCann 1983). Secondly, the difference between the metabolic clearance rates of LH and FSH and between the responses of peripheral levels of the gonadotrophins to LHRH pulses may influence the ratio of LH to FSH in peripheral plasma: a high pulse frequency causes a relative increase of LH, while a low pulse frequency favours the secretion of FSH (Lincoln 1979). A third possibility might be that a gonadal steroid or a combination of gonadal steroids can suppress LH or FSH selectively (Plant, Hess, Hotchkiss, and

Knobil 1978). Finally, differential changes of peripheral levels of LH and FSH might be explained by the existence of inhibin, a gonadal protein hormone, which has a specific or selective effect on the secretion of FSH, while not or only marginally influencing LH secretion. Although between 1932 and 1975 many authors considered the inhibin-concept as an artifact, more recent data provided a more factual background for this concept. It is the aim of this chapter to review the present knowledge on the estimation, occurrence, physicochemical properties and physiological actions of inhibin. The sections on these aspects will be preceded by a brief recording of the history of the inhibin-concept.

II   History of the inhibin-concept

The historical background of the inhibin-concept has been reviewed by Setchell and Main (1974), who made an extensive bibliography of the literature up to 1973, and later by Baker, Bremner, Burger, de Kretser, Dulmanis, Eddie, Hudson, Keogh, Lee, and Rennie (1976); Setchell, Davies, and Main (1977); de Jong (1979) and Baker, Eddie, Higginson, Hudson, and Niall (1982a). In the following sections, an outline of the various phases of the development of this concept will be described: it will be shown that after an initial period of indirect indications for the production of a hormone in the seminiferous tubules of the testes which might influence pituitary function, a period of disbelief ensued. It is only during the last 10 years that more direct indications for the existence of inhibin were obtained, and the final doubts have been removed during the last year, when several groups isolated pure preparations of inhibin and the nucleotides of the cDNAs of porcine, bovine and human inhibin were sequenced (see IV 3).

1   FIRST SUGGESTIONS ON THE EXISTENCE OF INHIBIN (1923–1940)

The first evidence for the existence of inhibin came from experiments in which the germinal epithelium of rat testes was damaged by irradiation, and subsequently changes in the histology of the anterior pituitary gland were observed (Mottram and Cramer 1923). These morphological changes resembled those found after castration and the authors postulated that the irradiation-induced absence of 'an internal secretion of the seminal epithelium' was responsible for the pituitary changes.

In 1931, Martins and Rocha published the results of a series of ingenious experiments, in which parabiotic pairs of castrated and intact male and female rats were used to investigate the relationship between gonadal secretions and pituitary function by injecting testicular homogenates into the castrated parabiont. Martins and Rocha (1931) concluded that the testis secretes at least two hormones: one stimulating the growth of the accessory sex organs and another, water soluble hormone, which can suppress the secretion of hormones from the pituitary gland, as evidenced by the disappearance of castration cells in the hypophysis and also by the absence of signs of precocious puberty in the intact immature female parabiont. Testicular homogenates from both immature and mature rats exerted this action in male castrated rats, while no such action was found in female animals after the injection of testicular homogenates.

Similar results were obtained by McCullagh (1932), who injected watery extracts of bull testes into castrated rats, and observed a disappearance of the 'castration cells' in the pituitary gland, without effect on the weight of the accessory sex organs. Benzene extracts of the testes, however, caused an increase of the weight of the accessory sex organs, but did not change the morphological appearance of the pituitary gland. McCullagh coined the names 'inhibin' and 'androtin' for the two separate activities which could be extracted from the testis. Since that time, the structure of 'androtin' was elucidated within a relatively short period, when David, Dingemanse, Freud, and Laqueur (1935) isolated testosterone from bulls' testes. However, it took about 40 years before the search for the structure of inhibin was resumed, and from that moment more than 10 years before this search was successfully concluded.

In a later study, McCullagh and Walsh (1934) showed that treatment of long-term castrated male rats with androtin for a 3 week period did not reverse the pituitary hypertrophy caused by castration, suggesting again that the hypertrophic growth of the pituitary cells after castration is due to a factor different from androgens. No search for inhibin in extracts from ovaries was made in this period, although McCullagh and Schneider (1940) showed that aqueous testicular extracts could disrupt the oestrous cycle in rats.

During the same period, Nelson (1934), Nelson and Gallagher (1935), Vidgoff and Vehrs (1940) and Rubin (1941) published results of experiments in which McCullagh's original data could not be confirmed and which were conflicting with his inhibin hypothesis. From these experiments it was concluded that steroids alone provide the negative feedback-action of the testis on the hypophysis, and that the aqueous extracts might have inhibited pituitary cell hypertrophy by non-specific toxic effects rather than by the action of a seminiferous tubular hormone on the pituitary gland.

2   DENIAL OF THE EXISTENCE OF INHIBIN (1940–1965)

Meanwhile, the development of bioassays for the urinary gonadotrophic hormones LH and FSH, which had been recognized as separate entities since the work of Fevold, Hisaw, and Leonard (1931), offered new possibilities for the investigation of the gonadal control of gonadotrophin secretion in men. When it became clear that the urinary excretion of FSH, but not LH, correlated with the degree of damage of spermatogenesis in azoo- or oligozoospermic men (Klinefelter, Reifenstein, and Albright 1942; Heller and Nelson 1945; del Castillo, Trabucco, and de la Balze 1947; McCullagh and Schaffenburg 1952), and that normal FSH levels could be found in men with damaged or absent Leydig cells (McCullagh and Schaffenburg 1952), these observations did not lead to a revival of the inhibin-concept. Instead, the 'utilization theory' was put forward by Heller and Nelson (1948) and Heller, Paulsen, Mortimore, Junck, and Nelson (1952). According to the latter theory, the increases of urinary FSH without concomitant rise of LH should be explained on basis of non-utilization or non-inactivation of FSH by the cells in the damaged seminiferous tubules. An alternative possibility was suggested by Howard, Sniffen, Simmons, and Albright (1950), who rejected the utilization theory because it did not explain the increased pituitary gonadotrophin levels in castrated rats, and because of the fact that, for the utilization theory to be valid, most of the blood should go through the gonads before being cleared of gonadotrophins in the kidney. These authors suggested that oestrogens might play an important role in regulation of FSH secretion. Similarly, McCullagh and Schaffenburg (1952), Johnsen (1964) and Lacy and Pettitt (1970) thought that steroids, most probably oestrogens which had been produced in the seminiferous tubules might be involved in the negative feedback of FSH secretion.

3   INDIRECT EVIDENCE FOR THE EXISTENCE OF SEPARATE FEEDBACK MECHANISMS FOR LH AND FSH (1965–1975)

Apart from the clinical observations of increases of FSH, not accompanied by increases of LH in urine or serum from azoo- or oligozoospermic men (see Setchell *et al.* 1977), a large number of experimental studies were aimed at the specific destruction of the spermatogenic epithelium in experimental animals, followed by detailed studies on the regulation of gonadotrophin secretion. From the results of these studies, a number of cell types was suggested as the source for the FSH-inhibiting substance(s) (Table 1.1): any tubular cell type, including 'an independent testicular station' (Leonard, Leach, Couture, and Paulsen 1972) was included. Other clinical observations also indicated that separate feedback mechanisms for LH and FSH must exist: in testicular feminization, a syndrome in which androgen receptors are defective, increased concentrations of LH are found in the presence of normal FSH levels, which increase after castration (Judd, Hamilton, Barlow, Yen, and Kliman 1972, Tremblay, Foley, Corvol, Park, Kowarski, Blizzard, Jones, and Migeon 1972). A similar picture of high LH, but normal FSH was reported in a patient with congenital absence of Leydig cells by Berthezene, Forest, Grimaud, Claustrat, and Mornex (1976); the testes of this patient consisted of Sertoli cells and gonocytes,

Table 1.1

Testicular cell types which have been suggested as possible sources of inhibin on basis of in vivo *experiments*

Cell type	Observations	Reference
Sertoli cells	– Increased FSH levels in Sertoli cell-only patients; further increase after castration	– Christiansen (1975)
	– Negative correlation between numbers of Sertoli cells and peripheral FSH in neonatally irradiated male rats	– de Jong and Sharpe (1977)
	– Negative correlation between testicular ABP and peripheral ABP after destruction of germinal epithelium	– Rich and de Kretser (1977)
Sertoli cells and/or spermatogenesis	– Increased FSH levels in men after chemotherapy	– van Thiel *et al.* (1972)
	– Increased FSH levels in men with insufficient numbers of spermatids	– Franchimont *et al.* (1972)
	– Increased FSH levels in vitamin A deficient male rats	– Krueger *et al.* (1974)
Spermatocytes and/ or early spermatids	– Increased FSH and LH levels in busulphan–treated male rats	– Debeljuk *et al.* (1973); Gomes *et al.* (1973)
Mature spermatids	– Increased FSH levels in cryptorchid rats of various ages	– Swerdloff *et al.* (1971)
Residual bodies	– Increased FSH levels in men with incomplete spermiogenesis	– Johnsen (1964)
'Independent testicular station'	– Increased FSH levels in oliogozoospermia, not correlated with specific damage of spermatogenesis	– Leonard *et al.* (1972)

implicating these cells types as the possible source of production of the FSH-regulator. However, from these studies it is not possible to decide if (a combination of) special tubular steroids or another hormonal principle was required for suppression of FSH secretion. Conclusions from experiments, in which steroids were injected and stronger suppression of LH than of FSH was measured (androgens: Goh, Karim, and Ratnam 1981; oestrogens: Goh, Chew, Karim, and Ratnam 1980) might be invalid because it is possible that feedback inhibition of FSH requires a more subtle rythmicity of steroid levels, as indicated by Plant *et al.* (1978), who gave a combination of implanted and injected testosterone in castrated male rhesus monkeys, and suppressed both LH and FSH concentrations in serum to control levels. Finally, Lincoln (1979) suggested that a change in the frequency of

LHRH-pulses might change the ratio between peripheral levels of LH and FSH.

4 DIRECT EVIDENCE FOR THE EXISTENCE OF INHIBIN (1963–1975)

A more direct approach to the investigation of the question if non-steroidal substances may exert a selective suppression of FSH-levels is the study of concentrations of FSH in blood of experimental animals injected with steroid-free gonadal preparations. Using this type of experiment, Fachini, Toffoli, Gaudiano, Marabelli, Polizzi, and Mangili (1963), Fachini and Ciaccolini (1966), Lugaro, Giannattasio, Ciaccolini, Fachini, and Gianfranceschi (1969), and Lugaro, Carrea, Casellato, Mazzola, and Fachini (1973) indicated that FSH-inhibiting activity might be present in semen of various species, in bovine spermatozoa, in urine and in spermatic venous blood. Lugaro, Casellato, Mazzola, Fachini, and Carrea (1974) also partially purified a peptide from bovine spermatozoa, which exerted this action. Furthermore, Setchell and Wallace (1972) and Setchell and Sirinathsinghji (1972) indicated that rete testis fluid could suppress FSH, using the suppression of uterine weight in hCG-injected mice as an endpoint. At the same time, Franchimont, Millet, Vendrely, Letawe, Legros, and Netter (1972) showed that injection of human seminal plasma into castrated male rats suppressed FSH levels: similar experiments with ovine rete testis fluid were reported by Setchell and Jacks (1974), who also indicated that steroids could not be present in the injected material.

Since 1975, inhibin activity has been detected in testicular fluids from many species. These experiments will be reviewed in section IV.1.

5 EVIDENCE FOR THE PRESENCE OF INHIBIN IN THE FEMALE

Indirect evidence for a separate regulatory system for FSH in women was first provided by Sherman and Korenman (1975) and Sherman (1976), who observed that at the beginning of the menopausal transition increased peripheral levels of FSH were found, concomitant with normal LH concentrations. Boyar, Wu, Liao, and Finkelstein (1977) found selectively increased FSH concentrations in a woman who had normal levels of LH and oestradiol, but no follicles in her ovaries.

Direct evidence for the presence of inhibin in the ovary was provided by de Jong and Sharpe (1976), who prevented the postcastration increase of FSH, but not LH, in adult male rats by the injection of steriod-free ovarian follicular fluid. This observation has been amply confirmed afterwards (see section IV.1.). The first indication of secretion of the FSH-suppressing activity from the ovary was described by Uilenbroek, Tiller, de Jong, and

Vels (1978), who transplanted rat ovaries into the spleen of castrated male or female rats. In these animals, FSH-levels were suppressed, while LH levels were not affected. This phenomenon can be explained on the basis of inactivation of steroids in the liver of these animals before they could reach the peripheral circulation, while inhibin-activity is not destroyed during passage through the liver.

6   CONCLUSIONS

Indirect evidence for the existence of inhibin was reported as early as 1923. Probably due to non-specificity or bad reproducibility of detection systems for inhibin activity, this concept had been dismissed until the late sixties, when it became clear that manipulation of spermatogenesis could lead to specific or preferential increases of FSH. The final proof for the existence of inhibin was obtained after the demonstration that testicular, but also ovarian, protein preparations can cause a specific suppression of the secretion of FSH from the pituitary gland.

III   Detection and estimation of inhibin

Inhibin activity can be detected using a number of *in vivo* and *in vitro* bioassay systems, while a small number of radioimmunological methods for the estimation of inhibin has also been reported. Here, the main techniques which have been used will be reviewed and their relative merits will be discussed. Earlier reviews on this subject were published by Hudson, Baker, Eddie, Higginson, Burger, de Kretser, Dobos, and Lee (1979), Baker, Eddie, Higginson, Hudson, Keogh, and Niall (1981) and Baker *et al.* (1982*a*).

1   *IN VIVO* BIOASSAYS

i   Assays based on decrease of organ weight

After the experiments with parabiotic rats, from which Fachini *et al.* (1963) and Lugaro *et al.* (1969, 1973) concluded that inhibin-activity is present in semen and extracts of spermatozoa, the observation by Setchell and Wallace (1972) of decreases of ovarian weight in the Steelman-Pohley assay for FSH activity after injection of concentrated ovine rete testis fluid was the first direct suggestion for the presence of an 'antigonadotrophic activity' in gonadal fluids. This observation was repeated by Setchell and Sirinathsinghji (1972), who measured uterine rather than ovarian weight and also observed a decrease of the hCG-induced increase of the weight of this organ after injection of rete testis fluid. A number of authors developed assays for

Table 1.2

Inhibin bioassays, based on suppression of organ weights in in vivo *systems*

Method	Precision index (λ)	Reference
Inhibition of hCG-effect on growth of		
uterus (rat)	–	Setchell and Sirinathsinghji (1972)
(mouse)	0.104	Ramasharma et al. (1979)
ovary (rat)	0.21	Chari et al. (1976)
(mouse)	–	Sheth et al. (1979)

the estimation of inhibin using the same principles, measuring ovarian or uterine weights (Table 1.2).

Since the mechanism of the increase of the weight of ovaries or uterus in immature hCG-injected mice or rats depends on an increase of endogenous FSH (Ramasharma, Shashidara Murthy, and Moudgal 1979), suppression of the increase of organ weights might be due to suppression of FSH. However, an alternative explanation might be that the injected, putatively inhibin-containing materials interact with the binding of FSH to its receptor on the granulosa cell. The presence of such 'FSH-binding inhibitors' in gonads or gonadal fluids has been amply documented (Reichert and Abou-Issa 1977; Dias and Reichert 1984; Sluss and Reichert 1984), and its presence in partly purified inhibin preparations has been reported recently (Mohapatra, Duraiswami, and Chari 1985). For this reason, results obtained on the basis of inhibin-estimations, in which only organ weights were used, should be considered with caution.

Furthermore, the results of this type of assay appear to depend very much on age of the animals, timing of the injections and the dose of hCG used: using protocols similar to those of Chari, Duraiswami, and Franchimont (1976) and Ramasharma et al. (1979), Hudson et al. (1979) and de Jong, Welschen, Hermans, Smith, and van der Molen (1979) reported negligible suppression of organ weights with inhibin-preparations which were shown to contain bioactivity using other methods. Finally, the precision of this type of method is relatively low, while the sensitivity is much lower than that observed using *in vitro* techniques (Hudson *et al.* 1979). This combination of reservations makes it extremely difficult to interpret results obtained with this type of assay.

ii Assays based on suppression of serum FSH concentrations

Inhibin-activity has been detected using a large number of systems, using the suppression of FSH in peripheral serum of intact or gonadectomized,

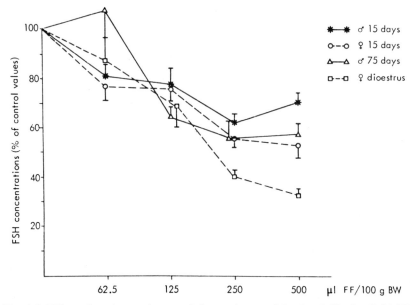

Fig. 1.1 Effect of various volumes of charcoal-treated bovine follicular fluid (FF) on serum concentrations of FSH in gonadectomized male and female rats 8 h after injection (mean ± s.e.m). Concentrations have been expressed as percentage of values found in bovine plasma injected controls. (From de Jong *et al.* 1978 by permission).

prepubertal or mature, male or female experimental animals as an end-point. The specificity of this type of assay depends on the possibility to exclude effects of steroids and to show that LH concentrations are not affected.

In none of the models employed have FSH levels been suppressed to values lower than 25 per cent of those observed in control animals (Fig. 1.1). The FSH-suppressing effect of inhibin preparations generally becomes apparent within a few hours after injection, and starts to disappear within 6 to 16 h (Nandini, Lipner, and Moudgal 1976; Hermans, Debets, van Leeuwen, and de Jong 1981, 1982).

Male animals. Adult intact male rats (Hermans *et al.* 1980; Lorenzen, Dworkin, and Schwartz 1981) and rats castrated 12 to 48 h previously (Nandini *et al.* 1976; Hermans *et al.* 1980; Lorenzen *et al.* 1981) or castrated 2 to 3 weeks before injection of inhibin-containing preparations (Lugaro *et al.* 1974; Setchell and Jacks 1974; Franchimont, Chari, and Demoulin 1975; Hopkinson, Daume, Sturm, Fritze, Kaiser, and Hirschhäuser 1977; Le Lannou and Chambon 1977; Peek and Watkins 1979; Vijayalakshmi, Bandivdekar, Joshi, Moodbidri, and Sheth 1980) all appeared to be suitable models for detection of inhibin activity. However, hardly any experiments were reported in which this type of assay was optimized by varying the interval between the injection of inhibin-containing preparations and the collection of blood for estimation of gonadotrophins. Neither were dose-response curves constructed in most of these studies, making it difficult to assess the value of this type of assay in the quantitative measurement of inhibin-activity. Furthermore, the paradoxical increases of FSH reported to occur after injection of larger doses (Davies, Main, and Setchell 1978; Franchimont, Demoulin, Verstraelen-Proyard, Hazee-Hagelstein, and Tunbridge 1979; Hudson et al. 1979; Rush and Lipner 1979) and the suppression of LH after injection of gonadal preparations observed by several authors (Hodgen, Nixon, and Turner 1974; Lee, Pearce, and de Kretser 1977; Braunstein and Swerdloff 1977; Davies et al. 1978), make it difficult to interpret results in terms of specific FSH-suppression.

The suppression of the immediate post-castration rise of FSH after administration of inhibin-containing preparations is greater in absolute, but not in relative terms. This system has been used in adult rats (de Jong and Sharpe 1976; Hermans et al. 1980; Lorenzen et al. 1981) and in immature animals (Nandini et al. 1976; Daume, Chari, Hopkinson, Sturm, and Hirschhäuser 1978; Davies et al. 1978; Peek and Watkins 1979; Rush and Lipner 1979; Hermans et al. 1980). The larger suppressive effect obtained in immature rats, when compared with that in adult rats, appears to be related to the faster increase of peripheral FSH levels after castration in the prepubertal animal (Hermans et al. 1980).

Comparisons of sensitivity between models using acutely versus long-term castrated rats are scarce: Hermans et al. (1980) suggested that the response of FSH diminishes with increasing time after castration.

Inhibin-containing preparations have also been injected or infused into larger male animals, especially in experiments in which the time course of the response of FSH-levels was studied. Keogh, Lee, Rennie, Burger, Hudson, and de Kretser (1976) infused ovine rete testis fluid and testicular lymph into castrated rams over 26 h, and observed a suppression of FSH, but not of LH, from the sixth hour of the infusion onwards. After stopping the infusion, baseline levels of FSH were reached at 24 to 48 h. Similar experiments were reported by Cahoreau, Blanc, Dacheux, Pisselet, and Courot (1979), who injected crude ovine rete testis fluid preparations into cryptorchid rams. Surprisingly, these injections resulted in the immediate abolishment of LH-pulses, while it took much longer before FSH suppression was observed. Baker, Eddie, Higginson, Hudson, and Niall (1980) suggested that this suppression of LH may have been due to non-specific effects, since infusion with lymph obtained from non-testicular tissue can cause similar effects.

Female animals. Female animals have also been used extensively for the detection of inhibin-activity, using the suppression of peripheral concentrations of FSH as an endpoint. Animals were always bled within 3 to 24 h after the last injection of the inhibin-containing preparations. Most experiments have been performed in rats, although mice (Bronson and Channing 1978), long-term ovariectomized rhesus monkeys (Chappel, Holt, and Spies 1980), hamsters (Chappel 1979) and mares (Miller, Wesson, and Ginther 1979) have also been used.

Inhibin-containing preparations have been injected into adult female rats on various days of the oestrous cycle (Marder, Channing, and Schwartz 1977; Schwartz and Channing 1977; DePaolo, Wise, Anderson, Barraclough, and Channing 1979; Hoffmann, Lorenzen, Weil, and Schwartz 1979; Hermans et al. 1981, 1982). The timing of the collection of blood after the injection is rather critical: the initial suppression of FSH seen after the administration of inhibin-containing preparations decreases with time and is followed by an increase of FSH levels, which leads to concentrations which are significantly larger than the levels before the injection (DePaolo, Hirschfield, Anderson, Barraclough, and Channing 1979; Hermans et al. 1981, 1982). This 'rebound phenomenon' may be due to decreased concentrations of endogenous inhibin, as a result of the suppression of FSH levels (see section IV2(i)).

In acutely ovariectomized rats, the response of FSH to the administration of preparations with inhibin-activity varies with the day of the cycle on which the operation took place. For this reason, the use of acutely ovariectomized rats is less practical than the use of animals which had been ovariectomized earlier (Marder et al. 1977; Hermans et al. 1980): the influence of the cycle is abolished in rats which had been ovariectomized for 2 days (Hermans et al. 1980), for 7 days (Campbell and Schwartz 1979), or for 12 or 14 d (Hopkinson et al. 1977; Welschen, Hermans, Dullaart, and de Jong 1977), and no rebound of FSH after the initial suppression is found.

Immature female rats offer a less suitable model for detection of inhibin-activity: in these animals the duration of the suppression of FSH is shorter and the relative decrease of FSH levels is less than in mature animals (Hermans et al. 1981).

iii   Conclusions

In vivo methods for the detection of inhibin-activity using suppression of organ weight as an endpoint lack specificity. If suppression of FSH is used as response-parameter, qualitative data may be obtained. However, in order to obtain quantitative data, large groups of rats and large amounts of inhibin-containing material must be used, due to the relative insensitivity of the system and the large interindividual differences in response. Therefore, the use of in vitro bioassays, as discussed in the next section, has definite advantages.

2   IN VITRO BIOASSAYS

In vitro methods for the detection and estimation of inhibin activity (see Table 1.3) do have a number of advantages over the in vivo techniques described in

Table 1.3

Inhibin bioassays, based on estimation of suppression of FSH in in vitro *systems*

Method	Reference
Incubation of rat hemipituitaries	
Inhibition of LHRH-stimulated	Davies *et al.* (1978)
FSH secretion	Hudson *et al.* (1979)
Inhibition of spontaneous	Jenner *et al.* (1982)
FSH secretion	
Rat pituitary cell cultures	
Inhibition of spontaneous FSH release	Steinberger and Steinberger (1976); de Jong *et al.* (1978, 1979*a*)
Inhibition of LHRH-stimulated FSH release	Labrie *et al.* (1978)
	Eddie *et al*(1979)
Suppression of cell content of FSH	Scott *et al.* (1980)
Hamster pituitary cell cultures	
Inhibition of LHRH-stimulated FSH release	Chappel *et al.* (1980)
Ovine pituitary cell culture	Huang and Miller (1984)
Inhibition of spontaneous FSH release	Tsonis *et al.* (1986)

the previous section: these methods estimate direct effects on FSH production or release from the pituitary gland, have a better precision index (see Baker *et al.* 1981) and need a smaller number of experimental animals, when compared with the *in vivo* methods. On the other hand, the *in vitro* system might be more sensitive to non-specific or toxic effects of added preparations: agents which damage the incubated hemipituitary glands or cultured pituitary cells and therefore suppress the amount of FSH present in the system might erroneously be regarded as containing inhibin activity.

Assay systems using incubated hemipituitary glands are fast and simple to perform (Davies *et al.* 1978), but have been reported to be less sensitive and to have a lower precision index than systems in which dispersed pituitary cells are used (Franchimont, Verstraelen-Proyard, Hazee-Hagelstein, Renard, Demoulin, Bourguignon, and Hustin 1979; Hudson *et al.* 1979; Baker *et al.* 1981). The spontaneous release of FSH from incubated hemipituitary glands is only slightly reduced after addition of inhibin-con-taining preparations (Davies *et al.* 1978). For this reason a system in which the release of gonadotrophins was stimulated by addition of LHRH has been used. In these LHRH-stimulated glands, however, LH release was also suppressed after addition of inhibin-containing preparations from testicular sources. On the other hand, Jenner, de Koning, and van Rees (1982) indicated the hemipituitary glands *in vitro* secrete relatively large amounts of FSH, even in the absence of LHRH; this release of FSH can be prevented by the addition of steroid-free follicular fluid to the incubation medium. Sairam, Ranganathan, Ramasharma, and Lamothe (1981) reported that fractions of bovine seminal plasma could selectively inhibit the LHRH-stimulated release of FSH from incubated mouse pituitaries.

At present *in vitro* bioassays using dispersed pituitary cells are most widely used for the detection and estimation of inhibin activity. A number of end points have been described: the selective suppression of the spontaneous release of FSH (Fig. 1.2a; Steinberger and Steinberger 1976; de Jong, Welschen, Hermans, Smith, and van der Molen 1978; Erickson and Hsueh, 1978; de Jong, Smith, and van der Molen 1979; DePaolo, Shander, Wise, Barraclough, and Channing 1979; Lagacé, Labrie, Lorenzen, Schwartz, and Channing 1979; Tsonis, McNeilly, and Baird 1986), of the cellular content of FSH (Scott, Burger, and Quigg 1980; Scott and Burger 1981), or the suppression of the LHRH-stimulated release of FSH after preculture in the presence of inhibin-containing preparations (Fig. 1.2b; de Jong *et al.* 1978, 1979*a*; Eddie, Baker, Dulmanis, Higginson, and Hudson 1978; Labrie, Lagacé, Ferland, Kelly, Drouin, Massicotte, Bonne, Raynaud, and Dorrington, 1978; Eddie, Baker, Higginson, and Hudson 1979; Franchimont *et al.* 1979). In the latter system, the release of LH is usually also suppressed by the addition of inhibin-containing preparations (Fig. 1.2b). This means that one of the criteria for intactness of the pituitary cells, that is, non-changed release of LH after addition of inhibin-containing preparations, cannot be used in the LHRH-stimulated system.

This introduces the problem of the negative response, by which inhibin bioactivity has to be measured. Apart from the negative effect of steroids (Drouin and Labrie 1976, 1981; Eddie *et al.* 1978, 1979; de Jong *et al.* 1979*a*), it may be that toxic or non-specific effects influence negatively the biosynthetic or secretory activity of the pituitary gonadotrophic cells. Such effects may be exerted by as yet undefined principles in ethanol extracts of rete testis fluid (Baker, Burger, de Kretser, Eddie, Higginson, Hudson, and Lee 1978; Baker *et al.* 1980), acid ethanol extracts of bovine follicular fluid (de Jong *et al.* 1979*b*) and crude human seminal plasma (Scott, Burger, Quigg, Dobos, Robertson, and de Kretser 1982), but also by serum proteins (Baker *et al.* 1978, 1980), concanavalin A (Ponsin, Khar, Kunert-Radek, Bennardo, and Jutisz 1980), and changes in the ionic composition of the culture medium (McCann 1974; Kao, Gunsalus, Williams, and Weisz 1977). Indications about the presence of toxic substances may be obtained by estimation of release of radiolabelled chromate after prelabelling of the cells with this material (Robertson, Au, and de Kretser 1982; Scott *et al.* 1982). Although it is not likely that subtle changes in the gonadotrophic cells will be detected by this method, Scott *et al.* (1982) showed that increases of chromium release paralleled decreases of the cellular content of LH and morphological changes in the total pituitary cell population. A final point which may be of importance in this discussion of specificity of the suppression of LHRH-induced gonadotrophin release from pituitary cells is the presence in gonadal extracts of

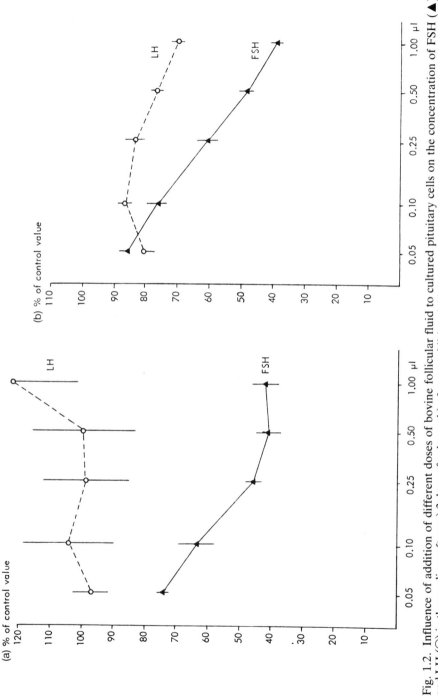

Fig. 1.2. Influence of addition of different doses of bovine follicular fluid to cultured pituitary cells on the concentration of FSH (▲) and LH (○) in the medium after: a) 3 days of culture; b) after an additional 6 h in the presence of 50 ng LHRH per ml. Values are mean ± s.e.m, data from 4 different experiments, each with 4 dishes per dose, were combined. (From de Jong *et al.* 1979*b* by permission).

endopeptidases, which may metabolize LHRH (Vale, Rivier, Brown, Leppaluoto, Ling, Monahan, and Rivier 1976; Baker *et al.* 1982*a*). The breakdown of LHRH may lead to spurious overestimation of inhibin-activity in LHRH-stimulated systems. On the other hand, peptides with an LHRH-like activity, which have been reported to be present in gonadal fluids or tissue (Sharpe 1983), might exert effects on basal release of gonadotrophins, while their action in the presence of relatively high concentrations of LHRH is less likely.

Results obtained for the same samples using two or more of these assay systems have been compared in a limited number of studies. Comparison between *in vitro* and *in vivo* bioassays have been reported by Hudson *et al.* (1979), who found a good correlation over a thousand-fold range of activity. De Jong *et al.* (1979*b*) indicated that ethanol extracts of bovine follicular fluid may contain a principle which cannot suppress basal release of FSH, while the LHRH-stimulated release of both FSH and LH is inhibited after addition of this fraction. De Jong, Jansen, Steenbergen, van Dijk, and van der Molen (1983) reported that the relative bioactivity of inhibin in a protein preparation from ovine testicular lymph was significantly higher when compared in an LHRH-stimulated system with a standard preparation from bovine follicular fluid, than in an assay based on spontaneous release of FSH. Finally, Scott and Burger (1981) also reported differences between inhibin potencies of preparation, obtained in assays performed in the presence or absence of LHRH, or in assays based on estimation of the FSH content of cells.

The sensitivity of these *in vitro* bioassays using rat pituitary cells is sufficient to estimate inhibin concentrations in gonadal fluids, but it is not possible to measure inhibin in peripheral blood. However, Tsonis *et al.* (1986) reported that the sensitivity of an inhibin bioassay system based on the suppression of spontaneous release of FSH from ovine pituitary cells is thirty- to forty-fold higher than that of a rat pituitary cell bioassay, thus making peripheral levels of inhibin-activity detectable.

3   IMMUNOLOGICAL METHODS AND ANTIBODIES AGAINST INHIBIN

The earliest publication, in which the production of antibodies against inhibin was mentioned, was that of Franchimont, Chari, Hagelstein, and Duraiswami (1975), who used fractions of human seminal plasma for the immunization of rabbits. After injection of serum from these animals, increased concentrations of FSH were found in rats, suggesting immuno-neutralization of inhibin. However, an alternative explanation of these observations might be that the rabbit immunoglobulins, which were present in the rat serum, displaced the (rat FSH-rabbit anti rat FSH)-complex from the second antibody in the radioimmunoassay for rat FSH, leading to a

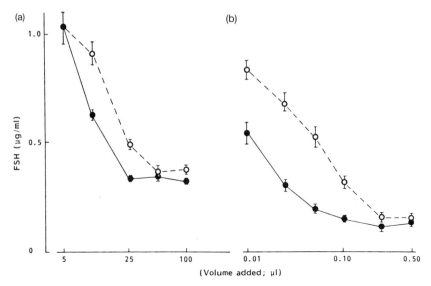

Fig. 1.3. Influence of addition of different doses of ovine follicular fluid (a), and ovine testicular lymph (b), with (○) or without (●) 1 μl of ovine antiserum against bovine follicular fluid inhibin, to cultured pituitary cells on the concentration of FSH in the medium (mean ± s.e.m, n = 4). The immunoneutralization of inhibin in the ovine follicular fluid was significantly more effective than that of inhibin in the ovine testicular lymph. (From van Dijk *et al.* 1986 by permission).

decrease of bound radioactivity and, therefore, to seemingly increased FSH levels (Baker *et al.* 1982*a*). Vaze, Thakur, and Sheth (1979) reported similar data, using a preparation of human seminal plasma as an antigen.

Immunoneutralization of inhibin *in vitro* has been described by Channing, Tanabe, Turner, and Hodgen (1982), de Jong, Jansen, Hermans, and van der Molen (1982*a*), Rivier, Spiess, McClintock, Vaughan, and Vale (1985), Lee, Kraft, Atkins, and Burger (1986), and van Dijk, Steenbergen, Gielen, and de Jong (1986). In this kind of system the specificity of the neutralization can be investigated in more detail by testing for the absence of influence of the antiserum on the gonadotrophin secretion when no inhibin-containing preparations are added. Channing *et al.* (1982) obtained an antiserum against inhibin after repeated injection of porcine follicular fluid in monkeys; after a number of injections no FSH suppression was obtained and the possibility of immunoneutralization was ascertained *in vitro*, using dispersed pituitary cells as an assay system. De Jong *et al.* (1982*a*) and van Dijk *et al.* (1986) immunized rabbits and sheep with partially purified bovine follicular fluid preparations. The latter authors detected immunological differences between inhibin from male and female sources within the same species (Fig. 1.3). Rivier *et al.* (1985) showed immunoneutralization of inhibin *in vitro*, using an antiserum against the N-terminal hexapeptide of purified, sequenced inhibin from porcine follicular fluid. Finally, Lee *et al.*

(1986) described immunoneutralization of inhibin-activity from various species by addition of monoclonal antibodies, raised against fluid from the ovaries of rats injected with pregnant mare serum gonadotrophin, while Miyamoto, Hasegawa, Fukuda, and Igarashi (1986) described the detection of various forms of bovine inhibin using monoclonal antibodies against purified bovine inhibin (see Section IV3(i)).

Radioimmunoassays for inhibin have been described by Sheth, Vaze, and Thakur (1978) and Vaze *et al.* (1979), who used antibodies against an inhibin preparation from human seminal plasma. However, these authors did not report on the correlation between concentrations of inhibin, measured in various samples using this radioimmunological technique, and those obtained in a fully characterized bioassay system. On the other hand de Jong, Steenbergen, van Dijk, and van der Molen (1983b) described a highly significant correlation between these parameters for various inhibin preparations, obtained during the purification of inhibin from bovine follicular fluid, using an antiserum which was preabsorbed with bovine peripheral plasma. Due to technical difficulties no further data were provided.

Using an immobilized immunoglobulin fraction from the same antiserum, de Jong, Jansen, Hermans, and van der Molen (1982b) purified internally labelled inhibin from the culture medium of bovine granulosa cells; these authors indicated that the apparent molecular weight of the isolated protein was 65 kD. The isolated molecule contained ^{3}H-atoms, which were incorporated after culture of the cells in the presence of tritiated sugars, suggesting that the protein is glycosylated.

The only well-characterized radioimmunoassay system for the estimation of inhibin described until now is that of McLachlan, Robertson, Burger, and de Kretser (1986a). These authors found a good correlation between bio-and immuno-potencies of a number of preparations of bovine and human origin, and indicate a high cross-reactivity between the various molecular forms of inhibin (see Section IV3(i)). Using this method, McLachlan, Robertson, Healy, de Kretser, and Burger (1986b) described inhibin concentrations in peripheral plasma of women, who underwent gonadotrophin-induced ovarian hyperstimulation for *in vitro* fertilization (see Section IV2).

Finally, Sairam, Ranganathan, and Ramasharma (1981) described binding of a radioiodinated inhibin preparation from bull seminal plasma to pituitary membranes, and Seethalakshmi, Steinberger, and Steinberger (1984) found binding to pituitary cells of 'internally' labelled proteins which had been secreted by Sertoli cells cultured in the presence of tritiated amino acids. However, these systems have not been fully characterized and indisputable evidence that they do indeed measure inhibin is not available.

Antibodies against inhibin have also been used to affect ovulation rates in female animals. This subject will be discussed in Section V2ii.

4   MODE OF ACTION OF INHIBIN

From the data in Section III2, it appears that inhibin can act directly at the level of the pituitary gonadotroph to suppress the production and secretion of FSH. Additional data on the direct effect of inhibin on the biosynthesis of FSH in the pituitary gland have been provided by Chowdhury, Steinberger, and Steinberger (1978), who cultured pituitary cells in the presence of radiolabelled amino acids and observed a decrease of the amount of protein-incorporated radioactivity, which could be immuno-precipitated using an anti-FSH antiserum.

Little is known about the biochemical mechanism of action of inhibin. Jenner, de Koning, Tijssen, and van Rees (1985) suggested that agents, which stimulate cyclic AMP levels in pituitary cells, have a suppressive action on the secretion of FSH from hemipituitary glands *in vitro*, while LH release is stimulated by addition of these compounds. On the other hand, Franchimont, Lecomte-Yerna, Henderson, Verhoeven, Hazee-Hagelstein, Jaspar, Charlet-Renard, and Demoulin (1983) observed no change in the amount of endogenous cyclic AMP, but a dose-dependent increase of cyclic GMP after addition of an inhibin-containing preparation to cultured pituitary cells. No indications exist for a possible influence of inhibin on the type of FSH which is produced in pituitary gonadotrophs (unpublished results), while de Greef, de Jong, de Koning, Steenbergen, and van der Vaart (1983) showed that injection of inhibin-containing preparations into female rats did not influence the metabolic clearance rate of FSH or LH; this indicates that injection of inhibin does not cause a decrease of peripheral FSH levels by increasing the metabolism of FSH.

Extra-pituitary actions of inhibin have been described by a number of authors: they postulated actions of inhibin-containing preparations on the hypothalamic release of LHRH *in vitro* (Demoulin, Bourguignon, and Franchimont 1979) and *in vivo* (de Greef *et al.* 1983). However, the latter authors also showed that administration of more purified fractions of bovine follicular fluid had no effect on the LHRH-concentration in the portal blood collected from the pituitary stalk. Lumpkin, Negro-Vilar, Franchimont, and McCann (1981) and Condon, Leipheimer, and Curry (1983) also described effects on FSH secretion after intraventricular administration of inhibin-containing material, and postulated effects of inhibin on the hypothalamic release of FSHRH. A possible direct effect of inhibin preparations from ovine testis fluid on the incorporation of radiolabelled thymidine in testicular desoxyribonucleic acid was reported by Franchimont, Croze, Demoulin, Bologne, and Hustin (1981). Finally, direct effects of inhibin on the ovary have been described by Chari, Aumüller, Daume, Sturm, and Hopkinson (1981), who observed diminished follicular development in rats and by Chari, Daume, Sturm, Vaupel, and Schüler (1985), who described inhibitory effects of inhibin-containing fractions of human follicular fluid on aromatization in porcine granulosa cells. Ying, Becker, Ling, Ueno, and Guillemin (1986b) confirmed this effect, by showing that purified porcine inhibin suppressed FSH-induced aromatase activity of rat granulosa cells *in vitro*, while transforming growth factor-β (TGF-β) had the opposite effect (see Section IV3ii). In contrast, Hillier, van Hall, van den Boogaard, de Zwart, and Keijzer (1982) showed that effects of follicular fluid on aromatization of testosterone by granulosa cells were not found when more purified, inhibin-containing fractions were used. The interpretation of most of these observations is difficult, because purified preparations of inhibin were not used in all of these studies and usually a correlation between the size of the extrapituitary effects and the stage of purification of inhibin was not established.

IV   Sources and physicochemical characterization of inhibin

1   PRESENCE OF INHIBIN IN TISSUES AND BODY FLUIDS

Inhibin-like activity has been detected in gonadal extracts and fluids from a large number of species, using an enormous variety of methods for qualitative or quantitative estimation of the activity. Furthermore, activity has been claimed to be present in non-gonadal, 'reproduction-related' organs and fluids like seminal plasma, prostate tissue and placenta. Results of these experiments have been summarized in Table 1.4, in which detection methods have also been indicated. Often, the inhibin-like activity has been detected using single-dose assays, that is, without demonstration of parallellism of dose-response curves with those obtained using preparations in which the presence of inhibin-activity has been ascertained using a number of *in vivo* and *in vitro* methods. Apart from the studies in which bioactivity of inhibin was estimated, the presence of the cDNA for the α-subunit of inhibin (see Section IV3i) in a human placental cDNA library was detected by Mayo, Cerelli, Spiess, Rivier, Rosenfeld, Evans, and Vale (1986). However, Davis, Dench, Nikolaidis, Clements, Forage, Krozowski, and Burger (1986) did not detect mRNA for this subunit in rat placental tissue, while they could find such mRNA in rat ovaries.

The importance of establishing the presence of inhibin-activity by the use of well defined methods becomes clear from the results of Scott *et al.* (1982) and Robertson *et al.* (1982), who indicated that the FSH-suppression observed after addition of human seminal plasma to cultured pituitary cells, was largely due to toxic effects of compounds in the seminal plasma, as revealed by concomitant suppression of LH and release of radiolabelled chromate ions from the cells.

2   REGULATION OF INHIBIN PRODUCTION

i   In vivo *experiments*

There is only a small number of experiments in which the regulation of inhibin production and secretion has been studied in male animals without resorting to

Table 1.4
Sources of inhibin, and methods used for the detection of inhibin in these materials

Source	Method (Suppression of)	Reference
Ovarian follicular fluid		
– bovine	FSH in castrated male rats	de Jong and Sharpe (1976)
– porcine	FSH in intact/castrated female rats	Welschen et al. (1977) Marder et al. (1977)
– ovine	FSH secretion from hemipituitaries	Davies et al. (1978)
– rat	FSH secretion from pituitary cells	Fujii et al. (1983)
– mare	FSH in ovariectomized mares	Miller et al. (1979)
– human	FSH in castrated male rats	Daume et al. (1978)
– monkey	FSH secretion from pituitary cells	Channing et al. (1981)
Testes extracts		
–bovine	FSH in castrated rams	Keogh et al. (1976)
	FSH in castrated male rats	Nandini et al. (1976)
– ovine	uterine weight in mice	Shashidhara Murthy et al. (1979)
	FSH in castrated male rats, ovarian weight in mice	Sheth et al. (1979)
– rat	FSH in castrated male rats	Davies et al. (1978)
Rete testis fluid		
– ovine	FSH in male rats	Setchell and Jacks (1974)
– porcine	uterine weight	Setchell and Sirinath-singhji (1972)
Testicular lymph		
– ovine	FSH secretion from pituitary cells	Baker et al. (1978)
Placenta		
– rabbit	FSH secretion from pituitary cells	Hochberg et al. (1981)
– human	FSH content of pituitary cells, radioimmuno-assay	McLachlan et al. (1986c)

estimation of peripheral levels of FSH as a parameter for inhibin secretion. Testicular levels of inhibin activity appear not to be indicative for inhibin secretion (Au, Robertson, and de Kretser 1984a), which can only be estimated after ligation of the efferent ducts, causing intratesticular collection of the tubular fluid. This procedure will measure the amount of inhibin, secreted into the luminal compartment of the tubule, while the secretion to the outside of the tubule will not be measured. This may be important, since differential secretion of compounds to intra- and extralumi-nal compartments has been described (Gunsalus, Musto, and Bardin 1980), and the ligation of the tubules does not affect FSH secretion significantly within 72 h (Main, Davies, and Setchell 1978; Au, Robertson, and de Kretser 1984b).

Using the method of efferent duct ligation, Au, Robertson, and de Kretser (1985) observed that the secretion of inhibin was decreased after hypophysec-tomy. The authors also described stimulation of inhibin secretion after administration of FSH to hypophysectomized rats, while administration of testosterone or human chorion gonadotrophin had no effect.

In female animals, Lee, McMaster, Quigg, Findlay, and Leversha (1981) and Lee, McMaster, Quigg, and Leversha (1982) showed that the injection of pregnant mare serum gonadotrophin into immature rats caused a spectacular increase in peripheral levels of inhibin, probably as a result of an increase of the amount of mRNA for inhibin in the ovaries of these animals (Davis *et al*. 1986). No further *in vivo* studies have been published on the regulation of the secretion of inhibin by exogenous hormones. The influence of the cyclic growth of ovarian follicles on the levels of inhibin in the follicular fluid and ovarian vein plasma will be discussed in Section V.

ii   In vitro *experiments*

Inhibin is produced in the Sertoli cells in the testis and in granulosa cells in the ovary. Apart from the indirect evidence, supporting the role of these cells as origin of inhibin (see Section II3), more direct indications for the production of inhibin in these cell types were obtained from studies in which isolated testicular and ovarian cell types were cultured *in vitro*, and in which the spent media from these cells were assayed for inhibin-activity. Since the amounts of inhibin which are produced *in vitro* are relatively small, this type of study has only been performed using cultured pituitary cells as assay system for inhibin secretion.

The first direct evidence for the production of inhibin by Sertoli cells was published by Steinberger and Steinberger (1976), who observed a decrease of the release of FSH from pituitary cells, which were cultured together with Sertoli cells or in the presence of spent medium from cultured Sertoli cells. In contrast, de Jong *et al*. (1979a) found increased secretion of gonadotrophins in cocultures of pituitary and Sertoli cells, especially when the cells were cultured in the absence of fetal calf serum. However, the latter authors and many others (see Table 1.5) confirmed that spent medium from Sertoli cells

Table 1.5
In vitro *secretion of inhibin by Sertoli and granulosa cells*

Inhibin production influenced by	effect*	Reference
Sertoli cells		
age of animal	−	Steinberger (1981)
	−	Massicotte *et al.* (1984)
	−	Ultee-van Gessel and de Jong (1987)
temperature	−	Steinberger (1979)
	+	Ultee-van Gessel *et al.* (1986)
FSH	+	Steinberger *et al.* (1981)
	=	Verhoeven and Franchimont (1983)
	+	Le Gac and de Kretser (1982)
	+	Ultee-van Gessel *et al.* (1986)
testosterone	+	Verhoeven and Franchimont (1983)
	−	Ultee-van Gessel *et al.* (1986)
spermatogenic cells	+	Steinberger (1979)
	−	Ultee-van Gessel *et al.* (1986)
Granulosa cells		
androgens	+	Henderson and Franchimont (1981, 1983)
atresia	−	Henderson *et al.* (1984*a*)
duration of culture	−	Hermans *et al.* (1982*b*)
day of cycle	+	Sander *et al.* (1984)
gonadotrophins	−	Henderson and Franchimont (1981)
FSH	+	Henderson *et al.* (1984*a*)
addition of serum	+	Croze and Franchimont (1984)

* +: stimulating effect
 −: inhibiting effect
 =: no effect

suppresses FSH release in *in vitro* bioassays for inhibin activity; dose response curves are parallel to those obtained after addition of follicular fluid from various species to pituitary cells in culture, indicating similar actions on the pituitary of inhibin from these sources.

Results of experiments designed to reveal the factors which regulate the secretion of inhibin from Sertoli cells are also summarized in Table 1.5. Hormonal effects have been described by a large number of authors, who obtained varying results. Franchimont *et al.* (1983) and Verhoeven and Franchimont (1983) reported stimulated secretion of inhibin after addition of testosterone or 5α-dihydrotestosterone to the medium; this effect could be blocked by addition of the antiandrogen cyproterone acetate. On the other hand, Ultee-van Gessel, Leemborg, de Jong, and van der Molen (1986) reported that in their system, where hormones were added sooner after the isolation of the Sertoli cells, testosterone suppressed the release of inhibin. Stimulatory effects of FSH on the secretion of inhibin from cultured

Sertoli cells have been described by Steinberger (1981), le Gac and de Kretser (1982) and Ultee-van Gessel *et al.* (1986), while Franchimont *et al.* (1983) and Verhoeven and Franchimont (1983) could not detect stimulatory effects of FSH on the production of inhibin by Sertoli cells *in vitro*. Finally, Ultee-van Gessel *et al.* (1986) reported that the addition of testosterone and FSH to Sertoli cells suppressed the secretion of inhibin, when compared with the effect of addition of FSH alone. The cause for the differences in the observations of the various groups of investigators is not clear; it might be that the presence or absence of peritubular cells, which may play an important role in the mediation of effects of androgens on Sertoli cell function (Skinner and Fritz 1986) can explain these results. Furthermore, the use of fetal calf serum in the medium of the Sertoli cells may play a role: Le Gac and de Kretser (1982) indicated that the medium of Sertoli cells cultured in the presence of fetal calf serum may have toxic effects on pituitary cells in culture.

Effects of temperature on the secretion of inhibin by Sertoli cells *in vitro* have been studied because of the stimulatory effect of increased testicular temperature on peripheral levels of FSH *in vivo*. Steinberger (1981) indicated that after increasing the culture temperature of the Sertoli cells a diminished secretion of inhibin into the medium was observed, while Ultee-van Gessel *et al.* (1986) found a positive influence of increased temperature on the secretion of inhibin. The latter observation is in keeping with experiments on the effect of increased temperature on the release of oestradiol and ABP by Sertoli cells (Rommerts, de Jong, Grootegoed, and van der Molen 1980) and suggests that it is not the high temperature *per se* which impairs the secretion of inhibin *in vivo*.

Little certainty exists on the influence of spermatogenic cells on the production of inhibin by Sertoli cells. The experiments of Steinberger (1981) suggest that the presence of spermatogenic cells (spermatocytes or spermatids) does not affect the secretion of inhibin by Sertoli cells *in vitro*. However, in these studies the amount of inhibin added to the pituitary cells caused maximal suppression of the secretion of FSH, thus making it impossible to judge if the addition of the spermatogenic cells increased, or even decreased, the amount of inhibin secreted. In similar experiments Ultee-van Gessel *et al.* (1986) indicated that at 32° C the recombination of spermatogenic cells with Sertoli cells suppressed the amount of inhibin in the medium of these cells, while the addition of thymocytes had no such negative influence.

Granulosa cells are the source of inhibin from the ovary, as shown first by Erickson and Hsueh (1978), who cultured rat granulosa cells, and added the medium from these cells to cultured pituitary cells. This observation has been amply confirmed (Table 1.5). From these studies it becomes clear that FSH but also androgens and LH may be major factors in the regulation of inhibin secretion by granulosa cells from bovine ovaries (Henderson and

Franchimont 1981, 1983) and human infant ovaries (Channing, Tanabe, Chacon, and Tildon 1984). Henderson, Franchimont, Charlet-Renard, and McNatty (1984) indicated that FSH could only stimulate inhibin production by granulosa cells from healthy bovine follicles, while testosterone caused an increase of inhibin production by granulosa cells from both healthy and atretic follicles. This effect of testosterone could also be provoked by synthetic androgens, and could be blocked by addition of anti-androgens (Henderson and Franchimont 1983). Finally, Sander, van Leeuwen, and de Jong (1984) showed that the production of inhibin by rat granulosa cells depends on the stage of the oestrous cycle in which granulosa cells are harvested. These authors were not able to show stimulatory effects of FSH on inhibin secretion *in vitro*.

3 PURIFICATION AND CHARACTERIZATION OF INHIBIN

The subject of purification and characterization of inhibin has been reviewed extensively during the last few years (de Jong, Jansen, and van der Molen 1981; Baker *et al.* 1982a; Channing, Gordon, Liu, and Ward 1985b; de Jong and Robertson 1985). The apparent physicochemical characteristics of inhibin, as published before 1985, have been tabulated by de Jong and Robertson (1985). These data have been incorporated into Table 1.6, which also includes more recent information. In the next sections, only these more recent data will be discussed.

i Gonadal inhibin

Several groups of investigators have isolated proteins with inhibin activity from porcine and bovine ovarian follicular fluid. Most of these investigators reported apparent molecular weights of 32,000 for the biologically active protein (see Table 1.6), while Robertson, Foulds, Leversha, Morgan, Hearn, Burger, Wettenhall, and de Kretser (1985) found a molecular weight of 58,000. The latter authors also indicated that this 58 kD protein could be transformed into a 32 kD protein after inclusion of an acid precipitation step in the purification procedure (Robertson, de Vos, Foulds, McLachlan, Burger, Morgan, Hearn, and de Kretser 1986). The 32 and 58 kD inhibins consist of two dissimilar α and β subunits, as becomes apparent after reduction with mercaptoethanol; the isolated subunits have no biological activity. Amino acid sequences of the N-terminal ends of the subunits of the isolated 32 kD inhibin molecules are very similar for porcine and bovine material.

On basis of the information obtained from these amino acid sequences, Mason, Hayflick, Ling, Esch, Ueno, Ying, Guillemin, Niall, and Seeburg (1985), Forage, Ring, Brown, McInerney Cobon, Gregson, Robertson, Morgan, Hearn, Findlay, Wettenhall, Burger and de Kretser (1986), and Mason, Niall, and Seeburg (1986) were able to isolate and sequence the

Table 1.6

Reported molecular weights for inhibin from testicular and ovarian sources[*].
(Partly from de Jong and Robertson, 1985)

Source	Molecular weight (kD)	Estimated using	Inhibin activity measured by suppression of	Reference
Testis				
Human testis	'low'	gel filtr.	ovarian weight + FSH *in vivo*	Krishnan et al. (1982)
Ovine RTF	30	SDS-PAGE	FSH *in vitro*	Baker et al. (1982b)
Ovine testis	20	gel filtr.	Uterine weight	Moudgal et al. (1984)
Ovine testis	<1.5	gel filtr.	Ovarian weight + FSH *in vitro*	Vijayalakshmi et al. (1980)
Rat testis	50–60	gel filtr.	FSH *in vitro*	Au et al. (1983)
Ovarian follicular fluid				
Bovine	32	SDS-PAGE	FSH *in vitro*	Fukuda et al. (1986)
	32	SDS-PAGE	FSH *in vitro*	Robertson et al. (1986)
	55	SDS-PAGE	FSH *in vitro*	Robertson et al. (1985)
	65	SDS-PAGE	FSH *in vitro*	Jansen et al. (1981)
	65	SDS-PAGE	FSH *in vitro*	van Dijk et al. (1984)
	32, 55, 65, 88, 108, 120	blotting with monoclonal antibodies after SDS-PAGE		Miyamoto et al. (1986)
Human	23	SDS-PAGE	Ovarian weight + FSH *in vivo*	Chari et al. (1979)
Ovine	<1.5	Gel filtr.	Ovarian weight + FSH *in vivo*	Vijayalakshmi et al. (1980)
	80	Gel filtr.	FSH *in vitro*	Dobos et al. (1983)
Porcine	10–12	Gel filtr.	FSH *in vitro*	Rivier et al. (1984)
	10–35	Gel filtr.	FSH *in vivo*	Williams et al. (1979)
	25–30, 43	Gel filtr.	FSH *in vitro*	Sairam et al. (1984)
	32	SDS-PAGE	FSH *in vitro*	Rivier et al. (1985)
	32	SDS-PAGE	FSH *in vitro*	Ling et al. (1985)
	32, 55, 80, 100	SDS-PAGE	FSH *in vitro*	Miyamoto et al. (1985)
	194	Rad. inact.	FSH *in vivo*	Ward et al. (1983)

[*]gel filtr: gel filtration
SDS-PAGE: sodium dodecyl sulphate – polyacrylamide gel electrophoresis
Rad. inact.: radiation inactivation

nucleotides in the cDNA for the inhibin subunits from porcine, ovine and human granulosa cells. Mayo *et al.* (1986) published a nearly identical structure for the α-subunit of porcine inhibin. The following data can be derived from these nucleotide sequences: the subunits are separately synthesized in a pre-pro form. The α-subunit, of which the pre-pro form consists of about 360 amino acids, contains pairs of basic amino acids which precede the position of the amino acids, which were found to be present in the N-terminal sequence of the α-subunits of the 58 kD and 32 kD inhibins. Similarly, the proform of the β-subunit contains 5 basic amino acids, preceding the amino acids found to be present in the N-terminal sequence of the isolated β-subunit, for which Mason *et al.* (1985, 1986) published two isoforms. The α-subunit of porcine, bovine and human inhibin contain two, two and three glycosylation sites, respectively. This is consistent with reports on the glycoprotein character of gonadal inhibin (Jansen, Steenbergen, de Jong, and van der Molen 1981; de Jong *et al.* 1982*b*; Lecomte-Yerna, Hazee-Hagelstein, Charlet-Renard, and Franchimont 1984; van Dijk, Steenbergen, de Jong, and van der Molen 1985).

These results can partly explain the confusion in the literature about the physicochemical characteristics of inhibin: cleavage of the complete pre-pro-hormone, which would have a molecular weight of about 100,000, leads to a molecular weight around 60 K as suggested by a number of authors (see Table 1.6). This 60 kD molecule might be split up peripherally into the 32 kD protein, which apparently has biological activity. The reason why a number of authors detected 32 kD inhibin in follicular fluid extracts may have been because of specific steps in the isolation procedures: Channing *et al.* (1985*b*) indicated that acid conditions may inactivate protease inhibitors, which are present in follicular fluid together with a number of proteases. The various sizes of the inhibin molecules, which may be predicted on basis of this model, have been shown to be present in bovine follicular fluid by Miyamoto *et al.* (1986), who detected 32 kD, 55 kD and 65 kD molecules using monoclonal antibodies against the α- and β-subunits after polyacrylamide gel electrophoresis and subsequent blotting experiments. These authors also described 120, 108 and 88 kD molecules, detected by this technique, which they propose to be trimers of one α-subunits, and 2 β-subunits of different length.

Finally, in order to explain the very high molecular weights postulated by Ward, Glenn, Liu, and Gordon (1983); Channing, Liu, Gordon, Xue, and Ward (1984) and Channing *et al.* (1985*b*), one has to postulate aggregation of the – relatively hydrophobic (Jansen *et al.* 1981) – molecule.

ii Structural relationships between gonadal inhibin and other regulatory proteins

After the structure of porcine inhibin had been elucidated, Mason *et al.* (1985) noticed a significant homology between the β-subunit of inhibin and transforming growth factor-β (TGF-β). Subsequent experiments indicated that the secretion of FSH from pituitary cells can be stimulated 1.5-fold by addition of TGF-β. This effect was specific: the secretion of other pituitary hormones was not affected (Ying, Becker, Baird, Ling, Ueno, Esch, and Guillemin 1986a). These authors also found that a fraction of porcine follicular fluid could exert a similar stimulatory effect on FSH secretion, and postulated the presence of a TGF-β-like molecule in the follicular fluid. The molecule with this TGF-β-like activity was identified as a homo- or heterodimer of the inhibin-β-subunit(s) by Ling, Ying, Ueno, Shimasaki, Esch, Hotta, and Guillemin (1986a, b) and Vale, Rivier, Vaughan, McClintock, Corrigan, Woo, Karr, and Spiess (1986). These authors coined the names 'activin' and FSH-releasing protein (FRP), respectively, for the active principle. The question if these dimers of the inhibin β-subunit, produced in the ovary, can reach the circulation in quantities which are large enough to affect pituitary FSH secretion has not yet been resolved. However, the possibility of inhibin-antagonizing effects of these substances after the addition of gonadal preparations to the culture medium of pituitary cells cannot be excluded. This may make it mandatory to separate inhibin from FRP in order to make it possible to estimate bioactivity of inhibin in gonadal preparations.

Finally, a homology between the amino-acid sequence of the β-subunit of inhibin and the C-terminal end of Müllerian inhibiting substance (MIS) has been found by Cate, Mattaliano, Hession, Tizard, Farber, Cheung, Ninfa, Frey, Gash, Chow, Fisher, Bertonis, Torres, Wallner, Ramachandran, Ragin, Manganaro, MacLaughlin, and Donahoe (1986). These findings suggest that the genes for inhibin, TGF-β and MIS belong to one gene family. Especially the observation that various combinations of subunits in one molecule may cause opposing biological activities of the resulting hormones makes the study of these molecules very interesting (Tsonis and Sharpe 1986).

iii   Non-Gonadal inhibin-like molecules

Two peptides with inhibin-like activity in *in vivo* and *in vitro* assay systems have been isolated from human seminal plasma. The amino acid sequences of these peptides were elucidated by Ramasharma, Sairam, Seidah, Chrétein, Manjunath, Schiller, Yamashiro, and Li (1984) and Seidah, Ramasharma, Sairam, and Chrétien (1984), who described a 32 amino acid peptide (α-inhibin), and by Sheth, Arabatti, Carlquist, and Jörnvall (1984) and Seidah, Arbatti, Rochemont, Sheth and Chrétien (1984), who described a larger, 94-amino acid peptide (β-inhibin). Later, Arbatti, Seidah, Rochemont, Escher, Sheth, and Chrétien (1985) indicated that the biological activity of the 94 amino acid peptide resided in the 28 C-terminal amino acids, while Li, Hammonds, Ramasharma, and Chung (1985) claimed that the amino acid sequence of the 32 amino acid peptide described by Ramasharma *et al*. (1984) is also found in a 52 and a 92 amino acid peptide which can be isolated from human seminal plasma.

There is no homology between α- and β-inhibin. The amino acid sequence of β-inhibin was also described by Johansson, Sheth, Cederlund, and Jörnvall (1984), who suggested possible identity of β-inhibin with a sperm-coating antigen, which is synthesized in prostatic epithelium, and by Akiyama, Yoshioka, Schmid, Offner, Troxler, Tsuda, and Hara (1985), who described the peptide as β-microseminoprotein. Both authors mention homology with the peptide PDC-109, which was isolated from bovine seminal plasma by Esch, Ling, Böhlen, Ying, and Guillemin (1983), and which showed – albeit at high concentrations – FSH suppressing activity (Esch, Ling, Ying, and Guillemin 1983). Finally, Lilja and Jeppson (1985) indicated that α-inhibin might be of seminal vesicle origin.

The peptide described by Ramasharma *et al*. (1984) has been synthesized, and was shown to possess biological activity *in vivo*, and also *in vitro* in a system with incubated LHRH stimulated mouse pituitaries (Yamashiro, Li, Ramasharma, and Sairam, 1984). However, other authors have not been able to ascertain the biological activity of this material (de Jong, Sander, Ultee-van Gessel, and van der Molen 1985; Liu, Booth, Merriam, Barnes, Sherins, Loriaux, and Cutler 1985). Similarly, Kohan, Fröysa, Cederlund, Fairwell, Lerner, Johansson, Khan, Ritzen, Jörnvall, Cekan, and Diczfalusy (1986) showed that synthetic β-inhibins did not possess inhibin-bioactivity in an *in vitro* bioassay system with dispersed pituitary cells.

4 CONCLUSIONS

The recent data on the isolation and characterization of inhibin from ovarian follicular fluid show extensive homology between the molecules from various species. Bioactivities of the isolated proteins have been adequately characterized, while the physiological meaning of the two differing β-subunits remains to be elicidated. The significant homologies between the β-subunit of inhibin, TGF-β and MIS, suggest that these proteins are all derived from one gene family. The opposing biological effects of α-β and β-β dimers indicate that combination of varying subunits may provide an important mechanism for determining the biological activity of these hormones.

The difficulty in confirming bioactivity of seminal plasma inhibins is striking. It is most likely that these peptides have a function different from the suppression of peripheral levels of FSH *in vivo*.

V Physiological significance of inhibin

Experiments are scarce in which peripheral concentrations of inhibin have been estimated using adequate methods (Lee *et al*. 1981, 1982; McLachlan *et al*. 1986*b*). This means that most of the available evidence about the conditions in which inhibin might play a role in the regulation of the secretion of FSH, and therefore in the regulation of fertility, is circumstantial. This indirect evidence will be reviewed in the following sections.

1   INHIBIN IN MALE ANIMALS

i   Prepubertal male animals

The vast majority of data of the possible role of inhibin in the regulation of FSH secretion in immature male animals has been derived from experiments with rats. In this species, testicular development starts during the perinatal period, and it has been shown that this process can be influenced by FSH: this hormone stimulates the mitosis of Sertoli cells (Orth 1984), influences the microenvironment of spermatogenic cells by stimulating the production and secretion of a number of substances in Sertoli cells (Jutte, Jansen, Grootegoed, Rommerts, and van der Molen 1983) and might influence the maturation of Leydig cells in immature rats (Kerr and Sharpe 1985*a, b*).

These observations lead to the suggestion that a specific feedback on FSH secretion – that is, inhibin – should be present during the prepubertal period; this assumption is supported by the following experimental evidence obtained after bilateral or unilateral orchidectomy, testicular irradiation and estimation of inhibin in testes of rats of various ages.

Bilateral orchidectomy in immature male rats leads to a relatively fast increase in peripheral concentrations of FSH, while the response of LH to this operation takes much more time (Hermans *et al*. 1980). Eight hours after castration of 15 day old male rats FSH levels increased to approximately 300 per cent of control values, while in adult animals FSH levels amounted to only 120 per cent of those in sham-operated controls (Fig. 1.4). These increases could not be prevented by the administration of amounts of testosterone which resulted in normal peripheral levels of this steroid. However, administration of steroid-free bovine follicular fluid at the time of operation could prevent the increase of FSH levels at both ages. This suggests that inhibin plays a larger role in FSH feedback in immature than in adult animals.

Similarly, the compensatory testicular hypertrophy, which occurs in male rats after hemicastration early in life, and which is thought to be due to increased levels of FSH as a result of impaired feedback action by the remaining testis, can only be evoked if the hemicastration takes place before the age of 21 days, and can be prevented by the injection of inhibin-containing preparations (Selin and Moger 1979). After hemicastration on the first day of life, peripheral FSH levels are selectively increased on day 21, but on days 42 and 63 of life the stimulation of FSH levels has been

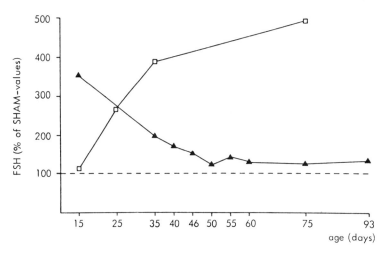

Fig. 1.4. Mean changes in concentrations of FSH 8 h after bilateral castration of male (▲) or female (□) rats of different ages. Results are expressed as percentages of the values in sham-operated control animals (broken line). (From Hermans *et al.* 1980 by permission).

abolished, although the compensatory weight increase of the remaining testis is similar at all ages studies. Similar data were found for the release of FSH from the pituitary glands of these animals *in vitro* (Ultee-van Gessel, Leemborg, de Jong, and van der Molen 1985).

Neonatal irradiation resulted in decreased numbers of Sertoli and spermatogenic cells in the testes of rats 21, 51 and 81 days old, and in significantly increased concentrations of FSH in peripheral plasma from the 21 days old animals (de Jong and Sharpe 1977). Significant negative correlations were found between Sertoli cell numbers and FSH levels at all ages studied. However the slope of these regression lines decreased from day 21 to day 81 of age, which again suggests that the Sertoli cell plays a more important role in FSH feedback in the immature animal than in the adult male rat.

More direct evidence about the importance of inhibin in FSH feedback in immature animals can be derived from experiments in which testicular content of inhibin in testes of rats of various ages was estimated. Ultee-van Gessel and de Jong (1987) showed that the concentration of inhibin in testes from immature rats is significantly higher than that in testis from mature rats, although the total amount of inhibin per testes was larger in adult animals. A confounding factor in this comparison is the changing architecture of the testis: the numbers of spermatogenic cells increase progressively, while around day 21 the blood testis barrier is formed (Dym and Fawcett 1970). Furthermore, results from experiments in which Sertoli cells from rats of different ages were cultured in short-term experiments show that the amount of inhibin activity produced per immature Sertoli cell is larger than that produced by Sertoli cells from adult testes (Massicotte, Lagacé, Labrie, and Dorrington 1985; Ultee-van Gessel and de Jong 1987). The fact that this effect of age was not detected by Steinberger (1979) is probably due to the dose of Sertoli cell culture medium used by this author to assess inhibin activities: at all ages tested a maximal suppression of *in vitro* FSH release was found.

Finally, injection of inhibin-containing material into immature rats causes a decrease of testicular weight and temporary arrest of the development of the spermatogenic epithelium (de Jong *et al.* 1978). However, these experiments should be viewed with caution because of the possibility of the presence of non-inhibin-related factors in the follicular fluid used, which might impair the interaction of FSH with its receptor (Sluss and Reichert 1984).

Extrapolation of these experimental data, obtained in rats, to the situation in other experimental animals or man is difficult. However, the different ratios between LH and FSH in pre- and neonatal male and female pigs (Colenbrander, MacDonald, Elsaesser, Parvizi, and van de Wiel 1982; Colenbrander, Meijer, MacDonald, van de Wiel, Engel, and de Jong 1987) and in boys and girls going through puberty (see Jenner *et al.* 1972) suggest that similar mechanisms might also play a role in other species.

ii   Adult male animals

The results of most studies on the regulation of testicular function in adult animals indicate that FSH plays only a minor, if any, role in the maintenance of spermatogenesis (Cunningham, and Huckins 1979). This may be the reason for the difficulties which were encountered in establishing the role of inhibin in the regulation of FSH in adult males: its significance appears to be limited or modulatory in species which do not have a clear seasonal rhythm in their reproductive cycle (Plant *et al.* 1978; Lipner and Dhanarajan 1984; Jones, Wood, and Rush 1985).

Administration of exogenous inhibin has been reported not to have profound effects in the adult rat (Davies, Main, Laurie, and Setchell 1979), and the only study in which effects in adult animals have been reported was that by Moudgal, Murthy, Murthy, and Rao (1984), who showed that repeated injection of a purified inhibin preparation in monkeys might impair spermatogenesis.

The regulation of the production of inhibin in the adult testis has been studied by Au *et al.* (1985), who reported an increase of the content of inhibin in the ligated testes of hypophysectomized rats, which were treated with FSH. LH preparations did not have any effects in this model.

Finally, in animals in which reproduction has a seasonal rhythm, inhibin might be important for the regulation of FSH during the period of growth of the testis: it has been shown that hemicastration of rams causes a compensatory hypertrophy of the remaining testis (Hochereau de Reviers, Loir, and Pelletier 1976), indicating that in this situation specific regulation of FSH, presumably by inhibin, plays a role in the regulation of testis function.

2   INHIBIN IN FEMALE ANIMALS

i   Prepubertal female animals

The role of inhibin in the feedback regulation of FSH secretion has been studied in immature female rats by a variety of techniques, including unilateral and bilateral ovariectomy, injection of inhibin-containing preparations and estimation of ovarian inhibin content at various stages of development.

The high level of FSH found in the peripheral blood of 15 day old female rats is not affected at 8 hours after bilateral ovariectomy, suggesting that at this age no inhibin feedback exists (Fig. 1.4). At the age of 25 days, a clear increase of FSH is found after the same period, which is even larger at later ages. These results correlate well with the observations of Sander, Meijs-Roelofs, Kramer, and van Leeuwen (1985), who measured endogenous inhibin in ovarian homogenates of rats of various ages. These authors found no detectable inhibin at 13 days of age, very low levels at 18 days and increasing amounts of inhibin from that day onwards, until peak levels were observed on the day of first ovulation. Ovarian inhibin levels were negatively correlated with peripheral FSH concentrations from 5 days before the first ovulation until the day of first ovulation (Sander, Meijs-Roelofs, van Leeuwen, Kramer, and van Capellen 1986). It remains unclear if inhibin also plays a role in the steady decline of peripheral FSH levels before that time (Meijs-Roelofs, Osman, and Kramer 1982), since no data exist about the relationship between ovarian and circulating levels of inhibin.

Unilateral ovariectomy in immature female rats leads to an increase of FSH levels within 5 to 8 hours from the age of 23 days onwards (Sander *et al*. 1986). This increase disappears after 24 h, when the number of developing follicles in the remaining ovary has been adjusted. Further support for the idea that inhibin feedback plays a role in the regulation of FSH secretion in prepubertal female rats has been obtained by injection of exogenous inhibin, in the form of steroid-free bovine follicular fluid. After an initial suppression of FSH, a clear overshoot ensues, indicating a lack of endogenous inhibin after the period of diminished FSH levels (Hermans *et al*. 1981). This phenomenon may be observed in rats of 25 days old.

The regulation of inhibin secretion in immature rats has been studied by Lee *et al*. (1981, 1982), who showed that peripheral levels of inhibin, as measured using dispersed pituitary cells, increase dramatically after injection of PMSG.

Finally, inhibin activity has been detected in ovarian follicular fluid from girls and monkeys during the neonatal period (Channing, Chacon, Tanabe, Gagliano, and Tildon 1984; Channing, Tanabe, Hahn, Phillips, and Carraher 1984). The physiological meaning of this observation is not clear.

ii   Adult female animals

Studies on the possible physiological role of inhibin in adult female animals have been reviewed earlier by Welschen, Hermans, and de Jong (1980), Grady, Charlesworth, and Schwartz (1982) and Channing et al. (1985b).

Bilateral ovariectomy of adult rats on the second day of dioestrus leads to a rapid specific increase in peripheral levels of FSH, which may reach a five-fold augmentation in 8 hours (Fig. 1.4). This increase cannot be suppressed by injection of 'physiological' amounts of steroids (Hermans et al. 1980), but is readily reversible by injection of steroid-free bovine follicular fluid, which contains inhibin-like activity.

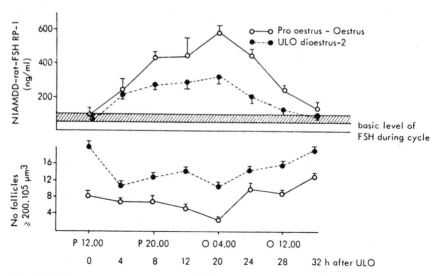

Fig. 1.5. Plasma concentrations of FSH and numbers of large follicles per ovary during the prooestrus-oestrus peak of FSH and after unilateral ovariectomy (ULO) on dioestrus-2. (From de Jong et al. 1980 by permission).

A similar increase of FSH occurs after unilateral ovariectomy (Welschen, Dullaart, and de Jong 1978): plasma FSH levels increase three- to fourfold within a 4 to 8 h period, and are only suppressed to control levels after the number of large follicles in the remaining ovary equals that in both ovaries at the time of operation (Fig. 1.5). This increase of FSH is not followed by an increase of LH levels, cannot be suppressed by steroids, and is reversible by injection of follicular fluid. Finally, the secondary peak of FSH, which is found after ovulation in the rat and lasts until a new cohort of large ovarian follicles is present in the ovary (Fig. 1.5, Welschen *et al.* 1980), is a third example of an increase of FSH levels, which is not accompanied by a rise of LH concentrations and cannot be suppressed by injecting steroids, but which can be prevented by the injection of follicular fluid (Schwartz and Channing 1977). The inverse relationship between the number of large ovarian follicles, which is directly related to the number of granulosa cells, and peripheral levels of FSH in these adult female rats strongly resembles the relationship between the numbers of Sertoli cells and peripheral FSH in neonatally irradiated immature male rats. When expressed in terms of the latter model, one might argue that 'puberty' occurs in the female at the development of each new cohort of follicles.

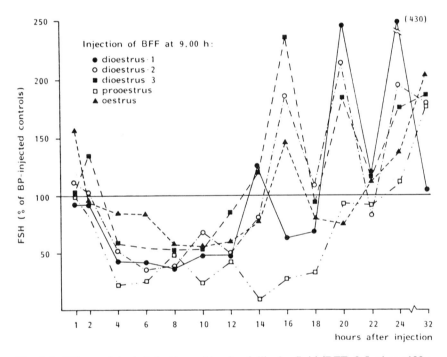

Fig. 1.6. Effects of single injections of bovine follicular fluid (BFF, 0.5 ml per 100 g body weight) at various days of the oestrous cycle on FSH concentrations in adult female rats. Results are shown as percentage of FSH concentrations in bovine plasma (BP) injected controls. (From de Jong *et al.* 1983*a* by permission).

This counterbalancing system probably becomes most clear when one looks at the effects of single injections of steroid-free bovine follicular fluid at various days of the oestrous cycle: initially FSH levels are reduced, but after 16 to 24 h a compensatory, apparently oscillating overshoot of FSH levels occurs (Fig. 1.6, Hermans *et al.* 1982). This overshoot may be the reason for ineffectiveness of daily injections of follicular fluid in intact female rats: the animals went on having normal cycles during 4 weeks of injections, and even showed increased FSH levels after this period (de Jong *et al.* 1978). In ewes, however, daily administration of a large dose of bovine follicular fluid was effective in suppressing FSH and preventing ovulation for a 27-day treatment period (Henderson, Prisk, Hudson, Ball, McNatty, Lun, Heath, Kieboom, and McDiarmid 1986). A similar overshoot of FSH levels is likely to be the cause of the increased ovulation rate, which was observed in ewes after administration of follicular fluid (Wallace and McNeilly 1985; Wallace, McNeilly, and Baird 1985; Henderson *et al.* 1986). These data contrast with the results of Stouffer and Hodgen (1980), who observed decreased preovulatory oestradiol levels and midluteal progesterone levels after repeated injections of porcine follicular fluid to rhesus monkeys on days 1 to 3 of the menstrual cycle, and concluded that FSH-dependent events during the early follicular phase are important determinants of subsequent function of the corpus luteum.

Active immunization against inhibin also leads to increased ovulation rates in ewes (Henderson, Franchimont, Lecomte-Yerna, Hudson, and Ball 1984; Bindon, Piper, Cahill, Driancourt, and O'Shea 1986), and it is probable that a genetic defect in the biosynthesis of inhibin, resulting in low follicular fluid levels of the hormone, may be responsible for the increased fecundity of Booroola Merino ewes (Cummins, O'Shea, Bindon, Lee, and Findlay 1983; cf. Bindon and Piper 1986).

The regulation of inhibin secretion throughout the oestrous cycle has not been completely elucidated yet. Sander *et al.* (1984) showed that granulosa cells, harvested from rat ovaries at different days of the cycle, produced increasing amounts of inhibin towards the day of prooestrus, when highest production was estimated. This observation is in agreement with data of Chappel (1979), who measured inhibin in hamster ovarian homogenates, and of Sander *et al.* (1985) for rat ovarian homogenates. Fujii, Hoover, and Channing (1983) also estimated highest levels of inhibin activity in rat follicular fluid on the day of prooestrus, while DePaolo *et al.* (1979*b*) showed highest levels of inhibin in ovarian venous plasma at this time of the cycle. However, these increasing levels of inhibin in ovarian venous plasma are not reflected in decreasing levels of FSH in peripheral plasma.

Studies on relationships between the size of the follicles, and the number of granulosa cells per follicle on the one hand, and follicular fluid inhibin levels on the other, revealed an increase in inhibin content with the size of the follicle in the ewe (Tsonis, Quigg, Lee, Leversha, Trounson, and Findlay 1983), the cow (Henderson *et al.* 1984*a*) and the pig (van de Wiel, Bar-Ami, Tsafriri, and de Jong 1983). Henderson *et al.* (1984*a*) described that small healthy and atretic follicles contained similar concentrations of inhibin activity in the follicular fluid, while in fluid from large atretic follicles lower inhibin levels were found compared to the concentrations in large healthy follicles. The *in vitro* production of inhibin by granulosa cells from healthy, but not from atretic, follicles could be stimulated after addition of FSH, in parallel with the stimulation of aromatase activity in these cells.

Also in the human the amount of inhibin in follicular fluid varies with the stage of the menstrual cycle: highest levels are found in preovulatory follicles, while little inhibin was detected in follicles harvested during the luteal phase (Chappel *et al*. 1980; Lefevre, Kraiem, Epstein, Lunenfeld, Gougeon, Thebault, Bomsel-Helmreich, Frydman, Papiernik, Saltarelli, Brailly, and Milgrom 1981; Channing, Gagliano, Hoover, Tanabe, Batta, Sulewski, and Lebech 1981; Channing, Tanabe, Seegar Jones, Jones, and Lebech 1984*e*; Channing, Gagliano, Tanabe, Fortuny, and Cortes-Prieto 1985*a*). Treatment with human menopausal gonadotrophins in an *in vitro* fertilization programme resulted in increased inhibin levels in follicular fluid (Channing *et al*. 1984*e*). Similar results were obtained in rhesus monkeys (Schenken, Anderson, and Hodgen 1984). The levels of inhibin in follicular fluid were positively correlated with those of oestrogen, but negatively with progesterone levels; inhibin levels decreased as soon as luteinization of the granulosa cells started.

Inhibin might also play a role in the pathogenesis of polycystic ovarian (PCO) disease. In this condition the concentration of FSH in peripheral plasma is low relative to that of LH. This may be caused by the increased secretion of inhibin activity in ovarian venous blood (Tanabe, Gagliano, Channing, Nakamura, Yoshimura, Iizuka, Fortuny, Sulewski, and Rezai 1983), although the intrafollicular inhibin concentration was normal. Furthermore, the lack of developing follicles during the early menopause may be the reason for the specific increase in circulating concentrations of FSH (Sherman and Korenman 1975). Another situation in which inhibin might play a role in human physiology arises after luteinization of unruptured follicles, where Koninckx, Brosens, Verhoeven, and de Moor (1981) suggested that peripheral FSH levels remain high through lack of inhibin, which may be resorbed from peritoneal fluid after normal ovulation but not after luteinization of unruptured follicles. However, Dhont, Serreyn, Duvivier, Van Luchene, de Boever, and VandeKerckhove (1984) could not confirm these data on increased FSH concentrations in women with unruptured follicles. Finally, increased FSH levels were found in women who gave birth to dizygotic twins (Martin, Olsen, Theile, El Beaini, Handelsman, and Bhatnagar 1984). This observation suggests that circulating levels of inhibin in these women are relatively low. This might cause a disregulation of the pituitary-ovarian axis which might be equivalent to that observed in high-fecundity Booroola Merino ewes (Cummins *et al*. 1983).

Summarizing, it appears that inhibin is at least one of the factors which determines the number of follicles which is destined to ovulate. In this way, interference with the action of inhibin might contribute to the regulation of fertility.

3   CONCLUSIONS

Because of the lack of reliable methods for estimation of peripheral levels of inhibin in man and experimental animals, all evidence for the physiological importance of inhibin in the regulation of reproductive processes is derived from circumstantial evidence. From these indirect results, it appears likely that inhibin plays an important role in the feedback regulation of peripheral concentrations of FSH during the period in which Sertoli cells and granulosa cells – the target cells for FSH – divide, that is, during puberty in male animals and during the development of ovarian follicles in female animals. In this way, inhibin may be an important regulator of the length of the seminiferous tubules in the testis and of the number of developing follicles in the ovary.

References

Akiyama, K., Yoshioka, Y., Schmid, K., Offner, G. D., Troxler, R. F., Tsuda, R., and Hara, M. (1985). The amino acid sequence of human β-microseminoprotein. *Biochim. Biophys. Acta* **829**, 288–294.

Arbatti, N. J., Seidah, N. G., Rochemont, J., Escher, E., Sheth, A. R., and Chrétien, M. (1985). β₂-Inhibin contains the active core of human seminal plasma β-inhibin: synthesis and bioactivity. *FEBS Lett.* **181**, 57–63.

Au, C. L., Robertson, D. M., and de Kretser, D. M. (1983). *In vitro* bioassay of inhibin in testes of normal and cryptochid rats. *Endocrinology* **112**, 239–244.

—— —— —— (1984*a*). An *in vivo* method for estimating inhibin production by adult rat testes. *J. Reprod. Fert.* **71**, 259–265.

—— —— —— (1984*b*). Relationship between testicular inhibin content and serum FSH concentrations in rats after bilateral efferent duct ligation. *J. Reprod. Fert.* **72**, 351–356.

—— —— —— (1985). Effects of hypophysectomy and subsequent FSH and testosterone treatment of inhibin production by adult rat testes. *J. Endocr.* **105**, 1–6.

Baker, H. W. G., Bremner, W. J., Burger, H. G., de Kretser, D. M., Dulmanis, A., Eddie, L. W., Hudson, B., Keogh, E. J., Lee, V. W. K., and Rennie, G. C. (1976). Testicular control of follicle-stimulating hormone secretion. *Recent Prog. Horm. Res.* **32**, 429–476.

—— Burger, H. G., de Kretser, D. M., Eddie, L. W., Higginson, R. E., Hudson, B., and Lee, V. W. K. (1978). Studies on purification of inhibin from ovine testicular secretions using an *in vitro* bioassay. *Int. J. Androl.* Suppl. **2**, 115–124.

—— Eddie, L. W., Higginson, R. E., Hudson, B., Keogh, E. J., and Niall, H. D. (1981). Assays of inhibin. In *Intragonadal regulation of reproduction* (ed. P. Franchimont and C. P. Channing) pp. 193–228. Academic Press, London.

—— —— —— —— and Niall, H. D. (1980). Testicular inhibin. In *Endocrinology 1980* (ed. I. A. Cumming, J. W. Funder and F. A. O. Mendelsohn) pp. 251–254. Australian Academy of Science, Canberra.

—— —— —— —— —— (1982*a*). Clinical context, neuroendocrine relationships, and nature of inhibin in males and females. In *Clinical neuroendocrinology*, Vol. II (ed. G. M. Besser and L. Martini) pp. 283–330. Academic Press, London.

———— ——— ——— ——— ——— (1982b). Studies on the purification of ovine inhibin. *Ann. N.Y. Acad. Sci.* **383**, 329–342.

Berthezene, F., Forest, M. G., Grimaud, J. A., Claustrat, B., and Mornex, R. (1976). Leydig-cell agenesis. A cause of male pseudohermaphroditism. *New Engl. J. Med.* **295**, 969–972.

Bindon, B. M. and Piper, L. R. (1986). The reproductive biology of prolific sheep breeds. In *Oxford Reviews of Reproductive Biology*, Vol. 8 (ed. J. R. Clarke) pp. 414–451. Clarendon Press, Oxford.

Bindon, B. M., Piper, L. R., Cahill, L. P., Driancourt, M. A., and O'Shea, T. (1986). Genetic and hormonal factors affecting superovulation. *Theriogenology* **25**, 53–70.

Boyar, R. M., Wu, R. H. K., Liao, P. Y., and Finkelstein, J. W. (1977). Endocrine and cytogenetic studies in a patient with Turner's phenotype and a ring chromosome. *J. clin. Endocr. Metab.* **44**, 340–345.

Braunstein, G. D. and Swerdloff, R. S. (1977). Effect of aqueous extracts of bull and rat testicles on serum FSH and LH in the acutely castrate male rat. In *The testis in normal and infertile men* (ed. P. Troen and H. R. Nankin) pp. 281–291. Raven Press, New York.

Bronson, F. H. and Channing, C. P. (1978). Suppression of serum follicle-stimulating hormone by follicular fluid in the maximally estrogenized, ovariectomized mouse. *Endocrinology* **103**, 1894–1898.

Cahoreau, C., Blanc, M. R., Dacheux, J. L., Pisselet, Cl., and Courot, M. (1979). Inhibin activity in ram rete testis fluid: depression of plasma FSH and LH in the castrated and cryptorchid ram. *J. Reprod. Fert.* Suppl. **26**, 97–116.

Campbell, C. S. and Schwartz, N. B. (1979). Time course of serum FSH suppression in ovariectomized rats injected with porcine follicular fluid (folliculostatin): effect of estradiol treatment. *Biol. Reprod.* **20**, 1093–1098.

Cate, R. L., Mattaliano, R. J., Hession, C., Tizard, R., Farber, N. M., Cheung, A., Ninfa, E. G., Frey, A. Z., Gash, D. J., Chow, E. P., Fisher, R. A., Bertonis, J. M., Torres, G., Wallner, B. P., Ramachandran, K. L., Ragin, R. C., Manganaro, T. F., MacLaughlin, D. T., and Donahoe, P. K. (1986). Isolation of the bovine and human genes for Müllerian inhibiting substance and expression of the human gene in animal cells. *Cell* **45**, 685–698.

Channing, C. P., Chacon, M., Tanabe, K., Gagliano, P., and Tildon, T. (1984a). Follicular fluid inhibin activity and steroid levels in ovarian tissue obtained at autopsy from human infants from 18 to 200 days of age. *Fert. Steril.* **42**, 861–869.

—— Gagliano, P., Hoover, D. J., Tanabe, K., Batta, S. K., Sulewski, J., and Lebech, P. (1981). Relationship between human follicular fluid inhibin F activity and steroid content. *J. clin. Endocr. Metab.* **52**, 1193–1198.

—— —— Tanabe, K., Fortuny, A., and Cortes-Prieto, J. (1985a). Demonstration of a gradient in inhibin activity, estrogen, progesterone, and Δ^4-androstenedione in follicular fluid, ovarian vein blood, and peripheral blood of normal women. *Fert. Steril.* **43**, 142–145.

—— Gordon, W. L., Liu, W. K., and Ward, D. N. (1985b). Physiology and biochemistry of ovarian inhibin. *Proc. Soc. exp. Biol. Med.* **178**, 339–361.

—— Liu, W. K., Gordon, W. L., Xue, Y. F., and Ward, D. N. (1984b). Porcine inhibin: initial fractionation as a high molecular weight complex. *Arch. Androl.* **13**, 219–231.

—— Tanabe, K., Chacon, M., and Tildon, J. T. (1984c). Stimulatory effects of follicle-stimulating hormone and luteinizing hormone upon secretion of progesterone and inhibin activity by cultured infant human ovarian granulosa cells. *Fert. Steril.* **42**, 598–605.

40   F. H. de Jong

——— ——— Hahn, D., Phillips, A., and Carraher, R. (1984d). Inhibin activity and steroid hormone levels in ovarian extracts and ovarian vein plasma of female monkeys during postnatal development. *Fert. Steril.* **42**, 453–458.

——— ——— Seegar Jones, G., Jones, H. W., Jr., and Lebech, P. (1984e). Inhibin activity of preovulatory follicles of gonadotropin-treated and untreated women. *Fert. Steril.* **42**, 243–248.

——— ——— Turner, C. K., and Hodgen, G. D. (1982). Antisera to porcine follicular fluid in monkeys: neutralization of human and pig inhibin *in vivo* and *in vitro*. *J. clin. Endocr. Metab.* **55**, 481–486.

Chappel, S. C. (1979). Cyclic fluctuations in ovarian FSH-inhibiting material in golden hamsters. *Biol. Reprod.* **21**, 447–453.

——— Holt, J. A., and Spies, H. G. (1980). Inhibin: differences in bioactivity within human follicular fluid in the follicular and luteal stages of the menstrual cycle. *Proc. Soc. exp. Biol. Med.* **163**, 310–314.

Chari, S., Aumüller, G., Daume, E., Sturm, G., and Hopkinson, C. (1981). The effects of human follicular fluid inhibin on the morphology of the ovary of the immature rat. *Arch. Gynec.* **230**, 239–249.

——— Daume, E., Sturm, G., Vaupel, H., and Schüler, I. (1985). Regulators of steroid secretion and inhibin activity in human ovarian follicular fluid. *Mol. Cell. Endocr.* **41**, 137–145.

——— Duraiswami, S., and Franchimont, P. (1976). A convenient and rapid bioassay for inhibin. *Horm. Res.* **7**, 129–137.

——— Hopkinson, C. R. N., Daume, E., and Sturm, G. (1979). Purification of "inhibin" from human ovarian follicular fluid. *Acta endocr. Copenh.* **90**, 157–166.

Charlesworth, M. C., Grady, R. R., Shin, L., Vale, W. W., Rivier, C., Rivier, J., and Schwartz, N. B. (1984). Differential suppression of FSH and LH secretion by follicular fluid in the presence or absence of GnRH. *Neuroendocrinology* **38**, 199–205.

Chowdhury, M., Steinberger, A., and Steinberger, E. (1978). Inhibition of *de novo* synthesis of FSH by the Sertoli Cell Factor (SCF). *Endocrinology* **103**, 644–647.

Christiansen, P. (1975). Urinary gonadotrophins in the Sertoli-cell-only syndrome. *Acta endocr. Copenh.* **78**, 180–191.

Colenbrander, B., MacDonald, A. A., Elsaesser, F., Parvizi, N., and van de Wiel, D. F. M. (1982). Response of luteinizing hormone and follicle-stimulating hormone to luteinizing hormone releasing hormone in the fetal pig. *Biol. Reprod.* **27**, 556–561.

——— Meijer, J. C., MacDonald, A. A., van de Wiel, D. F. M., Engel, B., and de Jong, F. H. (1987). Feedback regulation of gonadotrophic hormone secretion in neonatal pigs. *Biol. Reprod.* in press.

Condon, T. P., Leipheimer, R. E., and Curry, J. J. (1983). Preliminary evidence for a CNS site of action for ovarian inhibin. *Life Sci.* **32**, 1691–1698.

Croze, F. and Franchimont, P. (1984). An *in vitro* model for the study of inhibin production by rat ovarian cells. *J. Reprod. Fert.* **72**, 249–260.

Cummins, L. J., O'Shea, T., Bindon, B. M., Lee, V. W. K., and Findlay, J. K. (1983). Ovarian inhibin content and sensitivity to inhibin in Booroola and control strain Merino ewes. *J. Reprod. Fert.* **67**, 1–7.

Cunningham, G. R. and Huckins, C. (1979). Persistence of complete spermatogenesis in the presence of low intratesticular concentrations of testosterone. *Endocrinology* **105**, 177–186.

Currie, B. L., Warberg, J., Folkers, K., Bowers, C. Y., Paulsen, E., Mulder, J., and Paulsen, F. (1977). Presence of peptide hormones that control gonadotropin secretion in extrahypothalamic areas of the brain. *Biochem. Biophys. Res. Commun.* **79**, 1207–1211.

Daume, E., Chari, S., Hopkinson, C. R. N., Sturm, G., and Hirschhäuser, C. (1978). Nachweis von Inhibin-Aktivität in der Follikelflüssigkeit menschlicher Ovarien. *Klin. Wschr.* **56**, 369–370.

David, K., Dingemanse, E., Freud, J., and Laqueur, E. (1935). Über krystallin-isches männliches Hormon aus Hoden (Testosteron), wirksamer als aus Harn oder aus Cholesterin bereitetes Androsteron. *Hoppe-Seylers Z. physiol. Chem.* **233**, 281–282.

Davies, R. V., Main, S. J., Laurie, M. S., and Setchell, B. P. (1979). The effect of long-term administration of either a crude inhibin preparation or an antiserum to FSH on serum hormone levels, testicular function and fertility of adult male rats. *J. Reprod. Fert.* Suppl. **26**, 183–191.

—— —— and Setchell, B. P. (1978). Inhibin: evidence for its existence, methods of bioassay and nature of the active material. *Int. J. Androl.* Suppl. **2**, 102–114.

Davis, S. R., Dench, F., Nikolaidis, I., Clements, J. A., Forage, R. G., Krozowski, Z. and Burger, H. G. (1986). Inhibin A-subunit gene expression in the ovaries of immature female rats is stimulated by pregnant mare serum gonadotrophin. *Biochem. Biophys. Res. Commun.* **138**, 1191–1195.

Debeljuk, L., Arimura, A., and Schally, A. V. (1973). Pituitary and serum FSH and LH levels after massive and selective depletion of the germinal epithelium in the rat testis. *Endocrinology* **92**, 48–54.

de Greef, W. J., de Jong, F. H., de Koning, J., Steenbergen, J., and van der Vaart, P. D. M. (1983). Studies on the mechanism of the selective suppression of plasma levels of follicle-stimulating hormone in the female rat after administration of steroid-free bovine follicular fluid. *J. Endocr.* **97**, 327–338.

de Jong, F. H. (1979). Inhibin – fact or artifact. *Mol. Cell. Endocr.* **13**, 1–10.

—— Hermans, W. P., Jansen, E. H. J. M., Steenbergen, J., and van der Molen, H. J. (1980). Biochemical and physiological aspects of inhibin-like activity. In *Endocrinology 1980* (ed. I. A. Cumming, J. W. Funder and F. A. O. Mendelsohn) pp. 255–258. Australian Academy of Science, Canberra.

—— Jansen, E. H. J. M., Hermans, W. P., and van der Molen, H. J. (1982*a*). Purification, characterization, and *in vitro* production of inhibin. In *Intraovarian control mechanisms* (ed. C. P. Channing and S. J. Segal) pp. 37–52. Plenum Press, New York.

—— —— —— —— (1982*b*). Purification, characterization and physiological significance of inhibin from ovarian follicular fluid. In *Non-steroidal regulators in reproductive biology and medicine* (ed. T. Fujii and C. P. Channing) pp. 73–82. Pergamon Press, Oxford.

—— —— Steenbergen, J., van Dijk, S., and van der Molen, H. J. (1983*a*). Assay and purification of inhibin. In *Role of peptides and proteins in control of reproduction* (ed. S. M. McCann and D. S. Dhindsa) pp. 257–273. Elsevier, New York.

—— —— and van der Molen, H. J. (1981). Purification and characterization of inhibin. In *Intragonadal regulation of reproduction* (ed. P. Franchimont and C. P. Channing) pp. 229–250. Academic Press, London.

—— and Robertson, D. M. (1985). Inhibin: 1985 update on action and purification. *Mol. Cell. Endocr.* **42**, 95–103.

—— Sander, H. J., Ultee-van Gessel, A. M., and van der Molen, H. J. (1985). Specific regulation of the secretion of follicle-stimulating hormone from the pituitary gland: the inhibin concept. In *Frontiers of hormone research*, Vol. 14 (ed. Tj.B. van Wimersma Greidanus) pp. 53–69. Karger, Basel.

—— and Sharpe, R. M. (1976). Evidence for inhibin-like activity in bovine follicular fluid. *Nature, Lond.* **263**, 71–72.

—— —— (1977). Gonadotrophins, testosterone and spermatogenesis in neonatally irradiated male rats: evidence for a role of the Sertoli cell in follicle-stimulating hormone feedback. *J. Endocr.* **75**, 209–219.

—— Smith, S. D., and van der Molen, H. J. (1979a). Bioassay of inhibin-like activity using pituitary cells *in vitro*. *J. Endocr.* **80**, 91–102.

—— Steenbergen, J., van Dijk, S., and van der Molen, H. J. (1983b). Purification and characterization of inhibin from bovine ovarian follicular fluid. In *Hormones and cell regulation*, Vol. 7 (ed. J. E. Dumont, J. Nunez, and R. M. Denton) pp. 203–216. Elsevier, Amsterdam.

—— Welschen, R., Hermans, W. P., Smith, S. D., and van der Molen, H. J. (1978). Effects of testicular and ovarian inhibin-like activity, using *in vitro* and *in vivo* systems. *Int. J. Androl.* Suppl. 2, 125–138.

—— —— —— —— —— —— (1979b). Effects of factors from ovarian follicular fluid and Sertoli cell culture medium on *in vivo* and *in vitro* release of pituitary gonadotrophins in the rat: an evaluation of systems for the assay of inhibin. *J. Reprod. Fert.* Suppl. **26**, 47–59.

del Castillo, E. B., Trabucco, A., and de la Balze, F. A. (1947). Syndrome produced by absence of the germinal epithelium without impairment of the Sertoli or Leydig cells. *J. clin. Endocr.* **7**, 493–502.

Demoulin, A., Bourguignon, J. P., and Franchimont, P. (1979). '*In vitro*' inhibition of hypothalamic LH-RH by human inhibin. *Acta endocr.* Suppl. 225, 227.

DePaolo, L. V., Hirshfield, A. N., Anderson, L. D., Barraclough, C. A., and Channing, C. P. (1979a). Suppression of pituitary secretion of follicle-stimulating hormone by porcine follicular fluid during pro-oestrus and oestrus in the rat: effects on gonadotrophin and steroid secretion, follicular development and ovulation during the following cycle. *J. Endocr.* **83**, 355–368.

—— Shander, D., Wise, P. M., Barraclough, C. A., and Channing, C. P. (1979b). Identification of inhibin-like activity in ovarian venous plasma of rats during the estrous cycle. *Endocrinology* **105**, 647–654.

—— Wise, P. M., Anderson, L. D., Barraclough, C. A., and Channing, C. P. (1979c). Suppression of the pituitary follicle-stimulating hormone secretion during proestrus and estrus in rats by porcine follicular fluid: possible site of action. *Endocrinology* **104**, 402–408.

Dhont, M., Serreyn, R., Duvivier, P., Van Luchene, E., de Boever, J., and Vandekerckhove, D. (1984). Ovulation stigma and concentration of progesterone and estradiol in peritoneal fluid: relation with fertility and endometriosis. *Fert. Steril.* **41**, 872–877.

Dias, J. A. and Reichert, Jr., L. E. (1984). Evidence for a high molecular weight follicle-stimulating hormone binding inhibitor in bovine testis. *Biol. Reprod.* **31**, 975–983.

Dobos, M., Burger, H. G., Hearn, M. T. W., and Morgan, F. J. (1983). Isolation of inhibin from ovine follicular fluid using reversed-phase liquid chromatography. *Mol. Cell. Endocr.* **31**, 187–198.

Drouin, J. and Labrie, F. (1976). Selective effect of androgens on LH and FSH release in anterior pituitary cells in culture. *Endocrinology* **98**, 1528–1534.

—— —— (1981). Interactions between 17β-estradiol and progesterone in the control of luteinizing hormone and follicle-stimulating hormone release in rat anterior pituitary cells in culture. *Endocrinology* **108**, 52–57.

Dym, M. and Fawcett, D. W. (1970). The blood-testis barrier in the rat and the physiological compartmentation of the seminiferous epithelium. *Biol. Reprod.* **3**, 308–326.

Eddie, L. W., Baker, H. W. G., Dulmanis, A., Higginson, R. E., and Hudson, B. (1978). Inhibin from cultures of rat seminiferous tubules. *J. Endocr.* **78**, 217–224.

—— —— Higginson, R. E., and Hudson, B. (1979). A bioassay for inhibin using pituitary cell cultures. *J. Endocr.* **81**, 49–60.

Erickson, G. F. and Hsueh, A. J. W. (1978). Secretion of 'inhibin' by rat granulosa cells *in vitro*. *Endocrinology* **103**, 1960–1963.

Esch, F. S., Ling, N. C., Böhlen, P., Ying, S. Y., and Guillemin, R. (1983). Primary structure of PDC-109, a major protein constituent of bovine seminal plasma. *Biochem. Biophys. Res. Commun.* **113**, 861–867.

—— —— Ying, S. Y., and Guillemin, R. (1983). Peptides of gonadal origin involved in reproductive biology. In *Role of peptides and proteins in control of reproduction* (ed. S. M. McCann and D. S. Dhindsa) pp. 275–290. Elsevier, New York.

Fachini, G. and Ciaccolini, C. (1966). Pituitary inhibition by effluent blood from the testis. *Endokrinologie* **50**, 79–82.

—— Toffoli, C., Gaudiano, A., Marabelli, M., Polizzi, M., and Mangili, G. (1963). Nemasperm and pituitary inhibition. *Nature, Lond.* **199**, 195–196.

Fevold, H. L., Hisaw, F. L., and Leonard, S. L. (1931). The gonad stimulating and the luteinizing hormones of the anterior lobes of the hypophysis. *Am. J. Physiol.* **97**, 291–301.

Forage, R. G., Ring, J. M., Brown, R. W., McInerney, B. V., Cobon, G. S., Gregson, R. P., Robertson, D. M., Morgan, F. J., Hearn, M. T. W., Findlay, J. K., Wettenhall, R. E. H., Burger, H. G., and de Kretser, D. M. (1986). Cloning and sequence analysis of cDNA species coding for the two subunits of inhibin from bovine follicular fluid. *Proc. natn. Acad. Sci. U.S.A.* **83**, 3091–3095.

Franchimont, P., Chari, S., and Demoulin, A. (1975). Hypothalamus-pituitary-testis interaction. *J. Reprod. Fert.* **44**, 335–350.

—— —— Hagelstein, M. T., and Duraiswami, S. (1975). Existence of a follicle-stimulating hormone inhibiting factor 'inhibin' in bull seminal plasma. *Nature, Lond.* **257**, 402–404.

—— Croze, F., Demoulin, A., Bologne, R., and Hustin, J. (1981). Effect of inhibin on rat testicular desoxyribonucleic acid (DNA) synthesis *in vivo* and *in vitro*. *Acta endocr. Copenh.* **98**, 312–320.

—— Demoulin, A., Verstraelen-Proyard, J., Hazee-Hagelstein, M. T., and Tunbridge, W. M. G. (1979). Identification in human seminal fluid of an inhibin-like factor which selectively regulates FSH secretion. *J. Reprod. Fert.* Suppl. **26**, 123–133.

—— Lecomte-Yerna, M. J., Henderson, K., Verhoeven, G., Hazee-Hagelstein, M. T., Jaspar, J. M., Charlet-Renard, C., and Demoulin, A. (1983). Inhibin: mechanisms of pituitary action and regulation of secretion. In *Role of peptides and proteins in control of reproduction* (ed. S. M. McCann and D. S. Dhindsa) pp. 237–255. Elsevier, New York.

—— Millet, D., Vendrely, E., Letawe, J., Legros, J. J., and Netter. A. (1972). Relationship between spermatogenesis and serum gonadotropin levels in azoospermia and oligospermia. *J. clin. Endocr. Metab.* **34**, 1003–1008.

—— Verstraelen-Proyard, J., Hazee-Hagelstein, M. T., Renard, Ch., Demoulin, A., Bourguignon, J. P., and Hustin, J. (1979). Inhibin: from concept to reality. *Vitams. Horm.* **37**, 243–302.

Fujii, T., Hoover, D. J., and Channing, C. P. (1983). Changes in inhibin activity, and progesterone, oestrogen and androstenedione concentrations, in rat follicular fluid throughout the oestrous cycle. *J. Reprod. Fert.* **69**, 307–314.

Fukuda, M., Miyamoto, K., Hasegawa, Y., Nomura, M., Igarashi, M., Kangawa, K., and Matsuo, H. (1986). Isolation of bovine follicular fluid inhibin of about 32 kDa. *Mol. Cell. Endocr.* **44**, 55–60.

Gay, V. L., Midgley, A. R., and Niswender, G. D. (1970). Patterns of gonadotrophin secretion associated with ovulation. *Fedn. Proc.* **29**, 1880–1887.

44   F. H. de Jong

Goh, H. H., Chew, P. C. T., Karin, S. M. M., and Ratnam, S. S. (1980). Control of gonadotrophin secretion by steroid hormones in castrated male transsexuals. I. Effect of oestradiol infusion on plasma levels of follicle-stimulating hormone and luteinizing hormone. *Clin. Endocr.* **12**, 165–175.

—— Karim, S. M. M., and Ratnam, S. S. (1981). Control of gonadotrophin secretion by steroid hormones in castrated male transsexuals. II. Effects of androgens alone and in combination with oestradiol on the secretions of FSH and LH. *Clin. Endocr.* **15**, 301–312.

Gomes, W. R., Hall, R. W., Jain, S. K., and Boots, L. R. (1973). Serum gonadotropin and testosterone levels during loss and recovery of spermatogenesis in rats. *Endocrinology* **93**, 800–809.

Grady, R. R., Charlesworth, M. C., and Schwartz, N. B. (1982). Characterization of the FSH-suppressing activity in follicular fluid. *Recent Prog. Horm. Res.* **38**, 409–456.

Gunsalus, G. L., Musto, N. A., and Bardin, C. W. (1980). Bidirectional release of a Sertoli cell product, androgen binding protein, into the blood and seminiferous tubule. In *Testicular development, structure and function* (ed. A. Steinberger and E. Steinberger) pp. 291–297. Raven Press, New York.

Heller, C. G. and Nelson, W. O. (1945). Hyalinization of the seminiferous tubules associated with normal or failing Leydig cell function. Discussion of the relationship to eunuchoidism, gynecomastia, gonadotrophins, depressed 17-keto steroids and estrogens. *J. clin. Endocr.* **5**, 1–12.

—— —— (1948). The testis-pituitary relationship in man. *Recent Prog. Horm. Res.* **3**, 229–255.

—— Paulson, C. A., Mortimore, G. E., Junck, E. C., and Nelson, W. O. (1952). Urinary gonadotrophins, spermatogenic activity, and classification of testicular morphology – their bearing on the utilization hypothesis. *Ann. N.Y. Acad. Sci.* **55**, 685–702.

Henderson, K. M. and Franchimont, P. (1981). Regulation of inhibin production by bovine ovarian cells *in vitro*. *J. Reprod. Fert.* **63**, 431–442.

—— —— (1983). Inhibin production by bovine ovarian tissues *in vitro* and its regulation by androgens. *J. Reprod. Fert.* **67**, 291–298.

—— —— Charlet-Renard, Ch., and McNatty, K. P. (1984*a*). Effect of follicular atresia on inhibin production by bovine granulosa cells *in vitro* and inhibin concentrations in the follicular fluid. *J. Reprod. Fert.* **72**, 1–8.

—— —— Lecomte-Yerna, M. J., Hudson, N., and Ball, K. (1984*b*). Increase in ovulation rate after active immunization of sheep with inhibin partially purified from bovine follicular fluid. *J. Endocr.* **102**, 305–309.

—— Prisk, M. D. Hudson, N., Ball, K., McNatty, K. P., Lun, S., Heath, D., Kieboom, L. E. and McDiarmid, J. (1986). Use of bovine follicular fluid to increase ovulation rate or prevent ovulation in sheep. *J. Reprod. Fert.* **76**, 623–635.

Hermans, W. P., Debets, M. H. M., van Leeuwen, E. C. M., and de Jong, F. H. (1981). Time-related secretion of gonadotrophins after a single injection of steroid-free bovine follicular fluid in prepubertal and adult female rats. *J. Endocr.* **90**, 69–76.

—— —— —— —— (1982). Effects of single injections of bovine follicular fluid on gonadotrophin concentrations throughout the oestrous cycle of the rat. *J. Endocr.* **92**, 425–432.

—— van Leeuwen, E. C. M., Debets, M. H. M., and de Jong, F. H. (1980). Involvement of inhibin in the regulation of follicle-stimulating hormone concentrations in prepubertal and adult, male and female rats. *J. Endocr.* **86**, 79–92.

————— ——— Sander, H. J., and de Jong, F. H. (1982b). Estimation of inhibin-like activity in spent medium from rat ovarian granulosa cells during long-term culture. *Mol. Cell. Endocr.* **27**, 277–290.

Hillier, S. G., van Hall, E. V., van den Boogaard, A. J. M., de Zwart, F. A., and Keijzer, R. (1982). Activation and modulation of the granulosa cell aromatase system: experimental studies with rat and human ovaries. In *Follicular maturation and ovulation* (ed. R. Rolland, E. V. van Hall, S. G. Hillier, K. P. McNatty, and J. Schoemaker) pp. 51–70. Excerpta Medica, Amsterdam.

Hochberg, Z., Weiss, J., and Richman, R. A. (1981). Inhibin-like activity in extracts of rabbit placentae. *Placenta* **2**, 259–264.

Hochereau de Reviers, M. T., Loir M., and Pelletier, J. (1976). Seasonal variations in the response of the testis and LH levels to hemicastration of adult rams. *J. Reprod. Fert.* **46**, 203–209.

Hodgen, G. D., Nixon, W. E., and Turner, C. K. (1974). Suppression of serum gonadotrophins in castrated rats by testis homogenates. *IRCS, Med. Sci.* **2**, 1233.

Hoffmann, J. C., Lorenzen, J. R., Weil, T., and Schwartz, N. B. (1979). Selective suppression of the primary surge of follicle-stimulating hormone in the rat: further evidence for folliculostatin in porcine follicular fluid. *Endocrinology* **105**, 200–203.

Hopkinson, C. R. N., Daume, E., Sturm, G., Fritze, E., Kaiser, S., and Hirschhäuser, C. (1977). Inhibin-like activity of bovine ovarian extracts in male and female rats. *J. Reprod. Fert.* **50**, 93–96.

Howard, R. P., Sniffen, R. C., Simmons, F. A., and Albright, F. (1950). Testicular deficiency: a clinical and pathologic study. *J. clin. Endocr.* **10**, 121–186.

Huang, E. S. R. and Miller, W. L. (1984). Porcine ovarian inhibin preparations sensitize cultured ovine gonadotrophs to luteinizing hormone-releasing hormone. *Endocrinology* **115**, 513–519.

Hudson, B., Baker, H. W. G., Eddie, L. W., Higginson, R. E., Burger, H. G., de Kretser, D. M., Dobos, M., and Lee, V. W. K. (1979). Bioassays for inhibin: a critical review. *J. Reprod. Fert.* Suppl. **26**, 17–29.

Jansen, E. H. J. M., Steenbergen, J., de Jong, F. H., and van der Molen, H. J. (1981). The use of affinity matrices in the purification of inhibin from bovine follicular fluid. *Mol. Cell. Endocr.* **21**, 109–117.

Jenner, A. A. J, de Koning, J., Tijssen, A. M. I., and van Rees, G. P. (1985). Divergent effects on LH and FSH synthesis and release from intact female rat pituitary glands *in vitro* by methylxanthines, cyclic AMP derivatives and sodium fluoride. *Acta endocr. Copenh.* **109**, 315–319.

———— ——— and van Rees, G. P. (1982). Time course of inhibition of synthesis and basal release of FSH from pituitary glands of female rats by bovine follicular fluid *in vitro*. *Acta endocr. Copenh.* **101**, 501–506.

Jenner, M. R., Kelch, R. P., Kaplan, S. L., and Grumbach, M. M. (1972). Hormonal changes in puberty: IV. Plasma estradiol, LH, and FSH in prepubertal children, pubertal females, and in precocious puberty, premature thelarche, hypogonadism, and in a child with a feminizing ovarian tumor. *J. clin. Endocr. Metab.* **34**, 521–530.

Johansson, J., Sheth, A., Cederlund, E., and Jörnvall, H. (1984). Analysis of an inhibin preparation reveals apparent identity between a peptide with inhibin-like activity and a sperm-coating antigen. *FEBS Lett.* **176**, 21–26.

Johnsen, S. G. (1964). Studies on the testicular-hypophyseal feed-back mechanism in man. *Acta endocr. Suppl.* **90**, 99–124.

Jones, H. M., Wood, C. L., and Rush, M. E. (1985). A role for inhibin in the control of follicle-stimulating hormone secretion in male rats. *Life Sci.* **36**, 889–899.

Judd, H. L., Hamilton, C. R., Barlow, J. J., Yen, S. S. C., and Kliman, B. (1972). Androgen and gonadotropin dynamics in testicular feminization syndrome. *J. clin. Endocr. Metab.* **34**, 229–234.

Jutte, N. H. P. M., Jansen, R., Grootegoed, J. A., Rommerts, F. F. G., and van der Molen. H. J. (1983). FSH stimulation of the production of pyruvate and lactate by rat Sertoli cells may be involved in hormonal regulation of spermatogenesis. *J. Reprod. Fert.* **68**, 219–226.

Kao, L. W. L., Gunsalus, G. L., Williams, G. H., and Weisz, J. (1977). Response of the perifused anterior pituitaries of rats to synthetic gonadotrophin releasing hormone: a comparison with hypothalamic extract and demonstration of a role for potassium in the release of luteinizing hormone and follicle-stimulating hormone. *Endocrinology* **101**, 1444–1454.

Keogh, E. J., Lee, V. W. K., Rennie, G. C., Burger, H. G., Hudson, B., and de Kretser, D. M. (1976). Selective suppression of FSH by testicular extracts. *Endocrinology* **98**, 997–1004.

Kerr, J. B. and Sharpe, R. M. (1985*a*). Follicle-stimulating hormone induction of Leydig cell maturation. *Endocrinology* **116**, 2592–2604.

—— —— (1985*b*). Stimulatory effect of follicle-stimulating hormone on rat Leydig cells. *Cell Tissue Res.* **239**, 405–415.

Klinefelter, H. F., Reifenstein, E. C., and Albright, F. (1942). Syndrome characterized by gynecomastia, aspermatogenesis without aleydigism and increased excretion of follicle-stimulating hormone. *J. clin. Endocr.* **2**, 615–627.

Kohan, S., Fröysa, B., Cederlund, E., Fairwell, Th., Lerner, R., Johansson, J., Khan, S., Ritzen, M., Jörnvall, H., Cekan, S., and Diczfalusy, E. (1986). Peptides of postulated inhibin activity. Lack of *in vitro* inhibin activity of a 94-residue peptide isolated from human seminal plasma, and of a synthetic replicate of its C-terminal 28-residue segment. *FEBS Lett.* **199**, 242–248.

Koninckx, P. R., Brosens, I. A., Verhoeven, G., and de Moor, P. (1981). Increased postovulatory plasma follicle stimulating hormone levels in the luteinized unruptured follicle syndrome: a role for inhibin? *Br. J. Obstet. Gynaec.* **88**, 525–530.

Krishnan, K. A., Panse, G. T., and Sheth, A. R. (1982). Comparative study of inhibin from human testis, prostate and seminal plasma. *Andrologia* **14**, 409–415.

Krueger, P. M., Hodgen, G. D., and Sherins, R. J. (1974). New evidence for the role of the Sertoli cell and spermatogonia in feedback control of FSH secretion in male rats. *Endocrinology* **95**, 955–962.

Labrie, F., Lagacé, L., Ferland, L., Kelly, P. A., Drouin, J., Massicotte, J., Bonne, C., Raynaud, J.-P., and Dorrington, J. H. (1978). Interactions between LHRH, sex steroids and 'inhibin' in the control of LH and FSH secretion. *Int. J. Androl.* Suppl. **2**, 303–318.

Lacy, D., and Pettitt, A. J. (1970). Sites of hormone production in the mammalian testis, and their significance in the control of male fertility. *Br. med. Bull.* **26**, 87–91.

Lagacé, L., Labrie, F., Lorenzen, J., Schwartz, N. B., and Channing, C. P. (1979). Selective inhibitory effect of porcine follicular fluid on follicle stimulating hormone secretion in anterior pituitary cells in culture. *Clin. Endocr.* **10**, 401–405.

Lecomte-Yerna, M. J., Hazee-Hagelstein, M. T., Charlet-Renard, C., and Franchimont, P. (1984). Effect of neuraminidase on inhibin activity *in vivo* and *in vitro. Horm. Res.* **20**, 277–284.

Lee, V. W. K., Kraft, N. C., Atkins, R. C., and Burger, H. G. (1986). Monoclonal antibody to rat ovarian inhibin. *J. Endocr.* **109**, 379–383.

—— McMaster, J., Quigg, H., Findlay, J., and Leversha, L. (1981). Ovarian and peripheral blood inhibin concentrations increase with gonadotropin treatment in immature rats. *Endocrinology* **108**, 2403–2405.

—— —— —— and Leversha, L. (1982). Ovarian and circulating inhibin levels in immature female rats treated with gonadotropin and after castration. *Endocrinology* **111**, 1849–1854.

—— Pearce, P. T., and de Kretser, D. M. (1977). The assessment of rodent models in evaluating the capacity of bovine testes extracts to suppress FSH levels. In *The testis in normal and infertile men* (ed. P. Troen and H. R. Nankin) pp. 293–303. Raven Press, New York.

Lefevre, B., Kraiem, Z., Epstein, Y., Lunenfeld, B., Gougeon, A., Thebault, A., Bomsel-Helmreich, O., Frydman, R., Papiernik, E., Saltarelli, D., Brailly, S., and Milgrom, E. (1981). Inhibin-like activity and steroid content in human follicular fluid during the periovulatory period. *Int. J. Fert.* **26**, 295–296.

Le Gac, F. and de Kretser, D. M. (1982). Inhibin production by Sertoli cell cultures. *Mol. Cell. Endocr.* **28**, 487–498.

Le Lannou, D. and Chambon, Y. (1977). Présence dans l'épididyme d'un facteur abaissant fortement le taux sanguin de FSH chez le rat. *C.r. Séanc. Soc. Biol.* **171**, 636–638.

Leonard, J. M., Leach, R. B., Couture, M., and Paulsen, C. A. (1972). Plasma and urinary follicle-stimulating hormone levels in oligospermia. *J. clin. Endocr. Metab.* **34**, 209–214.

Li, C. H., Hammonds, R. G., Ramasharma, K. and Chung, D. (1985). Human seminal α inhibins: isolation, characterization and structure. *Proc. natn. Acad. Sci. U.S.A.* **82**, 4041–4044.

Lilja, H. and Jeppsson, J. O. (1985). Amino acid sequence of the predominant basic protein in human seminal plasma. *FEBS Lett.* **182**, 181–184.

Lincoln, G. A. (1979). Differential control of luteinizing hormone and follicle-stimulating hormone by luteinizing hormone releasing hormone in the ram. *J. Endocr.* **80**, 133–140.

Ling, N., Ying, S. Y., Ueno, N., Esch, F., Denoroy, L., and Guillemin, R. (1985). Isolation and partial characterization of a M_r 32,000 protein with inhibin activity from porcine follicular fluid. *Proc. natn. Acad. Sci. U.S.A.* **82**, 7217–7221.

—— —— —— Shimasaki, S., Esch, F., Hotta, M., and Guillemin, R. (1986a). Pituitary FSH is released by a heterodimer of the β-subunits from the two forms of inhibin. *Nature, Lond.* **321**, 779–782.

—— —— —— —— —— —— —— (1986b). A homodimer of the β-subunits of inhibin A stimulates the secretion of pituitary follicle stimulating hormone. *Biochem. Biophys. Res. Commun.* **138**, 1129–1137.

Lipner, H. and Dhanarajan, P. (1984). The role of inhibin in adult male rats: the effect of inhibin deficiency and androgen antagonism on serum and pituitary gonadotrophins. *Acta endocr. Copenh.* **105**, 302–307.

Liu, L., Booth, J., Merriam, G. R., Barnes, K. M., Sherins, R. J., Loriaux, D. L., and Cutler, G. B. (1985). Evidence that synthetic 31-amino acid inhibin-like peptide lacks inhibin activity. *Endocr. Res.* **11**, 191–197.

Lorenzen, J. R., Dworkin, G. H., and Schwartz, N. B. (1981). Specific FSH suppression in male rat by porcine follicular fluid. *Am. J. Physiol.* **240**, E209–E215.

Lugaro, G., Carrea, G., Casellato, M. M., Mazzola, G., and Fachini, G. (1973). Evidence for a peptidic factor in spermatozoa inhibiting the ovarian maturation. *Biochim. Biophys. Acta* **304**, 719–724.

—— Casellato, M. M., Mazzola, G., Fachini, G., and Carrea, G. (1974). Evidence for the existence in spermatozoa of a factor inhibiting the follicle stimulating hormone releasing hormone synthesis. *Neuroendocrinology* **15**, 62–68.

—— Giannattasio, G., Ciaccolini, C., Fachini, G., and Gianfranceschi, G. L. (1969). Biochemical study on the pituitary inhibition of gonadal origin. *Experientia* **25**, 147–148.

Lumpkin, M., Negro-Vilar, A., Franchimont, P., and McCann, S. (1981). Evidence for a hypothalamic site of action of inhibin to suppress FSH release. *Endocrinology* **108**, 1101–1104.

Main, S. J., Davies, R. V., and Setchell, B. P. (1978). Feedback control by the testis of gonadotrophin secretion: an examination of the inhibin hypothesis. *J. Endocr.* **79**, 255–270.

Marder, M. L., Channing, C. P., and Schwartz, N. B. (1977). Suppression of serum follicle stimulating hormone in intact and acutely ovariectomized rats by porcine follicular fluid. *Endocrinology* **101**, 1639–1642.

Martin, N. G., Olsen, M. E., Theile, H., El Beaini, J. L., Handelsman, D., and Bhatnagar, A. S. (1984). Pituitary-ovarian function in mothers who have had two sets of dizygotic twins. *Fert. Steril.* **41**, 878–880.

Martins, T. and Rocha, A. (1931). The regulation of the hypophysis by the testicle, and some problems of sexual dynamics. *Endocrinology* **15**, 421–434.

Mason, A. J., Hayflick, J. S., Ling, N., Esch, F., Ueno, N., Ying, S. Y., Guillemin, R., Niall, H., and Seeburg, P. H. (1985). Complementary DNA sequences of ovarian follicular fluid inhibin show precursor structure and homology with transforming growth factor-β. *Nature, Lond.* **318**, 659–663.

—— Niall, H. D., and Seeburg, P. H. (1986). Structure of two human ovarian inhibins. *Biochem. Biophys. Res. Commun.* **135**, 957–964.

Massicotte, J., Lagacé, L., Labrie, F., and Dorrington, J. H. (1984). Modulation of gonadotropin secretion by Sertoli cell inhibin, LHRH, and sex steroids. *Am. J. Physiol.* **247**, E495-E504.

Mayo, K. E., Cerelli, G. M., Spiess, J., Rivier, J., Rosenfeld, M. G., Evans, R. M., and Vale, W. (1986). Inhibin A-subunit cDNAs from porcine ovary and human placenta. *Proc. natn. Acad. Sci. U.S.A.* **83**, 5849–5853.

McCann, S. M. (1974). Regulation of secretion of follicle-stimulating hormone and luteinizing hormone. In *Handbook of physiology*, Sect. 7 Vol. 4 Pt 2, (ed. E. Knobil and W. H. Sawyer) pp. 489–517. American Physiological Society, Washington, D.C.

McCullagh, D. R. (1932). Dual endocrine activity of the testes. *Science, N.Y.* **76**, 19–20.

—— and Schneider, I. (1940). The effect of a non-androgenic testis extract on the estrous cycle in rats. *Endocrinology* **27**, 899–902.

—— and Walsh, E. L. (1934). Further studies concerning testicular function. *Proc. Soc. exp. Biol. Med.* **31**, 678–680.

McCullagh, E. P. and Schaffenburg, C. A. (1952). The role of the seminiferous tubules in the production of hormones. *Ann. N.Y. Acad. Sci.* **55**, 674–684.

McLachlan, R. I., Healy, D. L., Robertson, D. M., Burger, H. G. and de Kretser, D. M. (1986c). The human placenta: a novel source of inhibin. *Biochem. Biophys. Res. Commun.* **140**, 425–430.

—— Robertson, D. M., Burger, H. G., and de Kretser, D. M. (1986a). The radioimmunoassay of bovine and human follicular fluid and serum inhibin. *Mol. Cell. Endocr.* **46**, 175–185.

—— —— Healy, D. L., de Kretser, D. M., and Burger, H. G. (1986b). Plasma inhibin levels during gonadotropin-induced ovarian hyperstimulation for IVF: A new index of follicular function? *Lancet i*, 1233–1234.

Meijs-Roelofs, H. M. A., Osman, P., and Kramer, P. (1982). Ovarian follicular development leading to first ovulation and accompanying gonadotrophin levels as studied in the unilaterally ovariectomized rat. *J. Endocr.* **92**, 341–349.

Miller, K. F., Wesson, J. A., and Ginther, O. J. (1979). Changes in concentrations of circulating gonadotropins following administration of equine follicular fluid to ovariectomized mares. *Biol. Reprod.* **21**, 867–872.

Miyamoto, K., Hasegawa, Y., Fukuda, M., and Igarashi, M. (1986). Demonstration of high molecular weight forms of inhibin in bovine follicular fluid (bFF) by using monoclonal antibodies to bFF 32K inhibin. *Biochem. Biophys. Res. Commun.* **136**, 1103–1109.

—— —— —— Nomura, M., Igarashi, M., Kangawa, K., and Matsuo, H. (1985). Isolation of porcine follicular fluid inhibin of 32K daltons. *Biochem. Biophys. Res. Commun.* **129**, 396–403.

Mizunuma, H., Samson, W. K., Lumpkin, M. D., and McCann, S. M. (1983). Evidence for an FSH-releasing factor in the posterior portion of the rat median eminence. *Life Sci.* **33**, 2003–2009.

Mohapatra, S. K., Duraiswami, S., and Chari, S. (1985). On the identity of bovine seminal plasma inhibin. *Mol. Cell. Endocr.* **41**, 187–196.

Moore, C. R. and Price, D. (1932). Gonad hormone functions, and the reciprocal inference between gonads and hypophysis with its bearing on the problem of sex hormone antagonism. *Am. J. Anat.* **50**, 13–72.

Mottram, J. C. and Cramer, W. (1923). On the general effects of exposure to radium on metabolism and tumour growth in the rat and the special effects on testis and pituitary. *Q. Jl. exp. Physiol.* **13**, 209–229.

Moudgal, N. R., Murthy, H. M. S., Murthy, G. S., and Rao, A. J. (1984) Regulation of FSH secretion in the primate by inhibin: studies in the bonnet monkey (*M. radiata*). In *Gonadal proteins and peptides and their biological significance* (ed. M. R. Sairam and L. E. Atkinson) pp. 21–37. World Scientific Publishing Co., Singapore.

Nandini, S. G., Lipner, H., and Moudgal, N. R. (1976). A model system for studying inhibin. *Endocrinology* **98**, 1460–1465.

Nelson, W. O. (1934). Effect of oestrin and gonadotrophic hormone injections upon hypophysis of adult rats. *Proc. Soc. exp. Biol. Med.* **32**, 452–454.

—— and Gallagher, T. F. (1935). Studies on the anterior hypophysis. 4. The effect of male hormone preparations upon the anterior hypophysis of gonadectomized male and female rats. *Anat. Rec.* **64**, 129–145.

Orth, J. M. (1984). The role of follicle-stimulating hormone in controlling Sertoli cell proliferation in testes of fetal rats. *Endocrinology* **115**, 1248–1255.

Peek. J. C. and Watkins, W. B. (1979). Gonadotrophin-inhibiting activity in seminal plasma from intact and vasectomized bulls. *J. Reprod. Fert.* **57**, 281–285.

Plant, T. M., Hess, D. L., Hotchkiss, J., and Knobil, E. (1978). Testosterone and the control of gonadotrophin secretion in the male rhesus monkey (*Macaca mulatta*). *Endocrinology* **103**, 535–541.

Ponsin, G., Khar, A., Kunert-Radek, J., Bennardo, T., and Jutisz, M. (1980). Concanavalin A inhibits the basal and stimulated hormone release from pituitary cells in monolayer cultures. *FEBS Lett.* **113**, 331–334.

Ramasharma, K., Sairam, M. R., Seidah, N. G., Chrétien, M., Manjunath, P., Schiller, P. W., Yamashiro, D., and Li, C. H. (1984). Isolation, structure, and synthesis of a human seminal plasma peptide with inhibin-like activity. *Science, N.Y.* **223**, 1199–1202.

—— Shashidara Murthy, H. M., and Moudgal, N. R. (1979). A rapid bioassay for measuring inhibin activity. *Biol. Reprod.* **20**, 831–835.

Reichert, L. E. and Abou-Issa, H. (1977). Studies on a low molecular weight testicular factor which inhibits binding of FSH to receptor. *Biol. Reprod.* **17**, 614–621.

Rich, K. A. and de Kretser, D. M. (1977). Effect of differing degrees of destruction of the rat seminiferous epithelium on levels of serum follicle stimulating hormone and androgen binding protein. *Endocrinology* **101**, 959–968.

Rivier, J., McClintock, R., Spiess, J., Vaughan, J., Dalton, D., Corrigan, A., Azad, R., and Vale, W. (1984). Isolation from porcine follicular fluid of a protein exhibiting potent inhibin like biological activity. In *Endocrinology* (ed. F. Labrie and L. Proulx) pp. 1141–1144. Excerpta Medica, Amsterdam.

—— Spiess, J., McClintock, R., Vaughan, J., and Vale, W. (1985). Purification and partial characterization of inhibin from porcine follicular fluid. *Biochem. Biophys. Res. Commun.* **133**, 120–127.

Robertson, D. M., Au, C. L., and de Kretser, D. M. (1982). The use of ^{51}Cr for assessing cytotoxicity in an *in vitro* bioassay for inhibin. *Mol. Cell. Endocr.* **26**, 119–127.

—— de Vos, F. L., Foulds, L. M., McLachlan, R. I., Burger, H. G., Morgan, F. J., Hearn, M. T. W., and de Kretser, D. M. (1986). Isolation of a 31 kDa form of inhibin from bovine follicular fluid. *Mol. Cell. Endocr.* **44**, 271–277.

—— Foulds, L. M., Leversha, L., Morgan, F. J., Hearn, M. T. W., Burger, H. G. Wettenhall, R. E. H., and de Kretser, D. M. (1985). Isolation of inhibin from bovine follicular fluid. *Biochem. Biophys. Res. Commun.* **126**, 220–226.

Rommerts, F. F. G., de Jong, F. H., Grootegoed, J. A., and van der Molen, H. J. (1980). Metabolic changes in testicular cells from rats after long-term exposure to 37° C *in vivo* or *in vitro*. *J. Endocr.* **85**, 471–479.

Rubin, D. (1941). The question of an aqueous hormone from the testicle. *Endocrinology* **29**, 281–287.

Rush, M. E. and Lipner, H. (1979). Effect of bovine testicular extracts on plasma gonadotrophins of X-irradiated rats. *Proc. Soc. exp. Biol. Med.* **162**, 85–89.

Sairam, M. R., Kato, K., Manjunath, P., and Ramasharma, K. (1984). Isolation and characterization of a protein with inhibin like activity from pig follicular fluid. In *Gonadal proteins and peptides and their biological significance* (ed. M. R. Sairam and L. E. Atkinson) pp. 65–83. World Scientific Publishing, Singapore.

—— Ranganathan, M. R., and Ramasharma, K. (1981). Binding of an inhibin-like protein from bull seminal plasma to ovine pituitary membranes. *Mol. Cell. Endocr.* **22**, 251–264.

—— —— —— and Lamothe, P. (1981). Isolation and characterization of a bovine seminal plasma protein inhibiting pituitary FSH secretion. *Mol. Cell. Endocr.* **22**, 231–250.

Sander, H. J., Meijs-Roelofs, H. M. A., Kramer, P., and van Leeuwen, E. C. M. (1985). Inhibin-like activity in ovarian homogenates of prepubertal female rats and its physiological significance. *J. Endocr.* **107**, 251–257.

—— —— van Leeuwen, E. C. M., Kramer, P., and van Capellen, W. A. (1986). Inhibin increases in the ovaries of female rats approaching first ovulation: relationships with follicle growth and serum FSH concentrations. *J. Endocr.* **111**, 159–166.

—— van Leeuwen, E. C. M., and de Jong, F. H. (1984). Inhibin-like activity in media from cultured rat granulosa cells collected throughout the oestrous cycle. *J. Endocr.* **103**, 77–84.

Schenken, R. S., Anderson, W. H., and Hodgen, G. D. (1984). Follicle-stimulating hormone increases ovarian vein nonsteroidal factors with gonadotropin-inhibiting activity. *Fert. Steril.* **42**, 785–790.

Schwartz, N. B. and Channing, C. P. (1977). Evidence for ovarian 'inhibin': Suppression of the secondary rise in serum follicle stimulating hormone levels in proestrous rats by injection of porcine follicular fluid. *Proc. natn. Acad. Sci. U.S.A.* **74**, 5721–5724.

Scott, R. S. and Burger, H. G. (1981). Mechanism of action of inhibin. *Biol. Reprod.* **24**, 541–550.

—— —— and Quigg, H. (1980). A simple and rapid *in vitro* bioassay for inhibin. *Endocrinology* **107**, 1536–1542.

—— —— —— Dobos, M., Robertson, D. M., and de Kretser, D. M. (1982). The specificity of inhibin bioassay using cultured pituitary cells. *Mol. Cell. Endocr.* **27**, 307–316.

Seethalakshmi, L., Steinberger, A., and Steinberger, E. (1984). Pituitary binding of ³H-labeled Sertoli cell factor *in vitro*: a potential radioreceptor assay for inhibin. *Endocrinology* **115**, 1289–1294.

Seidah, N. G., Arbatti, N. J., Rochemont, J., Sheth, A. R., and Chrétien, M. (1984). Complete amino acid sequence of human seminal plasma β-inhibin. *FEBS Lett.* **175**, 349–355.

—— Ramasharma, K., Sairam, M. R. and Chrétien, M. (1984). Partial amino acid sequence of a human seminal plasma peptide with inhibin-like activity. *FEBS Lett.* **167**, 98–102.

Selin, L. and Moger, W. H. (1979). Effects of hemicastration and porcine follicular fluid (inhibin) on testicular function in the immature rat. *Fert. Steril.* **32**, 248.

Setchell, B. P., Davies, R. V., and Main, S. J. (1977). Inhibin. In *The testis*, Vol. 4 (ed. A. D. Johnson and W. R. Gomes) pp. 189–238. Academic Press, New York.

—— and Jacks, F. (1974). Inhibin-like activity in rete testis fluid. *J. Endocr.* **62**, 675–676.

—— and Main, S. J. (1974). Inhibin. *Bibliography of Reproduction* **24**, 245–252; 361–367.

—— and Sirinathsinghji, D. J. (1972). Antigonadotrophic activity in rete testis fluid, a possible 'inhibin'. *J. Endocr.* **53**, lx–lxi.

—— and Wallace, A. L. C. (1972). The penetration of iodine-labelled follicle-stimulating hormone and albumin into the seminiferous tubules of sheep and rats. *J. Endocr.* **54**, 67–77.

Sharpe, R. M. (1983). Local control of testicular function. *Q. Jl. exp. Physiol.* **68**, 265–287.

Shashidhara Murthy, H. M., Ramasharma, K., and Moudgal, N. R. (1979). Studies on purification of sheep testicular inhibin. *J. Reprod. Fert.* Suppl. 26, 61–70.

Sherman, B. M. (1976). The menopausal transition: hormonal evidence for independent ovarian control of pituitary FSH secretion. *University Michigan Medical Center Journal* **42**, 33–37.

—— and Korenman, S. G. (1975). Hormonal characteristics of the human menstrual cycle throughout reproductive life. *J. clin. Invest.* **55**, 699–706.

Sheth, A. R., Arbatti, N., Carlquist, M., and Jörnvall, H. (1984). Characterization of a polypeptide from human seminal plasma with inhibin (inhibition of FSH secretion)-like activity. *FEBS Lett.* **165**, 11–15.

—— Joshi, L. R., Moodbidri, S. B., and Rao, S. S. (1979). Characterization of a gonadal factor involved in the control of FSH secretion. *J. Reprod. Fert.* Suppl. **26**, 71–85.

—— Vaze, A. Y., and Thakur, A. N. (1978). Development of a radioimmunoassay for inhibin. *Ind. J. exp. Biol.* **16**, 1025–1026.

Skinner, M. K. and Fritz, I. B. (1986). Identification of a non-mitogenic paracrine factor involved in mesenchymal-epithelial cell interactions between testicular peritubular cells and Sertoli cells. *Mol. Cell. Endocr.* **44**, 85–97.

Sluss, P. M. and Reichert, L. E. (1984). Porcine follicular fluid contains several low molecular weight inhibitors of follicle-stimulating hormone binding to receptor. *Biol. Reprod.* **30**, 1091–1104.

Steinberger, A. (1979). Inhibin production by Sertoli cells in culture. *J. Reprod. Fert.* Suppl. **26**, 31–45.

Steinberger, A. (1981). Regulation of inhibin secretion in the testis. In *Intragonadal regulation of reproduction* (ed. P. Franchimont and C. P. Channing) pp. 283–298. Academic Press, London.

—— and Steinberger, E. (1976). Secretion of an FSH-inhibiting factor by cultured Sertoli cells. *Endocrinology* **99**, 918–921.

Stouffer, R. L. and Hodgen, G. D. (1980). Induction of luteal phase defects in rhesus monkeys by follicular fluid administration at the onset of the menstrual cycle. *J. clin. Endocr. Metab.* **51**, 669–671.

Swerdloff, R. S., Walsh, P. C., Jacobs, H. S., and Odell, W. D. (1971). Serum LH and FSH during sexual maturation in the male rat: effect of castration and cryptorchidism. *Endocrinology* **88**, 120–128.

Tanabe, K., Gagliano, P., Channing, C. P., Nakamura, Y., Yoshimura, Y., Iizuka, R., Fortuny, A., Sulewski, J., and Rezai, N. (1983). Levels of inhibin-F activity and steroids in human follicular fluid from normal women and women with polycystic ovarian disease. *J. clin. Endocr. Metab.* **57**, 24–31.

Tremblay, R. R., Foley, T. P., Corvol, P., Park, I. J., Kowarski, A., Blizzard, R. M., Jones, H. W., and Migeon, C. J. (1972). Plasma concentration of testosterone, dihydrotestosterone, testosterone-oestradiol binding globulin, and pituitary gonadotrophins in the syndrome of male pseudo-hermaphroditism with testicular feminization. *Acta endocr. Copenh.* **70**, 331–341.

Tsonis, C. G., McNeilly, A. S., and Baird, D. T. (1986). Measurement of exogenous and endogenous inhibin in sheep serum using a new and extremely sensitive bioassay for inhibin based on inhibition of ovine pituitary FSH secretion *in vitro*. *J. Endocr.* **110**, 341–352.

—— Quigg, H., Lee, V. W. K., Leversha, L., Trounson, A. O., and Findlay, J. K. (1983). Inhibin in individual ovine follicles in relation to diameter and atresia. *J. Reprod. Fert.* **67**, 83–90.

—— and Sharpe, R. M. (1986). Dual gonadal control of follicle-stimulating hormone. *Nature, Lond.* **321**, 724–725.

Uilenbroek, J. Th, J., Tiller, R., de Jong, F. H., and Vels, F. (1978) Specific suppression of follicle-stimulating hormone secretion in gonadectomized male and female rats with intrasplenic ovarian transplants. *J. Endocr.* **78**, 399–406.

Ultee-van Gessel, A. M. and de Jong, F. H. (1987). Inhibin-like activity in Sertoli cell culture media and testicular homogenates from rats of various ages. *J. Endocr.* **113**, 103–110

—— Leemborg, F. G., de Jong, F. H., and van der Molen, H. J. (1985). Influence of neonatal hemicastration on in-vitro secretion of inhibin, gonadotrophins and testicular steroids in male rats. *J. Endocr.* **106**, 259–265.

—— —— —— —— (1986). *In vitro* secretion of inhibin-like activity by Sertoli cells from normal and prenatally irradiated immature rats. *J. Endocr.* **109**, 411–418.

Vale, W., Rivier, C., Brown, M., Leppaluoto, J., Ling, N., Monahan, M, and Rivier, J. (1976). Pharmacology of hypothalamic regulatory peptides. *Clin. Endocr.* **5**, 261s–273s.

—— —— Vaughan, J., McClintock, R., Corrigan, A., Woo, W., Karr, D., and Spiess, J. (1986). Purification and characterization of an FSH releasing protein from porcine ovarian follicular fluid. *Nature, Lond.* **321**, 776–779.

van de Wiel, D. F. M., Bar-Ami, S., Tsafriri, A., and de Jong, F. H. (1983). Oocyte maturation inhibitor, inhibin and steroid concentrations in porcine follicular fluid at various stages of the oestrous cycle. *J. Reprod. Fert.* **68**, 247–252.

van Dijk, S., de Jong, F. H., and van der Molen H. J. (1984). Use of fast protein liquid chromatography in the purification of inhibin from bovine follicular fluid. *Biochem. Biophys. Res. Commun.* **125**, 307–314.

—— Steenbergen, C., de Jong, F. H., and van der Molen, H. J. (1985). Comparison between inhibin from human and bovine ovarian follicular fluid using fast protein liquid chromatography. *Mol. Cell. Endocr.* **42**, 245–251.

—— —— Gielen, J. Th., and de Jong, F. H. (1986). Sexual dimorphism in immunoneutralization of bioactivity of rat and ovine inhibin. *J. Endocr.* **111**, 255–261.

van Thiel, D. H., Sherins, R. J., Myers, G. H., and De Vita, V. T. (1972). Evidence for a specific seminiferous tubular factor affecting follicle-stimulating hormone secretion in man. *J. clin. Invest.* **51**, 1009–1019.

Vaze, A. Y., Thakur, A. N., and Sheth, A. R. (1979). Development of a radioimmunoassay for human seminal plasma inhibin. *J. Reprod. Fert.* Suppl. **26**, 135–146.

Verhoeven, G. and Franchimont, P. (1983). Regulation of inhibin secretion by Sertoli cell-enriched cultures. *Acta endocr. Copenh.* **102**, 136–143.

Vidgoff, B. and Vehrs, H. (1940). Studies on the inhibitory hormone of the testes. IV. Effect on the pituitary, thyroid and adrenal glands of the adult male rat. *Endocrinology* **26**, 656–661.

Vijayalakshmi, S., Bandivdekar, A. H., Joshi, L. R., Moodbidri, S. B., and Sheth, A. R. (1980). Isolation and characterization of ovine testicular and ovarian inhibin. *Arch. Androl.* **5**, 179–188.

Wallace, J. M. and McNeilly, A. S. (1985). Increase in ovulation rate after treatment of ewes with bovine follicular fluid in the luteal phase of the oestrous cycle. *J. Reprod. Fert.* **73**, 505–515.

—— —— and Baird, D. T. (1985). Ovulation rate and embryo survival in Damline ewes after treatment with bovine follicular fluid in the luteal phase of the oestrous cycle. *J. Reprod. Fert.* **75**, 101–109.

Ward, D. N., Glenn, S. D., Liu, W. K., and Gordon, W. L. (1983). Chemistry and physiology of gonadal peptides. In *Factors regulating ovarian function* (ed. G. S. Greenwald and P. F. Terranova) pp. 141–155. Raven Press, New York.

Welschen, R., Dullaart, J., and de Jong, F. H. (1978). Interrelationships between circulating levels of estradiol-17β, progesterone, FSH and LH immediately after unilateral ovariectomy in the cyclic rat. *Biol. Reprod.* **18**, 421–427.

—— Hermans, W. P., and de Jong, F. H. (1980). Possible involvement of inhibin in the interrelationship between numbers of antral follicles and peripheral FSH concentrations in female rats. *J. Reprod. Fert.* **60**, 485–493.

—— —— Dullaart, J., and de Jong, F. H. (1977). Effects of an inhibin-like factor present in bovine and porcine follicular fluid on gonadotrophin levels in ovariectomized rats. *J. Reprod. Fert.* **50**, 129–131.

Williams, A. T., Rush, M. E., and Lipner, H. (1979). Isolation and preliminary characterization of inhibin-f. In *Ovarian follicular and corpus luteum function* (ed. C. P. Channing, J. Marsh, and W. A. Sadler) pp. 429–435. Plenum Press, New York.

Yamashiro, D., Li, C. H., Ramasharma, K., and Sairam, M. R. (1984). Synthesis and biological activity of human inhibin-like peptide-(1–31). *Proc. natn. Acad. Sci. U.S.A.* **81**, 5399–5402.

Ying, S. Y., Becker, A., Baird, A., Ling, N., Ueno, N., Esch, F., and Guillemin, R. (1986a). Type beta transforming growth factor (TGF-β) is a potent stimulator of the basal secretion of follicle stimulating hormone (FSH) in a pituitary monolayer system. *Biochem. Biophys. Res. Commun.* **135**, 950–956.

—— —— Ling, N., Ueno, N., and Guillemin, R. (1986b). Inhibin and beta type transforming growth factor (TGFβ) have opposite modulating effects on the follicle stimulating hormone (FSH)-induced aromatase activity of cultured rat granulosa cells. *Biochem. Biophys. Res. Commun.* **136**, 969–975.

2   GnRH and ovarian hormone feedback

IAIN J. CLARKE

I Introduction

This review focuses on the synthesis, release and action of GnRH and the
way in which these are regulated by ovarian feedback hormones. Particular
reference is made to work done on female rats, sheep and monkeys. Neural
mechanisms that control GnRH neurones have not been considered in
detail and reference can be made to recent reviews by Barraclough and Wise
(1982), Kalra and Kalra (1983), Barraclough, Wise, and Selmanoff (1984)
and Ferin, Van Vugt, and Wardlaw (1984). Also not covered in this review is
an appraisal of the role of changing GnRH receptor populations as a
mechanism for altering GnRH action. This topic has been reviewed
extensively by Clayton and Catt (1981) and Conn (1986).

   The review begins with an appreciation of the arrangement of GnRH
neurones in the hypothalamus and then considers the synthesis of GnRH.
Studies on GnRH *release* are considered in terms of feedback regulation
and, finally, feedback effects on GnRH *action* are reviewed. I have
attempted to provide a synopsis of the hypothalamo-pituitary-gonadal axis

in the female so that the article is of broad general interest to all reproductive biologists.

II Hypothalamic GnRH

1 ANATOMICAL CONSIDERATIONS

Studies in a number of species have shown a broadly similar arrangement of GnRH localization in the brain, although there is some interspecific variation. In all species the median eminence (ME) contains the greatest amount of GnRH, this being the area in which the peptide is stored in neuronal terminals prior to release into hypophyseal portal blood (Wheaton, Krulich, and McCann 1975; Polkowska, Dubois, and Domanski 1980; Samson, McCann, Chud, Dudley, and Moss 1980; Selmanoff, Wise, and Barraclough 1980). The region containing the second highest level is the organum vasculosum of the lamina terminalis (OVLT), where the GnRH is also localized in neuronal cell terminals (Baker, Dermody, and Reel 1975; Barry and Carette 1975; Wenger and Leonardelli 1980; Samson et al. 1980; Selmanoff et al. 1980). Using microdissection and radioimmunoassay, GnRH has been measured in a number of intra- and extrahypothalamic locations (Samson et al. 1980; Selmanoff et al. 1980).

In initial immunohistochemical studies difficulties were encountered in the staining of GnRH perikarya (Baker, Dermody, and Reel 1974; King, Parsons, Erlandsen, and Williams 1974; Polkowska et al. 1980) but these problems have been overcome by improvements in fixing and embedding of tissues and the identification of suitable antisera (Barry, Dubois, and Poulain 1973; Setalo, Vigh, Schally, Arimura, and Flerko 1975; Vigh, Setalo, Arimura, Schally, and Flerko 1978; Marshall and Goldsmith 1980; Kawano and Daikoku 1981; Silverman, Autunes, Abrams, Nilaver, Thau, Robinson, Ferin, and Krey 1982). Complex networks of neurones project from the hypothalamus to various parts of the brain (Krey and Silverman 1978a; Silverman et al. 1982; Hoffman 1983; King and Anthony 1984; Lehman, Robinson, Karsch, and Silverman 1986) as well as the ME and OVLT. In the rat, GnRH cell bodies are found mainly in the preoptic/septal region but scattered cells can be found as far rostral as the olfactory bulb and caudally to the retrochiasmatic area. GnRH containing perikarya have been found in the rat arcuate nucleus (AN) also (Clayton and Hoffman 1979; King, Tobet, Snavely, and Arimura 1982; Witkin, Paden, and Silverman 1982), but these may be of minor significance, since horizontal cuts between the AN and ME do not affect the GnRH content of either region (Kawano and Daikoku 1981). Some studies of the rat deny the presence of GnRH containing perikarya in the AN (Flerko, Setalo, Vigh, and Merchenthaler 1980; Flerko 1981; Witkin et al. 1982). In reviewing the topic, Kalra and Kalra (1983) reached the conclusion that GnRH cells in the medio-basal hypothalamus might be of functional significance because elimination of more rostrally located GnRH containing areas could never totally eliminate gonadotrophin secretion (see below). An early study of the sheep (Dees, Sorensen, Kemp, and McArthur 1981) reported scattered GnRH-positive neurones throughout the AN and ventromedial nucleus, but the cells were not grouped into discrete nuclei. More recently, Lehman *et al.* (1986) have found that the AN of the sheep contains very few GnRH containing cell bodies. In other species such as the guinea pig, ferret, bat (Krey and Silverman 1978), rabbit (Flerko 1981), monkey (Silverman, Antunes, Ferin, and Zimmerman 1977; King and Anthony 1984; Ellinwood, Ronnekleiv, Kelly, and Resko 1985) and human (Barry 1977; King and Anthony 1984) GnRH containing cell bodies are relatively abundant in the AN-tuberal region. In the human and the monkey GnRH cell bodies are also found in the more rostrally located areas including the preoptic nuclei, bed nucleus of the stria terminalis and OVLT (Barry 1977; Silverman *et al.* 1982; Ellinwood *et al.* 1985). In fact, Silverman *et al.* (1982) stated that these rostral areas contained the largest number of GnRH containing perikarya.

GnRH containing cell bodies of the preoptic/anterior hypothalamic area are thought to be the most pertinent to the control of gonadotrophin secretion in rodent species. This concept is derived from immunohistochemical tracking of neuronal pathways, lesioning studies, and the use of the Halasz knife (Halasz and Pupp 1965) to intercept fibre tracts.

Immunohistochemical studies of the GnRH in rat hypothalamus have been performed by a number of groups (Ibata, Watanabe, Kinoshita, Kubo, Sano, Sin, Hashimura, and Imagawa 1979; Flerko *et al.* 1980; Merchenthaler, Kovacs, Lovaoz and Setalo 1980; Kawano and Daikoku 1981; Witkin *et al.* 1982; Jew, Leranth, Arimura, and Palkovits 1984). There is general agreement that, in the rat, cell bodies in the septal, preoptic and suprachiasmatic regions (excluding the suprachiasmatic nucleus: Rethelyi, Vigh, Setalo, Merchenthaler, and Flerko, 1981) give rise to neuronal tracts that project caudally into the hypothalamus. One representation of these preoptico-infundibular tracts was provided by Hoffman (1983) and is shown in Fig. 2.1. A peculiar feature that has been seen in the rat is that some fibres run ventral to the optic chiasma (Flerko *et al.* 1980).

Lesions in the preoptic nucleus, the periventricular area caudal to the OVLT, or in the suprachiasmatic nucleus (SCN) of the female rat brain lead to impairment of gonadotrophin secretion and loss of cyclic ovarian function (Clemens, Smalstig, and Sawyer 1976; Brown-Grant and Raisman 1977; Gray, Sodersten, Tallentine, and Davidson 1978; Kawakami, Yoshioka, Konda, Arita, and Visessuvan, and Akema 1978; Wiegand, Terasawa, and Bridson 1978; Samson and McCann 1979; Wiegand, Teresawa, Bridson, and Goy 1980). More caudally located lesions may damage neuronal tracts arising from rostrally located areas and traversing these regions. For

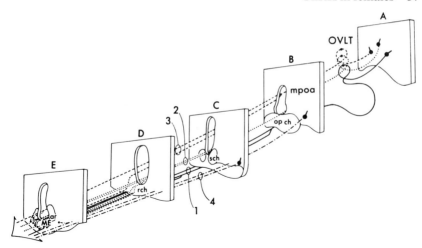

Fig. 2.1. GnRH tracts to the ME of the intact rat. Tracts of GnRH axons to the ME arise from a triangular field of cells whose apex lies in the caudal septum just rostral to the OVLT within the median preoptic nucleus or preoptic periventricular area, level 'A'. The rostral-most group of cells initiates three tracts: one which courses under the optic chiasm (1); one which traverses along the floor of the third ventricle (2); and a third that courses along the lateral walls of the third ventricle (3). The more lateral and caudal LHRH cells, seen in levels 'B' and 'C', give rise to the fourth tract that courses with the fibres of the medial forebrain bundle and enters the more caudal median eminence (ME). Fibres from all the tracts branch as they course to the ME and begin to merge at the retrochiasmatic area (rch), level 'D'. They then continue into the ME, level E. Because terminals may be found in the OVLT as GnRH fibres loop around the OVLT vessels, loops with dots are indicated. ar, arcuate nucleus; sch, suprachiasmatic nucleus, mpoa, medial preoptic area; op ch, optic chiasm. (From Hoffman (1983), by permission).

example the OVLT and the SCN do not contain GnRH containing perikarya (Wenger and Leonardelli 1980; Rethelyi *et al.* 1981) but lesions in this region do perturb the GnRH axis (Brown-Grant and Raisman 1977; Wenger, Kerdelhue, and Halasz 1979; Kawakami, Arita, and Yoshioka 1980). Thus some caution is required to interpret such studies. Nevertheless, these lesion studies provide support to the theory that GnRH is synthesized in rostrally located areas and transported along axons to the ME. Lesions in the septal and lateral preoptic area (POA) did not reduce GnRH levels in the ME whereas lesions in the medial POA did so (Ibata *et al.* 1979). This raises the question of the relevance of these most rostrally located GnRH containing neurones in the control of gonadotrophin secretion, and suggests that the medial POA is the most important source of GnRH.

Kalra and Kalra (1983) reviewed studies in the rat which showed that lesions in various rostrally located nuclei had very little impact on mediobasal hypothalamic GnRH and gonadotrophin secretion. As these reviewers pointed out, it is possible that GnRH output of neurones in the mediobasal hypothalamus may be increased after such insults, compensating for the removal of the normal inputs. For example, such an explanation might be applied to the observations by Soper and Weick (1980) that complete hypothalamic deafferentation did not block the pulsatile release of LH in long term ovariectomized (OVX) rats but lesions in the anterior AN did affect LH release. These authors were led to believe that there are two neural pathways that stimulate LH pulses—one being hypothalamic (AN) and the other being extrahypothalamic involving rostro-caudal pathways. It remains an open question as to whether or not the small numbers of cells in the mediobasal hypothalamus of the rat can compensate for the loss of more rostrally located nuclei that are more densely populated with GnRH containing neurons.

In the sheep, electrolytic lesions in the medio-basal hypothalamic region caused ovarian acyclicity (Clegg, Santalucito, Smith, and Ganong 1958; Clegg and Ganong 1960; Radford 1967; Przekop and Domanski 1970) whereas lesions in the anterior hypothalamus did not do so (Przekop and Domanski 1970). Lesions in the mediobasal hypothalamus also disrupted the oestrogen-induced positive feedback phenomenon (Radford 1979). These studies suggested that the anterior hypothalamus plays a minor role in sustaining gonadotrophin secretion in this species.

Halasz knife cuts between the anterior and mediobasal regions of the rat hypothalamus caused a build up of GnRH in the neurons proximal to the cut and a decrease in ME content of GnRH (Setalo et al. 1976; Weiner, Pattou, Kerdelhue, and Kordon 1975; Kalra 1976; Merchenthaler et al. 1980; Jew, Leranth, Arimura, and Palkovits 1984), suggesting that a substantial amount of GnRH travels in a rostro-caudal direction. In the sheep, Halasz knife cuts have produced equivocal results regarding the role of a preoptico-tuberoinfundibular GnRH tract. Jackson, Kuehl, McDowell, and Zaleski (1978) and Thiery, Pelletier, and Signoret (1978) found that anterior hypothalamic deafferentation would prevent the occurence of oestrogen-induced LH surges but basal LH secretion persisted. Post-suprachiasmatic cuts or cuts through the anterior part of the AN reduced basal LH output in ovariectomised ewes. Complete deafferentation of the mediobasal hypothalamus, with the anterior extent of these cuts passing through the anterior portion of the AN, eliminated LH secretion (Jackson et al. 1978). It was concluded that the area subserving tonic LH secretion resided anterior to the AN and posterior to the SCN (Jackson et al. 1978) and this was in accord with earlier observations with electrolytic lesions (see above). Pau, Kuehl, and Jackson (1982) substantiated the findings of Jackson et al. (1978) and Thiery et al. (1978), and found that that knife cuts at the level of the SCN reduced the size of oestrogen induced LH surges but did not affect basal LH secretion. They concluded that these results were '... consistent with the presence of luteinizing hormone releasing hormone (LH-RH) in the medio-basal hypothalamus of the ewe' and that '... these data suggest that in sheep a functional medio-basal hypothalamo-pituitary unit controls basal LH secretion.' It is possible that this interpretation is incorrect since complete deafferentation of the medio-basal hypothalamus abolished gonadotrophin secretion (Jackson *et al.* 1978). Furthermore, the area of the hypothalamus that Jackson *et al.* (1978) and Pau *et al.* (1982) regarded as the most pertinent to the maintenance of basal gonadotrophin output (post-SCN but pre-AN) is not well populated with GnRH containing cell bodies (Advis, Kuljis, and Dey 1985; Lehman *et al.* 1986). It is possible that a misinterpretation has arisen because the frontal cuts used in the sheep have not intercepted dorsal and lateral fascicles coursing into the ME (Lehman *et al.* 1986). The reduction in the oestrogen positive feedback response (Pau *et al.* 1982) could have been the result of the selective cuts through of medially orientated fibre bundles. Thus, for the sheep, it is still not clear which hypothalamic centres are most relevant to LH secretion, but the dense population of GnRH containing neurons in the medial POA suggest that this region is of major significance (Advis *et al.* 1985; Lehman *et al.* 1986).

A series of experiments in monkeys have suggested that GnRH neurones in the rostral hypothalamus do not play a major role in the control of gonadotrophin secretion. Krey, Butler, and Knobil (1975) created mediobasal hypothalamic 'islands' with the Halasz knife so that the AN, ventromedial nucleus, premammillary nucleus and parts of the mammillary bodies were isolated from the rest of the brain. These hypothalamic 'islands' did not prevent the occurrence of regular menstrual cycles in normal females or the secretion of LH and FSH in ovariectomised females. Furthermore, these monkeys had typical biphasic gonadotrophin responses to oestrogen challenges. On the basis of these experiments Krey *et al.* (1975) concluded that, in contrast to the rat, '... the central components of the neuroendocrine system which controls the gonadotropin surges, like those of the system which governs the tonic secretion of these hormones, are resident within the MBH (mediobasal hypothalamus)-hypophysial unit (of the monkey).' In contrast, Norman, Resko, and Spies (1976) found that lesions in the POA of the hypothalamus, extending into the suprachiasmatic area, interfered with ovulatory cycles in monkeys. Similar results were obtained with Halasz knife cuts (Norman *et al.* 1976) but some of the lesioned/cut animals did eventually ovulate. In order to resolve the discrepancy between the studies of Krey *et al.* (1975) and Norman *et al.* (1976), an experiment was carried out in which all brain structures anterior to the optic chiasma were removed from ovariectomised monkeys (Hess, Wilkins, Moossy, Chang, Plant, McCormack, Nakai, and Knobil 1977). Immediately following ablation, the monkeys were challenged with oestrogen and the biphasic plasma gonadotrophin responses were similar to those in unlesioned ovariectomised monkeys. This strongly suggested that areas rostral to the optic chiasma were not necessary for the induction of preovulatory-like LH surges in the monkey. Finally, Plant, Moosy, Hess, Nakai, McCormack, and Knobil (1979) repeated the studies of Norman *et al.* (1976), placing rostral hypothalamic radiofrequency lesions in four female monkeys, and found that ovulatory cycles continued in three animals. Plant *et al.* (1979) and Knobil (1980) have attempted to explain the discrepancy in these lesioning studies and have concluded that the differences in the results are qualitative and that anovulation obtained in the experimental series of Norman *et al.* (1976) could have been due to factors other than the lesioning procedure. In particular, Knobil (1980) reasoned that the monkeys in the study of Norman *et al.* (1976) may have become anovulatory for non-specific reasons. The data of Hess *et al.* (1977) and the fact that the basal hypothalamus of the monkey is rich in GnRH containing cell bodies certainly provides compelling evidence to support the interpretations of Krey *et al.* (1975). In spite of this, the more rostrally located nuclei contain the richest beds of GnRH containing cells in the monkey (Silverman *et al.* 1982), and it is difficult to dismiss their contribution to the ME and their role in the regulation of gonadotrophin secretion.

To investigate further the role of the anterior hypothalamus in the control of gonadotrophin secretion in the monkey, Cogen, Antunes, Louis, Dyrenfurth, and Ferin (1980) used a transorbital surgical approach to make horizontal cuts through the optic chiasma and into the hypothalamus, under the anterior hypothalamus and ventromedial nuclei and dorsal to the AN. A silastic barrier was inserted into the cut to separate the anterior and mediobasal hypothalamus. Complete bilateral cuts caused anovulation in the short term but cyclicity was restored later (120–210 days postoperatively). Also, these monkeys failed to show gonadotrophin surges in response to oestrogen 45–60 days after surgery but did so later (120–210 days). Thus, the anterior hypothalamic nuclei may play some role in controlling the menstrual cycle of the monkey but do not appear to be essential. Perhaps, as suggested by Norman *et al.* (1976) and Kalra and Kalra (1983) (see above), the mediobasal hypothalamus undergoes some neural reorganization after anterior hypothalamic deafferentation, and the GnRH containing cells, that are relatively abundant in this region, assume control of the cycle.

In an effort to localize further the medio-basal hypothalamic centre that controls the menstrual cycle of the monkey, Plant, Krey, Moossy, McCormack, Hess, and Knobil (1978) placed discrete lesions in the AN. Incomplete lesions, sparing ventral or caudal aspects of the AN, or being unilateral, did not eliminate gonadotrophin secretion, whereas destruction of the entire AN or caudal aspect of the AN/ME did so. Whilst this provides a useful 'hypophysiotropic clamp' model (Knobil 1980) (see below), it does not conclusively pinpoint the AN as the mediobasal nucleus responsible for elaborating GnRH signals. This is because effective AN lesions encroach on the ME and in destroying the AN they may also destroy neurones traversing the region into the ME. GnRH containing perikarya are present throughout the basal hypothalamus of the monkey (see above) and those relevant to gonadotrophin secretion may be outside the AN.

The general picture that has emerged from lesioning and deafferentation studies is that a preoptico-tuberoinfundibular GnRH axis does exist in the rat and appears to participate in the regulation of gonadotrophin secretion. The roles of GnRH neurones arising for example from extreme rostral areas, septum, or from the AN, medio-basal hypothalamus, are less certain (Kalra and Kalra 1983). In the sheep, it seems that rostro-caudal projections are vital for both basal and surge release of gonadotrophins. In the monkey, where a substantial proportion of the GnRH neurons are resident in the mediobasal hypothalamic region, the rostral areas may be of minor importance for both tonic and surge release of gonadotrophins.

2 GnRH SYNTHESIS

Because of the multidirectional aspects of GnRH neurones and storage in axonal terminals, it is difficult to make firm conclusions regarding the rate of synthesis of GnRH by merely measuring tissue content in various parts of the brain. However, for want of more sophisticated approaches many studies have relied upon macro- or micro-dissections of the brain and subsequent determination of tissue content. This may be reasonable for areas in which there is a predominance of perikarya or terminals, for example the medial POA and the ME respectively. On the other hand, problems can arise when areas containing both cell bodies and terminals are studied. Such a problem exists in some studies where preoptic/suprachias-matic areas are dissected and assayed (Wheaton 1979; Crowder and Nett 1984). These dissected areas presumably contained the GnRH-rich neuronal terminals of the OVLT.

A further consideration when measuring hypothalamic GnRH content by RIA is the specificity of the antiserum used. Some antisera are specific for the C- and N-terminal residues of the GnRH molecule and will not detect the peptide in its prohormone form. In fact, Millar and colleagues (Millar, Denniss, Tobler, King, Schally, and Arimura 1977; Millar, Wegener, and Schally 1981) used this to their advantage when postulating the existence of a prohormone for GnRH. Antiserum directed towards the central portion of GnRH detected larger amounts of high molecular weight GnRH (presumably prohormone) than antiserum directed towards the C- and N-terminals of GnRH. This led Millar and colleagues to the conclusion (later found to be correct-see below) that the prohormone contains GnRH with C- and N-terminus extensions. King and Anthony (1984) found that neuronal perikarya in the rat could not be visualized by immunohistochemical means when an antiserum (Arimura R422) was used that did not allow C- and N-terminal modifications or extensions. Interestingly, such antisera will detect perikarya as well as axons and terminals in other species (for example, humans, monkeys, ferrets and bats). These observations led King and Anthony (1984) to suggest that the prohormone is processed in the axons and/or terminals of the rat hypothalamus, whereas processing may occur in the perikarya of other species. Ellinwood *et al.* (1985) argued against this hypothesis since they used a similar antiserum (EL-14) and were able to immunostain GnRH containing perikarya in the rat. As they pointed out, one must be aware of the mixture of antibodies in polyclonal antisera and this may explain differences between EL-14 and R422.

The most exciting recent development in this area has been the discovery of the gene and amino-acid sequence for pro-GnRH. Evidence for pro-GnRH had been provided by chromatography (Millar *et al.* 1977) and cell free translation of poly A^+ enriched hypothalamic RNA (Curtis and Fink 1983). Utilizing recombinant DNA technology Seeburg and Adelman (1984) were able to isolate cloned human genomic DNA that encoded for GnRH. Using a restriction fragment of DNA that contained the GnRH coding sequence they probed a cDNA library that was constructed in λ phage using human placental tissue to derive the cDNA. Human placenta was used because GnRH synthesis had been shown in this tissue (Tan and Rousseau 1982) and human hypothalami were not available. Clones that hybridized with the cDNA inserts allowed the deduction of the amino-acid sequence for pro-GnRH which consists of a 23 amino-acid signal peptide, a decapeptide sequence for GnRH and a C-terminal extension of 56 amino acids. The GnRH sequence begins with a glutamic residue at the amino-terminal which undergoes cyclization to pyroglutamate. The sequence continues with glycine, which donates an amino group for C-terminal amidation, and then a lys-arg dibasic enzymatic cleavage point. The remaining C-terminal region has been called GnRH-associated peptide (GAP) being a potent inhibitor of prolactin secretion and also stimulates the release of gonadotrophins from cultured rat pituitary cells (Nikolics, Mason, Szonyi, Ramachandran, and Seeburg 1985). Complementary DNA clones encoding for pro-GnRH have now been isolated from human and rat hypothalamic mRNA (Adelman, Mason, Hayflick, and Seeburg 1986).

Pro-GnRH has been localized in the rat brain using an antiserum raised against pro-GnRH (40–53) (Phillips, Nikolics, Branton, and Seeburg 1985). Dual immunofluorescence staining was found with the anti-pro-GnRH (40–53) and anti-GnRH in cell bodies of the septal/preoptic region and various more rostrally located nuclei. The anti-pro-GnRH 40–53 serum also stained axons and terminals which suggests that some pro-GnRH may be transported along axons to be secreted from terminals into portal blood.

Using a probe consisting of an oligomer of 59 deoxyribonucleotides complementary to the sequence of GnRH mRNA coding for the amino acids −5−15 of the human preprohormone, Shivers, Harlan, Hejtmanick, Conn, and Pfaff (1986) have localized cells in the rat brain containing GnRH-like mRNA. These workers use the method of hybridization histochemistry to localize cells in the diagonal band of Broca and the preoptic region. The RNA species that hybridizes to this probe has not been characterized but it seems likely that it is the GnRH message. It is hoped that this technique will now allow us to quantify changes in GnRH synthesis in response to steroid feedback.

The isolation of the gene sequence for pro-GnRH will hopefully allow definitive studies on the synthesis of GnRH, using Northern blot hybridization or dot blot analysis. This represents a major advance since it overcomes the problem of interpreting changes in tissue content in terms of synthesis or release. Until the GnRH gene was discovered the only definitive method of measuring *in vivo* biosynthesis of GnRH was to inject tritiated amino-acids into the brain and then extract the tissue to quantify the degree of incorporation of the label (Krause, Advis, and McKelvy 1982). One problem with the application of this new technology might be the low abundance of GnRH message in discrete regions of the hypothalamus (Shivers, Harlan, Morrell, and Pfaff 1983*a, b*).

3   EFFECTS OF OVARIECTOMY AND STEROIDS ON GnRH CONTENT OF THE HYPOTHALAMUS

Ovariectomy causes a reduction in the mediobasal hypothalamic content of GnRH in rats (Piacsek and Meites 1966; Moguilevsky, Scacchi, Debeljuk, and Faigon 1975; Kalra 1976) and sheep (Wheaton, 1979). This is primarily due to the reduction in content of the ME (Kobayashi, Lu, Moore, and Yen 1978) and in female rats the effect can be reversed by the administration of oestrogen (Kalra 1976). The oestrogen effect can also be seen in castrated male rats and this can be blocked by an anti-oestrogen (Kalra and Kalra 1980). Also in male rat hypothalami, Millar *et al.* (1981) found tht high molecular weight species of GnRH (presumably prohormone) were reduced by castration and increased by oestrogen treatment of castrates, suggesting that the effect might be due to changes in rate of synthesis. Alternatively, this effect could have been due to changes in the rate of processing of the prohormone. In sheep, Wheaton (1979) was unable to reverse the effect of ovariectomy by administering oestrogen, which is somewhat puzzling given the consistent responses achieved in rats. Very little, if any, change was seen in the GnRH content of hypothalamic nuclei of rats after ovariectomy (Kobayashi *et al.* 1978).

Using male rats, Kalra and colleagues have examined the temporal changes in GnRH content of the hypothalamus and LH release. Castration reduced mediobasal hypothalamic GnRH stores but these were restored after 7 days by administration of testosterone. Whereas the change in

GnRH content was gradual after testosterone treatment, the fall in plasma LH levels was rapid. It was thus argued that the change in GnRH content of the mediobasal hypothalamus was not simply a result of decreased secretion, which would have caused a sharp rise in ME content, but was due to increased *de novo* synthesis (Kalra and Kalra 1980). Using similar reasoning, Kalra, Simpkins, and Kalra (1984) localized the effect to the ME. As these authors also pointed out, the replenishment of GnRH stores by steroidal treatment of gonadectomized animals could, in part, be due to an increase in the rate of processing from the prohormone and an alteration in the degradation of GnRH could also play a part. An alternative explanation is that the steroid treatment blocked LH secretion by direct action on the pituitary and that GnRH secretion continued at a reduced rate.

The role of progesterone *per se* in controlling GnRH stores has not been studied extensively. An early study (Piacsek and Meites 1966) using bioassay techniques showed no effect of progesterone in ovariectomised rats whereas, in the same study, oestrogen did have an effect.

4 ALTERATIONS IN HYPOTHALAMIC GnRH CONTENT DURING THE OESTROUS CYCLE.

Early studies (Barry and Dubois 1974) using immunofluorescence showed that GnRH content of the guinea pig ME increased during dioestrus and then fell suddenly during pro-oestrus/oestrus. This suggested a gradual accumulation throughout the cycle and a substantial discharge from ME terminals at the time of the preovulatory LH surge. Similar results had been obtained for the rat (Ramirez and Sawyer 1965) using bioassay methodology. More recently, radioimmunoassay studies have verified that the GnRH content of the rat mediobasal hypothalamus decreases around the start of the pro-oestrous LH surge (Araki, Ferin, Zimmerman, and Van de Wiele 1975; Asai and Wakabayashi 1975; Kalra and Kalra 1977; Barr and Barraclough 1978; Oshima, Morishita, Omura, and Saito 1978). Some studies showed that mediobasal hypothalamic GnRH levels rise during the LH surge (Barr and Barraclough 1976; Kalra and Kalra 1977) and this was taken as evidence that rapid *de novo* synthesis of GnRH occurs at this time.

The significance of changes in the hypothalamic content of GnRH during pro-oestrus in the rat must be viewed with caution since large daily fluctuations can also be found during dioestrus (Kalra and Kalra 1977). Also, nembutal treatment of pro-oestrous rats which blocks the LH surge, presumably by an effect on the brain, does not prevent the decrease in GnRH content of the mediobasal hypothalamus that normally occurs at this time (Barr and Barraclough 1978). If decreases in hypothalamic levels are taken as evidence of increased release, then these results are clearly anomalous. Furthermore, Snabes, Kelch, and Karsch (1977) found that hypothalamic GnRH levels decreased during the early afternoon in untreated OVX rats without an associated increase in plasma LH concentrations. Again, there is a clear discrepancy between the fall in hypothalamic GnRH content and the presumptive rise in GnRH secretion. One possible reason for fluctuations in ME content of GnRH could be that there are changes in degradation, but enzyme activity appears to be reduced during the pro-oestrous period (Advis, Krause, and McKelvy 1982; O'Connor, Lapp, and Mahesh 1984).

Using the micropunch method of Palkovits (1973) linked with radioimmuno-assay, one group has provided detailed information on cyclical changes in GnRH content of various nuclei in the rat hypothalamus (Rance, Wise, Selmanoff, and Barraclough 1979: Wise, Rance, Selmanoff, and Barra-clough 1979). In the ME, GnRH levels increased on the morning of pro-oestrus and decreased in the afternoon (during the commencement of the LH surge). Parallel changes were seen in the SCN, preoptic nucleus and retrochiasmatic area. Similar patterns of change were not seen during dioestrus. These studies, and those of Oshima *et al.* (1978) suggest that the preoptic-tuberoinfundibular GnRH tract acts as an integrated unit, a view that is contrary to that proposed by Araki *et al.* (1975) who found divergent alterations in the GnRH content of the anterior and mediobasal hypothalamus.

Wenger and Leonardelli (1980) have found cyclic variations in the GnRH content of the OVLT in the rat with the highest levels being seen during pro-oestrus. Dissection and immunoassay data were supported by fluores-cence studies and electron microscopy. Why GnRH levels should change in the neuronal terminals of the OVLT is not clear, since there is unlikely to be any vascular link with the pituitary gland (Ambach, Kovovics, and Palkovits 1978). Scott, Dudley, and Knigge (1974) have suggested that GnRH from the OVLT might reach the pituitary gland by a cerebrospinal route and via tanycytes in the infundibular recess, but this seems a very tortuous route given the simpler option of discharge from ME terminals directly into the hypophyseal portal blood. Thus any cyclic changes in GnRH content of the OVLT may be of little significance (Barraclough and Wise 1982).

A model that has been extensively used for the study of neural mechanisms regulating the pro-oestrous LH surge is the steroid-primed OVX rat. The original model was oestrogen- and progesterone-treated (Caligaris, Astrada, and Taleisnik, 1968). In fact, OVX rats given oestrogen alone will have daily LH surges (Legan and Karsch 1975) but progesterone treatment advances and amplifies the LH surge in oestrogen-primed OVX rats (Caligaris, Astrada, and Taleisnik 1971; De Paolo and Barraclough 1979). It is assumed that the neural mechanisms responsible for generating the LH surge—via GnRH—in the steroid-primed OVX rat are similar to those operating at the time of the cyclic preovulatory LH surge, but Barraclough and Wise (1982) give some reasons for caution in making this assumption. Kalra and Kalra (1979) examined hypothalamic GnRH content in OVX rats given either oestrogen alone (on day 0) or oestrogen on day 0 followed by progesterone at 10.00 h on day 2. The rats were then killed between 10.00 h and 19.00 h on that day. In the oestrogen treated rats, serum gonadotrophin levels were elevated in the afternoon of day 2; mediobasal hypothalamic GnRH levels changed very little during the early stages of the LH surge, but declined abruptly between 15.00 h and 18.00 h when serum LH levels were maximal and levels in the anterior hypothalamus showed no consistent pattern of change. In the oestrogen-primed, progesterone treated rats, mediobasal hypothalamic GnRH levels fell around the start of the LH surge and then rose during the surge. These results suggest two different mechanisms for the generation of the LH surge in the oestrogen alone versus the oestrogen-progesterone treated model. The latter seems more appropriate as a model for the pro-oestrous GnRH/LH surge since there is a rise in GnRH content of the mediobasal hypothalamus during the surge, as seen during pro-oestrus in intact females (Barr and Barraclough 1978; Kalra and Kalra 1977). Peduto and Mahesh (1985) have also reported that the appropriate progesterone treatment of oestrogen primed immature rats will cause changes in mediobasal hypothalamic GnRH content similar to those seen in the cyclic female.

In sheep, Jackson, Roche, Foster and Dzuik (1971) measured ME extracts using a rat bioassay and found that GnRH levels were low around day 5 of the oestrous cycle, rose around days 9–11 and fell through days 13–15 (ovulation is on day 17). Lowest levels were seen during pro-oestrus. Wheaton (1979) used radioimmunoassay to quantify GnRH in grossly dissected hypothalamic blocks. In the ME, levels rose progressively throughout the cycle to be highest during the pro-oestrous/oestrous period. In the preoptic/suprachiasmatic region, GnRH levels were higher during the later luteal phase (day 10–14) than during the early luteal phase (day 3–8) or during the pro-oestrous/oestrous period. Using qualitative immunohistochemical procedures, Polkowska et al. (1980) saw little change in the ME across the oestrous cycle, except for a reduction in GnRH content on day 1, after the LH surge.

Crowder and Nett (1984) conducted a more detailed study of the pro-oestrous period in ewes. In this case, sheep were given a prostaglandin injection on day 14 of the cycle to synchronise luteolysis and then given an oestrogen injection 30 h later. There were no significant changes in the GnRH content of the POA, the hypothalamus or the ME in the ensuing 96 h, over which time the preovulatory LH surge occurred. Although there was no statistical significance, the data do suggest that ME content fell slightly at the time of the LH surge. In spite of the pharmacological manipulations used in this experiment and the analysis of grossly dissected regions of the hypothalamus, the data are the best available for the sheep. There is clearly some conflict between the various reports of cyclic changes in hypothalamic GnRH content in the sheep but it does seem that alterations are minor. Again, we know nothing of the relative changes in synthesis, degradation and secretion. The use of more sophisticated methods, using cDNA probes and microdissection could greatly improve our knowledge in this area.

III GnRH secretion

It is now well established that GnRH is secreted in a pulsatile manner into the hypophyseal portal blood of rats (Sarkar and Fink 1980), monkeys (Carmel, Araki, and Ferin 1976; Carmel, Antunes, and Ferin 1979), sheep (Clarke and Cummins 1982) and humans (Carmel *et al.* 1976*a*). In the sheep and the monkey, conscious experimental models have been developed that allow the simultaneous measurement of GnRH and gonadotrophin secretion. One of these systems is the push-pull perfusion in which medium is pumped into and out of the ME region (Levine and Ramirez 1980). With careful placement of cannulae into discrete parts of the ME, and accurate regulation of the flow of medium, pulses of GnRH can be detected in perfusate. These pulses are believed to represent the pulsatile secretion of GnRH from neuronal terminals in the ME. This system has been developed in rats (Levine and Ramirez 1980, 1982), sheep (Levine, Pau, Ramirez, and Jackson 1982) and monkeys (Levine, Norman, Gliessman, Oyama, Bangsberg, and Spies 1985). A second model which allows the direct sampling of hypophyseal portal blood in sheep is one in which an artificial sinus is created anterior to the pituitary gland, in the sphenoid bone (Clarke and Cummins 1982). Access to the artifical sinus is provided by two needles that are introduced through the nose. With heparinization of the animal, and the puncturing of the hypophyseal portal blood vessels using an accurately made stillette introduced through one of the implanted needles, hypophyseal portal blood may be obtained from the artificial sinus. Most recently, Van Vugt, Diefenbach, Alston, and Ferin (1985) have shown that GnRH secretion from the hypothalamus is reflected by concentrations in the cerebrospinal fluid of the third ventricle in ovariectomized monkeys. One problems with this technique is that GnRH is not always detected during a sampling run. Another is that GnRH pulses do not always occur when there are LH pulses (Fig. 2.2*c*). However, this method does allow the simultaneous measurement of GnRH and LH in conscious animals and has the advantage, over other methods, that repeated observations may be made on the same individual. Figure 2.2 shows data from monkeys and from sheep which demonstrate the pulsatile nature of GnRH secretion and the close temporal relationship with LH secretion. The portal sampling technique indicates a more exact relationship between GnRH and LH in sheep than has been seen with the push-pull perfusion technique or cerebrospinal fluid sampling in monkeys. These models have provided conclusive proof that

RELATIONSHIP BETWEEN GnRH AND LH

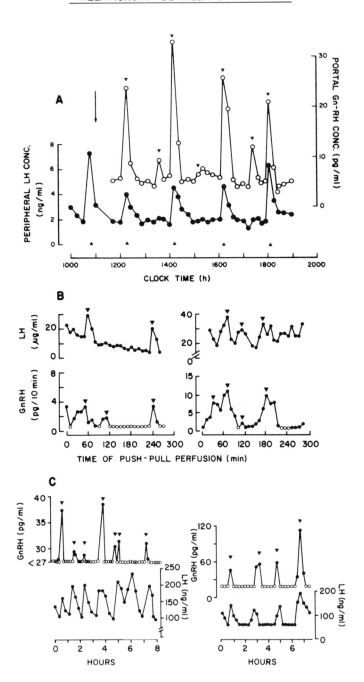

LH pulses are the direct reflection of GnRH pulses (Clarke and Cummins 1982; Levine *et al.* 1982, 1985).

In both the push–pull model and the portal access model, LH pulses are invariably associated with GnRH pulses but the converse is not always the case. A consistent feature of the studies in sheep (Clarke and Cummins 1982; Levine *et al.* 1982) and monkeys (Levine *et al.* 1985; Van Vugt *et al.* 1985) is that GnRH pulses may be seen in portal blood that are unaccompanied by LH Pulses (Fig. 2.2). These pulses are generally small and are more frequently observed in push–pull perfusates than in portal blood. It is possible that, in some circumstances, the measurement of LH pulses provides an accurate indication of the frequency of large GnRH pulses only. It has yet to be firmly established whether or not these small 'silent' GnRH pulses, if real, are physiologically meaningful.

1   DEFINITION OF FEEDBACK

In this and the following section on GnRH action it is necessary to consider three different types of steroidal feedback. These are best defined in ovariectomised animals and the example may be used of the changes that occur following oestrogen treatment of ovariectomised sheep (Clarke, Funder, and Findlay 1982) or monkeys (Yamaji, Dierschke, Hotchkiss, Bhattacharya, Surve, and Knobil 1971). The first and immediate effect is a fall in plasma LH levels; this may be called a *short-term negative feedback* effect. Following this, plasma LH levels rise well above pre-injection values (Clarke *et al.* 1982; Yamaji *et al.* 1971) by what may be called a *positive feedback* mechanism. Thereafter and with continued oestrogen treatment, plasma LH concentrations are held below those of untreated ovariectomised animals by a *long-term* (tonic) *negative feedback* effect (Diekman and Malven, 1973; Karsch, Dierschke, Weick, Yamaji, Hotchkiss, and Knobil 1973*a*; Goodman 1978). The positive feedback effect may also be obtained in chronically oestrogen treated OVX female monkeys (Karsch *et al.* 1973*a*) or in seasonally anoestrous female sheep (Goding, Catt, Brown, Kaltenbach, Cumming, and Mole 1969). Although these feedback parameters may change with the seasons (Legan, Karsch, and Foster, 1977), this issue will not be addressed here.

Fig. 2.2. Examples of the pulsatile secretion of GnRH in conscious animals. Panel A shows hypophysial portal GnRH levels in relation to jugular venous LH levels in a conscious OVX ewe; ▼ ▲ indicate GnRH and LH pulses respectively and the arrow indicates the time at which portal blood vessels were cut. Panel B shows data from push–pull perfusions in the rhesus monkey where the concordance between GnRH and LH pulses is less accurate than with portal sampling. Panel C shows data from cerebrospinal fluid sampling of rhesus monkeys; in this case some LH pulses occur without the concomitant appearance of GnRH pulses. ▲ or ▼ indicate the occurrence of pulses of LH and GnRH. (From Clarke and Cummins 1982; Levine *et al.* 1985, and Van Vugt *et al.* 1985, by permission).

Not only does this classification provide some method of systematic study but the three types of feedback can be related to normal cyclic events. In particular, the preovulatory LH surge might be generated by similar mechanisms in cyclic animals and oestrogen-treated OVX or anoestrous animals. Whereas this may be readily applicable to species such as monkeys, sheep and humans, it is not the case for rats in which a daily neural signal causes LH surges during oestrogen treatment (Legan, Coon, and Karsch 1975; Legan and Karsch 1975).

2   SHORT-TERM NEGATIVE FEEDBACK

Does oestrogen treatment cause an acute reduction in GnRH secretion? Using anaesthetized rats, Sarkar and Fink (1980) found that an intravenous injection of 1 µg oestradiol did result in an acute reduction of GnRH levels in hypophyseal portal blood, the effect being mainly due to a reduction in GnRH pulse amplitude. Using a different strain of rats, Sherwood and Fink (1980) found that a subcutaneous injection of 50 µg oestradiol benzoate had no effect on GnRH secretion but this may be of little relevance since, in the same study, no effect was observed on LH secretion. In OVX monkeys, Carmel et al. (1976) found no change in GnRH secretion into portal blood within 2 h of an intravenous injection of 1 µg oestradiol. In OVX sheep Clarke and Cummins (1985) found that an intramuscular injection of 50 µg oestradiol benzoate did not cause a short-term negative feedback effect on GnRH secretion but did inhibit LH secretion. This was taken as evidence that any short-term effect was probably due to pituitary feedback action, a conclusion that has been verified in an isolated pituitary model (see below).

With the exception of one study in rats, the available data suggest that short-term negative feedback effects on GnRH secretion are not the result of decreased GnRH secretion.

3   POSITIVE FEEDBACK

Original studies in anoestrous ewes (Goding et al. 1969) and in OVX ewes (Radford, Wheatley, and Wallace 1969) showed that oestrogen caused a preovulatory-like LH surge. This positive feedback mechanism has also been demonstrated in oestrogen-primed OVX rats (Caligaris, Astrada, and Taleisnik 1971b), OVX monkeys (Yamaji et al. 1972) and eugonadal women (Swerdloff and Odell 1969).

The fundamental question that arose from these early studies was whether or not the LH surge was triggered by a rise in GnRH secretion. This was initially addressed by Sarkar, Chiappa, Fink, and Sherwood (1976) who measured GnRH levels in the hypophyseal portal blood of cyclic rats. The technique used for this study did not allow synchronous collection of GnRH and LH because the pituitary stalk was cut to obtain the portal blood.

However, a marked increase in GnRH secretion was found at the time that the pro-oestrous LH surge was expected. Using the push–pull perfusion system, Levine and Ramirez (1982) also found an increase in GnRH secretion from the ME during pro-oestrus in rats.

Whereas data from animals suggest that GnRH secreted from the hypothalamus is 'diluted out' or degraded in the peripheral circulation and cannot be detected outside the portal capillaries, this may not be the case in humans. Early studies by Malacara, Seyler, and Reichlin (1972) in which GnRH was measured in the peripheral plasma of women by bioassay, indicated that there was increased GnRH secretion around the time of the mid-cycle LH surge. Subsequently, Miyake, Kawamura, Aono, and Kurachi (1980) and Elkind-Hirsch, Ravnikar, Schiff, Tulchinsky, and Ryan (1982) have confirmed this using radioimmunoassay. Although GnRH in peripheral plasma may not originate from the hypothalamus, the cyclic pattern observed is strongly suggestive of this. Nevertheless it remains possible that the immunoreactive material comes from organs such as the ovaries. Since the human is the only species in which GnRH can be detected in peripheral plasma, they may be the best subjects for studies of the mid-cycle surge. Because GnRH levels are low or undetectable at other times of the cycle, more detailed studies are not appropriate.

Evidence for a mid-cycle rise in GnRH secretion in monkeys was obtained by Neill, Patton, Dailey, Tsou, and Tindall (1977), who collected portal blood from sectioned pituitary stalks in anaesthetized animals. Most recently, a moderate rise in GnRH secretion has been measured in conscious sheep at the time of the cyclic preovulatory LH surge (Clarke, Thomas, Doughton, Burman, Yao, and Cummins 1986).

The next obvious question regarding the GnRH/LH surge is whether or not oestrogen causes the GnRH rise that is seen in cycling animals. Accordingly, Sarkar and Fink (1979) measured portal GnRH concentrations in rats that had been ovariectomized during dioestrus, immediately given oil or oestradiol benzoate, and then sampled on the day of expected pro-oestrus. The concentrations of GnRH in the oestrogen treated animals were similar to those in normal pro-oestrous rats, whereas concentrations of GnRH in oil treated rats were significantly lower. This provided direct evidence that oestrogen permits or facilitates the daily neuronal signal for the LH surge in this species and that this signal is transmitted to the pituitary gland in the form of a GnRH rise. To emphasize further that this effect was indeed due to oestrogen it was found that the injection of an antioestrogen at 17.00 h of dioestrus reduced portal GnRH concentrations between 16.00 h–18.00 h during the next pro-oestrus (Sarkar and Fink 1979). Further studies (Sarkar and Fink 1980) examined the effects of oestrogen in long-term (21–28 days) OVX rats and revealed that oestrogen injections given 72 h apart did not cause a rise in GnRH secretion, which was expected on the basis of a rise in plasma LH levels in similarly treated animals (Brown-Grant 1974). In fact, portal GnRH levels were significantly reduced in the oestrogen treated rats and the authors (Sarkar and Fink 1980) were led to the conclusion that the LH surge produced by oestrogen in long term OVX rats is due to increased pituitary responsiveness. A similar result was obtained in oestrogen-primed, progesterone treated animals (Sarkar and Fink 1980). These results were surprising for two reasons. Firstly, they differed markedly from the results obtained in acutely OVX and oestrogen treated rats (Sarkar and Fink 1979), suggesting that the long-term OVX rat is a bad model for the cyclic LH surge. Secondly, one might have anticipated that an increase in GnRH would occur, given the fact that the ME content of GnRH falls at the start of the LH surge in oestrogen-primed, progesterone treated rats (Kalra and Kalra 1977; see Section II.4). Using push–pull cannulae to sample ME perfusates of oestrogen-primed rats 10–22 days after ovariectomy, Levine and Ramirez (1980) found that progesterone injections did cause a rise in GnRH secretion at the expected time of the LH surge. These results are clearly difficult to reconcile with those of Sarkar and Fink (1980).

In women, GnRH and LH concentrations in peripheral plasma are increased by the injection of oestrogen in the mid-follicular phase of the menstrual cycle (Miyake, Tasaka, Sakumoto, Kawamura, and Aono 1983). This study also provided good evidence that the initial negative feedback component of the biphasic LH response to oestrogen is not accompanied by a fall in GnRH secretion (Fig. 2.3). It was notable that oestrogen induced a small (but not significant) rise in GnRH levels *before* the rise in LH secretion and during the short-term negative feedback phase but the major increment occurred at the start of the LH surge. More detailed studies might show the initial rise to be important as this could serve to prime the pituitary to respond maximally to a subsequent and more substantial rise in GnRH secretion (Aiyer, Chiappa, and Fink 1974, and Section IV.2).

In OVX ewes GnRH secretion has been studied following bolus intramuscular injection of 50 µg oestradiol benzoate, by push–pull perfusion (Schillo, Leshin, Kuehl, and Jackson 1985) and by direct portal sampling (Clarke and Cummins 1985). The study of Schillo *et al.* (1985) was complicated by the fact that GnRH could not be detected in ME perfusates of five of the nine sheep that were sampled during LH surges. In two sheep the LH surge was preceded by a large GnRH pulse. There was some indication that GnRH secretion increased in three out of the remaining four sheep sampled during the LH surge but the low GnRH values and high variability disallowed a significant result. This study highlighted some of the problems, especially cannula placement, that are encountered with the push–pull perfusion method. With direct portal sampling, it was found that oestrogen decreased the GnRH inter-pulse interval (that is, increased frequency) from 53.4 ± 8.7 min (mean $\pm$ s.e.m.) in controls to 26.8 ± 9.8 min during the LH surge (Clarke and Cummins 1985). An example of the pattern of response obtained in one ewe is shown in Fig. 2.4. It is perhaps significant that one

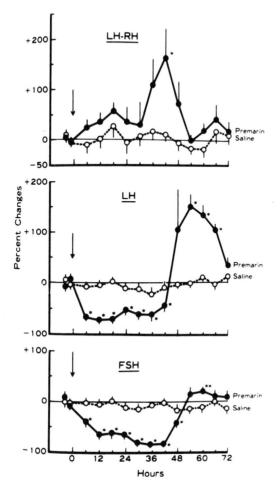

Fig. 2.3. Changes in the levels of GnRH (LHRH), LH and FSH in the peripheral plasma of women after injection (arrow) of oestrogen (premarin) in the mid-follicular phase of the menstrual cycle. (From Miyake *et al.* 1983, by permission).

large GnRH pulse occurred near the start of the LH surge consistent with the observations in two of the sheep in the study of Schillo *et al.* (1985).

As with the rat (see above), the long-term OVX ewe may not be the best animal model in which to study the positive feedback mechanism in view of the chronic deprivation of ovarian steroids and the inherently high GnRH/LH pulsatility (Levine *et al.* 1982; Clarke and Cummins 1985). A far better model might be the seasonally anoestrous ewe which has low frequency LH pulses and will show an LH positive feedback response to oestrogen (Goding *et al.* 1969). Indeed, recent studies by the author do show

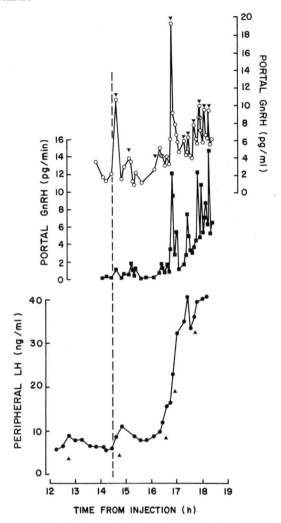

Fig. 2.4. GnRH levels in the hypophysial portal blood of an OVX ewe at the onset of an oestrogen-induced LH surge, showing the increase in GnRH pulse frequency that occurs at this time. The data are expressed as pg per ml (○) or pg per min (■) to indicate that changes are not related to altered blood flow. The dotted line indicates the start of the surge. ▼ and ▲ indicate GnRH and LH pulses respectively. (From Clarke and Cummins (1985), by permission).

that there is a GnRH surge associated with the oestrogen-induced LH surge in anoestrous ewes (unpublished observations).

The push–pull perfusion system has also been used in OVX monkeys to study the positive feedback mechanism in oestrogen-treated OVX monkeys (Levine *et al*. 1985). Once again, variable patterns of GnRH secretion were seen at the time of the LH surge with either short bursts of high amplitude

GnRH pulses or more prolonged periods of high frequency pulses. As similarly various patterns have been seen in cyclic ewes at the time of the LH surge (Clarke *et al*. 1986), it is becoming apparent that GnRH secretion at this time does not conform to a rigid pattern. To substantiate this it will be necessary to monitor GnRH profiles in a wider range of experimental models.

The frequency of LH pulses at the time of the LH surge is higher than at any other time of the cycle in rats (Gallo 1981), monkeys (Marut, Williams, Cowan, Lynch, Lerner, and Hodgen 1981), sheep (Karsch, Foster, Bittman, and Goodman 1983), cows (Rahe, Owens, Fleeger, Newton, and Harms 1980) and women (Djahanbakhch, Warner, McNeilly and Baird 1984). This has been taken to mean that GnRH frequency is maximal at this time. Although GnRH pulse frequency was increased in oestrogen-treated OVX sheep (Clarke and Cummins 1985), it must also be remembered that apparent LH pulses occur when cultured pituitary cells are given constant infusions of GnRH (Magness, Millar, and Michie 1981). This may mean that LH pulses occurring during the LH surge can occur as a consequence of oestrogen priming and an elevated but constant output of GnRH as well as discrete GnRH pulses.

Positive feedback effects on gonadotrophin secretion have been observed in various species when a progesterone injection is given after a period of oestrogen priming. This may be of physiological relevance in species such as rats, humans and monkeys where a rise in plasma progesterone levels precedes the cyclic LH surge (Weick, Dierschke, Karsch, Butler, Hotchkiss, and Knobil 1973; Hoff, Quigley and Yen 1983; Kalra and Kalra 1983), but may be of little relevance in species such as sheep in which the there is no preovulatory rise in plasma progesterone levels (Baird and Scaramuzzi 1976). In women, progesterone was found to amplify an oestrogen-induced LH surge in hypogonadal women (Liu and Yen 1983), but the site of progesterone action is not known. In fact, progesterone can induce an LH surge in oestrogen-treated, castrated or post-menopausal women (Nillius and Wide 1971). Since mid-cycle GnRH surges can be detected in the peripheral plasma of women (see above) it might be possible to determine whether or not progesterone acts on the hypothalamus to increase GnRH secretion by sampling at mid-cycle with appropriate overlap of progesterone treatment. This might allow a more definitive statement about effects of progesterone on GnRH secretion than has been forthcoming from rat studies (see above).

A preovulatory rise in GnRH secretion has been measured in the hypophyseal blood of rabbits after cupric acetate treatment (Tsou, Dailey, McLannan, Parent, Tindall, and Neill 1977). The rabbit is a reflex ovulator and cupric acetate treatment caused a rise in gonadotrophin secretion similar to that caused by mating. These findings are important because they suggest that a GnRH surge initiates the LH surge rather than accompanies it. Because of the rapid nature of the reflex response it seems unlikely that factors other than those arising from the hypothalamus (for example, steroidal priming of the pituitary gland), might be wholly responsible for initiating the LH surge.

4   LONG-TERM NEGATIVE FEEDBACK AND GnRH SECRETION

The tonic negative feedback effects are difficult to study in rats because the daily neural signal that causes a surge in gonadotrophin secretion is potentiated by steroids (see above). For this reason, tonic feedback effects are best studied in species where this diurnal component is not apparent. With one exception (Clarke, Karsch, and Cummins 1985) studies of GnRH secretion have not been done in relation to tonic negative feedback, but inferences may be drawn from studies on the pulsatile secretion of LH. However, since changes in LH pulse amplitude may be due to changes in hypothalamic GnRH pulse amplitude or in pituitary responsiveness to GnRH, we can use LH pulse frequency only, and not LH pulse amplitude, to make such inferences.

In rats, luteal phase plasma levels of progesterone had relatively little effect on the plasma LH rise seen after ovariectomy, whereas oestradiol levels equivalent to those seen during dioestrus II had a partial effect (Goodman 1978). When oestrogen and progesterone treatments were combined, the post-castration rise was virtually abolished. This study did not identify the site of action of the steroids in terms of long-term negative feedback.

In Suffolk sheep treated immediately after ovariectomy performed during the breeding season, follicular phase levels of oestrogen were found to reduce LH pulse amplitude but not frequency whereas luteal phase levels of progesterone reduced the frequency and increased amplitude (Goodman and Karsch 1980). Since oestrogen treatment reduced the pituitary response to GnRH but progesterone did not, these authors concluded that the progesterone effect was hypothalamic (on GnRH secretion) and the oestrogen effect was pituitary directed. In a further study, Goodman, Bittman, Foster, and Karsch (1981) used a low dose of progesterone which given alone had no effect on LH secretion. In the presence of oestrogen however, this low progesterone dose caused a reduction in LH pulse frequency. This could be taken as evidence that oestrogen can sensitize the central nervous system to respond to progesterone. Alternatively, oestrogen effects on GnRH/LH pulse amplitude might only be expressed in the presence of progesterone. Using lower dosages of oestrogen and progesterone than Goodman and Karsch (1980), neither steroid had a tonic negative feedback effect on LH secretion in OVX merino ewes during the breeding season (Martin, Scaramuzzi, and Henstridge 1983). However, combined treatment caused a reduction in LH pulse frequency.

A series of studies on sheep have emphasized the role of changing tonic negative feedback effects in the determination of oestrous cycle events. Thus, during the luteal phase of cycle oestrogen and progesterone may act synergistically to reduce GnRH/LH pulse frequency (Baird and Scaramuzzi 1976; Karsch, Legan, Hauger, and Foster 1977; Goodman, Legan, Ryan, Foster, and Karsch 1980). With the onset of luteolysis, oestrogen alone is unable to limit pulse frequency (Goodman and Karsch 1980; Goodman *et al.* 1981; Martin *et al.* 1983) and thus the LH pulse frequency increases (Hauger, Karsch, and Foster 1977; Baird 1978). Since each LH pulse provokes oestrogen secretion from the ovarian follicles (Baird 1978) the level of this steroid also rises (Baird and Scaramuzzi 1976; Baird 1978; Baird, Swanston, and McNeilly 1981). Other factors within the enlarging preovulatory follicle allow an increase oestrogen production per unit pulse of LH (Baird 1978). To add further impetus to the cascade that occurs at this time, oestrogen, by virtue of a 'positive' feedback effect, increases the LH pulse frequency (Karsch *et al.* 1983). Because oestrogen is the primary signal for the preovulatory surge, it is generally accepted that at some time in the follicular phase a critical plasma level of this steroid is reached that provides the appropriate neural signal for the LH surge. One fascinating feature of this oestrogen trigger is that some time delay is required for the LH surge to occur. To date no explanation for this delay has been forthcoming and studies of sub-cellular mechanisms within the hypothalamus and pituitary gland will be necessary to provide this.

In this scheme it is not necessary to regard oestrogen alone as a negative feedback factor. There are, however, some situations in which this effect assumes importance. For example, oestrogen is able to inhibit LH secretion in anoestrous ewes (Legan, Karsch, and Foster 1977). Martin *et al.* (1983) found this effect to be manifest by a reduction in pulse frequency. More recently, Clarke *et al.* (1985) showed that oestrogen eliminated GnRH secretion in Corriedale ewes OVX during the anoestrous season. How can such a negative feedback action be reconciled with the fact that a bolus injection of oestrogen elicits an LH surge (Goding *et al.* 1969)? The answer to this question might be found in an early study in monkeys where it was found that chronic oestrogen treatment did not prevent a positive feedback response to a sudden rise in oestrogen levels (Karsch *et al.* 1973*a*). From this it might be inferred that the positive feedback effect results from a sudden change in oestrogen levels (Karsch, Weick, Butler, Dierschke, Krey, Weiss, Hotchkiss, Yamaji, and Knobil 1973*b*). To add biological credence to this concept, the magnitude of the response is related to the size and duration of the oestrogen stimulus (Karsch *et al.* 1973*b*). Once again we are compelled to ask what subcellular machinery is participating in these responses in such a rigidly defined manner?

IV   GnRH action on the pituitary gland

The dissection of steroidal feedback effects into those manifest via actions on the hypothalamus and those directly on the pituitary gland demands suitable anatomical separation of the two targets. This can be simply achieved by cell culture or other *in vitro* systems. Pituitary cell perfusion systems allow short term experiments on gonadotrophin response to GnRH but do not permit extended studies even with pulsatile GnRH administration (Liu and Jackson 1984). Static cultures allow rudimentary conclusions to be drawn, but in most cases where gonadotrophin secretion has been studied this has been done with a constant background of GnRH. Another objection to the use of cultures is that the paracrine relationships between cells (Denef 1984) is lost after dispersion. The pattern of response to a steroid, particularly oestrogen, changes with time and it therefore seems appropriate that such responses are studied with the use of *in vivo* models that can be maintained for extended periods.

A number of *in vivo* models have been developed for the study of the pituitary gland in isolation from the hypothalamus. In most species separation by sectioning of the hypophyseal stalk causes infarction of the pituitary gland; this is almost total in the sheep (Clarke, Cummins, and de Kretser 1983). A viable anterior pituitary gland will be maintained after stalk-section in the monkey (Vaughan, Carmel, Dyrenfurth, Frantz, Antunes, and Ferin 1980) probably because of a substantial contribution to the vital blood supply via hypophyseal arteries which enter the pituitary vascular bed below the stalk (Bergland and Page 1978). Other methods of removing GnRH inputs to the pituitary gland include the lesioning of the arcuate nucleus in the rhesus monkey (Plant *et al.* 1978) and the hypothalamo-pituitary disconnection (HPD) procedure in the sheep (Clarke *et al.* 1983). Pentobarbitone anaesthesia can be used to block the LH surge mechanism (Radford and Wallace 1974) but this does not eliminate endogenous GnRH pulses (Wright and Clarke 1985).

This section will focus on studies in which GnRH action on the pituitary gland has been evaluated in terms of LH secretion *in vivo*. Studies of GnRH receptor dynamics and post-receptor mechanisms will not be considered.

1   SHORT-TERM NEGATIVE FEEDBACK

In monkeys with the arcuate nucleus lesioned (Nakai, Plant, Hess, Keogh, and Knobil 1978) and in HPD ewes (Clarke and Cummins 1984) where LH secretion was maintained with pulsatile GnRH replacement, the injection of oestrogen caused an acute reduction in plasma LH levels. This effect was due to a reduction in LH pulse amplitude (Clarke and Cummins 1984) in spite of constant GnRH stimulation, and thus the acute negative feedback effect can be explained by a pituitary action of oestrogen. These data, in combination with those on GnRH secretion (see above), clearly establish that the short-term effect of oestrogen is not mediated via the hypothalamus.

The role that an acute negative feedback effect might play during the oestrous/menstrual cycle is obscure. Perhaps it serves to prevent a premature LH surge during the follicular phase, holding LH secretion in abeyance (Baird *et al.* 1981; Scaramuzzi and Radford 1983) until follicular maturation is completed. Also it may allow time for GnRH 'self priming' and oestrogen priming of the pituitary gland (see below).

Progesterone alone had no acute negative feedback effect on GnRH responsiveness but could enhance the effects of oestrogen (Clarke and Cummins 1984).

2 POSITIVE FEEDBACK

Responsiveness to GnRH is enhanced at the time of the preovulatory LH surge in sheep (Reeves, Arimura, and Schally 1971), women (Yen and Lein 1976) and rats (Cooper, Fawcett, and McCann 1973; Aiyer, Fink, and Greig 1974). This increased responsiveness is also apparent when LH surges are produced by oestrogen treatment of sheep (Coppings and Malven 1976), rats (Libertun, Orias, and McCann 1974; Van Dieten, Steijsiger, Dullaart, and Van Rees 1974; Vilchez-Martinez, Arimura, Debeljuk, and Schally 1974) and women (Yen, Vandenberg, and Siler 1974).

The 'self-priming' effect of GnRH may also be relevant at the time of positive feedback. This is a phenomenon that became apparent when two equal doses of GnRH were injected 30–240 min apart to female rats during pro-oestrus (Aiyer, Chiappa, and Fink 1974). The response to the second injection was much greater than the response to the first, particularly during pro-oestrus and to a lesser extent during dioestrus. In women, infusions of GnRH that were so low as not to affect basal LH secretion were able to 'prime' the pituitary gland to respond to an injection of GnRH (Hoff, Lasley, and Yen 1979). Also in women, the 'self-priming' effect increases during the follicular phase of the menstrual cycle (Faure and Oliver 1978).

The 'self-priming' effect that is seen during pro-oestrus or in oestrogen-treated rats ceases with successive GnRH injections, but the heightened responsiveness to GnRH persists (Higuchi and Kawakami 1982). Thus, the 'self-priming' effect may be of importance in the initiation of the preovulatory LH surge.

A further mechanism that may play a role at this time has been described by Araki, Chikazawa, Motoyama, Ijima, Abe, and Tamada (1985). These authors showed that the desensitization phenomenon occurring during high-dose infusions of GnRH, is diminished in the follicular phase in women. Thus, during the oestrogen-dominant phase of the cycle, GnRH is able to exert more prolonged effects on LH and FSH secretion.

To what extent does an oestrogen-induced increase in pituitary responsiveness to GnRH contribute towards the LH surge? As outlined above, there is no doubt that oestrogen increases pituitary responsiveness. Many experiments have been conducted with either very high doses of GnRH (for example, Reeves *et al.* 1971; Crighton and Foster 1977) or with pulse frequencies that may be physiologically inappropriate; Aiyer *et al.* (1974), chose a 60 min interval between injections because this gave the best evidence of a priming effect. Now that we have estimates of GnRH/LH pulse frequency during the LH surge and we know portal blood GnRH concentrattions at this time (Sarkar *et al.* 1976; Clarke and Cummins 1985), we should be able to design experiments to define accurately the extent to which enhanced pituitary responsiveness contributes towards the LH surge. The degree to which it is important is probably species dependant because the incremental rise in LH secretion varies. In rats, the LH surge represents a 20–30 fold increase in peripheral plasma values (Naftolin, Brown-Grant, and Corker, 1972). Comparable values for sheep, monkeys and women are 30 to 50 fold (Scaramuzzi, Caldwell, and Moor 1970), 7 to 10 fold (Knobil 1974) and 10 to 15 fold (Landgren, Unden, and Diczfalusy 1980; Shi-fan, Zhong-xing, Yao-e, Sheng-min, Bei-zhu, Jin-zhi, Yi-e, and Xin-ying 1986) respectively. Thus, in species such as the monkey where the incremental rise is relatively small, it may be possible to generate an ovulatory LH surge by pituitary priming alone. This certainly appears to be the case from Knobil's experiments.

Using arcuate nucleus-lesioned OVX monkeys, maintained on hourly pulses of exogenous GnRH, it was found that oestrogen treatment caused a biphasic alteration in LH and FSH secretion similar to that seen in non-lesioned animals (Nakai *et al.* 1978). In arcuate-lesioned monkeys with ovaries, hourly GnRH pulses were able to restore gonadotrophin levels to the normal range which, in turn, produced increments in ovarian oestrogen that were sufficient to induce LH surges and ovulation (Knobil, Plant, Wildt, Belchetz, and Marshall 1980). From these studies, Knobil has proposed that 'the preovulatory gonadotropin surges does not require an increment in GnRH release by the hypothalamus' (Knobil 1980). Nevertheless, an increase in GnRH secretion does occur at the time of the preovulatory surge in monkeys and in other species (see above). Thus, the LH surge seems to depend on both increases in pituitary responsiveness to GnRH and an increase GnRH secretion.

In species such as sheep, an increase in pituitary responsiveness to GnRH cannot account for the LH surge (Coppings and Malven 1976). In OVX-HPD ewes receiving hourly GnRH pulses, an injection of oestrogen could effect a 2 to 3 fold rise in plasma LH values (Clarke and Cummins 1984). In anaesthetized OVX sheep given 20 min pulses of GnRH, a 50 µg intramuscular injection of oestradiol benzoate provoked an LH surge that was quantitatively similar but qualitatively different to the oestrogen-induced surge in conscious sheep (Wright and Clarke 1985). In the former, the surge was sinusoidal in shape rather than a single peak. The reason for this is obscure but the endogenous *pattern* of GnRH secretion is obviously important in the generation of a 'normal' surge of LH.

If it is possible to develop an LH surge in arcuate-lesioned monkeys with hourly pulses of GnRH then how is this effected? The first and obvious answer is that oestrogen increases the responsiveness of the pituitary gland to GnRH in such a way that LH pulse amplitudes are increased. In the studies so far reported it is unfortunate the blood samples were not collected at sufficiently frequent intervals to allow an analysis of pulse amplitude.

Another possible mechanism by which pituitary effects alone might cause an LH surge could be by the prolongation of responses of LH and FSH to individual GnRH pulses. If this were the case then plasma gonadotrophin values would increase in spite of a fixed GnRH pulse frequency and constant (or even reduced) pulse amplitude. The way in which a surge could be generated by this mechanism is shown in Figure 2.5. There is some suggestion that this does indeed occur in the oestrogen-treated, OVX-HPD ewe given pulses of GnRH every hour (Fig. 2.5).

There are many explanations and interpretations of the normal cyclic patterns that were obtained in the arcuate nucleus-lesioned rhesus monkey pulsed with GnRH. Firstly, the exogenous GnRH pulse regime of one pulse of 6 μg GnRH given every hour may provide a dose and frequency that would in any event lead to an LH surge. In fact, the pulses produce plasma GnRH levels of > 2000 pg ml (Wildt, Hausler, Marshall, Hutchison, Plant, Belchetz, and Knobil 1981) which are in considerable excess of those reported for GnRH in the portal blood of monkeys at mid-cycle (700 pg ml, Carmel *et al.* 1979), during the pre-ovulatory LH surge (approximately 100 pg ml; Neill *et al.* 1977), or following ovariectomy (up to 800 pg ml; Carmel *et al.* 1976; Neill *et al.* 1977). So the particular situation that is provided by the Knobil model may be unphysiological in that the system is being 'driven' by high doses of GnRH given at a frequency that predisposes to the LH surge. Nevertheless, these important studies in monkeys have led to the development of pulsatile GnRH therapies that have assisted the treatment of hypothalamic amenorrhea (Crowley and McArthur 1980) and hypo-gonadism in males (Jacobson, Seyler, Tamborlane, Gertner, and Genel 1979).

As mentioned previously, progesterone is able to amplify the action of oestrogen in promoting the LH surge. It was thus of interest that a progesterone background enhanced the effect of oestrogen during the LH 'surge' phase in OVX-HPD ewes (Clarke and Cummins 1984). This suggests that the effect of progesterone, at this time, is due to action on the pituitary rather than the hypothalamus.

3   LONG TERM NEGATIVE FEEDBACK

Whereas the tonic effect of long-term exposure to oestrogen on LH secretion has been described in various species including monkeys (Karsch *et al.* 1973a), humans (Wallach, Root, and Garcia 1970), rats (Goodman 1978) and sheep (Diekman and Malven 1973), it is not certain whether the

MODEL

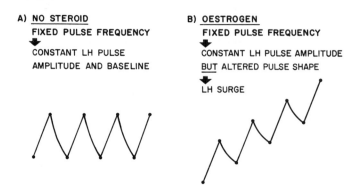

A) **NO STEROID**
FIXED PULSE FREQUENCY
↓
CONSTANT LH PULSE
AMPLITUDE AND BASELINE

B) **OESTROGEN**
FIXED PULSE FREQUENCY
↓
CONSTANT LH PULSE AMPLITUDE
BUT ALTERED PULSE SHAPE
↓
LH SURGE

EXPERIMENTAL

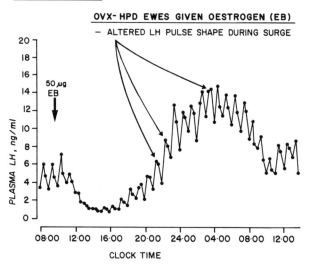

OVX-HPD EWES GIVEN OESTROGEN (EB)
– ALTERED LH PULSE SHAPE DURING SURGE

50 μg EB

Fig. 2.5. Possible mechanisms contributing to the induction of the LH surge. A: by a reduction in the rate of decay of LH pulses in plasma a simple model predicts that a 'surge' may occur without a change in the frequency or amplitude of the LH pulses. In model B, oestrogen causes a change in pulse shape and experimental demonstration of this is shown in the bottom half of the figure using data from an OVX-HPD ewe given hourly pulses of GnRH, and an injection of oestrogen (EB) (Data reproduced from Clarke and Cummins 1984, by permission).

suppression is due to hypothalamic or pituitary action. A case for tonic negative feedback was built upon observations that oestradiol reduced LH pulse amplitude and also reduced the LH responses to large bolus injections of GnRH to OVX ewes (Goodman and Karsch 1980). Although the amplitude of the LH response was reduced in the presence of oestrogen, the response phase was prolonged. Thus any pituitary action of oestrogen might produce qualitative as well as quantitative changes. In another study of ovariectomised ewes given a lower dose of oestrogen, Martin *et al.* (1983) found that, if anything, oestrogen increased LH pulse amplitude, particularly in the presence of progesterone. The clear discrepancy between these two studies makes it difficult to decide whether or not oestrogen has a tonic feedback effect on the pituitary gland. To resolve this issue, studies need to be conducted in animals with pituitaries isolated from endogenous GnRH inputs, and given exogenous GnRH pulses with appropriate steroidal overlays.

V   Conclusion

Since its isolation in 1970, studies on the synthesis, secretion and action of GnRH has yielded considerable material benefits. The original experiments by Knobil and co-workers, showing that menstrual cycles could be maintained in arcuate nucleus-lesioned monkeys (Knobil *et al.* 1980), have led to the use of pulsatile GnRH therapy for various types of infertility. Other work on GnRH agonists and antagonists (Schally 1984) has led to a variety of clinical applications including treatment for some types of cancer.

These clinical applications of GnRH have arisen from basic scientific findings. Hopefully, further research on GnRH will improve current applications and indicate new ones. In the future we need to expand our understanding of the neural mechanisms involved in GnRH secretion and the subcellular events involved in GnRH action. To date we have learnt very little of the processes involved in steroidal feedback. When such data are to hand, from the basic research efforts currently under way in a number of centres around the world, we can expect that these too will find their application in the clinic, complementing the currently available GnRH treatment regimes.

References

Adelman, J. P., Mason, A. J., Hayflick, J. S., and Seeburg, P. H. (1986). Isolation of the gene and hypothalamic cDNA for the common precursor of gonadotropin-releasing hormone and prolactin release-inhibiting factors in human and rat. *Proc. nat. Acad. Sci. U.S.A.* **83**, 179–183.
Advis, J. P., Krause, J. E., and McKelvy, J. F. (1982). Luteinizing hormone-releasing hormone peptidase activities in discrete hypothalamic regions and anterior pituitary of the rat: apparent regulation during the prepubertal period and the first estrous cycle at puberty. *Endocrinology* **110**, 1238–1245.
—— Kuljis, R. O., and Dey, G. (1985). Distribution of luteinizing hormone-releasing hormone (LH-RH) content and total LHRH-degrading activity (LHRH-DA) in the hypothalamus of the ewe. *Endocrinology* **116**, 2410–2418.

84 Iain J. Clarke

Aiyer, M. S., Chiappa, S. A., and Fink, G. (1974). A priming effect of luteinizing hormone releasing factor on the pituitary gland in the female rat. *J. Endocr.* **62**, 573–588.

—— Fink, G., and Greig, F. (1974). Changes in the sensitivity of the pituitary gland to luteinizing hormone releasing factor during the oestrous cycle of the rat. *J. Endocr.* **60**, 47–64.

Ambach, G., Kivovics, P., and Palkovits, M. (1978). The arterial and venous blood supply of the preoptic region of the rat. *Acta. morph. Hung.* **26**, 21–41.

Araki, S., Chikazawa, K., Motoyama, M., Ijima, K., Abe, N., and Tamada, T. (1985). Reduction in pituitary desensitization and prolongation of gonadotropin release by estrogen during continuous administration of gonadotropin-releasing hormone in women: its antagonism by progesterone. *J. clin. Endocr. Metab.* **60**, 590–598.

—— Ferin, M., Zimmerman, E. A., and Van de Wiele, R. L. (1975). Ovarian modulation of immunoreactive gonadotrophin releasing hormone (Gn-RH) in the rat brain: evidence for a differential effect on the anterior and mid-hypothalamus. *Endocrinology* **96**, 644–650.

Asai, T. and Wakabayashi, K. (1975). Changes in hypothalamic LH-RF content during the rat estrous cycle. *Endocr. Japon.* **22**, 319–326.

Baird, D. T. (1978). Pulsatile secretion of LH and ovarian estradiol during the follicular phase of the sheep estrous cycle. *Biol. Reprod.* **18**, 359–364.

—— and Scaramuzzi, R. J. (1976). The source of ovarian oestradiol and androstenedione in the sheep during the luteal phase. *Acta endocr. Copenh.* **83**, 402–409.

—— Swanston, I. A., and McNeilly, A. S. (1981). Relationship between LH, FSH and prolactin concentration and the secretion of androgens and estrogens by the preovulatory follicle in the ewe. *Biol. Reprod.* **24**, 1013–1025.

Baker, B. L., Demody, W. C., and Reel, J. R. (1974). Localization of luteinizing hormone-release hormone in the mammalian hypothalamus. *Am. J. Anat.* **139**, 129–135.

—— —— —— (1975). Distribution of gonadotropin releasing hormone in the rat brain as observed with immunocytochemistry. *Endocrinology* **97**, 125–135.

Barr, G. and Barraclough, L. A. (1976). Temporal alterations in medial basal hypothalamic (MBH) LH-RH associated with the surge of plasma LH. *Fedn Proc.* **35**, 700.

—— —— (1978). Temporal changes in medial basal hypothalamic LH-RH correlated with plasma LH during the rat estrous cycle and following electrochemical stimulation of the medial preoptic area in pentobarbital-treated proestrous rats. *Brain Res.* **148**, 413–423.

Barraclough, C. A. and Wise, P. M. (1982). The role of catecholamines in the regulation of pituitary luteinizing hormone and follicle-stimulating hormone secretion. *Endocr. Rev.* **3**, 91–119.

—— —— and Selmanoff, M. K. (1984). A role for hypothalamic catecholamines in the regulation of gonadotropin secretion. *Recent Prog. Horm. Res.* **40**, 487–529.

Barry, J. (1977). Immunofluorescence study of LRF neurons in man. *Cell Tissue Res.* **181**, 1–14.

—— and Carette, B. (1975). Immunofluorescence study of LRF neurones in primates. *Cell Tissue Res.* **164**, 163–178.

—— and Dubois, M. P. (1974). Immunofluorescence study of the preoptico-infundibular LH-RH neurosecretory pathway of the guinea pig during the estrous cycle. *Neuroendocrinology* **15**, 200–208.

—— —— and Poulain, P. (1973). LRF producing cells of the mammalian hypothalamus. *Z. Zellforsch. Mikrosk. Anat.* **146**, 351–366.

Bergland, R. M. and Page, R. B. (1978). Can the pituitary secrete directly into the brain (affirmative anatomical evidence). *Endocrinology* **102**, 1325–1338.

Brown-Grant, K. (1974). Steroid hormone administration and gonadotrophin secretion in the gonadectomized rat. *J. Endocr.* **62**, 319–332.

—— and Raisman, G. (1977). Abnormalities in reproductive function associated with the destruction of the suprachiasmatic nuclei in female rats. *Proc. R. Soc. B.* **198**, 279–296.

Caligaris, L., Astrada, J. J. and Taleisnik, S. (1968). Stimulating and inhibiting effects of progesterone on the release of luteinizing hormone. *Acta endocr. Copenh.* **59**, 177–185.

—— —— —— (1971a). Release of luteinizing hormone induced by estrogen injection into ovariectomized rats. *Endocrinology* **88**, 810–815.

—— —— —— (1971b). Biphasic effect of progesterone on the release of gonadotrophin in rats. *Endocrinology* **89**, 331–337.

Carmel, P. W., Araki, S., and Ferin, M. (1976). Pituitary stalk portal blood collection in rhesus monkeys: evidence for pulsatile release of gonadotropin-releasing hormone (GnRH). *Endocrinology* **99**, 243–248.

—— Antunes, J. L. and Ferin, M. (1979). Collection of blood from the pituitary stalk and portal veins in monkeys, and from the pituitary sinusoidal system of monkeys and man. *J. Neurosurg.* **50**, 75–80.

Clarke, I. J. and Cummins, J. T. (1982). The temporal relationship between gonadotropin releasing hormone (GnRH) and luteinizing hormone (LH) secretion in ovariectomized ewes. *Endocrinology* **111**, 1737–1739.

—— —— (1984). Direct pituitary effects of estrogen and progesterone on gonadotropin secretion in the ovariectomized ewe. *Neuroendocrinology* **39**, 267–274.

—— —— (1985). Increased GnRH pulse frequency associated with estrogen-induced LH surges in ovariectomized ewes. *Endocrinology* **116**, 2376–2383.

—— —— and de Kretser, D. M. (1983). Pituitary gland function after disconnection from direct hypothalamic influences in the sheep. *Neuroendocrinology* **36**, 376–384.

—— Funder, J. W. and Findlay, J. K. (1982). Relationship between pituitary nuclear oestrogen receptors and the release of LH, FSH and prolactin in the ewe. *J. Reprod. Fert.* **64**, 355–362.

—— Karsch, F. J. and Cummins, J. T. (1985). Oestrogen and progesterone stop GnRH pulses in ovariectomized anoestrous ewes. In *Proc. Aust. Soc. reprod. Biol. 17th Ann. Confr.* Abstr. 28.

—— Thomas, G. B., Doughton, B. W., Burman, K. J., Yao, B. and Cummins, J. T. (1986). Gonadotrophin-releasing hormone secretion during the ovine estrous cycle. *Proc. 68th meeting Endocr. Soc., Anaheim, U.S.A.*

Clayton, C. J. and Hoffman, G. E. (1979). Immunocytochemical evidence for anti-LHRH and anti-ACTH activity in the 'F' antiserum. *Am. J. Anat.* **155**, 139–145.

Clayton, R. N. and Catt, K. J. (1981). Gonadotrophin releasing hormone receptors; characterization, physiological regulation and relationship to reproductive function. *Endocr. Rev.* **2**, 186–206.

Clegg, M. T. and Ganong, W. F. (1960). The effect of hypothalamic lesions on ovarian function in the ewe. *Endocrinology* **67**, 179–186.

—— Santalucito, J. A., Smith, J. D., and Ganong, W. F. (1958). The effect of hypothalamic lesions on sexual behaviour and estrous cycles in the ewe. *Endocrinology* **62**, 790–797.

Clemens, J. A., Smalstig, E. B., and Sawyer, B. D. (1976). Studies on the role of the preoptic area in the control of reproductive function in the rat. *Endocrinology* **99**, 728–735.

Clifton, D. K., Steiner, R. A., Resko, J. A., and Spies, H. G. (1975). Estrogen-induced gonadotropin release in ovariectomized rhesus monkeys and its advancement by progesterone. *Biol. Reprod.* **13**, 190–194.

Cogen, P. H., Antunes, J. L., Louis, K. M., Dyrenfurth, I., and Ferin, M. (1980). The effects of anterior hypothalamic disconnection on gonadotropin secretion in the female rhesus monkey. *Endocrinology* **107**, 677–683.

Conn, M. P. (1986). The molecular basis of gonadotropin-releasing hormone action. *Endocr. Rev.* **7**, 3–10.

Cooper, K. J., Fawcett, C. P., and McCann, S. M. (1973). Variations in pituitary responsiveness to luteinizing hormone releasing factor during the rat oestrous cycle. *J. Endocr.* **57**, 187–188.

Coppings, R. J. and Malven, P. V. (1976). Biphasic effect of estradiol on mechanisms regulating LH release in ovariectomized sheep. *Neuroendocrinology* **21**, 146–156.

Crighton, D. K. and Foster, J. P. (1977). Luteinizing hormone release after two injections of synthetic luteinizing hormone releasing hormone in the ewe. *J. Endocr.* **72**, 59–67.

Crowder, M. E. and Nett, T. M. (1984). Pituitary content of gonadotropins and receptors for gonadotropin-releasing hormone (GnRH) and hypothalamic content of GnRH during the periovulatory period of the ewe. *Endocrinology* **114**, 234–239.

Crowley, W. F. and McArthur, J. W. (1980). Stimulation of normal menstrual cycle in Kallman's syndrome by pulsatile administration of luteinizing hormone-releasing hormone (LHRH). *J. clin. Endocr. Metab.* **51**, 173–175.

Curtis, A. and Fink, G. (1983). A high molecular weight precursor of luteinizing hormone releasing hormone from rat hypothalamus. *Endocrinology* **112**, 390–392.

Dees, W. L., Sorensen, A. M., Kemp, W. M., and McArthur, N. H. (1981). Immunohistochemical localization of gonadotropin-releasing hormone (GnRH) in the brain and infundibulum of the sheep. *Cell. Tissue Res.* **215**, 181–191.

Denef, C. (1984). Paracrine interaction in rat anterior pituitary. In *Endocrinology* (ed. F. Labrie, and L. Proulx) pp. 495–498, Exerpta Medica, Amsterdam.

De Paolo, L. V. and Barraclough, C. A. (1979). Interactions of estradiol and progesterone on pituitary gonadotropin secretion: possible sites and mechanisms of action. *Biol. Reprod.* **20**, 1173–1185.

Diekman, M. A. and Malven, P. V. (1973). Effect of ovariectomy and estradiol on LH patterns in ewes. *J. Anim. Sci.* **37**, 562–567.

Djahanbakhch, O., Warner, P., McNeilly, A. S., and Baird, D. T. (1984). Pulsatile release of LH and oestradiol during the periovulatory period in women. *Clin. Endocr.* **20**, 579–589.

Elkind-Hirsch, K., Ravnikar, U., Schiff, I., Tulchinsky, D., and Ryan, K. J. (1982). Determinations of endogenous immunoreactive luteinizing hormone-releasing hormone in human plasma. *J. clin. Endocr. Metab.* **54**, 602–607.

Ellinwood, W. E., Ronnekleiv, O. K., Kelly, M. J., and Resko, J. A. (1985). A new antiserum with conformational specificity for LH-RH: usefulness for radioimmunoassay and immunocytochemistry. *Peptides* **6**, 45–52.

Faure, N. and Oliver, G. C. (1978). Hypersensitivity of the human gonadotrophs to the repeated administration of small doses of LH-RH. *Horm. Res.* **9**, 12–21.

Ferin, M., Van Vugt, D., and Wardlaw, S. (1984). The hypothalamic control of the menstrual cycle and the role of endogenous opioid peptides. *Recent Prog. Horm. Res.* **40**, 441–485.

Flerko, B. (1981). The LH-RH neuron system of the rat and rabbit and a new insight into the hypophysiotrophic area of the hypothalamus. *Folia Endocrinol. Jap.* **57**, 292–297.

—— Setalo, G., Vigh, S. and Merchenthaler, I. (1980). Recent immunohistological findings on the LH-RH neuron system of the rat. *Materia Medica Polona* **12**, 119–123.

Gallo, R. V. (1981). Pulsatile LH release during the ovulatory surge on proestrus in the rat. *Biol. Reprod.* **24**, 100–104.

Goding, J. R., Catt, K. J., Brown, J. M., Kaltenbach, C. C., Cumming, I. A., and Mole, B. J. (1969). Radioimmunoassay for ovine luteinizing hormone. Secretion of luteinizing hormone during estrus and following estrogen administration in the sheep. *Endocrinology* **85**, 133–142.

Goodman, R. L. (1978). A quantitative analysis of the physiological role of estradiol and progesterone in the control of tonic and surge secretion of luteinizing hormone secretion in the rat. *Endocrinology* **102**, 142–151.

—— Bittman, E. L., Foster, D. L., and Karsch, F. J. (1981). The endocrine basis of the synergistic suppression of luteinizing hormone by estradiol and progesterone. *Endocrinology* **109**, 1414–1417.

—— and Karsch, F. J. (1980). Pulsatile secretion of luteinizing hormone: differential suppression by ovarian steroids. *Endocrinology* **107**, 1286–1290.

—— Legan, S. J., Ryan, K. D., Foster, D. L., and Karsch, F. J. (1980). Importance of variations in behavioural and feedback actions of oestradiol to the control of seasonal breeding in the ewe. *J. Endocr.* **23**, 404–413.

Gray, G. D., Sodersten, P., Tallentire, D., and Davidson, J. M. (1978). Effects of lesions in various structures of the suprachiasmatic-preoptic region on LH regulation and sexual behaviour in female rats. *Neuroendocrinology* **25**, 174–191.

Halasz, B. and Pupp, L. (1965). Hormone secretion of the anterior pituitary gland after physical interruption of all nervous pathways to the hypophysiotrophic area. *Endocrinology* **77**, 553–562.

Hauger, R. L., Karsch, F. J., and Foster, D. L. (1977). A new concept for the control of the estrous cycle of the ewe based on the temporal relationships between luteinizing hormone, estradiol and progesterone in peripheral serum and evidence that progesterone inhibits tonic LH secretion. *Endocrinology* **101**, 807–817.

Hess, D. L., Wilkins, R. H., Moossy, J., Chang, J. L., Plant, T. M., McCormack, J. T., Nakai, Y., and Knobil, E. (1977). Estrogen-induced gonadotropin surges in decerebrated female rhesus monkeys with medial-based hypothalamic peninsulae. *Endocrinology* **101**, 1264–1271.

Higuchi, T. and Kawakami, M. (1982). Luteinizing hormone responses to repeated injections of luteinizing hormone releasing hormone in the rat during the oestrous cycle and after ovariectomy with or without oestrogen treatment. *J. Endocr.* **93**, 161–168.

Hoff, J. D., Lasley, B. L., and Yen, S. S. C. (1979). The functional relationship between priming and releasing actions of luteinizing hormone-releasing hormone. *J. clin. Endocr. Metab.* **49**, 8–11.

—— Quigley, M. E. and Yen, S. C. (1983). Hormonal dynamics at midcycle: a reevalution. *J. clin. Endocr. Metab.* **57**, 792–796.

Hoffman, G. E. (1983). LHRH neurons and their projections. In *Structure and function of peptidergic and aminergic neurons* (ed. Y. Sano, Y. Ibata, and E. A. Zimmerman) pp. 183–202. Japan Scientific Societies Press, Tokyo.

Ibata, Y., Watanabe, K., Kinoshita, H., Kubo, S., Sano, Y., Sin, S., Hashimura, E., and Imagawa, K. (1979). The location of LH-RH neurons in the rat hypothalamus and their pathways to the median eminence. *Cell Tissue Res.* **198**, 381–395.

Jackson, G. L., Kuehl, D., McDowell, K., and Zaleski, A. (1978). Effect of hypothalamic deafferentation on secretion of luteinizing hormone in the ewe.

Biol. Reprod. **17**, 808–819.

—— Roche, J. F., Foster, D. L., and Dziuk, P. J. (1971). Luteinizing hormone releasing activity in the hypothalamus of anoestrous and cyclic ewes. *Biol. Reprod.* **5**, 5–12.

Jacobson, R. I., Seyler, L. E., Tamborlane, W. U., Gertner, J. M., and Genel, M. (1979). Pulsatile subcutaneous nocturnal administration of GnRH by portable infusion pump in hypogonadotrophic-hypogonadism: initiation of gonadotropin responsiveness. *J. clin. Endocr. Metab.* **49**, 652–654.

Jew, J. Y., Leranth, C., Arimura, A., and Palkovits, M. (1984). Preoptic LH-RH and somatostatin in the rat median eminence. *Neuroendocrinology* **38**, 169–175.

Kalra, P. S. and Kalra, S. P. (1977). Temporal changes in the hypothalamic and serum luteinizing hormone-releasing hormone (LH-RH) levels and the circulating ovarian steroids during the rat oestrous cycle. *Acta endocr. Copenh.* **85**, 449–455.

—— —— (1980). Modulation of hypothalamic luteinizing hormone releasing hormone levels by intracranial and sub-cutaneous implants of gonadal steroids in castrated rats: effects of androgen and estrogen antagonists. *Endocrinology* **106**, 390–397.

—— Simpkins, J. W., and Kalra, S. P. (1984). Testosterone raises LHRH levels exclusively in the median eminence of castrated rats. *Neuroendocrinology* **39**, 45–48.

Kalra, S. P. (1976). Tissue levels of luteinizing hormone-releasing hormone in the preoptic area and hypothalamus, and serum concentrations of gonadotropins following anterior hypothalamic deafferentation and estrogen treatment of the female rat. *Endocrinology* **99**, 101–107.

—— (1979). Dynamic changes in hypothalamic LH-RH levels associated with the ovarian steroid-induced gonadotrophin surge. *Acta endocr. Copenh.* **92**, 1–7.

—— —— (1983). Neural regulation of luteinizing hormone secretion in the rat. *Endocr. Rev.* **4**, 311–351.

Karsch, F. J., Dierschke, D. J., Weick, R. F., Yamaji, T., Hotchkiss, J., and Knobil, E. (1973a). Positive and negative feedback control by estrogen of luteinizing hormone secretion in the rhesus monkey. *Endocrinology* **92**, 799–804.

—— Foster, D. L., Bittman, E. L., and Goodman, R. L. (1983). A role for estradiol in enhancing luteinizing hormone pulse frequency during the follicular phase of the estrous cycle of sheep. *Endocrinology* **113**, 1333–1339.

—— Legan, S. J., Hauger, R. L., and Foster, D. L. (1977). Negative feedback action of progesterone on tonic luteinizing hormone secretion in the ewe: dependence on the ovaries. *Endocrinology* **101**, 800–806.

—— Weick, R. F., Butler, W. R., Dierschke, D. J., Krey, L. C., Weiss, G., Hotchkiss, J., Yamaji, and Knobil, E. (1973b). Induced LH surges in the rhesus monkey: strength-duration characteristics of the estrogen stimulus. *Endocrinology* **92**, 1740–1747.

Kawakami, M., Arita, J., and Yoshioka, E. (1980). Loss of estrogen-induced daily surges of prolactin and gonadotropins by suprachiasmatic nucleus lesions in ovariectomized rats. *Endocrinology* **106**, 1087–1092.

—— Yoshioka, E., Konda, N., Arita, J., and Visessuvan, S. (1978). Data on sites of stimulatory feedback action of gonadal steroids indispensable for luteinizing hormone release in the rat *Endocrinology* **102**, 791–798.

Kawano, H. and Daikoku, S. (1981). Immunohistochemical demonstration of LHRH neurons and their pathways in the rat hypothalamus. *Neuroendocrinology* **32**, 179–186.

King, J. C. (1974). Luteinizing hormone-releasing hormone (LH-RH) pathway of the rat hypothalamus revealed by the unlabelled antibody peroxide-antiperoxidase method. *Cell Tissue Res.* **153**, 211–217.

—— and Anthony, E. L. P. (1984). LHRH neurons and their projections in human and other mammals: species comparison. *Peptides* **5**, 195–207.

—— Parsons, J. A., Erlandsen, S. L. and Williams, T. H. (1974). Luteinizing hormone-releasing hormone (LH-RH) pathway of the rat hypothalamus revealed by the unlabelled antibody peroxidase-antiperoxidase method. *Cell Tissue Res.* **153**, 211–217.

—— Tobet, S. A., Snavely, F. L. and Arimura, A. A. (1982). LHRH immunopositive cells and their projections to the median eminence and organum vasculosum of the lamina terminalis. *J. comp. Neurol.* **209**, 287–300.

Knobil, E. (1974). On the control of gonadotrophin secretion in the rhesus monkey. *Recent Prog. Horm. Res.* **30**, 1–37.

—— (1980). The neuroendocrine control of the menstrual cycle. *Recent Prog. Horm. Res.* **36**, 53–88.

—— Plant, T. M., Wildt, L., Belchetz, P. E., and Marshall, G. (1980). Control of the rhesus monkey menstrual cycle: permissive role of hypothalamic gonadotropin-releasing hormone. *Science, N.Y.* **207**, 1371–1373.

Kobayashi, R. M., Lu, K. H., Moore, R. Y., and Yen, S. S. C. (1978). Regional distribution of hypothalamic luteinizing hormone-releasing hormone in proestrous rats: effects of ovariectomy and estrogen replacement. *Endocrinology* **102**, 98–105.

Krause, J. E., Advis, J. P., and McKelvy, J. F. (1982). *In vivo* biosynthesis of hypothalamic luteinizing hormone releasing hormone in individual free-running female rats. *Endocrinology* **111**, 344–346.

Krey, L. C., Butler, W. R., and Knobil, E. (1975). Surgical disconnection of the medial basal hypothalamus and pituitary function in the rhesus monkey. 1. Gonadotropin secretion. *Endocrinology* **96**, 1073–1087.

—— and Silverman, A. J. (1978). The luteinizing hormone-releasing hormone (LH-RH) neuronal networks of the guinea pig brain II. The regulation of gonadotropin secretion and the origin of terminals in the median eminence. *Brain Res.* **157**, 247–255.

Landgren, B. M., Unden, A. L., and Diczfalusy, E. (1980). Hormonal profile of the cycle in 68 normally menstruating women. *Acta endocr. Copenh.* **94**, 89–98.

Legan, S. J., Coon, G. A., and Karsch, F. J. (1975). Role of estrogen as initiator of daily LH surges in the ovariectomised rat. *Endocrinology* **96**, 50–56.

—— and Karsch, F. J. (1975). A daily signal for the LH surge in the rat. *Endocrinology* **96**, 57–62.

—— —— and Foster, D. L. (1977). The endocrine control of seasonal reproductive function in the ewe: a marked change in the response to the negative feedback action of estradiol on LH secretion. *Endocrinology* **101**, 818–824.

Lehman, M. N., Robinson, J. E., Karsch, F. J., and Silverman, A. J. (1986). Immunocytochemical localization of luteinizing hormone-releasing hormone (LH-RH) pathways of the sheep brain during anestrus and the mid-luteal phase of the estrous cycle. *J. comp. Neurol.* **244**, 19–35.

Levine, J. E., Norman, R. L., Gliessman, P. M., Oyama, T. T., Bangsberg, D. R., and Spies, H. G. (1985). *In vivo* gonadotropin-releasing hormone release and serum luteinizing hormone measurements in ovariectomized, estrogen-treated rhesus monkeys. *Endocrinology* **117**, 711–721.

—— Pau, K.-Y., Ramirez, V. D. and Jackson, G. L. (1982). Simultaneous measurement of luteinizing hormone-releasing hormone and luteinizing hormone release in unanaesthetized, ovariectomized sheep. *Endocrinology* **111**, 1449–1455.

—— and Ramirez, V. D. (1980). *In vivo* release of luteinizing hormone-releasing hormone estimated with push–pull cannulae from the mediobasal hypothalamus of ovariectomized, steroid-primed rats. *Endocrinology* **107**, 1782–1790.

—— —— (1982). Luteinizing hormone-releasing hormone release during the rat estrous cycle and after ovariectomy, as estimated with push–pull cannulae. *Endocrinology* **111**, 1439–1448.

Libertun, C., Orias, R., and McCann, S. M. (1974). Biphasic effect of oestrogen on the sensitivity of the pituitary to luteinizing hormone-releasing factor (LRF). *Endocrinology* **94**, 1094–1100.

Liu, J. H. and Yen, S. S. C. (1983). Induction of midcycle gonadotrophin surge by ovarian steroids in women: a critical evaluation. *J. clin. Endocr. Metab.* **57**, 797–802.

Liu, T. C. and Jackson, G. L. (1984). Long term superfusion of rat anterior pituitary cells: effects of repeated pulses of gonadotropin-releasing hormone at different doses, durations and frequencies. *Endocrinology* **115**, 605–612.

Magness, V. J., Millar, R. P., and Michie, J. (1981). Luteinizing hormone-releasing hormone, calcium and cyclic nucleotide interactions in luteinizing hormone release from ovine pituitary cells in culture. *Neuropeptides* **21**, 190–198.

Malacara, J. M., Seyler, E. L. J. R., and Reichlin, S. (1972). Luteinizing hormone releasing factor activity in peripheral blood from women during the midcycle luteinizing hormone ovulatory stage. *J. clin. Endocr. Metab.* **34**, 271–278.

Marshall, P. E. and Goldsmith, P. C. (1980). Neuroregulatory and neuroendocrine GnRH pathways in the hypothalamus and forebrain of the baboon. *Brain Res.* **193**, 353–372.

Martin, G. B., Scaramuzzi, R. J., and Henstridge, J. D. (1983). Effects of oestradiol, progesterone and androstenedione on the pulsatile secretion of luteinizing hormone in ovariectomized ewes during spring and autumn. *J. Endocr.* **96**, 181–193.

Marut, E. L., Williams, R. F., Cowan, B. D., Lynch, A., Lerner, S. P., and Hodgen, G. D. (1981). Pulsatile pituitary gonadotropin secretion during maturation of the dominant follicle in monkeys: estrogen positive feedback enhances the biological activity of LH. *Endocrinology* **109**, 2270–2272.

Merchenthaler, I., Kovacs, G., Lovasz, G., and Setalo, G. (1980). The preoptico-infundibular LH-RH tract of the rat. *Brain Res.* **198**, 63–74.

Millar, R. P., Denniss, P., Tobler, C., King, J. C., Schally, A. V., and Arimura, A. (1977). Presumptive prohormonal forms of hypothalamic peptide hormones. In *Colloques Internatonaux du CNRS. No. 180—Biologie cellulaire des processus neurosecretoires hypothalamiques* (ed. J. D. Vincent and C. Kordon) pp. 487–510.

—— Wegener, I., and Schally, A. U. (1981). Putative prohormonal luteinizing hormone-releasing hormone. In *Neuropeptides: biochemical and physiological studies* (ed. R. P. Millar) pp. 111–130. Churchill Livingstone, Edinburgh.

Miyake, A., Kawamura, Y., Aono, T., and Kurachi, K. (1980). Changes in plasma LRH during the normal menstrual cycle in women. *Acta endocr. Copenh.* **93**, 257–263.

—— Tasaka, K., Sakumoto, T., Kawamura, Y., and Aono, T. (1983). Estrogen induces the release of luteinizing-hormone-releasing hormone in normal cyclic women. *J. clin. Endocr. Metab.* **56**, 1100–1102.

Moguilevsky, J. A., Scacchi, P., Debeljuk, L., and Faigon, M. R. (1975). Effect of castration upon hypothalamic luteinizing hormone releasing factor (LH-RF). *Neuroendocrinology* **17**, 189–192.

Naftolin, F., Brown-Grant, K., and Corker, C. S. (1972). Plasma and pituitary luteinizing hormone and peripheral plasma oestrogen concentrations in the oestrous cycle of the rat and after experimental manipulation of the cycle. *J. Endocr.* **53**, 17–30.

Naik, D. V. (1975*a*). Immuno-electron microscope localization of luteinizing hormone-releasing hormone in the arcuate nuclei and median eminence of the rat. *Cell Tissue Res.* **157**, 437–455.

——— (1975*b*). Immunoreactive LH-RH neurons in the hypothalamus identified by light and fluorescent microscopy. *Cell Tissue Res.* **157**, 423–436.

Nakai, Y., Plant, T. M., Hess, D. L., Keogh, E. J., and Knobil, E. (1978). On the sites of negative and positive feedback actions of estradiol in the control of gonadotropin secretion in the rhesus monkey. *Endocrinology* **102**, 1008–1014.

Neill, J. D., Patton, J. M., Dailey, R. A., Tsou, R. C., and Tindall, G. T. (1977). Luteinizing hormone releasing hormone (LH-RH) in pituitary stalk blood of rhesus monkeys: relationships to levels of LH release. *Endocrinology* **101**, 430–434.

Nikolics, K., Mason, A. J., Szonyi, E., Ramachandran, J., and Seeburg, P. H. (1985). A prolactin-inhibiting factor within the precursor for gonadotropin-releasing hormone. *Nature, Lond.* **316**, 511–517.

Nillius, S. J. and Wide (1971). Effect of progesterone on the serum levels of FSH and LH in postmenopausal women treated with estrogen. *Acta endocr. Copenh.* **67**, 367–370.

Norman, R. L., Resko, J. A., and Spies, H. G. (1976). The anterior hypothalamus: how it affects gonadotropin secretion in the rhesus monkey. *Endocrinology* **99**, 59–71.

O'Connor, J. L., Lapp, C. A., and Mahesh, V. B. (1984). Peptidase activity in the hypothalamus and pituitary of the rat: fluctuations and possible regulatory role of luteinizing hormone releasing hormone-degrading activity during the estrous cycle. *Biol. Reprod.* **30**, 855–862.

Odell, W. D. and Swerdloff, R. S. (1968). Progesterone-induced luteinizing and follicle-stimulating hormone surge in postmenopausal women: a simulated ovulatory peak. *Physiology* **68**, 529–536.

Oshima, I., Morishita, H., Omura, K., and Saito, S. (1978). Changes in hypothalamic LH-RH content and blood levels of LH-RH, gonadotropin and estradiol during the preovulatory stage of rat estrous cycle. *Endocr. Japan.* **25**, 607–611.

Palkovits, M. (1973). Isolated removal of hypothalamic or other brain nuclei of the rat. *Brain Res.* **59**, 449–450.

Pau, K.-Y. F., Kuehl, D. E., and Jackson, G. L. (1982). Effect of frontal hypothalamic deafferentation on luteinizing hormone secretion and seasonal breeding in the ewe. *Biol. Reprod.* **27**, 999–1009.

Peduto, J. C. and Mahesh, V. B. (1985). Effects of progesterone on hypothalamic and plasma LHRH. *Neuroendocrinology* **40**, 238–245.

Phillips, H. S., Nikolics, K., Branton, D., and Seeburg, P. H. (1985). Immuno-cytochemical localization in rat brain of a prolactin release-inhibiting sequence of gonadotropin-releasing hormone prohormone. *Nature, Lond.* **316**, 542–545.

Piacsek, B. E. and Meites, J. (1966). Effects of castration and gonadal hormones on hypothalamic content of luteinizing hormone releasing factor (LRF). *Physiologist* **8**, 232–439.

Plant, T. M., Krey, L. C., Moossy, J., McCormack, J. T., Hess, D. L., and Knobil, E. (1978). The arcuate nucleus and control of gonadotropin and prolactin secretion in the female rhesus monkey (*Macaca mulatta*). *Endocrinology* **102**, 52–62.

——— Moossy, J., Hess, D. L., Nakai, Y., McCormack, J. T., and Knobil, E. (1979). Further studies on the effects of lesions in the rostral hypothalamus on gonadotropin secretion in the female rhesus monkey (Macaca mulatta). *Endocrinology* **105**, 465–473.

Polkowska, J., Dubois, M. P., and Domanski, E. (1980). Immunocytochemistry of luteinizing hormone releasing hormone (LH-RH) in the sheep hypothalamus during various reproductive stages. Correlation with the gonadotropic hormones of the pituitary. *Cell Tissue Res.* **208**, 327–341.

—— and Jutisz, M. (1979). Local changes in immunoreactive gonadotropin releasing hormone in the rat median eminence during the estrous cycle. Correlation with the pituitary luteinizing hormone. *Neuroendocrinology* **28**, 281–288.

Przekop, F. and Domanski, E. (1970). Hypothalamic centres involved in the control of gonadotrophin secretion and ovulation in sheep. *Acta physiol. Polonica* **21**, 34–49.

Radford, H. M. (1967). The effect of hypothalamic lesions on reproductive activity in sheep. *Endocrinology* **39**, 415–422.

—— (1979). The effect of hypothalamic lesions on estradiol-induced changes in LH release in the ewe. *Neuroendocrinology* **28**, 307–312.

—— and Wallace, A. L. C. (1974). Central nervous blockade of oestradiol-stimulated release of luteinizing hormone in the ewe. *J. Endocr.* **60**, 247–252.

—— Wheatley, I. S., and Wallace, A. L. C. (1969). The effects of oestradiol benzoate and progesterone on the secretion of luteinizing hormone in the ovariectomized ewe. *J. Endocr.* **44**, 135–136.

Rahe, C. H., Owens, R. E., Fleeger, J. L., Newton, J. T., and Harms, P. G. (1980). Pattern of plasma luteinizing hormone in the cyclic cow: dependence upon the period of the cycle. *Endocrinology* **107**, 498–503.

Ramirez, V. D. and Sawyer, C. H. (1965). Fluctuations in hypothalamic LH-RH (luteinizing hormone-releasing factor) during the rat estrous cycle. *Endocrinology* **76**, 282–289.

Rance, N., Wise, P. M., Selmanoff, M. K., and Barraclough, C. A. (1981). Catecholamine turnover rates in discrete hypothalamic areas and associated changes in median eminence luteinizing hormone-releasing hormone and serum gonadotrophins on proestrous and diestrous day 1. *Endocrinology* **108**, 1795–1802.

Reeves, J. J., Arimura, A., and Schally, A. V. (1971a). Pituitary responsiveness to purified luteinizing hormone-releasing hormone (LH-RH) at various stages of the estrous cycle in sheep. *J. Anim. Sci.* **32**, 123–126.

—— —— —— (1971b). Changes in pituitary responsiveness to luteinizing hormone-releasing hormone (LH-RH) in anestrous ewes pretreated with estradiol benzoate. *Biol. Reprod.* **4**, 88–92.

Rethelyi, M., Vigh, S., Setalo, G., Merchenthaler, I., and Flerko, B. (1981). The luteinizing hormone releasing hormone-containing pathways and their co-termination with tanycyte processes in and around the median eminence and in the pituitary stalk of the rat. *Acta morphol. Acad. Sci. Hung.* **29**, 259–283.

Samson, W. K. and McCann, S. M., (1979). Effects of lesions in the organum vasculosum of the lamina terminalis (OVLT) on the hypothalamic distribution of LHRH and gonadotropin secretion in the ovariectomized female rat. *Endocrinology* **105**, 939–946.

—— —— Chud, L., Dudley, C. A., and Moss, R. L. (1980). Intra- and extrahypothalmic luteinizing hormone-releasing hormone (LHRH) distribution in the rat with special reference to mesencephalic sites which contain both LHRH and single neurons responsive to LHRH. *Neuroendocrinology* **31**, 66–72.

Sarkar, D. K., Chiappa, S. A., Fink, G., and Sherwood, N. M. (1976). Gonadotrophin-releasing hormone surge in pro-oestrous rats. *Nature, Lond.* **264**, 461–463.

—— and Fink, G. (1979). Effects of gonadal steroids on output of luteinizing hormone releasing factor into pituitary stalk blood in the female rat. *J. Endocr.* **80**, 303–313.

—— —— (1980). Luteinizing hormone releasing factor in pituitary stalk plasma from long term ovariectomized rats: effects of steroids. *J. Endocr.* **86**, 511–524.

Scaramuzzi, R. J., Caldwell, B. V., and Moor, R. M. (1970). Radioimmunoassay of LH and estrogen during the estrous cycle of the ewe. *Biol. Reprod.* **3**, 110–119.

—— and Radford, H. M. (1983). Factors regulating ovulation rate in the ewe. *J. Reprod. Fert.* **69**, 353–367.

Schally, A. V. (1984). LH-RH analogs in contraception and cancer. In *LHRH and its analogs* (ed. F. Labrie, A. Belanger, and A. Dupont) pp. 3–15. Excepta Medica, Amsterdam.

Schillo, K. K., Leshin, L. S., Kuehl, D., and Jackson, G. L. (1985). Simultaneous measurement of luteinizing hormone-releasing hormone and luteinizing hormone during estradiol-induced luteinizing hormone surges in the ovariectomized ewe. *Biol. Reprod.* **33**, 644–652.

Scott, D. E., Dudley, G. K., and Knigge, K. M. (1974). The ventricular system in neuroendocrine mechanisms. II. *In vivo* monoamine transport by ependyma of the median eminence. *Cell Tissue Res.* **154**, 1–16.

Seeburg, P. H. and Adelman, J. P. (1984). Characterization of cDNA for precursor of human luteinizing hormone releasing hormone. *Nature, Lond.* **311**, 666–668.

Selmanoff, M. K., Wise, P. M., and Barrclough, C. A. (1980). Regional distribution of luteinizing hormone-releasing hormone (LH-RH) in the rat brain determined by microdissection and radioimmunoassay. *Brain Res.* **192**, 421–432.

Setalo, G., Vigh, S., Schally, A. V., Arimura, A., and Flerko, B. (1975). LH-RH-containing neural elements in the rat hypothalamus. *Endocrinology* **96**, 135–142.

—— —— —— and Flerko, B. (1976). Immunohistological study of the origin of LH-RH-containing nerve fibers of the rat hypothalamus. *Brain Res.* **103**, 597–602.

Sherwood, N. M. and Fink, G. (1980). Effect of ovariectomy and adrenalectomy on luteinizing hormone-releasing hormone in pituitary stalk blood from female rats. *Endocrinology* **106**, 363–367.

Shi-fan, L., Zhong-xing, W., Yao-e, Y., Sheng-min, B., Bei-zhu, Z., Jin-zhi, W., Yi-e, W., and Xin-ying, P. (1986)., Hormone changes during the menstrual cycle of Chinese women. *J. Reprod. Fert.* **76**, 43–52.

Shivers, B. D., Harlan, R. E., Hejtmancik, J. F., Conn, P. M., and Pfaff, D. W. (1986). Localization of cells containing LHRH-like mRNA in the rat forebrain using *in situ* tissue hybridization. *Endocrinology* **118**, 883–885.

—— —— Morrell, J. I. and Pfaff, D. W. (1983*a*). Absence of oestradiol concentration in the cell nuclei of LHRH-immunoreactive neurones. *Nature, Lond.* **304**, 345–347.

—— —— —— —— (1983*b*). Immunocytochemical localization of luteinizing hormone releasing hormone in male and female rat brains. *Neuroendocrinology* **36**, 1–12.

Silverman, A. J., Antunes, J. L., Abrams, G. M., Nilaver, G., Thau, R., Robinson, J. A., Ferin, M. and Krey, L. C. (1982). The luteinizing hormone-releasing hormone pathways in rhesus (*Macaca mulatta*) and pigtailed (*Macaca nemestina*) monkey: new observations on thick, unembedded sections. *J. comp. Neurol.* **211**, 309–317.

—— —— Ferin, M., and Zimmerman, E. A. (1977). The distribution of luteinizing hormone-releasing hormone (LH-RH) in the rat hypothalamus of the rhesus monkey. Light microscopic studies using immunoperoxidase technique. *Endocrinology* **101**, 134–142.

—— and Krey, L. C. (1978). The luteinizing hormone-releasing hormone (LH-RH) neuronal networks of the guinea pig brain. I. Intra- and extra-hypothalamic projections. *Brain Res.* **157**, 233–246.

Snabes, M. C., Kelch, R. P., and Karsch, F. J. (1977). A daily neural signal for luteinizing hormone release in the untreated ovariectomized rat: changes in gonadotropin-releasing hormone content of the preoptic area and hypothalamus throughout the day. *Endocrinology* **100**, 1521–1525.

Soper, B. D. and Weick, R. F. (1980). Hypothalamic and extrahypothalamic mediation of pulsatile discharges of luteinizing hormone in the ovariectomized rat. *Endocrinology* **106**, 348–355.

Swerdloff, R. S. and Odell, W. D. (1969). Serum luteinizing and follicle stimulating hormone levels during sequential and nonsequential contraceptive treatment of eugonadal women. *J. clin. Endocr. Metab.* **29**, 157–163.

Tan, L. and Rousseau, P. (1982). The chemical identity of the immunoreactive LHRH-like peptide biosynthesized in the human placenta. *Biochem. Biophys. Res. Commun.* **109**, 1061–1071.

Thiery, J. C., Pelletier, J., and Signoret, J. P. (1978). Effect of hypothalamic deafferentation on LH and sexual behaviour in ovariectomized ewe under hormonally induced oestrous cycle. *Annls. Biol. anim. Biochim. Biophys.* **18**, 1413–1426.

Tsou, R. C., Dailey, R. A., McLanan, C. S., Parent, A. D., Tindall, G. T., and Neill, J. D. (1977). Luteinizing hormone releasing hormone (LHRH) levels in pituitary stalk plasma during the preovulatory gonadotropin surge of rabbits. *Endocrinology* **101**, 534–539.

Van Dieten, J. A. M. J., Steijsiger, J., Dullaart, J., and Van Rees, G. P. (1974). The effect of estradiol benzoate on the pituitary responsiveness to LH-RH in male and female rats. *Neuroendocrinology* **15**, 182–188.

Van Vugt, D. A., Diefenbach, W. D., Alston, E., and Ferin, M. (1985). Gonadotropin-releasing hormone pulses in third ventricular cerebrospinal fluid of ovariectomized rhesus monkeys: correlation with luteinizing hormone pulses. *Endocrinology* **117**, 1550.

Vaughan, L., Carmel, P. W., Dyrenfurth, I., Frantz, A. G., Antunes, J. L., and Ferin, M. (1980). Section of the pituitary stalk in the rhesus monkey. I. Endocrine studies. *Neuroendocrinology* **30**, 70–75.

Vigh, S., Setalo, G., Arimura, A., Schally, A. V., and Flerko, B. (1978). LH-RH-containing nerve fibers in the brain of rats treated with sulpiride or reserpine. *Brain Res.* **152**, 401–405.

Vilchez-Martinez, J. A., Arimura, A., Debeljuk, L., and Schally, A. V. (1974). Biphasic effect of estradiol benzoate on the pituitary responsiveness of LH-RH. *Endocrinology* **94**, 1300–1303.

Wallach, E. E., Root, A. W., and Garcia, C.-R. (1970). Serum gonadotropin responses to estrogen and progestagen in recently castrated human females. *J. clin. Endocr. Metab.* **31**, 376–381.

Weick, R. F., Dierschke, D. J., Karsch, F. J., Butler, W. R., Hotchkiss, J., and Knobil, E. (1973). Periovulatory time courses of circulating gonadotropic and ovarian hormones in the rhesus monkey. *Endocrinology* **93**, 1140–1147.

Weiner, R. P., Pattou, E., Kerdelhue, B., and Kordon, C. (1975). Differential effects of hypothalamic deafferentation upon luteinizing hormone-releasing hormone in the median eminence and organum vasculosum of the lamina terminalis. *Endocrinology* **97**, 1597–1600.

Wenger, T., Kerdelhue, B., and Halasz, B. (1979). Short-term effect of the lesion of the organum vasculosum of the lamina terminalis on hypothalamic LH-RH and serum LH, FSH and prolactin in adult female rats. *Neuroendocrinology* **29**, 276–280.

—— and Leonardelli, J. (1980). Circadian and cyclic LHRH variations in the organum vasculosum of the lamina terminalis of female and male rats. *Neuroendocrinology* **31**, 331–337.

Wheaton, J. E. (1979). Regional brain content of luteinizing hormone-releasing hormone in sheep during the estrous cycle, seasonal anestrus, and after ovariectomy. *Endocrinology* **104**, 839–844.

—— Krulich, L., and McCann, S. M. (1975). Localisation of luteinizing hormone-releasing hormone in the preoptic area and hypothalamus of the rat using radioimmunoassay. *Endocrinology* **97**, 30–38.

Wiegand, S. J., Teresawa, E., and Bridson, W. E. (1978). Persistent estrus and blockade of progesterone-induced LH release follows lesions which do not damage the suprachiasmatic nucleus. *Endocrinology* **102**, 1645–1648.

—— —— —— and Goy, R. W. (1980). Effects of discrete lesions of preoptic and suprachiasmatic structures in the female rat. *Neuroendocrinology* **31**, 147–157.

Wildt, L., Hausler, A., Marshall, G., Hutchison, J. S., Plant, T. M., Belchetz, P. E., and Knobil, E. (1981). Frequency and amplitude of gonadotropin-releasing hormone stimulation and gonadotropin secretion in the rhesus monkey. *Endocrinology* **109**, 376–385.

Wise, P. M., Rance, N., Selmanoff, M., and Barraclough, C. A. (1979). Changes in radioimmunoassayable luteinizing hormone-releasing hormone in discrete brain areas of the rat at various times on proestrus, diestrus Day 1 and after phenobarbital administration. *Endocrinology* **108**, 2179–2185.

Witkin, J. W., Paden, C. M., and Silverman, A.-J. (1982). The luteinizing hormone-releasing hormone (LHRH) systems in the rat brain. *Neuroendocrinology* **35**, 429–438.

Wright, P. J. and Clarke, I. J. (1985). GnRH administered in a pulsatile regimen stimulates a surge release of LH in oestrogen treated, anaesthetized ovariectomized ewes. *Proc. endocr. Soc. Aust.* **28**, 2.

Yamaji, T., Dierschke, D. J., Bhattacharya, A. N., and Knobil, E. (1972). The negative feedback control by estradiol and progesterone of LH secretion in the ovariectomized rhesus monkey. *Endocrinology* **90**, 771–776.

—— —— Hotchkiss, J., Bhattacharya, A. N., Surve, A. H. and Knobil, E. (1971). Estrogen induction of LH release in the rhesus monkey. *Endocrinology* **89**, 1034–1041.

Yen, S. S. C. and Lein, A. (1976). The apparent paradox of the negative and positive feedback control system on gonadotrophin secretion. *Am. J. Obstet. Gynec.* **126**, 942–954.

—— VandenBerg, G., and Siler, T. M. (1974). Modulation of pituitary responsiveness to LRF by estrogen. *J. clin. Endocr. Metab.* **39**, 170–177.

3   Ovulation in the hen: neuroendocrine control

F. J. CUNNINGHAM

I   Introduction

The average Briton consumes 250 eggs per year and in the United Kingdom in 1985 more than 11,124 million eggs were produced resulting in total egg sales worth more than £900 million. Consequently because of its economic significance there has been a sustained research interest in the reproductive biology of poultry. Whilst the ovulatory cycles of birds and mammals share certain similarities, in some respects they are fundamentally different. Within the avian ovary, for example, follicles do not mature in synchrony:

instead a hierarchy of rapidly growing yolky follicles is established. The largest follicle ovulates at intervals, often daily, until a sequence is completed. In practical terms the sequence can be sustained if the newly laid eggs are removed and this fact has been much exploited by man. For economic reasons modern hybrid strains of domestic fowl do not have a reproductive cycle consisting of laying and incubating of eggs and brooding of the young. Instead selective breeding and efficient management have resulted in hens which lay an average 275 eggs during the first year of reproductive life, and it is not uncommon for individual hens to lay more than 300 eggs during the same period. Understanding the physiological basis of the mechanisms which, under standard conditions of light and darkness, determine the precise timing of a sequence of ovulations and restrict them to an 8–10 h period of the day remains a challenge.

The limitations to production imposed by the restriction of ovulation to an 8–10 h period of the day are not serious in young hens in which mid-sequence follicles mature and ovulate at about 24 h intervals: such hens may lay twenty or more eggs in a sequence. In older birds, however, follicular maturation occurs more slowly, for example 26 h, and ultimately this leads to a desynchronization of the rhythm of follicular maturation with the 8–10 h period of the day to which the occurrence of ovulation is normally restricted. Consequently sequences of only four to six eggs are laid, and the economic significance of this has encouraged research to examine the underlying physiology of the ovulation sequence.

II   Ovulation-oviposition sequence

Although the occurrence of ovulation (the release of the egg from the ovary) and oviposition (the laying of the hard shelled egg some 24–26 h later) in the domestic hen are normally restricted to an 8–10 h period of the day, ovulation can be induced experimentally outside this period. The position of the 8–10 h period of the day in relation to the cycle of light and darkness, that is, its phase angle, is regulated by the length of the periods of light and darkness and in the domestic hen is entrained primarily to dusk (Morris 1973).

Oviposition overtly expresses the occurrence of ovulation and follows ovulation by the time the egg spends in the oviduct which varies between 24–28 h. The time relationships between the two events are well established (Warren and Scott 1935; Fraps 1955; Melek, Morris, and Jennings 1973). Briefly, in a characteristic sequence of ovipositions (Fig. 3.1) of a hen maintained on a lighting schedule of 14 h light and 10 h of darkness (14L:10D) the first oviposition of a sequence occurs about 9–10 h after the onset of darkness. Subsequent eggs are laid progressively later each day by the period of lag, which is the interval between successive ovipositions less 24 h, until an egg laid about 17–18 h after the onset of

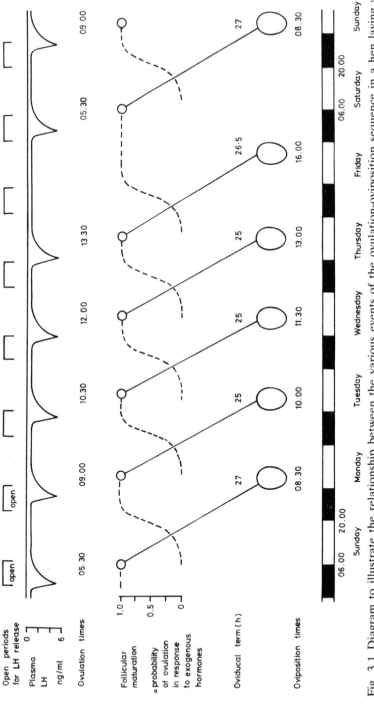

Fig. 3.1 Diagram to illustrate the relationship between the various events of the ovulation-oviposition sequence in a hen laying a sequence of five eggs. The open period for LH release refers to the 8–10 h interval of the day during which preovulatory releases of LH normally occur. (Courtesy of Professor T. R. Morris, University of Reading).

darkness occurs before the next sequence starts. Hens which lay sequences containing many eggs generally lay the first egg earlier in the day than do poor layers on the same lighting schedule. The period of lag is greatest between the last and penultimate ovipositions of a sequence, and in hens laying long sequences it is also greater between the first and second than between mid-sequence eggs. Ovulation usually occurs within 30–60 minutes of the oviposition of the previously ovulated egg except at the beginning and end of a sequence where these events are unrelated. There is evidence that hormonal changes associated with ovulation assist in bringing about oviposition. Thus when ovulation is induced prematurely, by treatment of hens with progesterone, the egg in the oviduct at the time may also be laid prematurely (Fraps 1942; Wilson and Sharp 1976a). Consequently it was suggested that the extended lag between the last and penultimate ovipositions may result from a delay in oviposition due to the absence of the oviposition-inducing stimulus associated with an accompanying ovulation (Fraps 1942).

III   Hypothalamus-pituitary-ovarian system

Generalisations about the neuroendocrine control of ovulation have largely come from studies using the rat, the ewe and the human female. The domestic hen, however, cannot be viewed from the mammalian perspective since ovulations occur in sequences and each ovulation occurs at a slightly later time on successive days until the sequence ends, when over a period of 40–44 h ovulation fails to occur. Because of this the ovulatory cycle of the domestic fowl was described as asynchronous (Bastian and Zarrow 1955) and is concerned with the interaction of two independent rhythms: namely, the process of follicular maturation; and the system which controls the luteinizing hormone (LH) release mechanism, which is believed to be circadian in nature.

1   HYPOTHALAMUS

The principal hypothalamic hormone affecting reproduction is luteinizing hormone-releasing hormone (LHRH). The structure of chicken(c) LHRH has been determined (King and Millar 1982a, b; Miyamoto, Hasegawa, Minegishi, Nomura, Takahashi, Igarashi, Kangawa, and Matsuo 1982; Miyamoto, Hasegawa, Igarashi, Chino, Sakakibara, Kangawa, and Matsuo 1983). Chicken LHRH is a decapeptide which differs from mammaliam LHRH in that glutamine is substituted for arginine at position 8 of the molecule. More recently a second form of LHRH was isolated from chicken hypothalamus (Miyamoto, Hasegawa, Igarashi, Kangawa, and Matsuo 1983). This latter peptide has some similarity of structure with salmon brain LHRH. The structural relationships between the various peptides are shown in Fig. 3.2. In view of the structural variations in chicken LHRH's it is

Porcine/Ovine LHRH

```
          1    2    3    4    5    6    7        8    9   10
     (pyro)Glu-His-Trp-Ser-Tyr-Gly-Leu- │Arg│ -Pro-Gly-NH₂
```

Chicken LHRH I

```
     (pyro)Glu-His-Trp-Ser-Tyr-Gly-Leu- │Gln│ -Pro-Gly-NH₂
```

Chicken LHRH II

```
     (pyro)Glu-His-Trp-Ser- │His│ -Gly- │Try│ - │Tyr│ -Pro-Gly-NH₂
```

Salmon LHRH

```
     (pyro)Glu-His-Trp-Ser-Tyr-Gly- │Try│ - │Leu│ -Pro-Gly-NH₂
```

Fig. 3.2. Comparison of the structures of vertebrate luteinizing hormone-releasing hormones.

feasible that changes also occurred in the chicken LHRH receptor during evolution. This could give rise to differences between species in the specificity of ligand receptor interactions which might account for the relatively poor specific activity of gonadotrophin stimulation by synthetic mammalian LHRH in birds (Johnson, Dickerman, and Advis 1984). The two forms of chicken LHRH, cLHRH I and cLHRH II, are biologically active, and the peptides are equipotent in stimulating the release of LH in cockerels *in vivo*. However, using a dispersed cockerel anterior pituitary cell assay system cLHRH II was found to be 4.7 times more potent than cLHRH I in stimulating LH release (Hsu-Fang Chou, Johnson, and Williams 1985). These authors suggested that cLHRH II may have greater affinity for the gonadotroph receptor, greater uptake by the cell, and/or that it may be more resistant to *in vitro* degradation than cLHRH I. On the other hand an extra pituitary site of degradation may be more effective in metabolizing cLHRH II, resulting in its equipotency with cLHRH *in vivo*.

i   Distribution of LHRH neurones in the hypothalamus

Immunocytochemical studies using antisera raised against synthetic mammalian LH (Sterling and Sharp 1982) and the measurement of immuno-

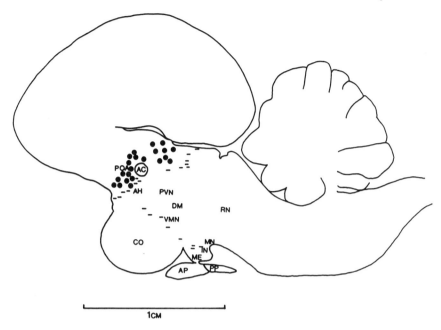

Fig. 3.3. Distribution of nerve cell bodies (●) and fibres (—) containing immunoreactive LHRH in the brain of the domestic hen. Anterior commissure (AC); anterior hypothalamus (AH); anterior lobe of the pituitary gland (AP); optic chiasma (CO); dorsomedial nucleus (DM); infundibular nucleus (IN); median eminence (ME); mammillary nucleus (MN); preoptic area (POA); posterior lobe of the pituitary gland (PP); paraventricular nucleus (PVN); nucleus ruber (RN); ventromedial nucleus (VMN). (Data are taken from Sterling and Sharp 1982).

reactive LHRH in discrete brain regions of the cockerel (Knight, Cunningham, and Gladwell 1983) have revealed that LHRH is found principally within the mediobasal and preoptic areas of the hypothalamus. Cell bodies containing LHRH lie within the preoptic area and adjacent anterior hypothalamic regions but not within the mediobasal hypothalamus (Fig. 3.3). This latter region contains LHRH axon terminals and stores about five times as much immunoreactive LHRH as does the preoptic area.

2   ANTERIOR PITUITARY GLAND

Progress in our knowledge of the chemistry and physiology of follicle-stimulating hormone (FSH) and luteinizing hormone (LH) stems from the first purification of chicken LH and FSH (Stockell-Hartree and Cunningham 1969) and the development of an homologous radioimmunoassay for chicken LH (Follett, Scanes, and Cunningham 1972). Since then chicken and turkey LH have been purified by various workers (see reviews by Licht, Papkoff, Farmer, Muller, Tsui, and Crews 1977; Ishii and

Sakai 1980). Both gonadotrophins are glycoproteins whose amino acid compositions resemble those of the mammaliam gonadotrophins (Ishii and Sakai 1980; Scanes 1981). Based on evidence from turkey gonadotrophins (Burke, Licht, Papkoff, and Bona Gallo 1979) avian LH and FSH are made up of two glycoprotein subunits each with a molecular weight of about 15,000. Both α and β subunits are required for the biological activity of the hormones. The anterior pituitary gland of the domestic fowl consists of well defined cephalic and caudal lobes. On the basis of immunocytochemical techniques LH containing cells are found in both lobes of the anterior pituitary gland (Mikami and Yamada 1984). As yet there have been no studies directed towards ascertaining the distribution of FSH containing cells and consequently there is no agreement as to whether FSH and LH are secreted from separate cells or from one type of cell.

Luteinizing hormone has long been known to stimulate ovulation (van Tienhoven 1961) and is generally considered to have steroidogenic properties (Wells, Gilbert, and Culbert 1980; Hammond, Burke, and Hertelendy 1981) whereas FSH is principally involved in ovarian growth, although it has been reported to have weak steroidogenic properties (Scanes and Fagioli 1980; Hammond *et al.* 1981).

3   OVARY

In birds only the left ovary is functional and in the adult may contain several million oocytes (Hutt 1949) all of which are formed during embryogenesis (Franchii, Mandl, and Zuckerman 1962; Hughes 1963). Many hundreds of these can be seen with the naked eye, though there is usually less than 100 with diameters greater than 1 mm. Most of the latter are usually described as "white" follicles since the yolk material they contain is pale in nature. The usually recognized follicular hierarchy consists of four to seven follicles with diameters greater than about 7 mm which contain mainly the "yellow" yolk. Other structures visible in the ovary are the postovulatory follicles of varying age and atretic follicles. In normal laying hens atresia is mainly associated with the smaller follicles. The larger follicles seldom become atretic except towards the end of the laying period or when a stressful stimulus has caused a sudden cessation of laying (Gilbert 1972).

The preovulatory follicles which form the hierarchy are identified according to size, with the largest (F_1) destined to ovulate first and the second largest (F_2) to ovulate the following day. Successive follicles are numbered in sequence and move one place up the hierarchy after ovulation of the preceding follicle. Ovulation occurs every 25–27 h with the exception of the day when there is no ovulation (Fig. 3.1).

The preovulatory follicle consists of an oocyte and its surrounding layers. The steroidogenic tissues are the granulosa layer and the theca layer containing nests of thecal cells, nerve terminals and blood vessels embedded in a collagen matrix (Gilbert 1965; Dahl 1970). Progesterone is the principal steroid produced by the granulosa layer especially that of the F_1 and F_2 follicles. On the other hand oestradiol is the major steroid secreted by the theca layer of the smaller and less mature follicles (Shahabi, Norton, and Nalbandov 1975; Huang and Nalbandov 1979; Culbert, Hardie, Wells, and Gilbert 1980; Wells, Gilbert, and Culbert 1980; Scanes and Fagioli 1980; Hammond et al. 1981; Bahr, Wang, and Calvo 1983). As follicles mature from F_5 to F_1 there is a large decrease in the concentration of oestradiol in the theca layer (Bahr et al. 1983). Measurements of the conversion of exogenous testosterone to oestradiol by isolated theca cells and of the activity of the aromatase enzyme system indicates that the decrease in oestradiol is a result of a reduction in aromatase activity (Wang and Bahr 1983; Armstrong 1984). As yet the mechanism by which oestradiol production is suppressed and that of progesterone increased has not been identified.

Apart from their role in sexual maturation and in the maintenance of reproductive tract tissues, the ovarian steroids are also important since a feedback loop exists between the ovary and the hypothalamus-pituitary gland system. For most of the ovulatory cycle a low level of tonic LH secretion is maintained by the negative feedback action of ovarian steroids. However, during the 8 h preceding an ovulation, one or more gonadal steroids exert a positive feedback action to bring about the preovulatory release of LH which leads to ovulation.

i Follicular maturation

In a comprehensive discussion of follicular maturation in the hen Etches (1984) considered the problem from three different standpoints, namely: follicular growth by which he meant changes in mass of the follicle and its contents; follicular competence which identified a follicle capable of being ovulated prematurely by trophic stimuli; and follicular maturity which referred to follicles that ovulate spontaneously.

ii Follicular growth

The entire follicle and its contained ovum increase in mass at a spectacular rate. For instance in the 9 days preceding ovulation the follicle and ovum increases in weight from 200 mg to 16 g. This is brought about largely by an accumulation of yolk accompanied during the last 4 days by an increase in mass of the thecal and granulosa cell layers, but the contribution of the follicle to the overall weight, as judged from the weight of the postovulatory follicle, is only about 1 g. The blood flow to the follicles could conceivably have a regulatory role in their development by altering the exposure of the follicle to gonadotrophic stimuli. However, although the follicle receives a larger proportion of the total ovarian blood flow as it increases in size, the amount of blood perfusing each gram of tissue remains unchanged (Scanes, Mozelic, Kavanagh, Merrill, and Rabii, 1982). Thus this mechanism would not seem to be significant with respect to the organisation of the hierarchy of follicles. Furthermore, the concentration of LH in all the follicles is identical and each follicle in the hierarchy, therefore, would appear to receive the same LH and presumably FSH stimulation (Etches, Croze, and Duke 1981).

iii   Follicular competence

It is well established that the follicle becomes more responsive to ovulation-inducing hormones as the time from the previous ovulation increases (Fig. 3.1). Such a relationship was first demonstrated by showing that the amount of chicken anterior pituitary powder required to induce ovulation in 50 per cent of hens declined as the time from the previous ovulation increased. Furthermore, the relationship was independent of the time of expected ovulation. Thus the follicle destined to become the first ovulation of a sequence acquired sensitivity to the treatment as quickly as any other follicle in the sequence (Fig. 3.1) (Fraps, 1955).

Fig. 3.4 shows the same relationship in response to treatment with LHRH. About 50 per cent of hens will ovulate in response to LHRH given 13–15 h after the previous ovulation (see also Fig. 3.1) and this relationship is identical for follicles destined to become mid-sequence or first ovulations of the sequence. This approach assumes that the ability of the anterior pituitary gland to secrete LH in response to LHRH between 8–16 h after a mid-sequence ovulation or between 8 and 30 h after a terminal ovulation is constant. The assumption is consistent with other findings (Bonney, Cunningham, and Furr 1974; Wilson and Sharp 1975) which showed either directly or indirectly that the response of the pituitary gland to exogenous stimuli is constant throughout the ovulatory cycle, except for a brief period after the preovulatory surge of LH. If, as it would appear, the F_1 follicle has acquired sensitivity to LH, the question is raised as to why the first egg of a sequence fails to ovulate at the expected time, that is, 24–27 h after its predecessor, and is held in abeyance until the following day. This question is central to our ideas about the physiological basis of the timing of ovulation and will be addressed later.

iv   Follicular maturity

Various studies have shown that, in the last few days before ovulation, the progesterone-producing capacity of the granulosa cells of the F_1 follicle

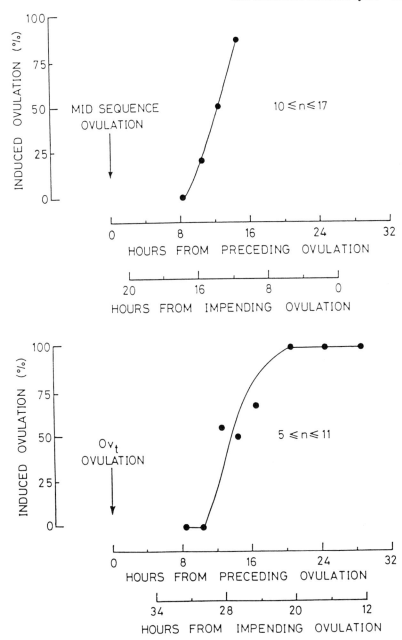

Fig. 3.4. Follicular maturation curve (cf Fig. 3.1) as determined by the ability of a follicle, destined to ovulate next, to be induced to ovulate prematurely in response to exogenous LHRH. The LHRH was administered either 8–16h after a mid-sequence ovulation (upper panel) or 8–28h after the terminal (Ov_t) of a sequence. (From Etches, 1984).

increases dramatically (for review see Etches 1984) whilst the synthesis of oestrogens and androgens by the thecal cells declines. The precise mechanism which causes maturation of the two cell types is unknown, although the increased LH responsiveness of the granulosa cells has been attributed to an increase in LH receptors. The decline in the capacity of the theca to produce androgens and oestrogens, however, is presumably associated with a decrease in the FSH binding activity of the tissue (Etches 1984). Ovulability, therefore, is closely associated with the capacity of the follicle to synthesize progesterone which, if this is associated with the acquisition of LH receptors, could lead to enhanced sensitivity of the follicle to an ovulation-inducing stimulus.

IV   Hormonal interrelationships during the ovulatory cycle

1   TEMPORAL RELATIONSHIPS BETWEEN PLASMA HORMONE CONCENTRATIONS

Since the advent of radioimmunoassay many independent studies have established the changes in the plasma concentrations of the reproductive hormones during the ovulatory cycle. Fig. 3.5, which is a diagrammatical representation of the pattern of hormone secretion before and after a mid-sequence ovulation, shows that peaks in the plasma concentrations of LH, progesterone, oestradiol and testosterone precede ovulation by 3–6 h (Furr, Bonney, England, and Cunningham 1973; Wilson and Sharp 1973; Senior and Cunningham 1974; Shahabi et al. 1975; Etches and Cunningham 1976a, 1977; Johnson and van Tienhoven 1980; Etches and Cheng 1981). Increases in the secretion of oestradiol and testosterone can be first observed as early as 18–14 h before ovulation, and this is consistent with the finding of peaks in the follicular concentrations of these steroids but not of progesterone at 20–18 h before ovulation (Shahabi et al. 1975). Plasma concentrations of testosterone start to increase 14–12 h before ovulation (Etches and Cunningham 1977; Johnson and van Tienhoven 1980), but a rise in progesterone is not seen until shortly before the preovulatory release of LH, that is, about 8 h before ovulation (Etches and Cunningham 1976a). After reaching a preovulatory maximum, plasma concentrations of gonadal steroids fall to basal values by the time of ovulation.

Changes in the plasma concentration of LH during the ovulatory cycle consist of cyclic preovulatory and postovulatory increases in LH superimposed on a diurnal rhythm of tonic LH secretion. The diurnal rhythm of tonic LH secretion is particularly evident in 4–15 week old hens, that is, before the onset of lay, as a 30–100 per cent increase in the concentration of LH at the onset of darkness (Fig. 3.6), which is maintained until 6 or 12 h

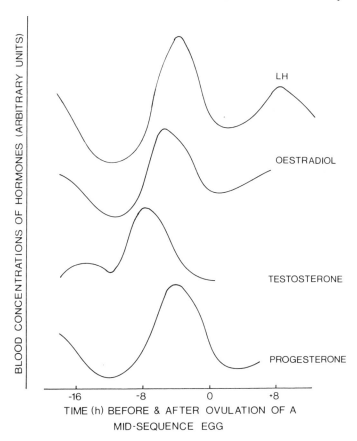

Fig. 3.5. Diagramatic representation of blood hormone profiles in hens in relation to a mid-sequence ovulation. Note particularly the preovulatory rises in the concentration of each of the hormones and also the preovulatory fall and postovulatory rise in LH.

after the onset of darkness in hens maintained on schedules of 16L:8D and 8L:16D respectively (Wilson, Jennings, and Cunningham 1983). By 19 weeks of age, shortly before the onset of lay, the diurnal rhythm of LH secretion changes to resemble the pattern observed during the day of "missed ovulation" in the adult hen, that is, a transient increase in the plasma concentration of LH at the onset of darkness followed by a decline in LH until a further increase occurs at about 14 h after the onset of darkness. The cyclic pattern of LH changes consists of a 1-3 fold preovulatory increase in the plasma concentration of LH, followed by a decline to the time of ovulation (Fig. 3.7a). Falling LH levels during the 3–4 h before ovulation are associated with a refractoriness of the pituitary gland to stimulation by

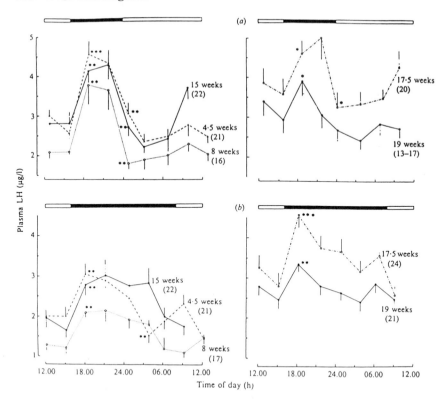

Fig. 3.6. Effect of age and the cycle of light and darkness on plasma LH concentrations in hens maintained in (a) 16h light : 8h darkness, or (b) 8h light : 16h darkness. The black horizontal bars indicate the periods of darkness. Values are means ± S.E.M. and the numbers in parentheses are the number of estimates per mean for each group. *P<0.05, **P<0.01, ***P<0.001 compared with the preceding value. (Reprinted from Wilson *et al.* 1983).

LHRH (Bonney *et al.* 1974). A further more gradual increase in the plasma concentration of LH occurs during the 8 h after ovulation (Fig. 3.7a) irrespective of the position of the ovulation in the sequence and is related to the time of ovulation rather than to the cycle of light and darkness. In hens which have ovulated an egg early in a sequence the postovulation increase in the plasma concentration of LH is followed by a significant decline in LH between 8 h after ovulation and the onset of darkness (Fig. 3.7a). Normally the preovulatory release of LH occurs subsequent to the period when plasma concentrations of LH are at their lowest in the cycle (Cunningham and Furr 1972; Etches and Cheng 1981; Wilson *et al.* 1983). In hens maintained on 16L:8D which have ovulated the terminal egg of a sequence,

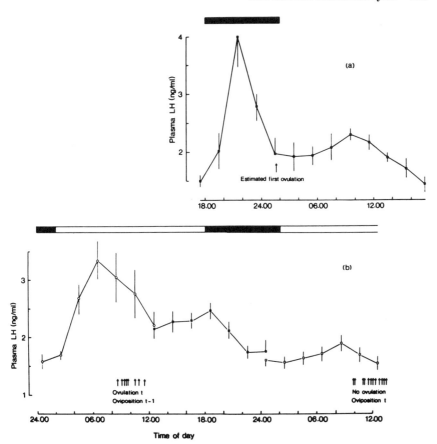

Fig. 3.7. Plasma LH concentrations (means ± S.E.M.) in hens in relation (a) to the time of the first ovulation of a sequence and (b) during the days of the terminal ovulation (ovulation t) and missed ovulation of a sequence (cf Fig. 3.1). The arrows indicate the times of oviposition in hens during the study period. (Modified from Wilson *et al*. 1983).

the postovulation increase in the plasma concentration of LH may not attain its maximum before being interrupted by the diurnal decline shortly after the onset of darkness (Fig. 3.7b). The possible significance of this in the termination of a sequence of ovulations is discussed later.

Because of difficulties in the establishment of a reliable assay system for FSH there is a dearth of information on the fluctuation of this hormone during the ovulatory cycle. Scanes, Godden, and Sharp (1977), however, found an increase in the plasma concentration of FSH between 16 and 14 h before ovulation. Using a bioassay procedure Imai and Nalbandov (1971) observed increases at 11 and 10 h before ovulation.

Although FSH is considered to be principally responsible for follicular growth there is evidence that it might also stimulate steroidogenesis (Scanes and Fagioli 1980; Hammond *et al.* 1981), possibly, as in mammals (Richards 1979; Leung and Armstrong 1980), by synergism with oestradiol which has been observed to increase both in plasma and in the wall of F_2 and F_3 follicles (Shahabi *et al.* 1975; Doi, Takai, Nakamura, and Tanabe 1980; Gulati, Nakamura, and Tanabe 1981), the interaction leading to the formation of LH receptors. There is evidence for a stimulatory action of FSH on the formation of gonadotrophin receptors in the male Japanese quail (Ishii, Tsuksui, and Adachi 1978), and this could explain the augmentative effect of exogenous FSH on LH-induced ovulation in the hen (Kamiyoshi and Tanaka 1972). Evidence for an increase in LH receptor content in the hen is provided by the observation that an extract of anterior pituitary tissue is more effective in stimulating the secretion of ovarian progesterone at 14 h than at 25 h before ovulation. Also, the injection of hens with LHRH or LH stimulates steroidogenesis at 12 but not at 6.5 h after the previous ovulation (Etches and Cunningham 1976*a*). A stimulation of the formation of LH receptors would be consistent with the observation that LH-stimulable adenylate cyclase activity increases in the largest ovarian follicle 18 and 12 h before ovulation (Calvo, Wang, and Bahr 1981).

2   GONADAL STEROID FEEDBACK ACTION ON LH RELEASE

The preovulatory increases in the plasma concentrations of oestradiol, progesterone and testosterone occur more or less simultaneously during the ovulation cycle. Consequently there are difficulties in assigning a role based solely on temporal relationships to any particular steroid in the LH release mechanism. Various approaches have been used to examine the question and these have involved treatment of intact or ovariectomized hens with gonadal steroids, passive immunization against steroids, or use of tamoxifen, a synthetic anti-oestrogen.

Although complete ovariectomy is difficult to perform in the hen, treatment of the oestrogen-progesterone primed ovariectomized hen (Wilson and Sharp 1976*c*) or the intact hen at an appropriate stage of the ovulatory cycle (Wilson and Sharp 1975; Etches and Cunningham 1976*a*) with a single injection of progesterone stimulates a release of LH comparable in magnitude and duration to the normal preovulatory release (Fig. 3.8). Testosterone only stimulates LH release if given at a time when the ovary contains a ripe follicle (Wilson and Sharp 1976*b*), and oestradiol has not been found to stimulate the release of LH under any circumstances (Wilson and Sharp 1976*b*, *c*). These results, together with the observation that hens which had been injected 14 h before with an antiserum raised

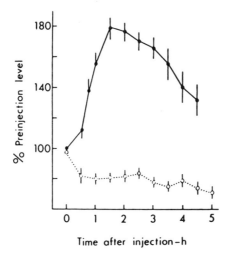

Fig. 3.8. Effect of a single injection of progesterone on plasma LH in ovariectomized hens (○) and in ovariectomized hens pretreated with a combinations of oestradiol and progesterone (●). The response in the oestrogen-progesterone treated hens resembles the normal preovulatory release. (From Wilson and Sharp 1976c).

against progesterone failed to ovulate (Furr and Smith 1975), suggest that progesterone is responsible for stimulating the preovulatory release of LH. Furthermore exogenous progesterone will stimulate a release of LH at any time during the ovulatory cycle except in the immediate preovulatory period, that is, after the surge of LH has occurred (Wilson and Sharp 1975) and the pituitary gland is refractory to LHRH (Bonney et al. 1974). Wilson and Sharp (1975) also found that, at a time when LH levels are depressed 12–18 h before a mid or late ovulation of a sequence, the LH response to progesterone was less than that observed when progesterone was given at other times of the cycle. The LH response to progesterone is most marked when hens are treated at the time of the preovulatory rises in LH and gonadal steroids. Another approach was used by A. L. Johnson, and van Tienhoven (1984), and Lang, Etches, and Walton (1984), who showed that progesterone could overcome the ovulation blocking action of aminoglutethimide, an inhibitor of steroidogenesis, and restore the preovulatory release of LH.

The preovulatory release of LH is produced by a "cascade effect" whereby progesterone stimulates the release of LH which in turn stimulates a further increase in the secretion of progesterone (Etches and Cunningham 1976a). Progesterone exerts its positive feedback effect on LH release principally by acting at receptor sites within the anterior hypothalamic-preoptic region (Opel 1979; van Tienhoven 1981). This action is through LHRH, since passive immunization against LHRH suppresses the progesterone-induced release of LH (Fraser and Sharp 1978). The progesterone-induced release of LH is associated with a significant decrease in the immunoreactive LHRH content of the posterior hypothalamus (Knight, Francis, Holman, and Gladwell 1982a). Presumably this reflects an hypothalamic site of action since receptors for progesterone have been detected in the hypothalamus as well as in the pituitary gland, and the receptor content of both tissues peaks at 18 h and 8 h before ovulation (Kawashima, Kamiyoshi, and Tanaka 1979). An important finding using a double immunohistochemical technique is that nuclear receptors for progesterone do not occur in LHRH neurones although they do occur in cells close to LHRH cell bodies and fibres (Sterling, Gase, Sharp, Tuohimaa, and Baulieu 1984). Thus it seems that progesterone, whilst being the positive feedback signal for LH release, does not act through a genomic mechanism within the LHRH neurone. Indeed the fact that exogenous progesterone can promote an LH response within 30 min suggests that the steroid may exert a direct effect on the membrane of LHRH or other neurones, and thereby modify their activity through a non-genomic mechanism.

Wilson and Sharp (1976c), in an impressive series of experiments using ovariectomized hens, showed that the capacity of progesterone to stimulate the release of LH is dependent on the attainment of critical blood concentrations of both oestradiol and progesterone. A single injection of progesterone fails to stimulate the release of LH in the unprimed ovariectomized hen or in the ovariectomized hen previously primed with either oestradiol or progesterone alone. However, if ovariectomized hens are given a treatment consisting of a combination of critical quantities of oestradiol and progesterone, LH concentrations in plasma are depressed to values comparable with those seen in laying hens. In these circumstances the further administration of progesterone provokes a release of LH similar in magnitude and duration to that seen in laying hens (Fig. 3.8). Oestradiol stimulates the production of progesterone receptors in the hypothalamus, and pituitary gland (Kawashima et al. 1979), and has been found to have a short term facilitatory effect on the responsiveness of the anterior pituitary gland of the immature hen to LHRH both in vivo and in vitro (Bonney and Cunningham 1977). It is likely that the enhanced release of LH in response to an injection of progesterone at a time when the preovulatory levels of LH are rising, is due to a sensitizing effect of the increased secretions of endogenous gonadal steroids.

The importance of oestradiol in the regulation of LH secretion has also been demonstrated in experiments involving the treatment of laying hens with the synthetic anti-oestrogen, tamoxifen. Since ovulation was not prevented either by a single injection of tamoxifen (Wilson and Cunningham 1981a), or by a passive immunization using an anti-oestradiol serum (Furr and Smith 1975), it would appear that oestradiol is not essential to the immediate ovulation. In contrast injection of higher dosages of tamoxifen on two successive days depressed or abolished the preovulatory release of LH and delayed or prevented the occurrence of ovulation. This was associated with a diminished response to progesterone. This same treatment also raised basal LH concentrations in association with an increased responsiveness of the pituitary gland to LHRH (Wilson and Cunningham 1981a). These findings using the anti-oestrogen suggest that oestradiol has at least two roles in the regulation of LH release in the hen. First it maintains a low basal level of LH in the blood by reducing the responsiveness of the pituitary gland to LHRH, and secondly it has a facilitative role in the mechanism by which progesterone stimulates the preovulatory release of LH. Progesterone also has a dual action in that it can depress tonic LH release while being able to induce cyclic LH release. Thus treatment of hens with progesterone initially depresses LH before exerting a stimulatory effect. There is also evidence that progesterone and oestradiol act synergistically to depress tonic LH secretion (Wilson and Sharp 1976c) not only at the pituitary gland but also at negative feedback sites within the hypothalamus to modify LHRH activity (Knight et al. 1983). The hypothalamic regions which may be involved in the control of tonic LH secretion include the paraventricular nucleus at the mediobasal hypothalamus, and catecholaminergic mechanisms may mediate this effect (Knight, Gladwell, and Cunningham 1981).

The role of testosterone during the ovulatory cycle has not been defined. The release of LH is stimulated by testosterone only when a ripe follicle is present in the ovary (Wilson and Sharp 1976b). The secretion of testosterone may be involved in changes within the ovary or the hypothalamus-pituitary system which set in train the hormonal events leading to ovulation. Supporting evidence for a role for testosterone in the hen is provided by the fact that passive immunization of laying hens against the steroid could delay or block ovulation (Furr and Smith 1975).

3   MODULATION OF LHRH NEURONAL ACTIVITY

The avian hypothalamus contains a network of catecholamine- and 5-hydroxytryptamine (5HT)-containing fibres and terminals (Sharp, Mac-Namee, Talbot, Sterling and Hall, 1984). In the preoptic area of the hypothalamus these correspond with the distribution of LHRH cell bodies, and terminals containing LHRH occur in the median eminence in close proximity to those containing catecholamines. The anatomical evidence, therefore, suggests that monoamines could regulate the release of LHRH. Indeed it has long been known that the adrenoceptor blocking drug dibenamine suppresses ovulation in the hen (van Tienhoven, Nalbandov and Norton, 1954). Furthermore, the progesterone-induced rise in plasma LH is accompanied by significant increases in the concentrations of

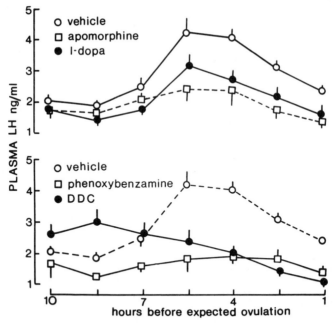

Fig. 3.9. Pharmacological manipulation of the preovulatory release of LH in hens using various drugs which affect central catecholaminergic neurotransmission (see text). Diethyldithiocarbamate (DDC) and α-methylparatyrosine (α-MPT) were administered 10.5 h before the expected time of ovulation, 1-dihydroxyphenylalanine (1-Dopa) 9.5 h before and apomorphine and phenoxybenzamine 8.5 h before. Values are means ± S.E.M. (n = 6). (Modified from Knight *et al.* 1982*b*).

dopamine and noradrenaline in the anterior hypothalamus-preoptic region (Knight *et al.* 1982*a*), which has been previously implicated in the neuroendocrine regulation of reproduction in birds (Opel 1980). Further evidence of a catecholaminergic involvement in the release of LHRH comes from the use of pharmacological precursors, synthesis blockers, antagonists and agonists (Buonomo, Rabii, and Scanes 1981; Knight, Wilson, Gladwell, and Cunningham 1982*b*). For instance, treatment with diethyl-dithiocarbamate, an inhibitor of dopamine β-hydroxylase, increased hypothalamic dopamine content two-fold and this was associated with a predominantly inhibitory action on LH secretion and blocked ovulation in four out of six hens. Similarly, treatments with the catecholamine precursor L-Dopa which also increased hypothalamic dopamine, significantly reduced the preovulatory release of LH but did not block ovulation (Fig. 3.9). Furthermore, stimulation of dopamine receptors with the agonist apomorphine markedly attenuated the preovulatory release of LH. Assuming that these inhibitory effects on LH secretion are acting centrally rather than at the level of the pituitary gland, then these findings are evidence for the existence of a dopaminergic system which is inhibitory to the release of LHRH. The failure of α-methyl-p-tyrosine, which caused a 60 per cent depletion of hypothalamic dopamine content, to modify the preovulatory LH surge suggests that decreased basal dopaminergic activity is not essential for mediation of the surge.

An involvement of an α-adrenergic system in mediating the preovulatory surge of LH is suggested by the observation that phenoxybenzamine, an α-adrenoceptor blocker, markedly inhibited the preovulatory release of LH (Fig. 3.9). This suggests that the positive feedback mechanism which leads to the preovulatory surge of LH may depend on the activation of α-adrenoceptors at central noradrenergic and/or adrenergic synapses. Surprisingly, although phenoxybenzamine markedly reduced the magnitude of the LH surge, five out of six hens ovulated at the expected time. This implies that the amount of LH normally released during the preovulatory surge exceeds the amount required for ovulation.

These conclusions about possible functional interactions between hypothalamic catecholaminergic neurones and LHRH producing neurones are strengthened by the finding that changes in the concentrations of dopamine, noradrenaline, adrenaline and LHRH in the hypothalamus can be related in time to the preovulatory release of LH (Knight, Wilson, Gladwell, and Cunningham 1984). Fig. 3.10 shows that, although no statistically significant fluctuations in the steady state concentrations in the anterior hypothalamus or posterior median eminence were detected during the ovulation cycle, the data indicate a role for catecholamines in modulating LHRH neuronal activity, since the concentrations of LHRH were significantly correlated with those of noradrenaline, adrenaline and dopamine. In both hypothalamic regions the preovulatory fall in LHRH content, which occurred simultaneously with the rise in the plasma LH concentration, was accompanied by simultaneous reductions in the contents of each of the three amines. Similarly, the subsequent rise in LHRH content was accompanied by increased amine contents. The fact that the contents of the three amines appear to change in synchrony during the study period suggests that the three catecholamines may coexist within the same population of neurones. The interpretation of steady-state concentrations, however, is complicated by the fact that fluctuations in levels may reflect alterations in the synthesis, storage and release and/or reuptake of transmitter. Ideally turnover studies of the amines are required, and in this regard Sharp et al. (1984) have reported that the turnover rates of dopamine and 5-HT decrease when preovulatory LH levels are at their highest. Recent findings implicate the natural opioid system in the mechanisms regulating LHRH secretion and hence LH release in the domestic fowl. The intracerebroventricular administration of a long acting analogue of metenkephalin significantly suppresses plasma concentrations of LH in laying hens, an effect which is reversible by naloxone. Also naloxone blocks

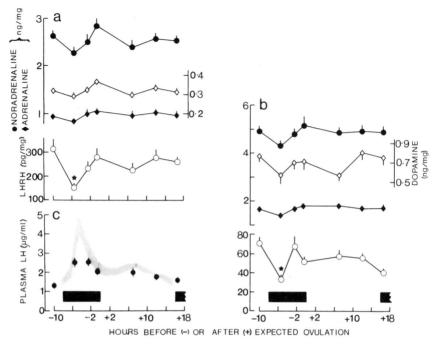

Fig. 3.10. Concentrations (means ± S.E.M. n = 12) of noradrenaline, adrenaline, dopamine and LHRH in (a) posterior hypothalamus-median eminence, (b) anterior hypothalamus-preoptic region, and (c) of LH in plasma of hens killed at different times before or after the expected time of ovulation. In (c) the hatched area represents the plasma LH concentrations of a group of hens sampled at 2 h intervals for 24 h. The black horizontal bars indicate the periods of darkness. *P<0.05 compared with the preceding value. (From Knight *et al.* 1984).

the inhibitory action of metenkephalin on the depolarization-induced release of LHRH in a hypothalamic superfusion system (S. C. Stansfield and F. J. Cunningham unpublished observations).

V   Control of ovulation rate

The rate of ovulation is dependent not only on the rate of follicular growth but also on the ability of a follicle to mature within the 8–10 h period of the day during which preovulatory releases of LH are normally initiated. The rate of follicular growth is affected to a large extent by age, being slower in older than in young hens, and by photoperiod. Whilst FSH is considered to be important in follicular growth (Hammond *et al.* 1981), a significant correlation between the rate of ovulation and plasma concentrations of LH has been reported both for hens maintained on the same photo-period (Wilson 1978), and for hens maintained on photoperiods known to have different gonad stimulating effects (Wilson 1982, Wilson and

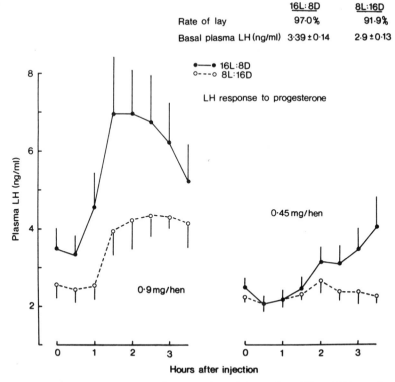

	16L:8D	8L:16D	
Rate of lay	97·0%	91·9%	n=40
Basal plasma LH (ng/ml)	3·39 ± 0·14	2·9 ± 0·13	

Fig. 3.11. Effect of long-days (16L : 8D) and short-days (8L : 16D) on the mean rate of lay (number of eggs laid from 173–197 days of age as a percentage of hen days) and on basal plasma LH concentrations at 188 days of age (means ± S.E.M.). The graphs show the changes in plasma concentrations of LH in six hens after an i.m. injection at 8 h after ovulation of 0.45 or 0.9 mg progesterone. (Reprinted from Wilson and Cunningham 1984).

Cunningham 1980*a*). The rate of lay in hens maintained on a schedule of 8L:16D is significantly less than in hens of a similar age maintained on 16L:8D (Fig. 3.11). This is associated with a significantly lower plasma concentration of LH and a reduced effectiveness of exogenous progesterone to stimulate the release of LH, which suggests that short photoperiods depress the functional activity of the hypothalamus-pituitary-gonad system. In immature hens transferred from 8L:16D to 16L:8D the plasma concentration of LH increases within 24 h, and since this is not associated with an increased responsiveness of the pituitary gland to exogenous LHRH, it must be assumed that the release of LH stimulated by increased daylength is mediated by an increased activity of LHRH secretory neurones.

    There is considerable evidence that the gonad stimulating effects of long photoperiods in the domestic hen (Wilson 1982) as well as in the turkey (Bacon and Nestor 1980) and Japanese quail (Follett 1981; Wada 1981) can be achieved by the substitution of a complete long photoperiod by short pulses of light at the appropriate time of the day. This can be explained by the fact that in the hen, as in many species of plants and animals, a circadian rhythm of sensitivity to light exists whereby the stimulatory effect of light varies according to the time of day when it is applied (Bünning 1936; Pittendrigh and Minis 1964). A physiological system which is circadian in nature must respond to changes in the light-dark cycle so that the phase of the environmental cue and of the physiological system, that is, the phase angle, retain a constant relationship. Experiments to determine the phase angle of the rhythm of sensitivity to light have involved a comparison of rates of lay in hens maintained on a skeleton photoperiod consisting of a main photoperiod and a further short period or pulse of light provided at different times during the hours of darkness (van Tienhoven and Ostrander 1976). The results of such experiments, however, have been confounded by the fact that the hens entrain their rhythm of photosensitivity to the pulse of light rather than to the main photoperiod, and the likelihood of this increases the later the pulse of light is applied relative to the onset of darkness in the main photoperiod. Wilson (1982) overcame the problem by studying the effects of skeleton photoperiods on the release of LH immediately after the imposition of the lighting treatments, that is, before re-entrainment can occur. The effect of photoperiod on the release of LH is most pronounced in the immature hen, the pituitary gland of which is more responsive to LHRH than is the laying hen (Bonney et al. 1974; Wilson and Sharp 1975). Thus if 18 week-old hens kept under a schedule of 8L:16D are transferred to 4L:16D with a further period of 4 h light provided at different times during the hours of darkness, then the 4 h light stimulus is most effective at releasing LH when given between 5 and 9 h after the end of the main photoperiod, that is, the point of maximum sensitivity under this schedule occurs 7 h after experimental dusk. The two 4 h periods of light provided in the most stimulatory skeleton schedule of 4L:5D:4L:11D are nearly as effective in stimulating LH release as a complete 16L:8D schedule (Wilson 1982). In hens subjected to a skeleton photoperiod consisting of a main period of 7.75 h light and a pulse of 0.25 h light provided at different times during the hours of darkness, again the period of maximum sensitivity to light occurs 6–8 h after the end of the main photoperiod even though its position in relation to the previous experimental dawn has been delayed. These relationships indicate that the circadian rhythm of sensitivity to light is entrained principally to dusk rather than to dawn. In contrast, the phase of maximum sensitivity in the Japanese quail occurs about 14 h after dawn and is entrained principally to dawn rather than to dusk (Follett 1981; Wada 1981).

VI   Timing of LH release and ovulation

The 8–10 h period of the day during which ovulations normally occur in hens maintained under standard conditions of light and darkness was referred to by

Fraps (1955) as the "open period". Since ovulation is preceded 8–4 h earlier by increases in the plasma concentrations of LH and gonadal steroids, the time during which the preovulatory release of LH normally occurs is also generally referred to as the "open period" (Fig. 3.1).

In hens maintained under continuous light, however, ovulations can occur at times throughout the 24 h day (Morris 1977; Wilson and Cunningham 1981*b*). The rate of lay of these hens, however, is not higher than those maintained on a standard photoperiod of 14L:10D (Morris 1973), since the interval between successive releases of LH in hens on continuous light is generally greater than the interval between mid-sequence LH releases in hens on a cycle of light and darkness. This perhaps suggests that under standard conditions of light and darkness, undefined hormonal events occur during the 8–10 h open period of the day which modify maturation of the developing follicle.

The open period possesses circadian properties in that its phase-angle within the cycle of light and darkness is determined by the duration of light and darkness. This is illustrated in Fig. 3.12 where the relationships of the open period of hens on different lighting schedules to the cycle of light and darkness is shown. There is a delay in relation to the onset of darkness of both the preovulatory release of LH associated with the first ovulation of the sequence, and the distribution of ovipositions as the period of darkness extends beyond 8 h. Releases of LH associated with subsequent ovulations of a sequence occur later in relation to the onset of darkness by the period of lag observed between successive ovulations (see Fig. 3.1).

In hens maintained on continuous light ovulation and oviposition occurs at intervals of 25–27 h throughout the 24 h solar day (Wilson and Cunningham 1981*b*), although for several days after the transfer of hens from a cycle of light and darkness to continuous light, ovulations continue to be synchronized within the flock. This is particularly noticeable in prolific layers and probably reflects the constancy between hens in the period of lag between successive ovulations. A completely uniform distribution of ovulations/ovipositions throughout the 24 h day in hens maintained on continuous light is not observed due to entraining influences of environmental stimuli other than the cycle of light and darkness (Morris 1977).

In the hen the open period is entrained primarily to dusk with a lesser influence exerted by dawn (Bhatti and Morris 1978). For example if hens maintained on 16L:8D are transferred to a schedule of 20L:4D in which dawn is advanced by 4 h, the onset of the open period is advanced in relation to dusk by only 30 min as judged by the mean time of oviposition. However, if the 20L:4D schedule is achieved by a 4 h delay of dusk, the timing of the open period is delayed by 3 h 8 min (Fig. 3.13).

As a further test of circadian rhythmicity the open period in the hen can be entrained to ahemeral cycles, that is, cycles less than or greater than 24 h, between 21 and 30 h (Biellier and Ostmann 1960), but on cycles outside this range becomes free running.

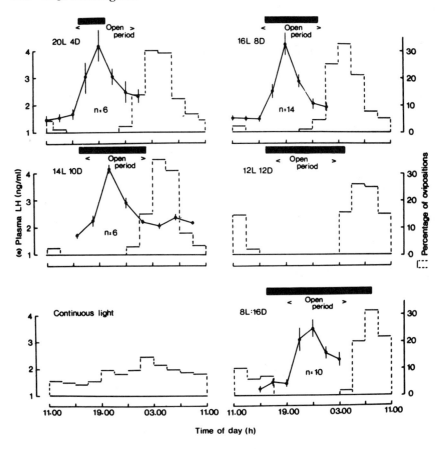

Fig. 3.12. Effect of six photoperiods on the timing of the first preovulatory LH release of a sequence in hens and on the distribution of ovipositions during 14 days in 24 hens per photoperiod. The black horizontal bars indicate the periods of darkness. LH concentrations are means ± S.E.M. (Reprinted from Wilson and Cunningham 1984).

VII   Physiological basis of the ovulation cycle

1   INTERACTION OF TWO INDEPENDENT RHYTHMS

A feature which distinguishes the reproductive physiology of the domestic hen is that any hypothesis which attempts to explain the neuroendocrine control of the ovulation-oviposition sequence must include considerations of the daily lag between ovulations, and the missed ovulation which terminates the clutch sequence (Fig. 3.1). Over the years various proposals have been made to account for these events. For instance Bastian and Zarrow (1955) postulated the existence of two independent rhythms which

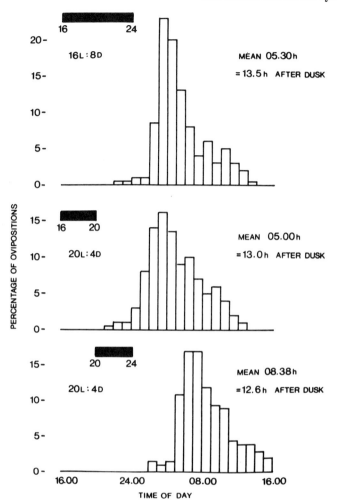

Fig. 3.13. Effect of lengthening the photoperiod from 16h : 8D to 20L : 4D by advancing dawn or by delaying dusk on the distribution of ovipositions and mean time of oviposition. Results are presented for ovipositions of 48 hens during days 8–10 after the change in lighting. The black horizontal bars indicate the periods of darkness. (Reprinted from Wilson and Cunningham 1984).

interact to produce the asynchronous ovulatory cycle. They considered that LH was released over a period of 8 h each night and that a follicle matured every 26 h. When these two events coincided ovulation occurred. According to this concept, lag is caused by successive follicles maturing later in the period of LH release until the two events are not coincident and ovulation fails to occur. The mature follicle in the ovary is duly ovulated during the period of LH release the next day. The development of radioimmunoassay procedures to measure plasma hormone concentrations, however, revealed that LH is not released for a prolonged period and furthermore a preovulatory release is not seen on the day of missed ovulation (Fig. 3.1).

2   DAILY VARIATION IN EXCITABILITY OF A NEURAL COMPONENT

The hypothesis which is still most frequently quoted to account for the events of the ovulatory cycle of the hen was proposed by Fraps (1954), who recognized the circadian or diurnal characteristics of the open period for LH release. He proposed a diurnal variation in the threshold of a neural component involved in the release of LH. He considered that the neural component was stimulated by a factor which was present, during the period of elevated threshold, in insufficient quantities to activate the LH release mechanism and so induce ovulation. The factor which Fraps termed the "excitation" hormone was thought to be secreted in increasing quantities by the maturing ovarian follicle. When the threshold of the neural component was sufficiently reduced, during the open period, to be activated by the "excitation" hormone, LH release was stimulated and ovulation followed. Since successive follicles mature at intervals of 24 h plus lag, the increase in the level of the "excitation" hormone would follow a similar time course. Consequently, the neural component would be stimulated at a progessively later time each day. Eventually the level of the "excitation" hormone would be maximal at a time when the threshold of the neural component was too high, and the release of LH would not be stimulated until the following day when the threshold decreased again. Progesterone was considered to be the likely "excitation" hormone since it was known to induce premature ovulation in laying hens (Fraps and Dury 1943; Rothchild and Fraps 1949).

The essence of this theory is that on the day of missed ovulation a gradual increase in the secretion of progesterone accompanies the cycle of follicular maturation. No such increase has been reported, although slightly higher concentrations of progesterone have been observed during the hours of darkness on the day of missed ovulation (Etches 1979).

Administration of progesterone outside the open period will induce an increase in the plasma concentration of LH, and in response to an adequate LH stimulus the ovary is able to synthesize and secrete steroids by 12 h after the previous ovulation (Shahabi et al. 1975; Etches and Cunningham 1976a). Thus the normal restriction of the spontaneous preovulatory release of LH to the open period of the day is not due to a failure of the positive feedback mechanism either at the hypothalamus-pituitary level or at the ovary. The findings, however, do not discount the possibility of the existence of diurnal changes in sensitivity to stimulation within the ovary or hypothalamus-pituitary system, since this has not been adequately tested. Indeed appropriate experimental models need to be devised to examine these possibilities.

3 RISE IN PLASMA LH AT THE ONSET OF DARKNESS

Before the first ovulation of a sequence a transient increase in plasma LH is seen at the onset of darkness in hens maintained under lighting schedules of 14h:10D (Wilson and Sharp 1973; P. A. Johnson and van Tienhoven 1984) and 16L:8D (Wilson and Cunningham 1981b). It has been claimed that this rise in LH, referred to as the crepuscular LH peak, is implicated in initiating the preovulatory events which culminate in ovulation (P. A. Johnson and van Tienhoven 1984). For this to hold true the beginning of the open period, during which the preovulatory releases of LH occur, might be expected to have the same relationship to the onset of darkness regardless of photoperiod. This is not the case since in hens on a short photoperiod (8L:16D) the beginning of the open period occurs some 3–4 h later than it does in hens kept under a photoperiod of 14L:10D (Fig. 3.12). Thus the small and transient change in LH at the onset of dark would appear to be unimportant, unless the delay in the onset of the open period in hens on a schedule of 8L:16D results from reduced activity of the hypothalamus-pituitary-ovary system which is reflected by a delay in the response to the crepuscular LH peak. This possibility is unlikely since, in hens on a schedule of 8L:16D, the long interval of 38–42 h between the LH release associated with the last ovulation of one sequence and the first of the next would tend to ensure that the follicle about to ovulate would be highly responsive to small changes in plasma LH. The possibility, however, cannot be excluded that the crepuscular peak is acting at hypothalamic-pituitary loci.

4 ROLE OF THE ADRENAL GLAND

The adrenal gland is known to be a circadian oscillator in many systems (Halberg, Halberg, Barnum, and Bitner 1959), and consequently much attention has been given to the possibility that it has a role in the timing of ovulation in the hen. Various anatomical and physiological observations support the proposition. The left adrenal gland is in close juxtaposition with the left ovary of the laying hen (Biswall 1954), and a dense network of adrenergic nerve fibres radiates from the adrenal-ovarian interface to the steroid producing cells within the thecal layer of the follicle (Unsicker, Seidel, Hofmann, Muller, Schmidt, and Wilson 1983). Furthermore, injections of either deoxycorticosterone or corticosterone will induce premature ovulation in hens (van Tienhoven 1961; Etches and Cunningham 1976b) through stimulation of LH release (Wilson and Sharp 1976b; Wilson and Lacassagne 1978). Treatment of hens with adrenocorticotrophic hormone (ACTH) can also induce premature ovulation (van Tienhoven 1961; Wilson and Lacassagne 1978), although when given at a time during which the largest ovarian follicle is too immature to be ovulated, ACTH causes a depression in the plasma concentration of LH associated with a reduction in responsiveness of the anterior pituitary gland to exogenous LHRH.

If corticosterone has a role in the control or timing of ovulation it might be expected that an increase in its secretion would precede the preovulatory release of LH. Such a temporal relationship in the preovulatory release of LH has been observed in the rat (Feder, Brown-Grant, and Corker 1971) and in some teleost fish (Cook, Stacey, and Peter 1980). In laying hens the secretion of corticosterone follows a daily circadian rhythm (Beuving and Vonder 1977; Etches 1979; Wilson and Cunningham 1980*b*; 1981*b*), the phase-angle of which is similar to that of the open period (Cunningham, Wilson, Knight and Gladwell 1984). In hens maintained on continuous light the circadian rhythm of corticosterone secretion is free running but again the timing of the preovulatory release of LH is related in time to the increase in the plasma concentration of corticosterone (Fig. 3.14). In a sequence of ovulations in hens kept under standard conditions of light and darkness the circadian rhythm in the secretion of corticosterone is modified during the sequence by the preovulatory changes in the secretion of progesterone, since during the period while the plasma concentrations of progesterone and LH are rising, concentrations of corticosterone decline (Wilson and Cunningham 1980*b*; 1981*b*). Increased corticosterone secretion also accompanies the occurrence of ovulation (Wilson and Cunningham 1980*b*) and oviposition (Beuving and Vonder 1977).

Experiments designed to modify adrenal steroid secretion in the hen have also led to the conclusion that corticosterone is probably involved in the timing of the preovulatory release of LH. Treatment of hens with dexamethasone, a synthetic corticosteroid which inhibits ACTH secretion and which leads to a reduction in the plasma concentration of corticosterone (Etches 1976; Wilson and Cunningham 1980*b*), blocks the preovulatory release of LH (Wilson and Lacassagne 1978) and ovulation (Soliman and Huston 1974; Wilson and Lacassagne 1978), an effect which can be overcome by an injection of ACTH. Metyrapone, a drug which inhibits 11β-hydroxylase activity in the adrenal gland (Liddle, Estep, Kendall, Carter Williams, and Townes, 1959; Kahnt and Neher 1962) and thus also reduces corticosterone production (Wilson and Cunningham 1980*c*), also modifies the timing of the preovulatory surge of LH such that LH is released at intervals of about 26–27 h at times through the 24 h day, instead of being confined to an 8–10 h period (Wilson and Cunningham 1980*c*). This could involve a modification of the diurnal rhythm of LH secretion and it was noticeable that in hens treated with either metyrapone or dexamethasone the plasma concentrations of LH did not fall as steeply during the hours of darkness as in untreated hens. This suggests that the

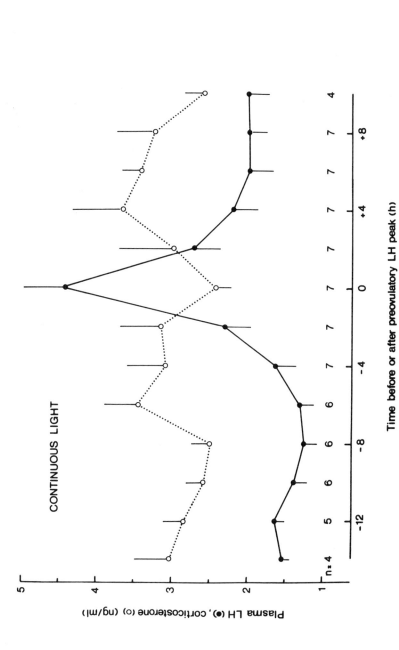

Fig. 3.14. Pattern of corticosterone secretion in relation to the preovulatory release of LH in hens maintained on continuous light. (Reprinted from Wilson and Cunningham 1984).

decline in the plasma concentrations of LH during the open period (Fig. 3.12) is of importance in the maturation of the follicle.

In a comprehensive study designed to establish the site of action of corticosterone in the hypothalamus-pituitary-ovary system Etches, Petite, and Anderson-Langmuir (1984) concluded that corticosterone had a number of actions. For instance the pituitary response to LHRH was not affected by infusions of corticosterone. However, corticosterone blocked the hypothalamic input to the pituitary gland, since exposure to corticosterone obliterated the photoperiodic response of hens transferred from short to long days. The steroid also had effects at the level of the ovary. This was concluded from the lack of ovarian growth in response to gonadotrophic stimulation in the presence of corticosterone. Further, in hens treated with aminoglutethimide, a drug which blocks steroidogenesis, ovulation occurred in the absence of preovulatory rises in LH and progesterone after treatment with corticosterone. The precise locus of action of corticosterone within the ovary however, remains unknown. Corticosterone does not alter the response of granulosa cells to LH. Assuming that ovulation occurred in aminoglutethimide-treated hens because the follicle was sensitized to the basal concentrations of LH, it would seem that in the short term corticosterone has the ability to enhance the sensitivity of the follicle to ovulation-inducing stimuli. This is consistent with the idea that falling concentrations of LH are of importance in the process of follicular maturation.

VIII Control of the ovulation cycle

Although the characteristic distribution of ovipositions of the domestic hen into a sequence has been well described, no entirely adequate explanation has been forthcoming for the physiological basis of the mechanism which restricts the occurrence of ovulation to an 8–10 h period of the 24 h day and which underlies the distribution of ovulations and ovipositions into sequences.

The fact that a distinct feature of the ovulation cycle is that the preovulatory release of LH nearly always occurs after a period of declining concentrations of LH in plasma, suggests that the diurnal rhythm in tonic LH secretion may be of significance in follicular maturation. Perhaps the fall in the plasma concentration of LH may sensitize the maturing follicle by increasing the availability of receptors for gonadotrophin, that is, "up-regulation" of LH receptors. The fall in the plasma concentration of LH is most pronounced towards the beginning of the open period and is related in time to the circadian increase in the plasma concentration of corticosterone (Wilson and Cunningham 1981b). A feature of the open period, therefore, may be an increase in the sensitivity of the maturing follicle to LH. This seems likely in view of the fact that the amount of chicken anterior pituitary extract required to induce premature ovulation at 17 and 8 h before the normally expected time decreased from 500 to 20 µg (Fraps 1955).

Perhaps, therefore, the increase in the plasma concentration of LH which occurs between 0 and 8 h after an ovulation stimulates the observed increase in synthesis and secretion of oestradiol from thecal cells (Shahabi *et al.* 1975). This may enhance the formation of granulosa cells and, in conjunction with an increase in the plasma concentration of FSH, stimulate the production of LH receptors on these cells. During the remaining period before the preovulatory release of LH associated with ovulation of the follicle the plasma concentration of LH falls and the consequent "up-regulation" of LH receptors may increase the sensitivity of the follicle to LH. At the beginning of the open period the plasma concentration of corticosterone rises in association with a steep fall in LH, and these changes perhaps increase the sensitivity of the maturing follicle to LH. Indeed Etches *et al.* (1984) concluded that, in the short term, corticosterone enhances the sensitivity of the ovary to stimulation. It is conceivable, therefore, that the sensitivity of the follicle is sufficiently increased during the open period that low circulating concentrations of LH are adequate to initiate an increase in steroidogenesis and hence trigger the cascade reaction which leads to the preovulatory release of LH.

It is possible that the facilitatory influences associated with the open period are graded, being minimal at the beginning and end and maximal at the mid-point. This would account for the extended lag between ovulations arising from LH releases occurring either early or late in the open period, and the comparatively small lag between ovulations resulting from LH releases initiated at the mid-point of the open period.

The distribution of ovulations and ovipositions in a sequence, however, can also be explained by the possibility that endocrine or neuroendocrine changes during the open period exert either a hastening or delaying effect on the maturation of the developing follicle, depending on the stage of maturity at which the follicle is exposed to these admittedly undefined hormonal changes.

This is suggested by experiments in which darkness was advanced between 4 and 13 h in hens maintained on a lighting schedule of 16L:8D and laying long sequences of eggs (Wilson, Jennings, and Cunningham 1985). The effects on subsequent ovulations and ovipositions were related to the degree of maturation of the largest follicle at the time of the advanced onset of darkness. In summary, LH release and ovulation were advanced if the onset of darkness occurred at greater than 7–8 h after the previous ovulation, and delayed or prevented if the onset of darkness occurred at less than 6–7 h after the previous ovulation. The mean times of oviposition in hens subjected to a 7 h advance in onset of darkness from lights out 16.00–24.00 h to lights out 09.00–17.00 h are given in Fig. 3.15. Thus if darkness occurred at greater than 7 h after the previous ovulation the next

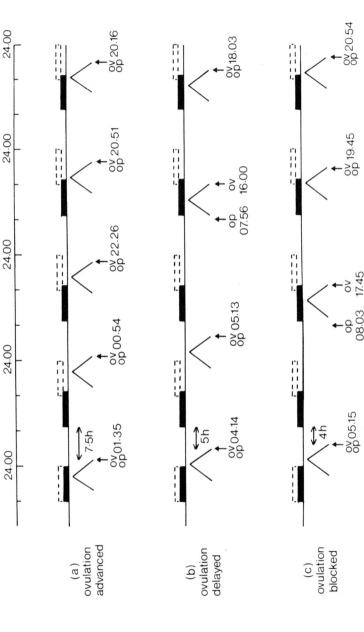

Fig. 3.15. Effect of advancing the onset of darkness by 7 h on the timing of the preovulatory release of LH (∧), ovulation (ov) and oviposition (op). Hens were laying long sequences of eggs and times of oviposition/ovulation were taken from hens which were ovulating mid-sequence eggs immediately before the change in lighting. In 12 hens (a) the advanced onset of darkness occurred 7–8 h after the previous ovulation and subsequent ovulations were advanced. When the advanced onset of darkness occurred less than 6 h after the previous ovulation the subsequent ovulation was delayed in 16 hens (b) and blocked in 11 hens as indicated by the day in which ovulation/oviposition failed to occur (c). (Reprinted from Wilson and Cunningham 1984; see also Wilson et al., 1985).

expected ovulation was marginally advanced and the subsequent ovulation was advanced by 2–3 h in accordance with the new position of the open period (Fig. 3.15a). On the other hand if darkness was advanced so that the onset occurred less than 6–7 h after the previous ovulation, the next expected LH increase and ovulation was delayed by 1 h and the subsequent LH release and ovulation prevented (Fig. 3.15b). This situation is reminiscent of the events which characterize the end of a normal sequence. Alternatively, an advance in the onset of the period of darkness exerted an immediate effect and blocked the next expected LH release and ovulation (Fig. 3.15c). In this case the preovulatory release of LH was delayed until the start of the next open period when the follicle would be sufficiently developed in its maturation to be facilitated rather than delayed by the hormonal changes during the open period. It may be significant that normally in hens maintained on 16L:8D the onset of darkness and the associated increase in plasma corticosterone, and decrease in plasma LH, occurs at less than 8 h after the final ovulation of a sequence (Wilson *et al.* 1983). It is, therefore, conceivable that the same hormonal changes associated with the open period which exert stimulatory effects on follicles late in their maturation also exert an inhibitory effect on follicles at a less mature stage. Again Etches *et al.* (1984) reported paradoxical actions of corticosterone in that the hormone either enhances or reduces the sensitivity of the follicle to stimulation. Thus endocrine or neuroendocrine factors associated with the open period would serve both to initiate and terminate a sequence of ovulations.

In conclusion just as the reproductive cycle in female mammals is the end result of a complex series of neuroendocrine interactions so is the sequence of ovulations and ovipositions in the hen. There are, however, a number of features which distinguish events in the hen from those in mammals. For instance in the mammalian ovary there is no hierarchy of yolk filled follicles such that those destined to ovulate two or three days hence can be easily identified. Another important ovarian structural and functional difference is that there is no comparable structure in birds to the mammalian corpus luteum. The avian post ovulatory follicle should not be considered as an alternative corpus luteum. Indeed in the hen the greatest amount of progesterone arises from the follicle about to ovulate. This has important implications with regard to the initiation of the ovulation-inducing surge of LH, since progesterone and LH interact in a cascade of events which lead to the preovulatory release of LH. In the mammal, by contrast, the preovulatory release of LH is brought about by an increased frequency of LH episodes and increased production of oestradiol by the preovulatory follicle. An episodic pattern of LH secretion has not been shown in the hen but perhaps this is a reflection of the lack of sensitivity of the assays for LH which are currently available. Also in notable contrast with mammals two distinct molecular species of LHRH have been characterized in chicken hypothalamic tissue. The two forms have similar biological potency with respect to the stimulation of LH secretion *in vivo* but the physiological significance of the duality of form has yet to be established.

Acknowledgements

I gratefuly acknowledge the financial support provided by both the Agricultural and Food Research Council and the British Egg Marketing Board Research and Education Trust for much of the research described here.

References

Armstrong, D. G. (1984). Ovarian aromatase activity in the domestic fowl. *(Gallus domesticus)*. *J. Endocr.* **100**, 81–86.

Bacon, W. L. and Nestor, K. E. (1980). Energy savings in turkey laying hens using an intermittent lighting program. *Poultry Sci.* **59**, 1953–1955.

Bahr, J. M., Wang, S. C., and Calvo, F. O. (1983). Sex steroid concentrations of isolated theca and granulosa layers of preovulatory follicles during the ovulatory cycle of the domestic hen. *Biol. Reprod.* **29**, 326–334.

Bastian, J. W. and Zarrow, M. X. (1955). A new hypothesis for the asynchronous ovulatory cycle of the domestic hen *(Gallus domesticus)*. *Poultry Sci.* **34**, 776–778.

Beuving, G. and Vonder, G. M. A. (1977). Daily rhythm of corticosterone in laying hens and the influence of egg laying. *J. Reprod. Fert.* **51**, 169–173.

Bhatti, B. M. and Morris, T. R. (1978). The relative importance of sunrise and sunset for entrainment of oviposition in the fowl. *Br. Poult. Sci.* **19**, 365–371.

Biellier, H. V. and Ostmann, A. W. (1960). Effect of varying day-length on time of oviposition in domestic fowl. *Bull. Mo. agric. Exp. Stn.* No. 747, pp. 1–52.

Biswall, G. (1954). Additional histological findings in the chicken reproductive tract. *Poultry Sci.* **33**, 843–851.

Bonney, R. C. and Cunningham, F. J. (1977). The sensitizing effect of oestradiol on the response of the anterior pituitary gland of the domestic fowl to luteinizing hormone releasing hormone *in vitro* and *in vivo*. *J. Endocr.* **72**, 16P.

—— —— and Furr, B. J. A. (1974). Effect of synthetic luteinizing hormone releasing hormone on plasma luteinizing hormone in the female domestic fowl *(Gallus domesticus)*. *J. Endocr.* **63**, 539–547.

Bünning, E. (1936). Die endogene Tagesrhytmik als Grundlage der photoperiodische Reaktion. *Ber. d. bot. Ges.* **54**, 590–607.

Buonomo, F., Rabii, J., and Scanes, C. G. (1981). Aminergic involvement in the control of luteinizing hormone secretion in the domestic fowl. *Gen. comp. Endocr.* **45**, 162–166.

Burke, W. H., Licht, P., Papkoff, H., and Bono Gallo, A. (1979). Isolation and characterization of luteinizing hormone and follicle-stimulating hormone from pituitary glands of the turkey *(Meleagris gallopavo)*. *Gen. comp. Endocr.* **37**, 508–520.

Calvo, F. O., Wang, S. C., and Bahr, J. M. (1981). LH stimulable adenlyl cyclase activity during the ovulatory cycle in granulosa cells of the three largest follicles and the postovulatory follicle of the domestic hen *(Gallus domesticus). Biol. Reprod.* **25**, 805–812.

Cook, A. F., Stacey, N. E., and Peter, R. E. (1980). Periovulatory changes in serum cortisol levels in the goldfish *(Carassius auratus). Gen. comp. Endocr.* **40**, 507–510.

Culbert, J., Hardie, M. A., Wells, J. W., and Gilbert, A. B. (1980). Effect of ovine LH on the progesterone content of granulosa cells in preovulatory follicles of the domestic fowl. *(Gallus domesticus). J. Reprod. Fert.* **58**, 449–453.

Cunningham, F. J. and Furr, B. J. A. (1972). Plasma levels of luteinizing hormone and progesterone during the ovulatory cycle of the hen. In *Egg formation and production* (ed. B. M. Freeman and P. E. Lake) pp. 51–64. British Poultry Science Ltd., Edinburgh.

—— Wilson, S. C., Knight, P. G. and Gladwell, R. T. (1984). Chicken ovulation cycle. *J. exp. Zool.* **232**, 485–494.

Dahl, E. (1970). Studies of the fine structure of ovarian interstitial tissue. 2. The ultra-structure of the thecal gland of the doemstic fowl. *Z. Zellforsch.* **109**, 195–211.

Doi, O., Takai, T., Nakamura, T., and Tanabe, Y. (1980). Changes in the pituitary and plasma LH and follicular progesterone and estradiol and plasma testosterone and esterone concentrations during the ovulatory cycle of the quail *(Coturnix coturnix japonica). Gen. comp. Endocr.* **41**, 156–163.

Etches, R. J. (1976). A radioimmunoassay for corticosterone and its application to the measurement of stress in poultry. *Steroids,* **28**, 763–773.

—— (1979). Plasma concentrations of progesterone and corticosterone during the ovulation cycle of the hen (*Gallus domesticus). Poultry Sci.* **58**, 211–216.

—— (1984). Maturation of ovarian follicles. In *Reproductive biology of poultry* (ed. F. J. Cunningham, P. E. Lake and D. Hewitt) pp. 51–73. British Poultry Science Ltd, Longman Group, Harlow.

—— and Cheng, K. W. (1981). Changes in the plasma concentrations of luteinizing hormone, progesterone, oestradiol and testosterone and in the binding of follicle-stimulating hormone to the theca of follicles during the ovulation cycle of the hen *(Gallus domesticus). J. Endocr.* **91**, 11–22.

—— Croze, F. and Duke, C. E. (1981). Plasma concentrations of luteinizing hormone, progesterone, testosterone and estradiol in follicular and peripheral venous plasma during the ovulation cycle of the hen. *Advances in Physiological Sciences.* Vol. 33. *Recent advances of avian endocrinology.* (eds G. Pethes, P. Péczely and P. Rudas) pp. 89–98. Pergamon Press, Akadémiai Kiadó, Hungary.

—— and Cunningham, F. J. (1976a). The interrelationship between progesterone and luteinizing hormone during the ovulation cycle of the hen *(Gallus domesticus). J. Endocr.* **71**, 51–58.

—— —— (1976b). The effect of pregneneolone, progesterone, deoxycorticosterone, or corticosterone on the time of ovulation and oviposition in the hen. *Br. Poult. Sci.* **17**, 637–642.

—— —— (1977). The plasma concentrations of testosterone and LH during the ovulation cycle of the hen *(Gallus domesticus). Acta endocr. Copenh.* **84**, 357–366.

—— Petite, J. N., and Anderson-Langmuir, C. E. (1984). Interrelationships between the hypothalamus, pituitary gland, ovary, adrenal gland and the open period for LH release in the hen *(Gallus domesticus). J. exp. Zool.* **232**, 501–511.

Feder, H. H., Brown-Grant, K., and Corker, C. S. (1971). Pre-ovulatory progesterone, the adrenal cortex and the 'critical period' for luteinizing hormone release in rats. *J. Endocr.* **50**, 29–39.

Follett, B. K. (1981). The stimulation of luteinizing hormone and follicle-stimulating hormone secretion in quail with complete and skeleton photoperiods. *Gen. comp. Endocr.* **45**, 306–316.

—— Scanes, C. G., and Cunningham, F. J. (1972). A radioimmunoassay for avian luteinizing hormone. *J. Endocr.* **52**, 359–378.

Franchii, L. L., Mandl, A. M., and Zuckerman, S. (1962). The development of the ovary and the process of oogenesis. In *The ovary* Vol. 1 (ed. S. Zuckerman, A. M. Mandl, and P. Eckstein) pp. 1–88. Academic Press, New York and London.

Fraps, R. M. (1942). Synchronized induction of ovulation and premature oviposition in the domestic fowl. *Anat. Rec.* **84**, 521.

—— (1954). Neural basis of diurnal periodicity in release of ovulation-inducing hormone in fowl. *Proc. natn. Acad. Sci. U.S.A.* **40**, 348–356.

—— (1955). Egg production and fertility in poultry. In *Progress in the physiology of farm animals*. Vol. 2 (ed. J. Hammond) pp. 661–740. Butterworths, London.

—— and Dury, A. (1943). Occurrence of premature ovulation in the domestic fowl following administration of progesterone. *Proc. Soc. exp. Biol. Med.* **52**, 346–349.

Fraser, H. M. and Sharp, P. J. (1978). Prevention of positive feedback in the hen *(Gallus domesticus)* by antibodies to luteinizing hormone releasing hormone. *J. Endocr.* **76**, 181–182.

Furr, B. J. A., Bonney, R. C., England, R. J., and Cunningham, F. J. (1973). Luteinizing hormone and progesterone in peripheral blood during the ovulatory cycle of the hen *(Gallus domesticus)*. *J. Endocr.* **57**, 159–169.

—— and Smith, G. K. (1975). Effects of antisera against gonadal steroids on ovulation in the hen *(Gallus domesticus)*. *J. Endocr.* **66**, 303–304.

Gilbert, A. B. (1965). Innervation of the ovarian follicle of the domestic hen. *Q. Jl. exp. Physiol.* **50**, 437–445.

—— (1972). The activity of the ovary in relation to egg production. In *Egg formation and production* (ed. B. M. Freeman and P. E. Lake) pp. 3–21. British Poultry Science Ltd., Edinburgh.

Gulati, D. P., Nakamura, T., and Tanabe, Y. (1981). Diurnal variations in plasma LH, progesterone, testosterone, estradiol and estrone in Japanese quail. *Poultry Sci.* **60**, 668–673.

Halberg, F., Halberg, E., Barnum, C. P., and Bitner, J. J. (1959). Physiological 24h periodicity in human beings and mice, the lighting regimen and daily routine. In *Photoperiodism and related phenomena in plants and animals* (ed. R. B. Withrow) pp. 803–878. American Association of Advanced Science, Publication Number 55, Washington, D.C.

Hammond, R. W., Burke, W. H., and Hertelendy, F. (1981). Influence of follicular maturation on progesterone release. *Biol. Reprod.* **24**, 1048–1055.

Hsu-Fang Chou, Johnson, A. L., and Williams, J. B. (1985). Luteinizing hormone releasing activity of [Gln^8]-LHRH and [His^5,trp^7,tyr^8]-LHRH in the cockerel, *in vivo* and *in vitro*. *Life Sciences* **37**, 2459–2465.

Huang, E. S. U. and Nalbandov, A. V. (1979). Steriodogenesis of chicken granulosa cells and theca cells: *In vitro* incubation system. *Biol. Reprod.* **20**, 442–453.

Hughes, G. C. (1963). The population of germ cells in the developing female chick. *J. Embryol. exp. Morph.* **11**, 513–536.

Hutt, F. B. (1949). *Genetics of the fowl.* McGraw-Hill, New York.

Imai, K. and Nalbandov, A. V. (1971). Changes in FSH activity of anterior pituitary glands and of blood plasma during the laying cycle of the hen. *Endocrinology* **88**, 1465–1470.

Ishii, S. and Sakai, H. (1980). Avian follicle-stimulating hormone: isolation and characterization. In *Biological rhythms in birds: Neural and endocrine aspects* (ed. T. Tanake, K. Tanaka and T. Ookawa) pp. 189–198. Springer Verlag, Berlin.

—— Tsutsui, K., and Adachi, T. (1978). Effects of gonadotrophins on elements of testes of birds. In *Comparative endocrinology* (ed. P. J. Gaillard and H. H. Boer) pp. 73–76. Elsevier, Amsterdam.

Johnson, A. L., Dickerman, R. W.,and Advis, J. P. (1984). Comparative ability of chicken and mammalian LHRH to release LH from rooster pituitary cells *in vitro. Life Sciences* **34**, 1847–1851.

—— and van Tienhoven, A. (1980). Plasma concentrations of six steroids and LH during the ovulatory cycle of the hen *(Gallus domesticus). Biol. Reprod.* **231**, 386–393.

—— —— (1984). Effects of aminoglutethimide on luteinizing hormone and steroid secretion, and ovulation in the hen. *Gallus domesticus. Endocrinology*, **114**, 2276–2283.

Johnson, P. A. and van Tienhoven, A. (1984). Investigations of the significance of the crepuscular LH peak in the ovulatory cycle of the hen *(Gallus domesticus). J. Endocr.* **100**, 307–313.

Kahnt, F. W. and Neher, R. (1962). On the specific inhibition of adrenal steroid biosynthesis. *Experientia* **18**, 499–501.

Kawashima, M., Kamiyoshi, M., and Tanaka, K. (1979). Cytoplasmic progesterone receptor concentrations in the hen hypothalamus and pituitary. Difference between laying and nonlaying hens and changes during the ovulatory cycle. *Biol. Reprod.* **20**, 581–585.

Kamiyoshi, M. and Tanaka, K. (1972). Augmentative effect of FSH on LH-induced ovulation in the hen. *J. Reprod. Fert.* **29**, 141–143.

King, J. A. and Millar, R. P. (1982a). Structure of avian hypothalamic gonadotrophin-releasing hormone. *S. Afr. J. Sci.* **78**, 124–125.

—— —— (1982b). Structure of chicken hypothalamic luteinizing hormone-releasing hormone 1. Structural determination on partially purified material. *J. biol. Chem.* **257**, 10722–10728.

—— —— (1982c). Structure of chicken hypothalamic luteinizing hormone-releasing hormone II. Isolation and characterization. *J. biol. Chem.* **257**, 10729–10732.

Knight, P. G., Cunningham, F. J., and Gladwell, R. T. (1983). Concentrations of immunoreactive luteinizing hormone releasing hormone in discrete brain regions of the cockerel: effects of castration and testosterone replacement therapy. *J. Endocr.* **96**, 471–480.

—— Francis, P. T., Holman, R. B., and Gladwell, R. T. (1982a). Changes in hypothalamic monoamine concentrations accompany the progesterone-induced release of luteinizing hormone in the domestic hen. *Neuroendocrinology* **35**, 359–362.

—— Gladwell, R. T., and Cunningham, F. J. (1981). Effect of gonadectomy on the concentrations of catecholamines in discrete areas of the diencephalon of the domestic fowl. *J. Endocr.* **89**, 389–397.

—— Wilson, S. C., Gladwell, R. T., and Cunningham, F. J. (1982b). Evidence for the involvement of central catecholaminergic mechanisms in mediating the preovulatory surge of luteinizing hormone in the domestic hen. *J. Endocr.* **94**, 295–304.

—— —— —— —— (1984). Hypothalamic content of LHRH and catecholamines during the ovulation cycle of the hen *(Gallus domesticus). J. Reprod. Fert.* **71**, 289–295.

Lang, G. F., Etches, R. J., and Walton, J. S. (1984). Effects of luteinizing hormone, progesterone, testosterone, estradiol and corticosterone on ovulation and luteinizing hormone release in hens treated with aminoglutethimide. *Biol. Reprod.* **30**, 278–288.

Leung, P. C. K. and Armstrong, D. T. (1980). Interactions of steroids and gonadotropins in the control of steroidiogenesis in the ovarian follicle. *A. Rev. Physiol.* **42**, 71–82.

Licht, P., Papkoff, H., Farmer, S. W., Muller, C. H., Tsui, H. W. and Crews, D. (1977). Evolution of gonadotropin structure and function. *Recent Prog. Horm. Res.* **33**, 169–248.

Liddle, G. W., Estep, H. L., Kendall, J. W., Carter Williams, W., and Townes, A. W. (1959). *J. clin. Endocr. Metab.* **19**, 875–894.

Melek, O., Morris, T. R., and Jennings, R. C. (1973). The time factor in egg formation for hens exposed to ahemeral light-dark cycles. *Br. Poult. Sci.* **14**, 493–498.

Mikami, S. and Yamada. S. (1984). Immunohistochemistry of the hypothalamic neuropeptides and anterior pituitary cells in the Japanese quail. *J. exp. Zool.* **232**, 405–417.

Miyamoto, K., Hasegawa, Y., Igarashi, M., Chino, N., Sakikibara, S., Kangawa, K., and Matsuo, H. (1983). Evidence that chicken hypothalamic luteinizing hormone-releasing hormone is [Gln^8]-LHRH. *Life Sciences.* **32**, 1341–1347.

—— —— —— Kangawa, K., and Matsuo, H. (1983). Structural determination of the second gonadotropin releasing hormone (GnRH II) in chicken hypothalamus. In *Peptide chemistry* (ed. E. Munekata) pp. 94–101. The Protein Research Foundation.

—— —— Minegishi, T., Nomura, M., Takahashi, Y., Igarashi, N., Kangawa, K., and Matsuo, H. (1982). Isolation and characterization of chicken hypothalamic luteinizing hormone-releasing hormone. *Biochem. Biophys. Res. Commun.* **107**, 820–827.

Morris, T. R. (1973). The effect of ahemeral light and dark cycles on egg production in the fowl. *Poultry Sci.* **52**, 423–445.

—— (1977). The clutch patterns of hens in constant illumination. *Br. Poult. Sci.* **18**, 397–405.

Opel, H. (1979). On hypothalamic control of ovulation in the turkey. *Poultry Sci.* **58**, 717–724.

—— (1980). The hypothalamus and reproduction in the female. *Poultry Sci.* **58**, 1607–1618.

Pittendrigh, C. S. and Minis, D. H. (1964). The entrainment of circadian oscillations by light and their role as photoperiodic clocks. *Am. Nat.* **98**, 261–294.

Richards, J. S. (1979). Hormonal control of ovarian follicular development: a 1978 perspective. *Recent Progr. Horm. Res.* **35**, 343–373.

Rothchild, I. and Fraps, R. M. (1949). The induction of ovulating hormone release from the pituitary of the domestic hen by means of progesterone. *Endocrinology* **44**, 141–149.

Scanes, C. G. (1981). Adenohypophysial hormones: their chemistry, physiology and control. *Advances in Physiological Sciences* Vol. 33. *Recent advances of avian endocrinology* (ed. G. Pethes, P. Péczely and P. Rudas) pp. 61–71. Pergamon Press, Akadémiai Kiadó, Hungary.

—— and Fagioli, J. T. (1980). Effects of mammalian and avian gonadotropins on *in vitro progesterone production by avian grnaulosa cells. Gen. comp. Endocr.* **41**, 1–7.

—— Godden, P. M. M., and Sharp, P. J. (1977). An homologous radioimmunoassay for chicken follicle-stimulating hormone: observations on the ovulatory cycle. *J. Endocr.* **73**, 413–481.

—— Mozelic, H., Kavanagh, E., Merrill, G. and Rabii, J. (1982). Distribution of blood flow in the ovary of domestic fowl *(Gallus domesticus)* and changes after prostaglandin F-2α treatment. *J. Reprod. Fert.* **64**, 227–231.

Senior, B. E. and Cunningham, F. J. (1974). Oestradiol and luteinizing hormone during the ovulatory cycle of the hen. *J. Endocr.* **60**, 201–202.

Shahabi, N. A., Norton, H. W., and Nalbandov, A. V. (1975). Steroid levels in follicles and plasma of hens during the ovulatory cycle. *Endocrinology* **96**, 962–968.

Sharp, P. J., MacNamee, M. C., Talbot, R. T., Sterling, R. J., and Hall, T. R. (1984). Aspects of the neuroendocrine control of ovulation and broodiness in the domestic hen. *J. exp. Zool.* **232**, 475–483.

Soliman, K. F. A. and Huston, T. M. (1974). Involvement of the adrenal gland in ovulation of the fowl. *Poultry Sci.* **53**, 1664–1667.

Sterling, R. J., Gase, J. M., Sharp, P. J., Tuohimaa, P., and Baulieu, E. E. (1984). Absence of nuclear progesterone receptor in LH-RH neurones in laying hens. *J. Endocr.* **102**, R5-R7.

—— and Sharp, P. J. (1982). The localization of LH-RH neurones in the diencephalon of the doemstic hen. *Cell Tissue Res.* **222**, 283–298.

Stockell-Hartree, A. and Cunningham, F. J. (1969). Purification of chicken pituitary follicle-stimulating hormone and luteinizing hormone. *J. Endocr.* **43**, 609–616.

Unsicker, K., Seidel, F., Hofmann, H. D., Muller, T. H., Schmidt, R., and Wilson, A. (1983). Catecholaminergic innervation of the chicken ovary. *Cell Tissue Res.* **230**, 431–450.

van Tienhoven, A. (1961). Endocrinology of reproduction in birds. In *Sex and internal secretions* **Vol. 2** (ed. W. C. Young) pp. 1088–1169. Williams and Wilkins, Baltimore.

—— (1981). Neuroendocrinology of avian reproduction with special emphasis on the reproductive cycle of the fowl *(Gallus domesticus)*. *World Poult. Sci. J.* **37**, 156–176.

—— Nalbandov, A. V., and Norton, H. W. (1954). Effect of dibenamine on progesterone-induced and spontaneous ovulation in the hen. *Endocrinology* **54**, 605–611.

—— and Ostrander, C. E. (1976). Short total photoperiods and egg production of White Leghorns. Egg production and shell breaking strength. *Poulty Sci.* **52**, 998–1001.

Wada, M. (1981). Photoinducible phase for gonadotrophin secretion entrained to dawn in Japanese quail. *Gen. comp. Endocr.* **43**, 227–233.

Wang, S. C. and Bahr, J. M. (1983). Estradiol secretion by theca cells of the domestic hen during the ovulatory cycle. *Biol. Reprod.* **28**, 618–624.

Warren, D. C. and Scott, H. M. (1935). The time factor in egg production. *Poultry Sci.* **14**, 195–207.

Wells, J. W., Gilbert, A. B., and Culbert, J. (1980). Effect of luteinizing hormone on progesterone secretion *in vitro* by the granulosa cells of the domestic fowl *(Gallus domesticus)*. *J. Endocr.* **84**, 249–254.

Wilson, S. C. (1978). Relationship between plasma concentration of luteinizing hormone and intensity of lay in the domestic hen. *Br. Poult. Sci.* **19**, 643–650.

—— (1982). Evidence of a photoinducible phase for the release of luteinizing hormone in the domestic hen. *J. Endocr.* **94**, 397–406.

—— and Cunningham, F. J. (1980a). Effect of increasing day length and intermittent lighting schedules in the domestic hen on plasma concentrations of luteinizing hormone (LH) and the LH response to exogenous progesterone. *Gen. comp. Endocr.* **41**, 546–553.

—— —— (1980b). Concentrations of corticosterone and luteinizing hormone in plasma during the ovulatory cycle of the domestic hen and after the administration of gonadal steroids. *J. Endocr.* **85**, 209–218.

—— —— (1980c). Modification by metyrapone of the open period for pre-ovulatory LH release in the hen. *Br. Poult. Sci.* **21**, 351–361.

136   F. J. Cunningham

—— —— (1981a). Effects of an anti-oestrogen, Tamoxifen (ICI 46,474) on luteinizing hormone release and ovulation in the hen. *J. Endocr.* **88**, 309–316.

—— —— (1981b). Effect of photoperiod on the concentrations of corticosterone and luteinizing hormone in the plasma of the domestic hen. *J. Endocr.* **91**, 135–143.

—— —— (1984). Endocrine control of the ovulation cycle. In *Reproductive biology of poultry* (ed. F. J. Cunningham, P. E. Lake and D. Hewitt) pp. 29–49. British Poultry Science Ltd., Longman Group, Harlow.

—— Jennings, R. C., and Cunningham, F. J. (1983). An investigation of diurnal and cyclic changes in the secretion of luteinizing hormone in the domestic hen. *J. Endocr.* **98**, 137–145.

—— —— —— (1985). Effects of the advance of darkness on the ovulatory cycle of the hen. *Br. Poult. Sci.* **26**, 83–96.

—— and Lacassagne, L. (1978). The effects of dexamethasone on plasma luteinizing hormone and oviposition in the hen *(Gallus domesticus)*. *Gen. comp. Endocr.* **35**, 16–26.

—— and Sharp, P. J. (1973). Variations of plasma LH levels during the ovulatory cycle of the hen *(Gallus domesticus)*. *J. Reprod. Fert.* **35**, 561–564.

—— —— (1975). Changes in plasma concentrations of luteinizing hormone after injection of progesterone at various times during the ovulatory cycle of the domestic hen. *(Gallus domesticus)*. *J. Endocr.* **67**, 59–70.

—— —— (1976a). The effects of progesterone on oviposition and ovulation in the domestic fowl *(Gallus domesticus)*. *Br. Poult. Sci.* **17**, 163–173.

—— —— (1976b). Effects of androgens, oestrogens and deoxycorticosterone acetate on plasma concentrations of luteinizing hormone in laying hens. *J. Endocr.* **69**, 93–102.

—— —— (1976c). Induction of luteinizing hormone release by gonadal steroids in the ovariectomized hen. *J. Endocr.* **71**, 87–98.

4 Chorion villus biopsy

CHARLES H RODECK

I Introduction

The last five years have seen a resurgence of interest in chorion villus biopsy (CVB) for fetal diagnosis. Since the chorionic villi are relatively accessible and are derived from the zygote, biochemical and chromosomal studies can give accurate information about the fetus. Recent advances in molecular genetics have provided impetus for the development of techniques for obtaining villi because gene analysis can be performed on the DNA of any tissue from an individual whereas phenotypic analysis of the gene product requires the specific tissue. Thus, diseases which have no effect on the chorion itself can be diagnosed from chorionic villi. Furthermore, CVB can be performed in the first trimester of pregnancy so that patients have less time to wait, receive an earlier result and if necessary, can have a simple aspiration termination. These are major advantages over all the other methods of fetal diagnosis, which can only be done in the second trimester when women are far more committed to their pregnancy and termination is more traumatic physically and emotionally.

II   Structure and development of chorionic villi

The chorionic villi constitute the bulk of the placenta and are fetal tissue bathed in the maternal blood of the intervillous space. They are covered by the trophoblast which is divided into an outer syncytial layer, the syncytiotrophoblast, and an inner layer of cytotrophoblast (Langhans cells). The core of the villi consists of loose mesenchymal tissue, macrophages (Hofbauer cells) and fetal capillaries. In the first trimester of pregnancy they are less numerous, larger in diameter (about $150\,\mu$), have thicker trophoblast layers and less prominent capillaries than later in gestation (Fig. 4.1).

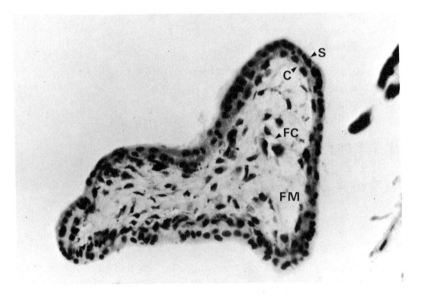

Fig. 4.1. Histology of a villus from a 10 week pregnancy ($\times$ 400). S, syncytiotrophoblast; C, cytotrophoblast; FC, fetal capillary; FM, fetal mesenchyme.

The villi begin to develop after the blastocyst has implanted in the decidua (6–10 days after fertilisation). On the 7th day, the trophoblast has differentiated into its two layers and the syncytiotrophoblast enlarges and erodes maternal capillaries. Lacunae appear which become confluent and fill with maternal blood (days 10–13). Trabecular columns of syncytio-trophoblast with a cytotrophoblast core (primary villus stems) are radially orientated between the chorionic plate (fetal surface) and the basal plate (maternal surface). Mesenchyme then extends into the villus stem from the extra-embryonic mesoderm and this later becomes vascularised (14–21 days). The maternal and fetal circulation of the placenta is thus established. Increasingly complex second- and third-order branching of villi occurs and is particularly extensive at the implantation site opposite the decidua basalis. This is the chorion frondosum which develops into the definitive placenta. The enlarging blastocyst, now called the gestation sac, is also covered by chorionic villi, which in turn are covered by the decidua capsularis. With continuing growth, the latter thins and the villi degenerate due to avascular necrosis, forming the chorion laeve. By 14 weeks post-menstrual age (about 12 weeks after fertilisation) the villi are confined to the placenta. The decidua capsularis gradually disappears as the uterine cavity is closed and the chorion laeve comes into contact with the decidua parietalis of the opposite wall of the uterus. Meanwhile, the amniotic cavity has grown and the amnion becomes apposed to the chorion laeve, obliterating the extra-embryonic coelom (Fig. 4.2) (Williams and Warwick 1980).

III Development of techniques for chorion villus biopsy

These have been critically reviewed elsewhere (Rodeck and Morsman 1983) and only a brief outline will be given here. The earliest reports came from Scandinavia (Hahnemann and Mohr, 1968; Kullander and Sandahl, 1973; Hahnemann, 1974). Quite large endoscopes (5–6 mm diameter) were introduced through the cervix immediately prior to termination of pregnancy, although in a proportion of patients the uterus was evacuated some days or weeks later to assess the complication rate. This was found to be quite high due to rupture of the amniotic sac and infection; and the success rate of chromosome analysis was rather low. At about this time it was becoming clear that amniocentesis early in the second trimester was entirely safe and reliable and interest in CVB waned.

An alternative approach was utilised by Rhine, Palmer and Thompson (1977): the lower part of the uterine cavity at the level of the internal os was lavaged with a small quantity of normal saline and the washings were cultured for chromosome analysis. A moderate success rate was obtained but this rate could not be confirmed by others (Goldberg, Chen, Young, and Reidy 1980). The method was based on the erroneous supposition that the villi on the gestation sac were exposed to the uterine cavity, into which they

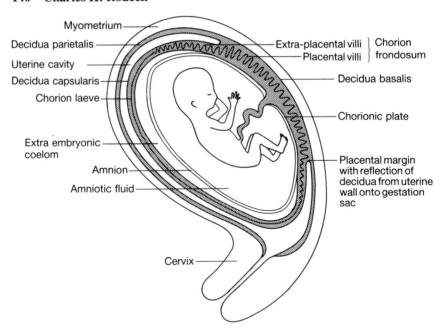

Fig. 4.2. Diagram of 10 week pregnancy *in utero*. The uterine cavity and extraembryonic coelom are exaggerated. The gestation sac is covered by decidua capsularis, chorion laeve and amnion and contains the embryo, the umbilical cord and amniotic fluid.

fell when they degenerated. This, of course, is not the case as they are covered by the decidua capsularis (Fig. 4.2) and the process of degeneration is one of necrosis and resorption.

A report from the hospital of the Anshan steelworks, Tietung, China, appeared in 1975. Transcervical needle aspiration was performed 'blindly', i.e. without endoscopic or sonographic guidance, with high sampling success. The aim was to sex the fetuses by Barr body counting and this was achieved. These were the first pregnancies that were allowed to go to term after CVB and in the majority there were no maternal, fetal or neonatal problems; some 6% of patients aborted after the procedure. This important paper was another factor stimulating the re-awakening of interest in CVB in Hungary, Russia and several centres in London. Kazy and Stigar (1980) and Kazy, Rozovsky, and Bakharev (1982) used two methods, both transcervical: ultrasound-guided biopsy forceps and endoscopically directed forceps. The results were very good in terms of pregnancy outcome, fetal sexing by X and Y fluorescence, and biochemical studies on normal villi.

Meanwhile, the β-globin gene had been identified in DNA extracted from placental tissue (Williamson, Eskdale, Coleman, Niazi, Loeffler, and Modell, 1981) indicating that villi could be used for fetal diagnosis of haemoglobinopathies. Further evaluation of 'blind' aspiration showed this to have too high a failure rate (Horwell, Loeffler, and Coleman, 1983, Liu, Mitchell, Johnson, and Wass, 1983). Endoscopically directed needle aspiration was found to be successful for fetal sexing using Y-specific DNA probes (Gosden, Mitchell, Gosden, Rodeck, and Morsman, 1982; Rodeck, Morsman, Gosden, and Gosden, 1983a) and was also tried by other groups (Kaplan, Dumez, and Goossens, 1983; Gustavii, 1983), but it was rather complex and difficult to teach. It also required the instillation of normal saline into the uterine cavity, which was a cause for worry in continuing pregnancies. Improvements in sonography made it possible to guide a cannula through the cervix (Old, Ward, Petrou, Karagozlou, Modell, and Weatherall, 1982; Ward, Modell, Petrou, Karagozlou, and Douratsos, 1983; Rodeck, Morsman, Nicolaides, McKenzie, Gosden, and Gosden, 1983b). This method was first used diagnostically for haemoglobinopathies (Old et al, 1982) and it became widespread when chromosome analysis proved to be possible. Earlier work suggested that villus culture was not easy (Niazi, Coleman, and Loeffler, 1981) but a reliable technique was established (Heaton, Czepulkowski, Horwell, and Coleman, 1984). Further impetus was provided by fast chromosome analysis on direct preparations (Simoni, Brambati, Danesino, Rosella, Terzoli, Ferrari, and Fraccaro, 1983)

Other methods that are preferred in some centres and which yield good results are sonographically guided transcervical biopsy forceps (Goossens, Dumez, Kaplan, Lupker, Chabret, Henrion, and Rosa, 1983) and transabdominal needle aspiration (Smidt-Jensen and Hahnemann, 1984). The latter, particularly is attracting interest. Very few have taken to the brush (Liu, Symonds, Jeavons, and Norman, 1985) for collecting villi, but the latest approach, transvaginal needle aspiration directed by a vaginal ultrasound probe, is receiving some attention (Popp, personal communication).

IV Clinical chorion villus biopsy

The large variety of sampling techniques that have been developed and are in use suggest that there is not one ideal method, although adequate information to compare one with another is not yet available. Clearly, there must, within the limits of safety, be room for operator preference. In this section, the main techniques for CVB will be described and their selection discussed, but firstly some wider aspects will be considered.

1 WHAT PATIENTS SHOULD HAVE CHORION VILLUS BIOPSY?

The immediate attractiveness of CVB to patients has led to a high degree of self referral. At the present state of knowledge, without long-term information being available and trials still underway, the procedure should not yet be offered as a routine and the clinician must be prepared to resist consumer pressure. Patients can be grouped into two broad categories, high and low risk. The former are those patients with a genetic history and therefore a 1:2 or 1:4 chance of an affected fetus, and those who are carriers of a chromosome translocation. The advantages of CVB seem to be so great in this group that they appear to outweigh any known (or unknown) disadvantages. The latter category which is by far the larger, has a slightly increased risk of a chromosomally abnormal fetus because the women are aged 35–40 years. Those that are older than 40 years, or have had a child with a chromosome aberration, or a previous second trimester termination may be placed in an intermediate risk category, although parental anxiety is a dominant influence in this group. Patients must be counselled in detail with a full explanation of the risks and options of first and second trimester diagnosis. Whenever possible, low risk patients should be invited to participate in a controlled trial comparing CVB with amniocentesis.

2   WHAT IS THE ROLE OF ULTRASONOGRAPHY?

Ultrasonography is an essential investigation before, during and after CVB (Table 4.1). Most workers agree that a real-time sector scanner gives the best pictures of pelvic organs although a high quality linear-array machine may also be suitable. It makes sense to perform a scan at the time of a preliminary counselling interview.

Table 4.1
Ultrasonography in chorion villus biopsy

1.	Confirmation of fetal life
2.	Confirmation of gestational age
3.	Number of gestation sacs and fetuses
4.	Placental site and cord insertion
5.	Position of uterus, presence of contractions and fullness of bladder
6.	Position of cervix
7.	Other abnormalities in pelvis
8.	Guidance of instrument
9.	Early warning of complications
10.	Follow-up

- The confirmation that the fetus is alive is clearly important. A CVB is unnecessary if there is an anembryonic pregnancy or an intrauterine death. The earlier in pregnancy that a patient is seen, the higher the chance of these being found. McFadyen (1985) found 7 per cent of pregnancies had empty sacs before 8 weeks' gestation and 4 per cent at 9–10 weeks, although higher figures are often quoted by other authors.

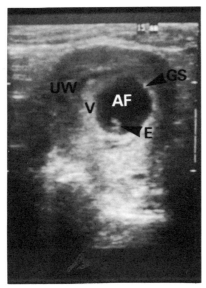

Fig. 4.3.   Ultrasound scan of 9 week pregnancy. The characteristic echogenic appearance of villi (V) extends over much of the gestation sac (GS) and the placental margins cannot yet be defined. AF, amniotic fluid; E, part of the embryo; UW, uterine wall.

• Gestational age must be confirmed so that the appropriate date for CVB can be arranged.

• Multiple pregnancy complicates CVB and counselling. Patients should be aware of the probable higher risk of abortion, possible lack of certainty that both (all) fetuses have been sampled and of the difficult decision that will have to be faced in the event of discordancy between the fetuses.

• The placental site is not easily defined by an ultrasound scan early in the first trimester because the extraplacental villi in the chorion of the gestation sac are also echogenic (Fig. 4.3). It becomes easier towards 12 weeks as the placenta grows and the chorion laeve forms. The area of implantation can be inferred from the thickest region of echogenicity and from the insertion of the umbilical cord which is usually visible after 9 weeks.

• The position of the uterus (anteversion, retroflexion or axial) and the relationship of the cervix and internal os to it are highly relevant to the performance of CVB. Focal uterine contractions are commonly seen and these may profoundly affect the topography, as may the capacity of the bladder.

• A search for uterine, adnexal or other pathology should not be omitted. Uterine or cervical fibroids may make CVB impossible.

• Use of ultrasound is integral with CVB and will be dealt with further below.

- Some complications can be detected with ultrasound, for example haematoma formation, rupture of the gestation sac and loss of amniotic fluid. Alteration in the fetal heart rate is extremely rare.
- Follow-up scans are valuable to ascertain whether development has continued normally or not. Policies vary from centre to centre, but at least one should be done 7–10 days after the CVB.

3   WHICH TECHNIQUE OF CHORION VILLUS BIOPSY?

Several of the most widely used procedures will be considered here. All are done on an out-patient basis and rarely, if ever, require local anaesthesia or sedation. An operating theatre is not necessary and a procedure room or a room in the ultrasound department is adequate. An operating table may be used although a colposcopy chair is more comfortable for transcervical procedures. Pre-sterilised packs containing drapes, swabs, pots, etc are convenient. Some instruments may be disposable (i.e. pre-sterilised) or a small autoclave in the room is suitable for others. A microscope is essential for the immediate selection of villi from the fragments of decidua and maternal blood clot that may be present in the sample. The individual doing this need not be a cytogeneticist, but must be trained and confident in recognising villi.

i   Transcervical aspiration

In conjunction with ultrasound guidance, this has become the most widely used technique: without ultrasound, 'blind' aspiration has an unacceptably low success rate.

The patient is scanned whether this has been done before or not. The information outlined above is then obtained or confirmed. The bladder is noted: it should not be empty or too full. Occasionally it may be helpful to modify its capacity. The patient is then positioned comfortably on a colposcopy chair which is preferable to the lithotomy position on an operating table. There will be at least a doctor, one nurse and possibly a cytogeneticist and an ultrasonographer present, and the woman will feel more at ease if she has already met one or more of these people; ideally one of them will have done the preoperative counselling. The presence of the partner may also promote relaxation.

The vulva and vagina are cleansed with an antiseptic solution and a vaginal speculum is passed to expose the cervix. We have found a Sims' speculum weighted at one end to be useful. A tenaculum is applied to the anterior cervical lip and the cervix and vagina are carefully inspected. If discharge is present, an endocervical swab should be taken. Florid infection should be regarded as a contraindication to transcervical CVB, until after a course of treatment. We believe that further meticulous cleansing is beneficial and the external os is swabbed several more times and then dried.

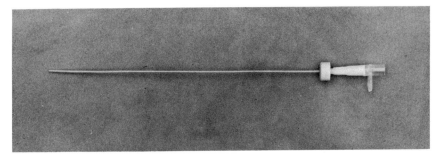

Fig. 4.4.   The Portex Cannula

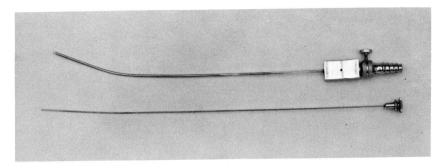

Fig. 4.5.   The Down's Cannula

The majority of centres use the Portex catheter (Fig. 4.4, Portex Ltd, England). It is disposable, soft polyethylene, 1.5 mm in diameter with an aluminium introducer which can be bent to the desired shape, depending on the relationship of cervix, internal os, uterus and placental site. An ultrasonographer helps the operator guide the cannula into the placenta. The introducer is removed, a syringe containing some heparinised culture medium is attached and suction is exerted while the cannula is slowly withdrawn. Involuntary movements of the catheter should be kept to a minium. After removal from the patient, the syringe and cannula are flushed into a petri dish for microscopic examination (Ward *et al.* 1983).

An alternative approach is the single-operator technique (Rodeck *et al.* 1983b). This utilises a 16 gauge malleable silver cannula which can be used many times (Fig. 4.5, Down's PLC). It also has an introducer and is bent into the required shape. When the tip is through the internal os, the operator puts down the tenaculum and picks up the ultrasound probe, thereafter guiding the highly echogenic cannula into the placental site him/herself. The introducer is removed and the cannula is connected via a neonatal mucus aspirator trap to a suction pump. A vacuum of 400–500 mg Hg is quite safe and can be controlled by the hole in the finger plate at the end of the cannula. Gentle 'hoovering' to and fro movements may help to dislodge villi, and the cannula is then slowly withdrawn whilst maintaining some vacuum, the operator meanwhile monitoring this continually with ultrasound. At this stage, the villi are usually still in the cannula; its tip is dipped into culture medium so that they are flushed into the trap. The contents can be inspected by naked eye, before microscopic examination.

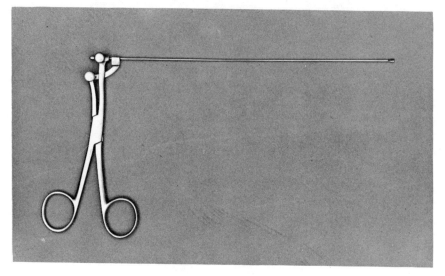

Fig. 4.6.   Biopsy forceps

ii   Transcervical biopsy forceps

A rigid, cupped or alligator-jaw biopsy forceps (Fig. 4.6) has been used by Dumez (Goossens *et al*. 1983) and by a number of other people. It is simple to use, does not rely on an aspiration system, and the operator can also guide himself with the ultrasound probe held in one hand and the biopsy forceps in the other. Early fears that such an instrument was too rigid, that the biopsies were too large and that there was a serious risk of tearing/rupturing the amnion do not appear to have been warranted.

iii   Transabdominal aspiration

Transabdominal ultrasound-guided needle aspiration has been described by Smidt-Jensen and Hahnemann (1984). A sector probe with a needle guide is aligned so that the dotted line on the ultrasound screen enters the placenta parallel to the chorionic plate. This is possible in most cases because a full bladder helps to elevate the uterus out of the pelvis, manual abdominal pressure may cause the uterus to rotate and because extensions of the

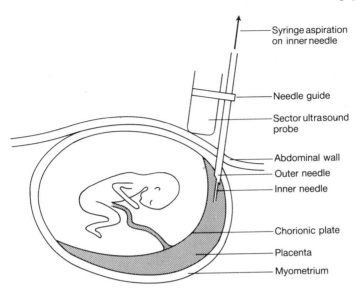

Fig. 4.7. Diagram of transabdominal needle aspiration showing a transverse section of the uterus with a posterior and right lateral placenta.

placental site onto more than one uterine wall increase the available choice of approaches. A 1.2 mm diameter needle is rapidly thrust through the abdominal wall and myometrium in the path of the needle guide (Fig. 4.7). Local anaesthetic is not necessary particularly if the needle entry is speedy. An inner needle with a syringe attached is passed down the first one into the placenta and villi are aspirated. Several re-insertions of the inner needle are usually required, depending on how much tissue is needed.

4   AT WHAT GESTATIONAL AGE?

Most centres are of the view that CVB should be performed between 8 and 12 weeks gestation and that 9–10 weeks is probably the optimum. Why not before 8 weeks? This would seem to be the ideal, that is as soon as a pregnancy has been confirmed. The main reason is technical—the intrauterine structures are so small, the ultrasound pictures are often not clear enough and consequently serious damage to the pregnancy is possible. This is only partly counterbalanced by the fact that viable villi are present not only in the placenta but are distributed all over the gestation sac, thus theoretically increasing the chances of successful biopsy. Further drawbacks of CVB before 8 weeks are the lack of time to organise referrals and the higher incidence of anembryonic pregnancies and subsequent spontaneous abortions than in the later first trimester, which render the procedure unnecessary.

Why not after 12 weeks? The received wisdom is that more fetal losses occur as a result of CVB. However, there do not appear to be any logical or biological reasons why this should be so. A possible explanation is that due to uterine growth and confinement of villi to the placenta, transcervical CVB is more difficult, particularly if the placenta is fundal, and that this leads to higher fetal losses. An additional factor may be the relative lack of experience of most operators in performing CVB after 12 weeks.

It would seem, however, that transabdominal CVB should be perfectly feasible in the second, and indeed the third trimester (Nicolaides, Soothill, Rodeck, Warren, and Gosden, 1986a). Much information needs to be collected on the rates of success and fetal loss as well as the mitotic index of villi at later stages of gestation. This might then become the procedure of choice for a woman at 13 weeks instead of asking her to wait several weeks for amniocentesis or fetoscopy. Should direct chromosomal preparations prove to be reliable in later pregnancy, this may be the best way of obtaining a karyotype when a fetus is found to have a non-lethal anomaly and when such further investigation is usually urgent (Nicolaides, Rodeck, and Gosden, 1986b).

At present, therefore, most CVB is transcervical and done at 9–11 weeks. In approximately 2 per cent of patients, there may be difficult access to the placental site, especially if this is posterior and the uterus is acutely retroflexed. The experienced operator will refrain from further attempts, or even from starting CVB, and re-schedule for one or two weeks time when the uterus will be more axial and the procedure may be entirely straightforward. Another situation where it may be wise to delay CVB, sometimes by no more than half an hour is due to a focal myometrial contraction. These may cause temporary distortion of the uterine cavity and should not be mistaken for uterine fibroids.

5   WHAT IS THE SUCCESS RATE OF CHORION VILLUS BIOPSY?

'Success' must be clearly defined. Here we are concerned with the rate at which adequate amounts of villi are obtained to perform a particular test; that can vary from 5 to 20 mg. The rates at which tests yield reliable results on adequate biopsies will be considered later.

The relationship of timing, type of procedure and pelvic anatomy to success rate has just been discussed. Another prime determinant of success is operator experience. The figures in Table 4.2 are from Jackson's WHO sponsored world registry. Centres with experience of less than 10 cases had a failure rate of 10 per cent whereas those with 300–500 cases failed in 2 per cent. The overall failure rate was 2.7 per cent and in the most recent total of 11,819 cases (CVS News, March 1986) it was 2.6 per cent. The degree of persistence in sampling has a bearing on this and most centres limit themselves to 2 or 3 transcervical passages of the instrument at one

Table 4.2
*Failure rate in obtaining an adequate sample of villi related to experience of centres.
(After Jackson, 1986)*

Experience of Centre (no. of cases)	No. of centres	Total no. patients	Failed Sample (%)
1–10	13	81	8 (9.9)
10–50	14	288	7 (2.4)
50–100	9	343	12 (3.5)
100–200	3	488	26 (5.3)
300–500	4	1697	32 (1.9)
TOTAL	43	3194	86 (2.7)

Table 4.3
*Percentage success rate of CVB and operator experience**

	Number of passages of cannula to obtain adequate CVB and percentage success rate				
	1	2	3	4 or more	Failed
First 50 cases	35	37(72)	16(88)	12(100)	—
Subsequent 150 cases	67	27(94)	4(98)	—	2

*(numbers in brackets are cumulative success rates in percent)

procedure. It is worth noting that Hahnemann, with the largest transabdominal experience, has a 0.5 per cent failure rate, possibly due to the ability to insert repeatedly the inner needle into the guide needle until an adequate quantity of villi has been aspirated.

The effect of operator experience is illustrated longitudinally by our own data (Warren and Rodeck, 1986). An increasing percentage of cases (Table 4.3) required fewer transcervical passages of the Down's Cannula to obtain an adequate CVB. There was a 2% failure rate in the latter 150 cases because a policy not to exceed 3 passages per procedure was adopted. In addition, the first 50 cases were performed by 3 operators whereas the subsequent 150 were virtually all done by one (RCW).

V   Risks and safety of chorion villus biopsy

The potential risks of CVB are shown in Table 4.4.

Table 4.4
Possible risks of CVB

Maternal	Discomfort
	Vaginal bleeding
	Trauma to organs, e.g. perforation of uterus
	Infection
Fetal	Spontaneous abortion
	Intrauterine death
	Perforation of membranes
	Infection
	Abnormal embryogenesis
	Amniotic bands
	Rhesus isoimmunisation
	Intrauterine growth retardation (IUGR)
	Abruptio placentae
	Premature rupture of membranes
Neonatal	Long term effects of: infection
	IUGR
	malformations

1 MATERNAL RISKS

Maternal discomfort is minimal; with transcervical CVB, some embarrassment may be suffered, and transabdominal CVB may cause abdominal pain. A few mothers experience mild abdominal pain for a day or two afterwards.

Vaginal bleeding is quite common for a few days and 10–40% incidences are quoted but it is usually no more than spotting. Most series have not differentiated between bleeding of intrauterine origin and bleeding from the cervix due to the use of a tenaculum. The latter is probably the cause in the majority of women and is therefore of no serious consequence.

Brambati, Oldrini, Ferrazi, and Lanzani (1986a) found intrauterine haematomas in 4.3 per cent of patients; all had disappeared by 16 weeks gestation. Other authors have not reported such findings.

Visceral trauma should be avoidable with careful transcervical ultrasound guided procedures. Transabdominally, the controlled puncture of the uterus should not be traumatic and the theoretical risks of damaging bowel or bladder are in practice extremely low.

A number of intrauterine bacterial infections have occurred, resulting in abortion or intrauterine death, and in a few, the maternal infection has been severe (CVS News, March 1986). They have been confined to transcervical procedures, which comprise the overwhelming majority of the 11,000 cases. The risks of infection may be lower transabdominally (although not absent) and none have yet been reported in the small number of cases (approximately 200) done in this way. It would seem prudent to exercise the most scrupulous cleansing procedure with transcervical CVB. In our own series, now over 400, no case appears to have become infected.

2   RISKS OF FETAL LOSS

The most pressing concern following the introduction of a new invasive intrauterine technique is the risk of immediate and short term complications, in particular of spontaneous abortion and intrauterine death. This problem is difficult to quantify because of the natural incidence of these complications and because of a lack of good information on them. It had generally been assumed that some 10 per cent of pregnancies were lost during the 1st trimester, but several recent studies have modified this view. Early ultrasonography (8–12 weeks) has shown that most of the losses in the first trimester are due to anembryonic pregnancies, while in others, a dead embryo is found; these pregnancies may abort spontaneously a few weeks later. Pregnancies with a live embryo also have a risk of aborting but this is remarkably low. For a woman of 25 years at 10 weeks' gestation, the chance of a spontaneous abortion still happening is under 2 per cent (McFadyen, 1985). This study also showed that the likelihood of spontaneously aborting a live fetus increases with maternal age and decreases with the gestational age at sonography. The probability of finding a dead fetus or an empty sac varies similarly.

These phenomena clearly make it difficult to compare different series and, in the absence of an adequate control group, to decide what the fetal mortality caused by CVB is or should be. Fetal losses up to 28 weeks' gestation occurred in 4.0 per cent of the 11,000 cases in the world registry, but highly experienced centres report rates such as 2.2 per cent (CVS News, March 1986). In our sequential cases, from number 51 to number 400 biopsied transcervically, 2.4 per cent lost their fetuses.

Claims have been made that the transabdominal technique is safer and less traumatic than the transcervical (Maxwell, Lilford, Czepulkowski, Heaton, and Coleman, 1986) although these claims were based on only 7 diagnostic patients. The centre with the most experience of this method reports 2.9 per cent fetal losses out of 218 cases (Hahnemann, CVS News, March 1986).

In terms of fetal mortality, therefore, there seems to be little difference between transcervical and transabdominal approaches and the same may be said, albeit with more hestitation, when comparing CVB with amniocentesis. The skill and experience of the operator are clearly of the greatest importance.

3   OTHER FETAL RISKS

Infection has been discussed above in the context of the mother. Perforation of the membranes, usually with loss of amniotic fluid, has happened in most centres but only during the learning phase, and should be avoidable. Amniotic bands and abnormal embryogenesis have not been reported.

Rhesus isoimmunisation is possible since several studies have shown a rise in maternal serum alpha-fetoprotein (MSAFP) after CVB in 50–70 per cent of patients (Warren, Butler, Morsman, McKenzie, and Rodeck, 1985; Brambati, Guercilina, Bonacchi, Oldrini, Lanzani, and Piceni, 1986b) suggesting that fetomaternal haemorrhage has occurred. The antigenic load must be too small for most women, but 'high responders' are likely to become isoimmunised. These are the patients who may develop particularly severe erythroblastosis fetalis, even in the same pregnancy. Anti-D immunoglobulin prophylaxis is strongly recommended in rhesus D negative women. Conversely, when anti-D antibodies are already present, CVB should be avoided because the antibody level may increase enormously and what could have been mild haemolytic disease may become severe and fatal. It is worth pointing out that the post-CVB rise in MSAFP does not cause false positives at the time when MSAFP is measured to screen for neural tube defects (16 weeks' gestation) (Brambati et al. 1986b).

Complications of later pregnancy require vigilant monitoring. In 1000 patients, 615 of whom had finished the pregnancy (Brambati et al. 1986a), abruptio placentae and placenta praevia occurred in 1.6 per cent (1 placenta accreta), premature rupture of membranes in 0.8 per cent, preterm delivery in 6.3 per cent and the perinatal mortality was 13/1000. Neither intrauterine growth measured by ultrasound nor birth weight deviated from expected values for a general population. While such findings are rather reassuring, they lack the support of a randomly assigned control group without which one cannot draw firm conclusions.

4 NEONATAL RISKS

There has so far been no suggestion of an increase in small for gestational age or low birth weight babies, nor that viral infections, which could affect brain development, are introduced by CVB.

Congenital defects were found in 2.6 per cent (16 cases) of 615 completed pregnancies in the series of Brambati et al. (1986a). Three fetuses had major anomalies (hydrops, diaphragmatic hernia and obstructive uropathy), 6 had club foot, and the remainder had minor anomalies. Hogge, Schonberg and Golbus (1985), in 600 cases, 25 per cent of whom had delivered, found 1 baby with Goldenhar syndrome and one craniosynostosis. Once again, no information is available on what the incidence of congenital malformation in these populations would have been, had they not had CVB. For this, and even more so for the assessment of the later development of these children, a suitable control group obtained by randomisation is essential.

VI Indications for Chorion Villus Biopsy

1 CHROMOSOME DISORDERS

The indications are similar to those for amniocentesis, i.e. advanced maternal age (greater than 35 years), previous child with a chromosome aberration, parental chromosomal translocation, and fetal sexing in X-linked diseases.

Chromosome analysis had been attempted on cultured villi obtained at abortion (Niazi *et al*. 1981) but was first performed diagnostically by the direct preparation method of Simoni *et al*. (1983). This was based on the presence of spontaneous mitoses in the cytotrophoblast of the villi. Subsequently, better culture techniques were developed (Heaton *et al*. 1984) which are also widely used. The advantages of direct preparations are the speed of the result (1–2 days) and the avoidance of some of the problems that may arise in cultures such as pseudomosaicism (the appearance of one or more extra cell lines with different karyotypes), bacterial infection and overgrowth of maternal cells. Their disadvantages are that good metaphase spreads may be few and banding is often poor. Many laboratories have not had sufficient success with direct preparations and use either culture alone or both methods in parallel. Preferably, an estimated weight of at least 10 mg of villi is obtained. Those with buds, sprouts and central (fetal) blood vessels are deemed to be viable and thought to have more mitosis, although this aspect has not yet been systematically studied. Villi with a smooth surface and no capillaries, resembling hydatidiform change, may have been aspirated from the extraplacental chorion which degenerates to form the chorion laeve. The importance of selecting healthy looking villi and excluding decidual debris and maternal blood is again emphasised.

The scale of some of the problems will now be considered:

i Failure to obtain a karyotype. With direct preparations, Simoni, Terzoli, and Romitti (1986) had a 0.7 per cent failure rate, although it is much higher in other laboratories. Culture failures occurred in 1.3 per cent of cases (Hogge *et al*. 1985) and 3.2 per cent (Czepulkowski, Heaton, Kearney, Rodeck, and Coleman, 1986). A great deal of practice and experienced laboratory personnel help to keep these failures to a minimum.

ii Maternal contamination. This should be negligible with direct preparations because decidua has no mitoses. No maternal contamination was found in 218 cultures by Czepulkowski *et al*. (1986) and it was identified twice in a series of 600 cultures by Hogge *et al*. (1985).

iii Discordance between direct preparations and cultured villi. This has been noted in a few cases in the world registry material and amniocentesis has confirmed the cultured (normal) result. On the other hand, two translocations found in cultured villi could not be detected in the direct preparations even after looking at them again, probably due to rather poor quality of the latter (Czepulkowski *et al*. 1986).

iv Discordance between villi and fetal tissue. In 6 cases with abnormal karyotypes on direct preparations (out of 250) Brambati, Simoni, Danesino, Oldrini, Ferazzi, Romitti, Terzoli, Rossella, Ferrari, and Fraccaro (1985) were not able to verify this on fetal tissue after termination of pregnancy. Hogge *et al*. (1985) also found a 2 per cent discrepancy between cultured villi and the fetus, most commonly involving mosaicism.

v The incidence of karyotypic abnormalities in villi. Several of the above authors found these in 5.9–7 per cent of cases, which is 2–3 times as high as in amniotic fluid in the second trimester and suggests that over half the affected fetuses abort before 16 weeks. Mosaicism comprises a quarter of the abnormalities; it is usually not present in the fetus, but even then there is a higher risk of spontaneous abortion (Hogge *et al.* 1985). How many of them are due to pseudomosaicism arising in culture is not known. Further studies are required comparing direct preparations with cultured villi, and the latter with placental and fetal cultures obtained from aborted material.

The mitotic and chromosomal properties of second and third trimester villi also need to be investigated, with a view to rapid cytogenetic diagnosis in fetal anomalies detected by ultrasound scanning (Nicolaides *et al.* 1986b).

2   METABOLIC DISEASES

Individually, these diseases are rare, but there are many of them and they are usually severe. The majority are autosomal recessive disorders, with a recurrence risk of 1 in 4 and most parents are aware of being carriers because they have already had an affected child. For these reasons, they form an important group in prenatal diagnosis and first trimester diagnosis is particularly beneficial for the patients at risk.

The first prerequisite for fetal diagnosis is to establish without doubt the genetic and biochemical diagnosis and to confirm carrier status by family studies. Then, reference ranges need to be established in direct assays on villi and in cultured villi from normal, heterozygous and affected pregnancies (Kazy *et al.* 1982; Simoni *et al.* 1983, Galjaard and Kleijer 1986). For some disorders, quantitative analysis of incorporated radioactivity is necessary by autoradiography, (for example, Lesch-Nyhan syndrome), or by scintillation counting, (for example, citrullinaemia, methylmalonic acidaemia). In most cases, 5–10 mg of villi is sufficient for diagnosis. Great care must be taken to avoid contaminating maternal tissues, as even tiny amounts can lead to a false diagnosis (Grabowski, Kruse, Goldberg, Chokkalingan, Gordan, Blakemore, Mahoney, and Desnick 1984).

A list of conditions in which first trimester fetal diagnosis has been made or excluded using chorionic villi is shown in Table 4.5. European data is being collected by Dr L Poenaru although not all countries have yet contributed. Information is available on 258 cases, 56 (22 per cent) of whom had affected fetuses. Confirmation of the diagnosis in abortus tissue was available in 53. Occasional difficulties have arisen because of inadequate samples or contamination with maternal tissue and in 4 cases there were discrepancies between direct assays on the villi and cultured villous and amniotic fluid cells. The presence of isoenzymes may also cause diagnostic inaccuracies as in metachromatic leukodystrophy (Sanguinetti, Marsh, Jackson, Fensom, Warren, and Rodeck 1986).

Table 4.5

Metabolic diseases diagnosed on chorionic villi

Adenosine deaminase deficiency	Mannosidosis
Argininosuccinicaciduria	Maple syrup urine disease
Citrullinaemia	Menke's syndrome
Chondrodysplasia punctata	Metachromatic leukodystrophy
Congenital adrenal hyperplasia	Methylmalonicacidaemia
Cystinosis	Mucolipidosis I
Fabry's disease	Mucolipidosis II (I cell disease)
Fanconi's disease	Mucopolysaccharidosis I (Hurler's disease)
Galactosaemia	Mucopolysaccharidosis II (Hunter's disease)
Gaucher's disease	Mucopolysaccharidosis III (San Fillipo disease)
Glutaricaciduria	Mucopolysaccharidosis VI
Glycogenosis II (Pompe's disease)	Mucopolysaccharidosis VII
Glycogenosis III	Multiple sulphatase deficiency
Gm$_1$ gangliosidosis (Landing)	Niemann-pick disease
Gm$_2$ gangliosidosis (Tay Sachs)	Propionicacidaemia
Gm$_2$ gangliosidosis (Sandhoff)	Refsum disease
Hypophosphatasia	Tyrosinaemia I
Krabbe's disease	Wolman's disease
Lesch-Nyhan disease	Zellweger syndrome

While it is reasonable to advise termination of pregnancy on the basis of a positive result, until more experience has been gained it is recommended that if the diagnosis indicates that the fetus is unaffected, this is later checked by amniocentesis.

3   CONDITIONS DIAGNOSED BY DNA ANALYSIS

The rapid progress in molecular genetics was one of the factors stimulating the development of CVB, and prenatal diagnosis was the first area in which this new technology had a clinical impact (see Gosden and Gosden 1985). The examination of the genotype both removes the need to acquire the phenotypic expression of the gene for analysis (for example, an enzyme or other protein) and also permits such studies to be made even if we do not know the cause of the disease, as in cystic fibrosis.

CVB has in turn facilitated these investigations as more DNA can be extracted from villi than from amniotic fluid; 20 mg of villi should yield about 20 μg of DNA.

If a gene has been identified and cloned (Table 4.6), this cloned DNA is used as a radioactive probe to search for the gene in the DNA from the sample, after it has been cut up by specific restriction endonucleases. Diseases caused by gene deletions can be diagnosed by direct DNA analysis, for example, sickle cell disease (Goossens *et al.* 1983). Where a point mutation has been identified, synthetic oligonucleotide probes may be used,

Table 4.6
Disease diagnosed by DNA analysis

Gene-specific probes	α-Thalassaemia
	β-Thalassaemia
	Sickle cell disease
	Haemophilia A
	Haemophilia B
	Antithrombin III deficiency
	α_1-Antitrypsin deficiency
	Phenylketonuria
Gene unknown	Cystic fibrosis
	Duchenne muscular dystrophy
	Huntington's chorea
Chromosome-specific probes	e.g. Y Chromosome
Viral c DNA	e.g. Rubella

as for α_1-antitrypsin deficiency (Kidd, Wallace, Itakura, and Woo 1983). At present, technical difficulties limit the use of this approach, although it has been successful for the β-thalassaemia mutation in Sardinia (Rosatelli, Falchi, Tuveri, Scalas, Di Tucci, Monni, and Cao 1985).

However, most diseases are very heterogenous; thirty different mutations are known to cause β-thalassaemia. For these, linkage analysis using restriction fragment length polymorphisms (RFLPs) is required. Even when the gene is not known (Table 4.6) extragenic probes are being used to identify RFLPs closely linked to the gene. Unfortunately, they are not informative in all families and the risk of recombination during meiosis, which is higher the further away the RFLP is from the gene, can lead to misdiagnosis (Winter, Harper, Goldman, Mibashan, Warren, Rodeck, Penketh, Ward, Hardisty, and Pembrey 1985).

The earliest experience of genotypic diagnosis in the first trimester was obtained with the haemoglobinopathies (Williamson *et al.* 1981; Old *et al.* 1982) and it is now the largest. In 200 pregnancies investigated there was only one failed diagnosis due to partial digestion of DNA, one misdiagnosis due to contamination with plasmid DNA (Old, Fitches, Heath, Thein, Weatherall, Warren, McKenzie, Rodeck, Modell, Petrou, and Ward 1986). Haemophilia A and B (Winter *et al.* 1985) and α_1-antitrypsin deficiency (Kidd, Golbus, Wallace, Itakura, and Woo 1984) are also being diagnosed increasingly in the first trimester and recently, the first few cases of cystic fibrosis have had fetal diagnosis (Farrall, Law, Rodeck, Warren, Stanier, Super, Lissens, Scambler, Watson, Wainwright, and Williamson 1986). Chromosome-specific probes have not yet proved to be useful with the exception of the Y chromosome for fetal sexing (Gosden *et al.* 1982). The presence of viral nucleic acid can be detected by hybridisation and fetal rubella infection has been diagnosed using a cDNA probe (Terry, Ho-Terry, Warren, Rodeck, Cohen, and Rees 1986).

These encouraging results clearly outweigh some of the limitations of DNA analysis. They include the expensive technology; the time consuming family studies that often have to be done, preferably before pregnancy, and which may show that the family is not informative for one or more DNA markers; and whether paternity is in question. Undoubtedly, however, cheaper, simpler methodology will be developed as well as new probes, leading to wider and further applications in prenatal diagnosis.

VII   Some consequences of chorion villus biopsy

1   FOR PATIENTS

Most patients see CVB as having clear advantages over 2nd trimester fetal diagnosis and its high degree of acceptability is altering their expectations: the feasibility of early diagnosis and early result is leading to greater haste and urgency and less tolerance of delay, factors that are intensified because many patients at the present are self-referred. These attitudes must still be tempered with caution until the results of longer term studies on the safety of CVB are obtained.

2   FOR THE MEDICAL PROFESSION

CVB is perceived by doctors as an exciting new technique which many will be attracted to and wish to provide. More opportunities for fetal diagnosis will arise, both for single gene defects and possibly for diseases with polygenic inheritance. Difficult ethical problems will have to be faced, such as with Huntington's chorea or requests for fetal sexing and selective abortion of fetuses of unwanted sex. These need to be considered in the context of the individuals involved, of patients' families and society.

3   FOR THE HEALTH SERVICES

In most Western countries, only 30–50 per cent of women with an indication for fetal diagnosis have it. If CVB is shown to have a sufficiently low risk, it will lead to an increase in uptake and a need for more resources. At the same time more genetic handicap will be prevented and the cost:benefit rate should improve. Antenatal care and counselling services will be particularly affected with greater emphasis being placed on pre-pregnancy counselling. Facilities for early, quick referral in pregnancy and well-trained clinical and laboratory staff will be required. Plans for their provision have scarcely begun. When heterozygote detection for the commonest gene in our population, cystic fibrosis, becomes possible, the organisational problems assume a different order of magnitude.

Acknowledgements

The author is grateful for the support of the Medical Research Council and the National Fund for Research into Crippling Diseases while he was Director of the Harris Birthright Research Centre for Fetal Medicine, King's College School of Medicine and Dentistry.

References

Brambati, B., Simoni, G., Danesino, C., Oldrini, A., Ferrazzi, G., Romitti, L., Terzoli, G. L., Rossella, F., Ferrari, M., and Fraccaro, M. (1985). First trimester fetal diagnosis of genetic disorders: clinical evaluation of 250 cases. *J. med. Genet.* **22**, 92–99.

—— Oldrini, A., Ferrazi, E., and Lanzani, A. (1986*a*) Chorionic villus sampling: an analysis of the obstetric experience of 1000 cases. *Prenat. Diagn.* (in press).

Brambati, G., Guercilina, S., Bonacchi, I., Oldrini, A., Lanzani, A., and Piceni, L. (1986*b*). Feto-maternal transfusion after chorionic villus sampling: clinical implications. *Human Reprod.* **1**, 37–40.

Czepulkowski, B. H., Heaton, D. E., Kearney, L. U., Rodeck, C. H., and Coleman, D. V. (1986). Chorionic villus culture for first trimester diagnosis of chromosome defects: evaluation by two London centres. *Prenat. Diagn.* In Press

Farrall, M., Law, H-Y., Rodeck, C. H., Warren, R., Stanier, P., Super, M., Lissens, W., Scambler, P., Watson, E., Wainwright, W., and Williamson, R. (1986). First trimester diagnosis of cystic fibrosis with linked DNA probes. *Lancet, i*, 1402–1405.

Galjaard, H., and Kleijer, W. J. (1986). Biochemical analysis of chorionic villi. In *Chorionic Villus Sampling* (eds. B. Brambati, G. Simoni, and S. Fabro) pp. 131–152. Marcel Dekker, Inc. New York and Basel.

Goldberg, M. F., Chen, A. T. L., Young, W. A., and Reidy, J. A. (1980). First trimester fetal chromosome diagnosis using endocervical lavage: a negative evaluation. *Am. J. Obstet. Gynecol.* **138**, 435–440.

Gosden, J. R. and Gosden, C. M. (1985). Recombinant DNA technology in prenatal diagnosis. In *Oxford Reviews of Reproductive Biology*, Vol. 7 (ed. J. R. Clarke) pp. 73–117 Clarendon Press, Oxford.

—— Mitchell, A. R., Gosden, C. M., Rodeck, C. H., and Morsman, J. M. (1982). Direct vision chorion biopsy and chromosome-specific DNA probes for determination of fetal sex in first trimester prenatal diagnosis. *Lancet ii*, 1416–1419.

Goossens, M., Dumez, Y., Kaplan, Y., Lupker, M., Chabret, C., Henrion, R., and Rosa, J. (1983). Prenatal diagnosis of sickle cell anaemia in the first trimester of pregnancy. *New Engl. J. Med.* **309**, 831–833.

Grabowski, G. A., Kruse, J. R., Goldberg, J. D., Chokkalingan, K., Gordan, R. E., Blakemore, K. J., Mahoney, M. J., and Desnick, R. J. (1984). First trimester prenatal diagnosis of Tay-Sachs disease. *Am. J. hum. Genet.* **36**, 1369–1378.

Gustavii, B. (1983). First trimester chromosomal analysis of chorionic villi obtained by direct vision technique. *Lancet, ii*, 507–508.

Hahnemann, N. (1974). Early prenatal diagnosis: a study of biopsy techniques and cell culturing from extra-embryonic membranes *Clin. Genet.* **6**, 294–306.

Hahnemann, N. and Mohr, J. (1968). Genetic diagnosis in the embryo by means of biopsy from extra-embryonic membranes. *Bull. Eur. Soc. human Genet.* **2**, 23.

Heaton, D. E., Czepulkowski, B. H., Horwell, D. H. and Coleman, D. V. (1984). Chromosome analysis of first trimester chorionic villus biopsies prepared by a maceration technique. *Prenat. Diagn.* **4**, 279–287.

Hogge, W. A., Schonberg, S. A., and Golbus, M. S. (1985). Prenatal diagnosis by chorionic villus sampling: lessons of the first 600 cases. *Prenat. Diagn.* **6**, 393–400.

Horwell, D. H., Loeffler, F. E., and Coleman, D. V. (1983). Assessment of a transcervical aspiration technique for chorionic villus biopsy in the first trimester of pregnancy. *Br. J. Obstet. Gynaecol.* **90**, 196–198.

Jackson, L. G. (1986). Surveillance of risks in first trimester prenatal diagnosis. In *Chorionic Villus Sampling* (eds. B. Brambati, G. Simoni, and S. Fabro) pp. 213–226. Marcel Dekker, Inc. New York and Basel.

Kaplan, L., Dumez, Y., and Goossens, M. (1983). A method for fetal tissue sampling by chorion biopsy: a new approach to first trimester prenatal detection of abnormal genes. *IRCS Med. Sci.* **11**, 85–86.

Kazy, A. and Stigar, A. M. (1980). Kozvetlin Real-time (ultrahang) kontroll melletti chorion biopsy. *Orv. Het.* **121**, 2765–2766.

Kazy, Z., Rozovsky, I. S., and Bakharev, V. A. (1982) Chorion biopsy in early pregnancy: a method of early prenatal diagnosis for inherited disorders. *Prenat. Diagn.* **2**, 39–45.

Kidd, V. J., Wallace, R. B., Itakura, K., and Woo, S. L. C. (1983). α_1-antitrypsin deficiency detection by direct analysis of the mutation in the gene. *Nature, Lond.* **304**, 230–234.

Kidd, V. J., Golbus, M. S., Wallace, R. B., Itakura, K., and Woo, S. L. C. (1984). Prenatal diagnosis of α_1-antitrypsin deficiency by direct analysis of the mutation site in the gene. *New Engl. J. Med.* **310**, 639–641.

Kullander, S., and Sandahl, B. (1973). Fetal chromosome analysis after transcervical placental biopsies in early pregnancy. *Acta obstet. gynecol. scand.* **52**, 355–359.

Liu, D. T. Y., Mitchell, J., Johnson, J., and Wass, D. M. (1983). Trophoblast sampling by blind transcervical aspiration. *Br. J. Obstet. Gynaecol.* **90**, 1119–1123.

—— Symonds, E. M., Jeavons, B., and Norman, S. (1985). Transcervical chorionic villus biopsy with a brush. *Prenat. Diagn.* **5**, 349–355.

McFadyen, I. R. (1985). Missed abortion and later spontaneous abortion in pregnancies clinically normal at 7–12 weeks. *Eur. J. Obstet, Gynaec and reprod. Biol.* **20**, 381–384.

Maxwell, D., Lilford, R., Czepulkowski, B., Heaton, D., and Coleman, D. (1986). Transabdominal chorionic villus sampling. *Lancet i*, 123–126.

Niazi, M., Coleman, D. V. and Loeffler, F. E. (1981). Trophoblast sampling in early pregnancy. Culture of rapidly dividing cells from immature placental villi. *Br. J. Obstet. Gynaec.* **88**, 1081–1085.

Nicolaides, K. H., Soothill, P. W., Rodeck, C. H., Warren, R. C. and Gosden, C. M. (1986a). Why confine chorionic villus (placental) biopsy to the first trimester? *Lancet i*, 543–544.

Nicolaides, K. H., Rodeck, C. H., and Gosden, C. M. (1986b). Rapid karyotyping in non-lethal malformations. *Lancet i*, 283–287.

Old, J. M., Fitches, A., Heath, C., Thein, S. L., Weatherall, D. J., Warren, R. C., McKenzie, C., Rodeck, C. H., Modell, B., Petrou, M., and Ward, R. H. T. (1986). First trimester fetal diagnosis for haemoglobinopathies: report on 149 cases. *Lancet ii*, 763–767.

—— Ward, R. H. T., Petrou, M., Karagozlou, F., Modell, B., and Weatherall, D. J.

(1982). First trimester fetal diagnosis for haemoglobinopathies: a report of 3 cases. *Lancet ii*, 1414–1416.

Rhine, S. A., Palmer, C. G. and Thompson, T. (1977). A simple first trimester alternative to amniocentesis for prenatal diagnosis. *Birth defects Orig. Art.* **Ser. XII 3D**, 231–247.

Rodeck, C. H., and Morsman, J. M. (1983). First trimester chorion biopsy. *Br. med. Bull.* **39**, 338–342.

—— —— Gosden, C. M., and Gosden, J. R. (1983*a*). The development of an improved technique for first trimester microsampling of chorion. *Br. J. Obstet. Gynaec.* **90**, 1113–1118.

—— —— Nicolaides, K. H., McKenzie, C. M., Gosden, C. M., and Gosden, J. R. (1983*b*). A single-operator technique for first trimester chorion biopsy. *Lancet ii*, 1340–1341.

Rosatelli, C., Falchi, A. M., Tuveri, T., Scalas, M. T., Di Tucci, A., Monni, G., and Cao, A. (1985). Prenatal diagnosis of β-thalassaemia with the synthetic oligomer technique. *Lancet i*, 241–243.

Sanguinetti, N., Marsh, J., Jackson, M., Fensom, A. H., Warren, R. C., and Rodeck, C. H. (1986). The arylsulphatases of chorionic villi: potential problems in the first trimester diagnosis of metachromatic leukodystrophy and Maroteaux-Lamy disease. *Clin. Genet.* in press.

Simoni, G., Brambati, B., Danesino, C., Rosella, F., Terzoli, G. L., Ferrari, M., and Fraccaro, M. (1983). Efficient direct chromosome analyses and enzyme determinations from chorionic villi samples in the first trimester of pregnancy. *Hum. Genet.* **63**, 349–357.

—— Terzoli, G., and Romitti, L. (1986). Fetal karyotyping by direct chromosome preparation. In *Chorionic Villus Sampling* (Eds. B. Brambati, G. Simoni and S. Fabro) pp. 99–117. Marcel Dekker, Inc. New York and Basel.

Smidt-Jensen, S. and Hahnemann, N. (1984). Transabdominal fine needle biopsy from chorionic villi in the first trimester. *Prenat. Diagn.* **4**, 163–169.

Terry, G. M., Ho-Terry, L., Warren, R. C., Rodeck, C. H., Cohen, A., and Rees, K. R. (1986). First trimester prenatal diagnosis of congenital rubella: a laboratory investigation. *Br. med. J.* **292**, 930–933.

Tietung Hospital of Anshan Iron and Steel Co., Department of Obstetrics and Gynaecology (1975) Fetal sex prediction by sex chromatin of chorionic villi cells during early pregnancy. *Chin. med. J.* **1**, 117–126.

Ward, R. H. T., Modell, B., Petrou, M., Karagozlou, F., and Douratsos, E. (1983). A method of chorionic villus sampling in the first trimester of pregnancy under real-time ultrasonic guidance. *Br. med. J.* **286**, 1542–1544.

Warren, R. C., Butler, J., Morsman, J. M., McKenzie, C. F., and Rodeck, C. H. (1985). Does chorionic villus sampling cause fetomaternal haemorrhage? *Lancet i*, 691.

Warren, R. C. and Rodeck, C. H. (1986). Experience with a silver cannula for transcervical chorion villus sampling. *Proceedings of meeting of First Trimester Fetal Diagnosis*, Lausanne 1985. In Press.

Williams, P. L. and Warwick, R. (1980). Gray's Anatomy, 36th Edition, pp. 88–107 Churchill Livingstone, Edinburgh.

Williamson, R., Eskdale, J., Coleman, D. V., Niazi, M., Loeffler, F. E., and Modell, B. (1981). Direct gene analysis of chorionic villi, a possible technique for the first trimester diagnosis of haemoglobinopathies. *Lancet ii*, 1125–1127.

Winter, R. M., Harper, K., Goldman, E., Mibashan, R. S., Warren, R. C., Rodeck, C. H., Penketh, R. J. A., Ward, R. H. T., Hardisty, R. M., and Pembrey, M. E. (1985). First trimester prenatal diagnosis and carrier detection of haemophilia A using the linked DNA probe DX13. *Br. med. J.* **291**, 765–769.

5   Human sex pre-selection
R. J. LEVIN

I   Introduction

This short review deals with the background and methods that have been developed to try to influence the fertilization of human ova by sperm, so that the parental choice of conceiving either a male or female child can be realised. As this human desire can be traced back at least to the Egyptian culture nearly fourteen centuries before the common era (BCE), the literature is huge and scattered. Much of the early work is now of little importance since serious studies only began when human mid-cycle homogametic ovulation and the chromosomal basis of sex via the heterogametic spermatozoa were established. The historical section is therefore brief: for those who wish to read about it in greater detail the popular but comprehensive account by Whelan (1977) is a useful source of references. Recent accounts of the prospects for sexing mammalian spermatozoa are in a conference proceedings edited by Amann and Seidel (1982).

1   HISTORICAL PERSPECTIVE

Because of the long interval between human conception and the final delivery, it is likely that primitive man took a very long time before human coitus was linked to the creation of the born child. Indeed some primitive people are said never to have made the connection. Once, however, the link was established the desire to control the sex of the offspring must surely have arisen. The Greek philosophers of antiquity (5th century BCE) speculated a great deal on the possible mechanism and gave us the recorded beginning of a series of contentious theories, hypotheses and biological myths in an effort to explain how the sex of the conceived child could be influenced (Edwards 1962).

Anaxagorus (500–428 BCE) suggested that the male determined embryonic sex by producing 'male' semen from the right testicle and 'female' from the left and as late as the 18th century French noblemen were told that removal of the left testicle would guarantee an heir. The womb was credited by Empedocles (494–434 BCE) and Democritus (470–370 BCE) to be the seat of sex determination, the former suggesting that the timing of coitus in the menstrual cycle affected the sex of the child. Aristotle (ca 350 BCE) reviewed all the then current theories and with his characteristic critical acumen showed their illogicalities and dismissed all the suggested mechanisms (overlooking, however, that of Hippocrates 420 BCE). Aristotle proposed his own theory that females were formed due to a parent's lack of 'heat', the deficiency of which caused the degeneration of reproductive fluids which became female-producing. Hippocrates, the father of modern medicine, had proposed that men and women secreted substances that had varying degrees of maleness and potency and thus when mixed together gave rise to males, females and hermaphrodites. Other early civilizations put forward their wisdom about the formation of males and females. The Talmud (completed between the fourth and sixth centuries C.E.) suggested that 'the determination of sex takes place at the moment of cohabitation. When the women emits her semen (orgasm?) before the man, the child will be a boy. Otherwise it will be a girl'. Focus on the orgasm as a mechanism influencing the sex of the child is also found in Tantric texts (7th to 17th century C.E.) that declared that if 'at the moment of orgasm the solar breath (sun is considered the masculine, fiery energy and solar breath is taken via the right nostril) dominates in man and the lunar breath (the moon is considered the feminine cooling energy and lunar breath is taken in via the left nostril) in woman and conception occurs the child will be male'. The opposite situation is said to produce female offspring while hermaphrodites were the result of the same dominant nostril breathing in both partners. Hindu love postures were created specifically to stabilise irregularities of breathing and create healthy, balanced children (Douglas and Slinger 1982).

Over the centuries remarkably little advance was made over such ideas but endless theories were constructed. Even by the 18th century something like 262 groundless hypotheses had been collected and by the beginning of the 20th century Thomson (1908) remarked that the number of speculations as to the nature of sex had doubled. However, by the end of the first decade of this century, the biological mechanism of sex determination had become established and its chromosomal mechanism described. Cytological confirmation of human sex chromosomes (X and Y) was published by Painter (1924). The identification of the sex-chromosome constitution of man is relatively recent, but it is now possible to type accurately the sex-chromosome constitution of individuals. Normal females are XX individuals while normal males are XY.

2   X AND Y SPERMATOZOA

Like other mammals human sex determination is brought about by the existence of two distinct populations of spermatozoa, one carrying an X chromosome and the other a Y chromosome. Rorvik and Shettles (1977) called these the gynosperm and the androsperm (presumably from the zoological literature on animal spermatozoa) but the nomenclature has never caught on and it is little used by the rest of the scientific community. The X and Y chromosomes are of unequal size, the former being larger and contains approximately 3.5 per cent more DNA. This weight difference, often the inspiration for techniques to separate the X and Y sperms, only becomes manifest as a change in specific gravity at the third or fourth decimal place (Lindahl 1971). Shettles (1961) claimed to have identified the two types of sperm by phase contrast microscopy. The Y chromosome was thought to reside in a small round head while the X chromosome was in an oval-shaped head. The finding was vigorously contested by a number of investigators (Bishop 1960; Rothschild 1960; Van Dujin 1960) and has not been subsequently confirmed even by scanning EM (Zaneveld, Gould, Humphrey and Williams 1971). In fact, morphological estimation of human spermatozoa is exceedingly complicated because there is a great variation in their shape even in normal, fertile men, as is true also of the gorilla.

X and Y spermatozoa are probably produced in the human seminiferous tubules in equal numbers, because they are created by division of the normal chromosome complement. There is no experimental evidence or observations that the maturation or survival rates of X and Y sperm normally differ in the male reproductive tract, but they may be influenced by specific conditions (see human sex ratio) giving rise to imbalances in their ratio in the ejaculate (Quinlivan and Sullivan 1974).

A major breakthrough in the study of human spermatozoa came about when it was found that the staining of one end of the Y chromosome by quinacrine, observed in human cells, was also present in the chromatin of the mature spermatozoa with the Y chromosome (Barlow and Vosa 1970). The part stained by quinacrine fluoresced (the F-body) and was claimed to be the distal segment of the Y chromosome. The staining technique has been used extensively by those investigators wanting to distinguish between the X and Y spermatozoa, especially when trying to show that their method enriched the ratio of Y and X spermatozoa. However, other investigators have suggested caution in assuming that the F-body fluorescence can always be equated with the number of Y spermatozoa. Beatty (1977) discussed the use of the F-body as the Y spermatozoa marker. He concluded that not all F-bodies are Y chromosomes and suggested that it should be standard practice to distinguish between F-bodies (that which is actually observed) and Y chromosomes (that which is inferred), since some Y chromosomes are not represented by F-bodies while some apparent F-bodies do not represent Y chromosomes. Gledhill (1983) has more recently questioned the sole use of the identification of Y spermatozoa by quinacrine since with this method up to 5.6 per cent of spermatozoa possess two spots. If these are always assumed to be two Y chromosomes then this represents a very high non-disjunction rate, which is not confirmed by modern cytogenetic studies of human spermatozoa (Martin, Balkan, Burns, Rademaker, Lin, and Rudd 1983). Corroboration of any enrichment procedures should, in Gledhill's view, be undertaken using other means of establishing the sex ratio of embryos in animals, by *in vitro* fertilization of zona-free hamster eggs by human sperm (Yanagimachi 1984) or by evaluating the claimed separation of X and Y by flow cytometry (Garner, Gledhill, Pinkel, Lake, Stephenson, Van Dilla, and Johnson 1983).

Because of the above experimental difficulties with F-body fluorescence and the effect of observer performance on counting rate (Roberts and Goodall 1976) other methods of idenfitying the Y spermatozoa have been investigated.

Recently Y sperm have been identified with phytohaemagglutinin (Hegde, Shastry, and Rao 1981) using an indirect immunofluorescent method that was claimed to have advantages over the quinacrine technique, but no studies have yet appeared using the method to identify sperm after application of a procedure to enrich the proportion of Y spermatozoa. Sarkar, Jolly, Friedman, and Jones (1984) have published a new development in the identification of Y from X spermatozoa which uses recombinant DNA clone specific to human male DNA (Deininger, Jolly, Rubin, Friedman, and Schmid 1981). The human male DNA-specific probe (pPD9) hydridizes to a 3.4 kb male-specific band in the enzyme Hae-111 digest but not to any specific band in female DNA. The densities of the bands were measured after running an agarose gel electrophoresis and staining with ethidium bromide. While this recombinant DNA clone test took nearly 2 weeks to complete, it is claimed to be more sensitive in quantifying Y enrichment than the F-body test. Sarkar (1984a) claims to have estimated the relative proportions of even the X-carrying spermatozoa by using an X chromosome-specific DNA clone.

3   THE HUMAN SEX RATIOS

The proportions of the sexes in populations is usually measured by the ratio of males to females, referred to as the 'sex ratio' (that is males/females $\times$ 100). The 'primary sex ratio' is defined as that ratio present at conception and is the proportion of ova that is fertilized by Y carrying spermatozoa. The 'secondary sex ratio' is defined as the ratio at birth and in essence is the modification of the primary sex ratio by the differential mortality rate *in utero*. It is obviously the more easily observed ratio and with a knowledge of the relevant mortality rates inferences are made about the practically unaccessable primary ratio. Many factors have been claimed to influence the human sex ratio at birth (Teitelbaum 1972; see Table 5.1). The difficulty with most of the studies of the secondary sex ratio is that while it may be shown statistically to shift, little can be gleaned from the data about any possible basic mechanisms that could bring about the changes. In many of the studies it is obvious that multiple factors are present that are unlikely to have all been controlled for and it often becomes impossible to isolate one particular factor as the others confound the data. Moreover, no satisfactory explanation exists as to how any one factor actually causes the shift. The factor that has stimulated much interest and has been extensively studied is the time of insemination during the menstrual cycle.

4   TIME OF INSEMINATION DURING THE MENSTRUAL CYCLE

The idea that the timing of insemination during the menstrual cycle could influence the sex of the conceived child probably stems from Hippocrates who suggested that male births were more likely from coitus in the days immediately after the end of menstrual flow. A number of studies have been reported where either natural or artificial insemination has been claimed to

Table 5.1.
Factors claimed to affect human sex ratio

Factor	Reference
Birth order	Russell 1936*
Family size	Russell 1936*
Sex of first born	Lancaster 1950*
Maternal age	Martin 1943*
Paternal age	Russell 1936*
Relative age of parents	Ciocco 1938*
Sex of child of prior pregnancy	Turpin & Schutzenburger 1948*
General genetic factors	Slater 1944*
Race and colour	Ciocco 1938*
Inbreeding and outbreeding	Schull 1958*
Ancestral longevity	Lawrence 1941*
Physique and temperament	Heath 1954*
Baldness of father	Damon & Nuttal 1965*
Blood groups	Cohen & Glass 1956*
Birth control	Winston 1932*
Hormonal induction of ovulation	James 1985
Artificial insemination	Jalavisto 1952*
Frequency of coitus	Jalavisto 1952*
Time of conception during menstrual cycle	Lawrence 1941*
Seasonal and monthly variation	Russell 1936*
Geographical and climatic conditions	Ciocco 1938*
	Lyster & Bishop 1965*
Illegitimacy	Ciocco 1938*
Multiple sclerosis	Alperovitch & Feingold 1981
Measles	Langaney & Pison 1979
Hepatitis B	Drew, London, Lustbader, Hesser & Bumberg 1978
Placenta praevia	MacGillivray, Davey & Isaacs 1986
Prostatic cancer	Hill, Fincham, Wijayasinghe, Haronga & Hendin 1985
Socio-economic status and conditions	Winston 1931*
Parental occupation	Russell 1936*
Urban/rural and other differences	Russell 1936*
	Ciocco 1938*
Cigarette smoking	Damon, Nuttall & Salber 1966*
Coffee drinking	Vogel, Kruger, Kurth & Schroder 1966*
High speed stress (aircraft pilots)	Snyder 1961*
Radiation damage	Schull & Neel 1958*
War and post-war conditions	Russell 1936*

*For full list of references, see Teitelbaum 1972

have differential effects on the sex ratio depending on the time of insemination in the cycle. Most of these studies suffer from poor methodology and their results conflict. Reviews of the influence of the time of insemination during the menstrual cycle have been published by Whelan (1974, 1977), Guerrero (1975), Rinehart (1975), Glass (1977), Harlap (1979a) and James (1971, 1976, 1980, 1983).

Modern attention to the old idea was reactivated by Kleegman's (1954) finding that artificial insemination (AI) immediately before ovulation was more likely to bring about a boy conception than insemination earlier or later in the cycle. Guerrero (1974), reporting on the sex ratio from natural (coital) insemination, found that the variation of sex ratio with the cycle day was U-shaped, male zygotes appear to be formed preferentially both at the early and late part of the fertile period while female zygotes were preferentially formed during the preovulatory period. Unexpectedly he obtained an inverted U-shaped distribution when artificial insemination was used. Thus, artificial insemination just prior to or at the time of ovulation should result in a preponderance of males while insemination 2 days or more prior to ovulation or some hours after should produce a prevalence of females. As AI procedures in clinics must try to bring the time of insemination as close as possible to the actual time of ovulation they all should, in theory, be showing a predominance of born males. Mortimer and Richardson (1982) surveyed the international AI literature and contacted major British centres providing AI and compared the sex ratio for fresh and for frozen semen with that found in normal, coital legitimate births during the period 1969–1978 in England, Scotland and Wales (51.5 per cent males). The results were complicated. The British data appeared to show that both types of AI significantly reduced the prevalence rate of male infants, but this was significantly lower than in the combined data from other European countries and U.S.A. Comparison of all the births after AI with fresh semen with data from British live births indicated that AI with fresh semen significantly increased the prevalence of male infants (51.5 to 57.7 per cent). This confirmed the data of Guerrero (1975). AI with cryostored semen slightly reduced the prevalence of males from that in the general population (51.5 to 49.7 per cent). Cryostorage thus appears to cause a reduction in the prevalence of male births of 8 per cent. Unfortunately a similar independent survey by Alfreddsson (1984) on the sex ratio of the births from AI with fresh and frozen semen did not confirm that cryostorage had any effect on the sex ratio at birth. Moreover, Alfreddsson could not find any increase in the sex ratio with fresh semen-AI when compared to that in the 10 year period of Icelandic births (1972–1981).

Harlap (1979a, b) investigating births to orthodox Jewish women who observed the religious ritual of coital abstinence for the week following the end of their menstruation (niddah), confirmed the findings of Guerrero (1974) in his study of natural (coital) insemination; namely, an excess of boys at the early and late time of the fertile period and an excess of girls in the middle of the fertile period. Alternative interpretations have been suggested by Vollman, James, Corson, and Rosner (1979). The usual criticisms offered against the Guerrero and Harlap studies are: 1.) that the experimental populations used were highly selected, in the case of Guerrero's, the women were those who had failed with a rhythm method of contraception or they could have been subfertile, while the Orthodox Jewish women studied by Harlap are obviously highly selected since they practised strict ritual periodic abstention from coitus; 2.) the time of ovulation was obtained by indirect methods either by analysis of the basal body temperature graphs (Guerrero 1975) with all its attendent difficulties (McCarthy and Rockette 1983) or by application of the equation $Y = N-14$ where Y is the expected day of ovulation and N is the reported mean cycle length for each woman (Harlap 1979a); 3.) data were obtained about coital behaviour of the conceiving cycle on the first day post partum by interview (Harlap 1979a). James (1983) recognizes these criticisms (and other minor ones) but does not think that they are very weighty. He argues that the likelihood that the results could have come about by chance are small but that it is not impossible that some artifact rather than the claimed phenomena could be responsible for the result.

Suggestions have been made about possible mechanisms allowing preferential formation or survival of male or female zygotes, but these are largely without any experimental evidence. Guerrero (1975) suggested that variations in vaginal and cervical pH might affect differentially the X and Y spermatozoa. The possible effect of pH on spermatozoa will be discussed in greater detail in the section on chemical factors affecting sperm functions and motility. Harlap (1979a) proposed that the differential survival of male and female conceptuses resulted from the fertilization of over-ripe ova by over-ripe spermatozoa. Over-ripeness of both gametes is associated with decreases in fertilization rates but increases in abnormal ova, chromosomal abnormalities and blastocyst deaths (Iffy 1963). In humans, only circumstantial evidence suggests that delayed fertilization causes chromosomal abnormalities and other congenital anomalies (Simpson 1979). Roberts (1978) examined the hypotheses designed to explain human sex ratio fluctuations and came to the conclusion that those based on the different maturation rates or effective lifetimes of X and Y spermatozoa could not account for many of the observations. He proposed a physiological model based on the supposed slightly greater motility of Y-bearing over X-bearing spermatozoa in the female reproductive tract. The cervical mucus was proposed as a site for this to take effect, operating for only a limited critical time before ovulation. Thereafter the decreasing viscosity of cervical (and uterine) fluids would reduce the advantages of the Y's motility. Another proposal invoked muscular activity of the tract which 'increases towards ovulation' possibly causing mixing of the contents. James (1983) proposed a different mechanism to explain the U-shaped curve of Guerrero (1974). The mother's gonadotrophin levels 'rise and fall very rapidly across the fertile period ... ' so that 'if these hormone levels were to affect the sex of the zygote then one might expect an excess of girls among the infants born following induction of ovulation by clomiphene or gonadotrophin (because the hormone surge seems to accompany an increased probability of female conceptions in normal cycles)'. James (1985) found that the lower proportion of male live-births following artificial induction of human ovulation was highly significant statistically. More recent data obtained by Ben-Rafael, Matalon, Blankstein, Serr, Lunenfield, and Mashiach (1986) do not support the reports of increased numbers of female infants in those conceived after induction of ovulation by human menopausal gonado-trophin and human chorionic gonadotrophin. The possible role of gonado-trophins acting via sex-selective abortion was ruled out by studies on dizygotic twinning (James 1983).

A number of side effects can be expected if the time of insemination during the menstrual cycle has a role in sex determination. James (1971) suggested that human coital rate could be causally associated with the sex of conceived children, high rates being associated with a prevalence of males. He further developed this idea (James 1983) to explain on the one hand the increase in the sex ratio during and just after wars, in the first few months of marriage and with younger marriage partners, and on the other hand the decrease associated with parity and with the presence of dizygotic twins. He concluded that the evidence was only decisive for the wartime effect. High coital rates would be expected in and around wartime associated with short home leaves and demobilization homecomings. As Bartos (1980) has pointed out the variations in human sex ratios can be explained by an entirely different hypothesis – that of coitus-induced ovulation (see Jöchle 1973, 1975, for discussion and references about such ovulations in women). He argued that coitus-induced ovulation could create the conditions appropriate for the increases in the sex ratio. If ova were immediately made available by the coital act inducing ovulation then any initial, temporary increase in Y sperm reaching the oviduct would significantly increase the chance of a male zygote being conceived. A number of studies have suggested that the Y spermatozoa are favoured by the female reproductive tract at specific times but the results have been contentious. They will be discussed in the next section on factors influencing sperm in the female reproductive tract.

i Prospective study of the preselection of the sex of the zygote by timing coitus relative to ovulation

Most studies quoted previously have been retrospective with all its attendant difficulties. France, Graham, Gosling, and Hair (1984) attempted to study the sex preselection theory of Shettles (1970) using data from 33 pregnancies in which the fertile coital act could be unequivocally identified. Couples were recruited who desired to preselect the sex of their child. If they desired a girl they were told to have coitus only 2–3 days before ovulation, or at ovulation if they desired a boy. The women were instructed to assess the 'peak day' of mucus symptoms, claimed to be a reliable indicator of ovulation (Billings, Billings, Brown, and Burger 1972; Hilgers, Abraham, and Cavanagh 1978), to take and chart their basal body temperature (BBT) and collect their urine for LH and creatinine assay. The pre-ovulatory surge of LH, peak mucus symptom and shift in BBT were used as indicators to define the time of ovulation in the cycle. Only 39 per cent of the couples obtained an infant of the sex they desired. Male infants were conceived by coitus occurring from 5 to 1 day before fertilization while female infants were conceived from coitus from 4 days before fertilization to 1 day after ovulation. Sixtyeight per cent of the male infants (total = 22) came from sperm survival of 2 d or more before fertilization but the majority of females (64 per cent) (total = 11) were conceived from sperm present for 1 day or less in the female reproductive tract. The findings, although based on a very small sample, were claimed to refute the theory of Kleegman (1954) and Shettles (1970) that a male child is more likely to be conceived from coitus close to the time of ovulation and to support the work of James (1971, 1980) and Guerrero (1974) that the sex ratio favours males the greater the interval between coitus and ovulation.

II   Putative factors affecting sperm functions in the female reproductive tract

Once the semen is ejaculated into the vagina and liquifaction occurs, the spermatozoa are able to leave the seminal fluid and come into contact with the fluids secreted by the human female genital tract. The chemical make-up of many of these secretions is only crudely known: indeed it was only recently that the concentrations of the major ions and the extraordinary high levels of some amino acids were determined (Wagner and Levin 1980; Levin and Wagner 1983). The inaccessibility of some of the secretions represents a major obstacle to our understanding of the roles of various constituents. It is becoming more obvious, however, that the luminal fluids of the female reproductive tract represent a series of complex environments that the fertile sperm has to pass through before it can reach the ovum and that many of the chemical constituents of these fluids are potent substances (ions, hormones and substrates) that are putative factors for activating or modulating spermatozoal functions and physiology (Farooqui 1983; Sarkar 1984a). The following sections are brief accounts of a selection of these chemical factors that may play a role in influencing spermatozoal activity.

There are a number of problems in relation to the studies of the actions of various genital tract factors on human spermatozoa. Firstly, there are strong species differences in relation to spermatozoal activity so that great care has to be taken if ever results from one species are extended to human spermatozoa. Second, practically all of the published work so far utilises the whole ejaculate containing both X and Y spermatozoa. Sarkar *et al.* (1984) have claimed that X and Y human sperm behave differently in a flow stream. We have hardly any knowledge about possible differences (if any) in activity of X and Y in the various female environments. Moreover, such measures as we have of spermatozoal physiology and biochemistry are the values of all the sperm present but the important ones to measure are the few hundred (at the most) that reach the Fallopian tubes. The physiological and biochemical parameters of all the others, while easy to measure, are perhaps of lesser interest. Our present knowledge is like describing the physiology and biochemistry of marathon winners by measurements on the tail-end runners and those who fail to finish the course – a salutary thought! A final problem is that in normal human sexual congress the ejaculate is deposited into the reproductive tract (vagina) of the sexually aroused woman. Sexual arousal induces some remarkable changes in the human female reproductive tract (Levin 1980). For example, under non-aroused conditions the vagina has a hypoxic lumen but on arousal the pO_2 rises quite dramatically (Wagner and Levin 1978) and this presumably allows any swimming ejaculated spermatozoa a much greater supply of energy, for aerobic metabolism creates 12 times more ATP than anaerobic metabolic pathways. While spermatozoa can function anaerobically, Bishop (1962) concluded that 'the energy available from respiration offers a wide thermodynamic safety factor in providing for the power requirements of the cell, whereas that gained from glycolysis alone may barely count for the energy needs'.

1 SODIUM, POTASSIUM AND CALCIUM IONS IN THE FEMALE REPRODUCTIVE FLUIDS

i *Sodium and potassium ions*

The concentrations of sodium and potassium in various human genital fluids and secretions are now known (see Wagner and Levin 1980). In brief the vagina, cervix and uterine fluids have much higher concentrations of K^+ and lower concentrations of Na^+ than plasma while oviductal fluid has a K^+ concentration only twice that of plasma. Possible roles that these ions play in human spermatozoal physiology are only just beginning to be explored. While the motility of mammalian spermatozoa is strongly influenced by the composition of their incubating media few systematic studies have been accomplished with human spermatozoa. The relationship between high potassium concentrations and human sperm motility is unclear and controversial (Sheth and Rao 1962; Battersby and Chandler 1977; Guerin and Czyba 1979). Lornage, Guerin, Czyba, and Menezo (1983) assessed the effects of potassium and various other ions on the motility of human ejaculated spermatozoa. They found that the optimal K^+ concentration was 11 mM, higher values decreasing motility. In general it appeared that human spermatozoal motility was less sensitive to changes in the ionic concentration of the medium than other species. The role of ions in capacitation and the acrosome reaction has been reviewed by Fraser (1984).

ii   Calcium ions

Although the functions of animal spermatozoa are known to be influenced by the environmental level of ionised calcium (Schackmann and Shapiro 1982) the picture is less clear with human spermatozoa. The importance of ionised calcium to human sperm is contentious because *in vitro* studies (Peterson and Freund 1976; Gorus, Finsy, and Pipeleers 1982) suggest that their motility is relatively independent of the external calcium concentration unless special techniques are used to change its entry into cells (Peterson and Freund 1976). However, as Levin and Wagner (1985) pointed out, these *in vitro* experiments were undertaken in the absence of cervical factors (prolactin: see below) that modulate spermatozoa calcium interactions. Reyes, Parra, Chavarria, Goicoechea, and Rosado (1979) reported that adding prolactin to ejaculated human spermatozoa induced a 60 per cent increase in their binding/or transport of Ca^{++}. Lornage, Guerin, Czyba, and Menezo (1983) found that the best survival of human spermatozoa when incubated *in vitro* was with an external medium concentration of 0.8 mM calcium and Sarkar *et al.* (1984) observed that the calcium concentration influences the curvature of the swimming tracks of human spermatozoa *in vitro*. Pilot experiments (Levin and Wagner 1985) indicate that the level of ionised Ca^{++} in human cervical fluid *in situ* after the induction of sexual arousal is 1.01 ± 0.3 mM ($\bar{x} \pm$ S.E.M.). The role of calcium in acrosome loss, motility and fertilization, *in vitro*, has been reviewed recently by Fraser (1984).

2   PROLACTIN

Prolactin is found in human seminal fluid (Sheth, Mugatwala, Shah, and Rao 1975) and in human cervical mucus (Sheth, Vaidya, and Raiker 1976). Its level in the plasma in both men and women increases significantly with sexual arousal, especially when breast and nipples are stimulated (Jacobs and Daughaday 1974) by a partner of opposite sex. In men however, masturbation is claimed not to cause any increases in either the blood or semen prolactin levels (Zavos and Albertson 1984). It is probable that the prolactin in male genital fluids originates from the cells of the male reproductive tract (Smith and Luqman 1982). Shah, Desai, and Sheth (1976) investigated the possible physiological role of prolactin in sperm metabolism and found that it stimulated human sperm adenyl cyclase, fructose utilization and glucose oxidation at physiological concentrations. They proposed that cervical prolactin may be beneficial in storage and survival of spermatozoa in the reproductive tract. Since these early studies a number of investigations have revealed that prolactin affects oxygen uptake, glycolysis, sperm motility and survival time, cAMP levels and initiation of an acrosomal reaction. It also increases maltase activity, decreases the binding of tetracycline to sperm and increases Ca^{++} uptake (Smith and Luqman 1982). Many of these changes are associated with capacitation of spermatozoa and it has been proposed that prolactin has a role in the capacitation process (Shah and Sheth 1979). The experiments of Reyes, Parra, Chavarria, Goicoechea and Rosado (1979) suggest that prolactin may act as a calcium ionophore for human spermatozoa. Definitive studies on the significance of prolactin and calcium for human sperm functions are urgently needed.

3 HYDROGEN ION CONCENTRATION (pH)

Unterberger (1930) reported that when 52 women he was treating for temporary sterility washed their strongly acid vaginas with sodium bicarbonate just before coitus the resulting children were all male. This suggested that an alkaline milieu was detrimental to the X-spermatozoa and led to a wave of interest and claims about the use of acid-alkali douches to irrigate the human vagina before coitus to influence the survival of either the X or the Y carrying spermatozoa. Cole and Johansson (1933) critically reviewed the studies and concluded that if Unterberger's reports (1930, 1932) were correct then sex can be controlled by such means. Although numerous studies have subsequently been carried out in animals and in man, no conclusive evidence for a differential effect of pH on X and Y spermatozoa has been reported. Downing and Black (1976) for example, found that the X and Y spermatozoa survived equally well in buffers of pH 5.2 and 8 while Diasio and Glass (1971) found no effect of pH on Y sperm motility. Despite repeated negative findings, the idea persists that alkaline or acid douches influence the selection of X and Y spermatozoa, probably via effects on motility and survival (Shettles 1970; Shettles and Rorvik 1985). Shettles propounded certain human coital principles that would favour the ascent of X or Y sperm to the ovum. Deep, rear entry coitus at the time of ovulation with female orgasm preceded by alkaline douche of the vagina, would favour the Y or androsperm. Gynosperm or X spermatozoa on the other hand would be favoured by shallow, face-to-face coitus 2–3 days before ovulation without female orgasm and preceded by a weak acid douche. These principles were given much publicity. Apart from brief reports that the migration of a small number of Y-bearing spermatozoa formed the frontal zone in an *in vitro* test using human mid-cycle cervical mucus (Rohde, Porstman, and Dörner 1973; Kaiser, Citoler, and Broer 1974), there is a paucity of scientific evidence to support any of the Shettle's principles (Levin 1982). A trial of the Shettle's techniques in a Singapore clinic with couples who desired sons reported that the Shettle's method was impractical on a mass basis and appeared to have questionable effectiveness (Williamson, Lean, and Vengadasalam 1978).

III   Techniques proposed to alter the chances of conceiving a male or female child

Because they overlap it is difficult to classify exactly the various methods but it is useful initially to divide them into preconception and postconception techniques.

1   PRECONCEPTION TECHNIQUES – SEPARATION OF X AND Y SPERM

Numerous attempts have been made to separate the X and Y spermatozoa to create enriched samples of either type. The separation techniques that have been applied are discussed below.

i   Columns and density gradients

By far the most successful and popular methods (in a field littered with temporary successes) are those employing the application of ejaculates to columns or density gradients. Early studies with immobilized animal spermatozoa had shown that when semen was placed on a column containing an approriate medium those sperm that had moved furthest down the column tended to be X-bearing while those that stayed at the top tended to be Y-bearing (see Rinehart 1975; Glass 1977 for references). Ericsson, Langevin, and Nishino (1973) in what could definitely be called a seminal paper introduced a fundamental change by using density gradients. They layered mobile human spermatozoa on columns of bovine serum albumin (BSA) or ovalbumin. Through their own motility Y-bearing spermatozoa moved progressively faster into increasingly dense BSA solutions (6, 10 and 20 per cent). The spermatozoa swimming into the BSA medium were recovered by centrifugation and then relayered over fresh BSA of higher density. By repeated application the final fraction obtained had greatly enhanced numbers of Y spermatozoa (estimated by quinacrine fluorescence). A simplification of the method was to manufacture a column with discontinuous density layers of BSA so reducing the three step procedure to one step. The 25–30 per cent sperm of the initial sample recovered from the last fraction of the column (highest density of BSA) were progressively motile and predominantly Y bearing. The success of the technique was claimed to be due to the Y spermatozoa having a greater ability to penetrate an interface between fluids and to swim faster than the X spermatozoa in fluids of high density and viscosity. This property of course has often been tacitly implied in relation to the timing of insemination during the menstrual cycle, and in the studies of the relationship between cervical mucus and the motility of Y spermatozoa by Rohde *et al.* (1973) and Kaiser *et al.* (1974). The layered separation technique has not always been successful (Glass 1977) but there is agreement that the sperm recovered from the distal portion of the column are progressively motile and of a more uniform morphology. Dmowski, Gaynor, Lawrence, Rao, and Scommegna (1979) used the modified technique to separate semen from debris, to improve motility and to decrease the number of abnormal forms from infertile men. Human serum albumin (HSA) was used instead of BSA. The real crux of the technique, however, was whether offspring conceived with spermatozoa subjected to the separation procedure were predominantly healthy boys. The first published trial using such supposedly enriched Y spermatozoa fractions was that of Dmowski, Gaynor, Rao, Lawrence, and Scommegna (1979). They obtained semen from 37 husbands requesting male child preselection. Semen was treated by a modification of the technique of Ericsson *et al.* (1973) using columns containing two discrete layers, a lower one of 20 per cent HSA and an upper one of 10 per cent HSA. The spermatozoa obtained from the lower layer after 30 minutes was collected by centrifugation and used for Artificial Insemination Husband (AIH). Ten conceptions were achieved and 7 pregnancies resulted in normal deliveries of 5 males and 2 females, one ended in a spontaneous abortion of a male fetus and two were still awaiting birth. The ratio of male to female conceptions in the small study was parallel to that of Y to X spermatozoa in the final specimen used for AIH assessed by F-body count. Obviously the numbers participating in this clinical trial were very small but it did show that the enriched sperm were fertile and that normal children could be conceived and born using such spermatozoa. The originator of the separation method also published his figures on the number of males born after using AIH with supposedly Y-enriched spermatozoa. Beernik and Ericsson (1982) reported briefly that of 84 deliveries from various centres using the method, 66 were males and 19 females, giving a statistically significant 79 per cent male birth ratio. One possible criticism of the significance of the result is that the patients are a 'pre-selected group' and thus cannot be thought of as typical of the population. The authors however, argued that the couples coming to the centres participating in the study had hitherto only had girls. Nothing definite can be said about the influence of any such bias in the sample of people (see section on human sex ratio).

Other materials have been used for the preparation of supposed X and Y subpopulations of human spermatozoa with varying degrees of success.

A previous attempt to separate X and Y spermatozoa using Sephadex gel-filtration (Steeno, Adimoelja, and Steeno 1975) resulted in a filtered fraction containing only 5 per cent of spermatozoa showing F-bodies. This contrasts with the Ericsson method. Two other groups could not repeat the separation (Downing, Black, Cary, and Delahanty 1976; Schilling, Lafrenz, and Klobasa 1978) but successful confirmation has been achieved by Adimoelja, Hariadi, Amitaba, Adisetya, and Soeharno (1977) and Quinlivan, Preciado, Long, and Sullivan (1982). The clinical application of the technique (for those couples who wished to avoid bearing sons) was published by Corson, Batzer, Alexander, Schlaff, and Otis (1984). Of the 19 women inseminated with supposed X-sperm separated by Sephadex, 12 conceived and up to the time of publication 7 female, 2 male and one set of mixed twins has been the result. The technique would, of course, be of great importance for those women who may be in danger of conceiving boys with male-linked genetic diseases.

Shastry, Hegde, and Rao (1977) employed Ficoll sodium metrizoate density gradients (63 per cent Y) while Kaneko, Yamaguchi, Kobayashi, and Izuki (1983) separated, without loss of motility, the X and Y by density gradient centrifugation in a Percoll and Ficoll-Paque medium. They found that one fraction had the highest concentration of Y spermatozoa (73 per cent) while the X-bearing spermatozoa was concentrated fourfold in another fraction. This technique offers the possibility of obtaining usable enriched X and Y fractions.

ii   Galvanic separation

Electrophoresis of animal spermatozoa has been attempted a number of times to separate X and Y spermatozoa on the assumption that they carry different charges or different amounts of net charge. Although varying degrees of success have been claimed (Glass 1977; Rinehart 1975) it has never resulted in such effective separations as have been obtained with columns. Fresh human semen from 11 people was subjected to electrophoresis in a specially designed cell by Shishito, Shirai, and Sasaki (1975). They found that Y-bearing spermatozoa (F-body stained) were predominantly attracted to the anode (61 per cent) and had an elevated motility while X-bearing spermatozoa were attracted to the cathode (89 per cent) and had a reduced motility. The spontaneous motility of the spermatozoa could dissipate the gradient within 15 minutes of switching the galvanic current off. No larger scale confirmatory investigation appears to have been published.

iii Counter current galvanic separation

The only other recently published study to use galvanic separation of human spermatozoa is that of Daniell, Herbet, Repp, Torbit, and Wentz (1982) who used a convection streaming galvanic cell apparatus developed by Bhattacharya, Shome, and Gunther (1977). As in the previous work by Shishito *et al.* (1975), the anode fraction attracted the Y spermatozoa (F-body count was enhanced from 48 to 77 per cent) while the cathode attracted the X spermatozoa (the fraction was enhanced from 52 to 77 per cent). Both fractions showed good motility but the mean recovery was only 15–16 per cent of the initial preseparation concentration. It is claimed that the great advantage of galvanic separation, as opposed to those of columns, is that both X and Y are separated at the same time and it only requires 3 hours of a trained technician's time. Obviously, until clinical insemination trials with the enriched fractions are undertaken and the altered birth ratios are obtained, without any reduction in pregnancy rates or increases in abnormal births, little further can be said about the long term feasibility of the technique.

iv Fractionation by laminar flow column

Sarkar (1984*b*) investigated the swimming patterns of human spermatozoa in a specially developed cylindrical flow column with laminar flow velocity gradients. The apparatus was initially designed to separate dead/non-motile spermatozoa from live swimming spermatozoa. Motile spermatozoa accumulated near solid wall surfaces and orientated against the direction of current, the dead cells were carried to the top fraction. Surprisingly the motile cells in the bottom fractions were found to be enriched for X sperm and the top fractions for Y sperm. This separation of X and Y could be greatly enhanced if the flow in the column was conically shaped (Sarkar, Jolly, Friedman, and Jones 1984). Most of the X spermatozoa remained in the bottom fractions (approximately 75–80 per cent X) while the Y was enriched to 75–80 per cent in the top fractions (estimated by F-body fluorescence). The yield of Y was only about 10 per cent of the total sample. The yield of X was much greater. Modifications of the technique (by the use of *in vitro* fertilization medium) by Sarkar (1984*a*) facilitated the separation of X and Y spermatozoa. This was because the medium had a special effect on the swimming behaviour of the spermatozoa: most of the spermatozoa swimming tracks were curved or circular (due to asymmetric flagellar beats), whereas in semen and saline the tracks are straight (due to symmetric flagellar beating). The fractionation technique relies on the effects of laminar flow amplifying swimming differences between X and Y sperm. The X spermatozoa accumulate at regions of steeper velocity gradient near walls, while the Y spermatozoa are transported to the middle of the flow stream. The technique allowed enrichment of Y up to 80 per cent. Sarkar *et al.* (1984) found that the ability to enrich X and Y spermatozoa was highly dependent on the initial individual semen. The basis for these different patterns in the ability of semen to be fractionated is unexplained. As previously mentioned, the level of calcium in the incubating medium regulated the degree of curvature of the spermatozoa swimming tracks. This effect may be important since swimming in circles could be advantageous to sperm enabling them to remain close to the egg. The calcium ions may be directly involved in the difference between the swimming of X and Y close to the walls of the streaming fluid. A further fascinating feature of spermatozoal motility is that addition of 1 μM 17β-oestradiol to the medium doubles the frequency of clockwise turning circular tracks (Sarkar 1984*a*). Sperm have been shown to have binding sites for the oestrogen on their outer surface and on the nucleus (Hernandez-Perez, Ballesteros, and Rosada 1979).

v Differential filtration

In a preliminary report Shettles (1976) claimed that when fresh semen samples were placed on one side of a millipore filter saturated with ovulatory cervical mucus (unstated origin but presumably human), after one hour at 37° C the spermatozoa that passed through were highly motile and apparently consisted practically entirely of what Shettles claimed was the small-headed Y carrying spermatozoa, apparently confirmed by F-body count. No further follow-up of this remarkable study has appeared and it remains a curiosity. The mechanism of separation presumably rested on the properties of the cervical mucus rather than of the pores of the filter. The previous reports by Rohde *et al.* (1973) and Kaiser *et al.* (1974) on the swimming separation of Y from X spermatozoa in human ovulatory mucus can be cited as evidence for such an effect. Hong, Chaput de Saintonge, and Turner (1981) showed that highly motile spermatozoa could pass through the pores (5 μ) of a millipore filter but no attempt has been made to assess whether the mechanical properties of the filter can bring about any significant Y spermatozoa enrichment.

2 DIET AND SEX SELECTION

A possible connection between diet and the sex of offspring has been suggested from animal studies (Schröder 1956; Foote 1977). From studies on the variation of male births in specific areas of Australia in relation to seasonal rainfall Lyster and Bishop (1965) speculated that the concentrations of K^+, Ca^{++}, Mg^{++} and other metallic ions could be the cause. Other experiments with a variety of animals and of concentrations of $K^+/Ca^{++}/Mg^{++}$ indicated high sex ratio (70 per cent males) with high K^+ and low sex ratio (70 per cent females) when Ca^{++} or Mg was high (see Stolkowski and Choukroun 1981 for references on frogs, guppies, marine worms (*Bonellia viridis*)). Further work led to the conclusion that it was not the actual concentration of the ions that was important but the ratio of potassium to the alkaline earths. Later experiments were conducted on rats and cows (Stolkowski and Choukroun 1981). With this background, Stolkowski hypothesised that the mineral intake of human mothers could influence the sex of the conceived child. A retrospective survey of the dietary habits of 100 couples suggested that the 40 women who had boys had had diets poor in Ca^{++} and/or rich in K^+ together with a high salt content, while 40 women with girls only had a diet low in K^+ and/or high in Ca^{++} and Mg^{++} with a low salt content. A trial programme of special diets to control the sex of children was begun in 1970. The diet for conceiving boys was rich in salt and K^+ while foods rich in Ca^{++} or Mg^{++} were excluded. The diet for conceiving girls was low in Na^+ and rich in alkaline earth elements. Each woman was given a diet sheet showing the food allowed and those excluded. For the girl diet, supplements were added to compensate for possible dietary deficiencies in the alkaline earths – namely Mg^{++} supplements, Ca^{++} syrup and vitamin D. For the boy diet, extra K^+ was given. The diets were administered for two menstrual cycles preceding that in which it was hoped fertilization would occur. The diets were abandoned if a pregnancy did not result after 3 cycles. Apparently various contraceptive measures (steroid pills, IUD) and infections were to be avoided, if possible for at least 3 months, before the start of the diet. By the time of publication (Stolkowski and Choukroun 1981) 47 recorded experiments had been undertaken giving rise as planned to 22 boys and 17 girls. In addition there was one set of boy/girl twins and 7 failures: 3 of these had been girls instead of boys and this was attributed to fathers with atypical spermatograms; a fourth failure was where a girl was required and a boy born; the remaining 3 failures were stated to be due to the women not having the correct dietary status. The claimed overall success rate was between 80–84 per cent but the occasional failures that occurred did so even in the most highly committed couples. A follow up for 'several' years of the children born in these trials revealed 'no special features' of the children; that is, they are presumed to have been normal. The composition of the diets was published first in French (Stolkowski 1977) but later were translated into English in a popular book by Papa and Labro (1984). The claim in that book is that of 155 women who used the 'diets', 123 successfully conceived a child with the sex of their choice (79 per cent). Lorrain and Gagnon (1975) reported that using the same diets with 50 couples trying to preselect the sex of their child, the efficacy of the method was some 80 per cent.

As yet no full scale clinical trial of the dietary technique has been published from any American or British source: all the data obtained stem from French or French–Canadian sources. The diets have been criticized because they can impose salt loads, they are restrictive and they are difficult to follow with accuracy. Contraindications are women with hypernervousness, hypercalcaemia, nephropathy and hypertension.

The outstanding question remains: how do the diets work? According to Stolkowski and to Lorrain (1980) the dietary excesses or deficiencies could affect the ionic exchanges of the ova with the bloodstream and change their potentialities (presumably meaning membrane resting potential?) which then could alter structures of the receptive sites in the pellucida membrana where spermatozoa bind. Such changes might affect the selection of the sperm which enters the ovum. However, Stolkowski also proposes that the diet may be responsible for destroying or inactivating the X or Y spermatozoa in the female tract. Lorrain, on the other hand, favours an immunological explanation whereby the spermatozoa are coated with antibody making them more susceptible to phagocytosis. The same speculations were advanced by Stolkowski and Choukroun (1981). The problem with such proposals is that with our present lack of knowledge about the chemical milieu of the female reproductive tract, a number of possible mechanisms could affect the X and Y spermatozoa following dietary manipulations. As detailed in the section on putative factors affecting spermatozoal functions, the levels of Na^+, K^+, Ca^{++} and of prolactin could be changed by the nutritional imbalances and affect differentially X and Y spermatozoa. The data on spermatozoal motility produced by Sarkar (1984a, b) and Sarkar et al. (1984) shows how complex the field is becoming.

A real difficulty, however, with any theory is our lack of basic facts about how the dietary alterations affect the levels of the ions under question in the various body fluids. Do the imbalances actually change the plasma levels of the ions, let alone at the level of the ovum? It would certainly be feasible for estimations to be made of vaginal and cervical Na^+, K^+, Ca^{++} and Cl^- before and during the consumptions of the diets to assess whether, compared to normal values, any significant changes occur (Wagner and Levin 1980). Moreover, it should not be too difficult nowadays to assess, *in vitro*, whether measured changes in the ionic milieu of ova can affect membrane potentials and even X and Y sperm binding. An *in vitro* model of the rabbit ovary has already been developed that allows the study of the ionic basis for ovulation (Kobayashi, Kitai, Santulli, Wright, and Wallach 1983).

Rorvik and Shettles (1977) had also suggested that certain aspects of a couple's diet should be modified if they were trying to conceive a boy or girl. Men were told to consult their acid/alkaline food list to make certain that a very marked alkaline milieu of their seminal fluid was not being encouraged by avoiding acidic foods although it was accepted that the importance of diet (especially acid-based foods) was unknown. Coffee drinking by the man before coitus was suggested to stimulate Y-bearing spermatozoa through the effect of caffeine (see Fraser 1984 for references). As pointed out previously there are no real data to test any of these suggestions that diet influences the sex of a conception.

IV   Postconception techniques

1   FETAL SEX SCREENING FOLLOWED BY SELECTIVE ABORTION

It is now possible to sex the fetus by a number of invasive techniques (amniocentesis, fetoscopy, chorionic villus biopsy) or a non-invasive technique (ultrasonic screening). Amniocentesis allows the collection of discarded fetal cells from the amniotic fluid directly by needle aspiration through the abdominal wall. The cells can be checked for F-body fluorescence (Y) or the Barr body (X) (Dewhurst 1956) (see section on X and Y spermatozoa). Amniocentesis however, carries risks (Philip and Bang 1978) and cannot normally be performed until the 15th week of pregnancy, at the earliest. Accurate sexing actually requires a proper chromosomal analysis (Anonymous 1980) which means at least 2–3 weeks cell culture to obtain sufficient DNA. Thus the sex of the fetus can only be known with reasonable accuracy using amniocentesis by week 18 of gestation. Termination of such pregnancies, at this stage, if unwanted, again carries its own risks and is mentally and physically taxing to the mother. Because of these risks and because many doctors refuse to undertake amniocentesis and abortion for couples who want a boy or girl for purely personal reasons not linked to any particular health or genetic hazard, selective termination is only offered (in Britain) when there is a risk of a serious sex-linked disorder (for example, muscular dystrophy). Recent advances with chorionic villus biopsy, however, allow enough cells to be obtained from the mesenchyme of only 2–3 chorionic villi to give a satisfactory amount of DNA in culture after only 10 days (Loeffler 1984). In fact, it is even possible to analyse first trimester chorionic villi directly without preliminary culture and obtain chromosomal evidence of fetal sex on the day of the biopsy. With the aid of real-time ultrasound, such sexing can be carried out at weeks 8–11 via the cervix. A cannula of 3 mm diameter has gained the most favour to obtain a suction sample. Success is improved by immediately examining the material obtained in theatre under the phase-contrast microscope and identifying the frond-like villi structures. The danger is that the amniotic sac will be punctured inducing an abortion. The possibility of damage to the fetus prevents the use of the method before week 8 of gestation (Loeffler 1984). Clearly, the earlier the detection of fetal sex, the easier is the abortion procedure and the mother's stress is greatly reduced (see Rodeck 1987).

Direct examination of the fetus by a fetoscope (fetoscopy) can obviously allow visualisation of the genital area so that the sex of the fetus can be determined, but there is a 5 per cent abortion rate associated with the procedure (Hobbins and Mahoney 1977).

With considerable experience ultrasound imaging, enabling genitalia to be identified, can be used as a non-invasive method for sexing the fetus. However, it is not easy to obtain a discriminating image of the small genitalia of a 16 week fetus (Hobbins 1983). It has been reported (Guardian, March 21st, 1986, Third World Review, page 11) that South Korean Health Authorities were investigating the abuse by couples of the results of ultrasound screening of their fetus (to monitor fetal growth and early detection of genetic/deformity problems); there was suspicion that if the fetal sex was wrong, abortion might be requested and carried out.

2   SEXING TECHNOLOGY RELATED TO *IN VITRO* FERTILIZATION (IVF)

i   Blastocyst sexing after IVF

Since the demonstration in rabbits (Edwards and Gardner 1967; Gardner and Edwards 1968) that the sex of a blastocyst can be checked by removing a few cells before implantation and then subsequently transferring the blastocyst (if of the required sex) to the recipient female for implantation and normal development, it is obvious that this technique could be employed in human IVF procedures. No human study describing such a procedure has been published. It should be remembered that sexing of the fetus, even by chromosome analysis, is not infallible for it is possible to have an XY fetus suffering from testicular feminization (Anonymous 1979).

ii   IVF with X or Y spermatozoa

Another possible technique to control the sex of the blastocyst would be to incubate unfertilized human ova with either X or Y enriched spermatozoal fractions. Under such circumstances even small numbers of spermatozoa may be effective. Of course there is always the chance that the sperm that fertilizes is a 'rogue' spermatozoa of the wrong sex so to be absolutely certain blastocyst sexing would have to be done. Such techniques demand high expertise. They would probably only be available to couples prepared to pay a premium for child sex selection. In Britain few couples can obtain even simple IVF on the National Health Service.

V   Social effects of human sex selection

Any review of current methods of human sex selection reveals the wide range of available techniques, some involving skilled technicians with elaborate laboratory equipment which are reputed to be highly effective while others can be used on a personal basis without any technical or medical involvement but have a lower or even questionable degree of selection effectiveness. The likelihood of a truly effective but simple method applicable at the 'mass' personal level is small but within a few years, even if the pace of research is maintained rather than accelerated, a laboratory-based method may allow parents to select the sex of their child with a reasonably high level of confidence (Gledhill 1983). Access will be limited by cost unless the number of people applying for sex preselection grows so large that Governments begin to worry about the possibility of an imbalanced sex ratio causing social problems. A concise account of the major arguments for and against free access to a technology of sex selection and of the possible social consequences is given by Largey (1978). While there have been numerous speculations about what would/could happen demographically to a society in which a highly effective sex selection technique were widely available (see Etzioni 1968; Largey 1973, 1978; Rinehart 1975), few real facts are known to make effective social models even in the developed nations where most studies have been undertaken. An obvious question to ask is – what is the justification for human sex selection? Reduction and elimination of sex-linked diseases is the major justification and impetus but the promotion of other aspects of human happiness also are recognized as playing a part. One argument is that parents would be happier because the child would be of the preferred sex and the child would be happier because it would not experience the rejection it might otherwise feel if it were of the undesired sex. However, husband and wife may not always be able to agree on the desired sex of the child. A less direct justification occurs in relation to the need to decrease the birthrate. It is assumed that reliable sex selection would reduce the total number of births as once the parents have children of the desired number and sex, no further attempts at 'trying' for a girl or a boy become necessary. Most of the articles written about sex preselection in the medical literature are very critical about a couple who want to select the sex of the child for only personal preference once the zygote has been created (Anonymous 1980). The issue however, is not so obviously clear-cut as many would have it seem and an excellent discussion on the ethics of the problem (especially in relation to wanting amniocentesis for initial sex identification) is given by Fletcher (1979). Counter arguments against allowing sex control have also been promoted. Because there is a general desire to want more boys than girls (see Etzioni 1968; Largey 1973; Rinehart 1975; Williamson 1983) a possible imbalance in the number of males could arise creating less mothers for further births. If such preferences for males were maintained for several generations the effects on the reproductive capacity of the population would be marked. However, others have thought that there would be a compensatory demand for girls as the shortage of females would increase their 'market' value. Such speculations may differ in different countries. What may not have great effects in Britain may create huge social problems in Asian countries where boys are very highly valued. Social problems such as increases in all-male juvenile gangs, violence and delinquency and an increase in homosexuality have been suggested (Etzioni 1968; Largey 1978) as possible sequels to a greater demand for boys.

In Great Britain a working party and a committee have looked at and reported on the question of allowing parental sex selection. The Council for Science and Society formed for the object of promoting the study of, and research into, the social effects of science and technology and of disseminating the results to the public set up an *ad hoc* working party to investigate the ethical aspects of the new techniques in human procreation. The report, (Human Procreation 1984), came to the conclusion that the 'new techniques should not be used to provide parents with children of desired sex, except for the purpose of avoiding sex linked disorders'. The Committee of Inquiry into Human Fertilization and Embryology (Warnock 1984) was established to examine the social, ethical and legal implications of recent and potential developments in the field of human assisted reproduction. Although its major objective was to discuss scientific and ethical problems bearing on the freezing, storage and use of human embryos and on *in vitro* fertilization, it did describe briefly, in a number of short sections, post fertilization and pre-implantation gender identification and possible future methods of sex selection, focussing mainly on separating male and female bearing sperm and pointing out that such spermatozoa could be used for fertilization either *in vitro* or *in vivo*. The conclusion was that while the Committee 'see no reason why, if a method of selecting the sex of a child before fertilization is developed, this should not be offered to couples who have good medical reasons for choosing the sex of their child, but if an efficient and easy method of ensuring the conception of a child of a particular sex became available, it is likely that some couples would wish to make use of it for purely social reasons ... It is impossible to predict, either in the long or the short term, the likely effects of such a practice on the ratio of males to females within society'. Nevertheless, the tone of the paragraph following implied that as the assumption is that the majority of couples would choose a boy as their first child, the social implications as to the position of women in society and the disadvantages of never being first born would be damaging to society's structures. However, no positive recommendation was made about sex selection except that it should be kept under review. The Committee did recommend that any 'do-it-yourself' kits or apparatus for self-administration for gender selection should come within the ambit of the Medicines Act (1986), with the aim of ensuring that such products are safe, efficacious and of an acceptable standard.

Acknowledgements

I would like to thank Mrs Jane Eborall for typing the manuscript.

References

Adimoelja, A., Hariadi, R., Amitaba, I. G., Adisetya, P., and Soeharno, (1977). The separation of X and Y spermatozoa with regard to the possible clinical application by means of artificial insemination. *Andrologia* **9**, 289–292.

Alfreddsson, J. H. (1984). Artificial insemination with frozen semen. Sex ratio at birth. *Int. J. Fert.* **29**, 152–155.

Alperovitch, A. and Feingold, N. (1981). Sex ratio in offspring of patients with multiple sclerosis. *New Engl. J. Med.* **305**, 1157.

Amann, R. P. and Seidel, G. E. (1982). *Prospects for sexing mammalian sperm.* Colorado Associated University Press, Boulder, Colorado.

Anonymous (1979). Antenatal prediction of sex. *Br. med. J.* **ii**, 754–755.

—— (1980). Choosing the baby's sex. *Br. med. J.* **i**, 272–273.

Barlow, P. and Vosa, C. G. (1970). The Y chromosome in human spermatozoa *Nature, Lond.* **226**, 961–962.

Bartos, L. (1980). Coitus-induced ovulation and sex ratio in man. *Med. Hypothesis* **6**, 899–906.

Battersby, S. and Chandler, J. A. (1977). Correlation between elemental composition and motility of human spermatozoa. *Fert. Steril.* **28**, 557–561.

Beatty, R. A. (1977). F-bodies as Y chromosome markers in mature human sperm heads: a quantitative approach. *Cytogenet. Cell Genet.* **18**, 33–49.

Beernik, F. J. and Ericsson, R. J. (1982). Male sex preselection through sperm isolation. *Fert. Steril.* **38**, 493–495.

Ben-Rafael, Z., Matalon, A., Blankstein, J., Serr, D. M., Lunenfeld, B. and Mashiach, S. (1986). Male to female ratio after gonadotrophin-induced ovulation. *Fert. Steril.* **45**, 36–40.

Bhattacharya, B. C., Shome, P., and Gunther, A. H. (1977). Successful separation of X and Y spermatozoa in human and bull semen. *Int. J. Fert.* **22**, 30–35.

Billings, E. L., Billings, J. J., Brown, J. B., and Burger, B. (1972). Symptoms and hormonal changes accompanying ovulation. *Lancet* **i**, 282–284.

Bishop, D. W. (1960). X and Y spermatozoa. *Nature, Lond.* **187**, 255–256.

—— (1962). Sperm motility. *Physiol. Rev.* **42**, 1–59.

Ciocco, A. (1938). Variation in the sex ratio at birth in the U.S. *Human Biol.* **10**, 36–64.

Cohen, B. H. and Glass, B. (1956). The ABO blood groups and the sex ratio. *Human Biol.* **28**, 20–42.

Cole, L. J. and Johansson, I. (1933). Sex control again. *J. Heredity* **24**, 265–274.

Corson, S. L., Batzer, F. R., Alexander, N. J., Schlaff, S., and Otis, C. (1984). Sex selection by sperm separation and insemination. *Fert. Steril.* **42**, 756–760.

Damon, A. and Nuttall, R. L. (1965). Ponderal index of fathers and sex ratio of children. *Human Biol.* **37**, 23–28.

—— —— and Salber, E. J. (1966). Tobacco smoke as possible genetic mutagen: parental smoking and sex of children. *Am. J. Epidem.* **83**, 520–536.

Daniell, J. F., Herbert, C. M., Repp, J., Torbit, D. A., and Wentz, A. C. (1982).

Initial evaluation of a convection counter streaming galvanisation technique of sex separation of human spermatozoa. *Fert. Steril.* **38**, 233–237.

Deininger, P. L., Jolly, D. J., Rubin, C. M., Friedman, T., and Schmid, C. W. (1981). Base sequence studies of 300 nucleotide renatured repeated human DNA clones. *J. mol. Biol.* **151**, 17–33.

Dewhurst, C. J. (1956). Diagnosis of sex before birth. *Lancet* i, 471–472.

Diasio, R. B. and Glass, R. H. (1971). Effects of pH on the migration of X and Y sperm. *Fert. Steril.* **22**, 303–305.

Dmowski, W. P., Gaynor, L., Lawrence, M., Rao, R., and Scommegna, A. (1979). Artificial insemination homologous with oligospermic semen separated on albumin columns. *Fert. Steril.* **31**, 58–62.

—— —— —— and Scommegna, A. (1979). Use of albumin gradients for X and Y sperm separation and clinical experience with male sex preselection. *Fert. Steril.* **31**, 52–57.

Douglas, N. and Slinger, P. (1982). *Sexual secrets – the alchemy of ecstasy*. Arrow Books, London.

Downing, D. C. and Black, D. L. (1976). Equality in survival of X and Y chromosome bearing human spermatozoa. *Fert. Steril.* **27**, 1191–1193.

—— —— Cary, W. H., and Delahanty, D. L. (1976). The effect of ion exchange column chromatography on separation of X and Y chromosome-bearing human spermatozoa. *Fert. Steril.* **27**, 1187–1190.

Drew, J. S., London, W. T., Lustbader, E. D., Hesser, J. E., and Blumberg, B. S. (1978). Hepatitis B virus and sex ratio of offspring. *Science, N.Y.* **201**, 687–692.

Edwards, A. W. F. (1962). Genetics and the human sex ratio. *Adv. Gen.* **11**, 239–272.

Edwards, R. G. and Gardner, R. L. (1967). Sexing of live rabbit blastocysts. *Nature, Lond.* **214**, 576–577.

Ericsson, R. J., Langevin, C. N., and Nishino, M. (1973). Isolation of fractions rich in human Y sperm. *Nature, Lond.* **246**, 421–424.

Etzioni, A. (1968). Sex control, science and society. *Science, N.Y.* **161**, 1107–1112.

Farooqui, A. A. (1983). Biochemistry of sperm capacitation. *Int. J. Biochem.* **15**, 463–468.

Fletcher, J. C. (1979). Ethics and amniocentesis for fetal sex identification. *New Engl. J. Med.* **301**, 550–553.

Foote, R. H. (1977). Sex ratios in dairy cattle under various conditions. *Theriogenology* **8**, 349–356.

France, J. T., Graham, F. M., Gosling, L., and Hair, P. I. (1984). A prospective study of the preselection of the sex of offspring by timing intercourse relative to ovulation. *Fert. Steril.* **41**, 894–900.

Fraser, L. R. (1984). Mechanisms controlling mammalian fertilisation. In *Oxford reviews of reproductive biology*, Vol. 6 (ed. J. R. Clarke) pp. 174–225. Clarendon Press, Oxford.

Gardner, R. L. and Edwards, R. G. (1968). Control of the sex ratio at full term in the rabbit by transferring sexed blastocysts. *Nature, Lond.* **218**, 346–349.

Garner, D. L., Gledhill, B. L., Pinkel, D., Lake, S., Stephenson, D., Van Dilla, M. A., and Johnson, L. A. (1983). Quantification of the X- and Y-chromosome bearing spermatozoa of domestic animals by flow cytometry. *Biol. Reprod.* **28**, 312–321.

Glass, R. H. (1977). Sex preselection. *Obstet. Gynec.* **49**, 122–126.

Gledhill, B. L. (1983). Control of mammalian sex ratio by sexing sperm. *Fert. Steril.* **40**, 572–574.

Gorus, F. K., Finsy, R., and Pipeleers, D. G. (1982). Effect of temperature, nutrients, calcium and cAMP on motility of human spermatozoa. *Am. J. Physiol.* **242**, C304–C311.

Guerin, J. F. and Czyba, J. C. (1979). Effects of ions and $K^+:Na^+$ ratio on motility and oxygen consumption of human spermatozoa. *Arch. Androl.* **2**, 295–300.

Guerrero, R. (1974). Association of the type and time of insemination within the menstrual cycle with the human sex ratio at birth. *New Engl. J. Med.* **291**, 1056–1059.

—— (1975). Type and time of insemination within the menstrual cycle and the human sex ratio at birth. *Stud. Fam. Plan.* **6**, 367–371.

Harlap, S. (1979a). Gender of infants conceived on different days of the menstrual cycle. *New Engl. J. Med.* **300**, 1445–1448.

—— (1979b). Gender of infants conceived on different days of the menstrual cycle. *New Engl. J. Med.* **301**, 1126.

Heath, C. W. (1954). Physique, temperament, and sex ratio. *Human Biol.* **26**, 337–342.

Hegde, U. C., Shastry, P. R., and Rao, S. S. (1981). Phytohaemagglutinin as a molecular probe to study the membrane constituents of human X- and Y-bearing spermatozoa. *J. reprod. Immunol.* **2**, 351–357.

Hernandez-Perez, O., Ballesteros, L. M., and Rosada, A. (1979). Binding of 17β-estradiol to the outer surface and nucleus of human spermatozoa. *Arch. Androl.* **3**, 23–29.

Hilgers, T. W., Abraham, G. E., and Cavanagh, D. (1978). Natural family planning 1. The peak symptom and estimated time of ovulation. *Obstet. Gynec.* **52**, 575–582.

Hill, G. B., Fincham, S. M., Wijayansinghe, C., Haronga, C. L., and Hendin, M. M. (1985). Sex ratio of offspring of patients with prostatic cancer. *Can. med. Ass.* **133**, 567–571.

Hobbins, J. C. (1983). Determination of fetal sex in early pregnancy. *New Engl. J. Med.* **309**, 979–980.

—— and Mahoney, M. J. (1977). Fetoscopy in continuing pregnancies. *Am. J. Obstet. Gynec.* **129**, 440–442.

Hong, C. Y., Chaput de Saintonge, D. M., and Turner, P. (1981). A simple method to measure drug effects on human sperm motility. *Br. J. clin. Pharmac.* **11**, 385–387.

Human procreation – ethical aspects of the new techniques (1984). Report of a working party of the Council for Science and Society. Oxford University Press, Oxford.

Iffy, L. (1963). The time of conception in pathological gestations. *Proc. R. Soc. Med.* **56**, 1098–1100.

Jacobs, L. S. and Daughaday, W. (1974). Physiologic regulation of prolactin secretion in man. In *Lactogenic hormones, fetal nutrition and lactation* (ed. J. B. Josimovich, M. Reynolds, and E. Cobo) pp. 351–378. Wiley, New York.

Jalavisto, E. (1952). Genealogical approach to factors influencing the sex ratio at birth. *Annls. Chir. Gynaec. Fenn.* **41**, 182–196.

James, W. H. (1971). Cycle day of insemination, coital rate and sex ratio. *Lancet* **i**, 112–114.

—— (1976). Timing of fertilisation and sex ratio of offspring – a review. *Ann. Hum. Biol.* **3**, 549–556.

—— (1980). Time of fertilisation and sex of infants. *Lancet* **i**, 1124–1126.

—— (1983). Timing of fertilisation and the sex ratio of offsprings. In *Sex selection of children* (ed. N. G. Bennett) pp. 73–79. Academic Press, New York.

—— (1985). The sex ratio of infants born after hormonal induction of ovulation. *Br. J. Obstet. Gynaec.* **92**, 299–301.

Jöchle, W. (1973). Coitus induced ovulation. *Contraception* **7**, 523–533.

—— (1975). Current research in coitus-induced ovulation: a review. *J. Reprod. Fert. Suppl.* **22**, 165–207.

Kaiser, R., Citoler, P., and Broer, K. H. (1974). Relative increase in Y chromatin bearing spermatozoa after *in vitro* penetration into human cervical mucus. *IRCS Medical Science* **2**, 1100.

Kaneko, S., Yamaguchi, J., Kobayashi, T., and Iizuku, R. (1983). Separation of human X- and Y-bearing sperm using Percoll density gradient centrifugation. *Fert. Steril.* **40**, 661–665.

Kleegman, S. J. (1954). Therapeutic donor insemination. *Fert. Steril.* **5**, 7–31.

Kobayashi, Y., Kitai, H., Santulli, R., Wright, K. H., and Wallach, E. (1983). The influence of a calcium- and magnesium-free environment on *in vitro* ovulation in the rabbit. *Fert. Steril.* **39**, 396A.

Lancaster, H. O. (1950). The sex ratios in sibships with special reference to Geissler's data. *Annals Eugenics* **15**, 153–158.

Langaney, A. and Pison, G. (1979). Rougeole et augmentation temporaire de la masculinité des naissances: coincidence ou causalité. *C.r. hebd. Séanc. Acad. Sci., Paris* **289**, 1255–1258.

Largey, G. (1973). Sex control and society: a critical assessment of sociological speculations. *Social Problems* **20**, 310–318.

—— (1978). Reproductive technologies – I. Sex Selection. In *Encyclopedia of bioethics* (ed. W. T. Reich) pp. 1429–1444. The Free Press, New York.

Lawrence, P. S. (1941). The sex ratio, fertility and ancestral longevity. *Q. Rev. Biol.* **16**, 35–79.

Levin, R. J. (1980). The physiology of sexual function in women. *Clinics Obstet. Gynec.* **7**, 213–252.

—— (1982). Current sex preselection methods. *IPPF Med. Bull.* **16**, 1–3.

—— and Wagner, G. (1983). Quantitative analysis of amino acids in human vaginal fluid during the menstrual cycle. *J. Physiol., Lond.* **343**, 87P.

—— —— (1985). Ionised calcium concentrations in human cervical fluid *in situ* before and after sexual arousal. *IRCS Medical Science* **13**, 406–407.

Lindahl, P. E. (1971). Centrifugation as a means of separating X- and Y-chromosome bearing spermatozoa. In *Sex ratio at birth – prospects for control.* (ed. C. A. Kiddy and H. D. Hafs) pp. 69–75. American Society for Animal Science, Illinois.

Loeffler, F. E. (1984). Prenatal diagnosis. Chorionic villus biopsy. *Br. J. Hosp. Med.* **31**, 418–420.

Lornage, J., Guerin, J. F., Czyba, J. C., and Menezo, Y. (1983). Influence of cations and albumin on human spermatozoa. *Arch. Androl.* **10**, 119–125.

Lorrain, J. and Gagnon, R. (1975). Selection preconceptionelle du sexe. *Union Med. Can.* **104**, 800–803.

Lyster, W. R. and Bishop, M. W. (1965). An association between rainfall and sex in man. *J. Reprod. Fert.* **10**, 35–47.

MacGillivray, I., Davey, D., and Isaacs, S. (1986). Placenta praevia and sex ratio at birth. *Br. med. J.* **292**, 371–372.

McCarthy, J. J. and Rockette, H. E. (1983). A comparison of methods to interpret the basal body temperature graph. *Fert. Steril.* **39**, 640–646.

Martin, R. H., Balkan, W., Burns, K., Rademaker, A. W., Lin, C. C., and Rudd, N. L. (1983). The chromosome constitution of 1000 human spermatozoa. *Human Gen.* **63**, 305–309.

Martin, W. J. (1943). Sex ratio during war. *Lancet* **ii**, 807.

Mortimer, D. and Richardson, D. W. (1982). Sex ratio of births resulting from artificial insemination. *Br. J. Obstet. Gynaec.* **89**, 132–135.

Painter, T. S. (1924). The sex chromosomes of man. *Am. Nat.* **58**, 506–524.

Papa, F. and Labro, F. (1984). *Boy or girl? Choosing your child through diet.* Souvenir Press, England.

Peterson, R. N. and Freund, M. (1976). Relationship between motility and the transport and binding of divalent cations to the plasma membrane of human spermatozoa. *Fert. Steril.* **27**, 1301–1307.

Philip, J. and Bang, J. (1978). Outcome of pregnancy after amniocentesis for chromosome analysis. *Br. med. J.* **ii**, 1183–1184.

Quinlivan, W. L. G., Preciado, K., Long, T. L. and Sullivan, H. (1982). Separation of human X and Y spermatozoa by albumin gradients and Sephadex chromatography. *Fert. Steril.* **37**, 104–107.

—— Sullivan, H. (1974). The ratios and separation of X and Y spermatozoa. *Fert. Steril.* **25**, 315–318.

Reyes, A., Parra, A., Chavarria, M. E., Goicoechea, B., and Rosado, A. (1979). Effect of prolactin on the calcium binding and/or transport of ejaculated and epididymal human spermatozoa. *Fert. Steril.* **31**, 669–672.

Rinehart, W. (1975). Sex preselection – not yet practical. *Population Reports* Series 1, no. 2 (May).

Roberts, A. M. (1978). The origins of fluctuations in the human secondary sex ratio. *J. biosoc. Sci.* **10**, 169–182.

—— and Goodall, H. (1976). Y chromosome visibility in quinacrine-stained human spermatozoa. *Nature, Lond.* **262**, 493–494.

Rodeck, C. H. (1987). Chorion villus biopsy. In *Oxford Reviews of Reproductive Biology*, Vol. 9 (ed. J. R. Clarke) pp. 137–160. Clarendon Press, Oxford.

Rohde, W., Porstman, T., and Dorner, G. (1973). Migration of Y-bearing human spermatozoa in cervical mucus. *J. Reprod. Fert.* **33**, 167–169.

Rorvik, D. M. and Shettles, L. (1977). *Choose your baby's sex – the one selection method that works.* Dodd, Mead and Company, New York.

Rothschild, V. (1960). X and Y spermatozoa. *Nature, Lond.* **187**, 253–254.

Russell, W. T. (1936). Statistical study of the sex ratio at birth. *J. Hyg.* **36**, 381–401.

Sarkar, S. (1984*a*). Motility, expression of surface antigen, and X and Y human sperm separation in *in vitro* fertilisation medium. *Fert. Steril.* **42**, 899–905.

—— (1984*b*). Human sperm swimming in flow. *Differentiation* **27**, 126–132.

—— Jolly, J., Friedman, T., and Jones, O. W. (1984). Swimming behaviour of X and Y sperm. *Differentiation* **27**, 120–125.

Schackman, R. W. and Shapiro, B. M. (1982). Calcium and the metabolic activation of spermatozoa. In *Calcium and cell function*, Vo. 2 (ed. W. Y. Cheung) pp. 339–353. Academic Press, New York.

Schilling, E., Lafrenz, R. and Oblasa, F. (1978). Failure to separate human X and Y chromosome bearing spermatozoa by Sephadex gel-filtration, *Andrologia* **10**, 215–217.

Schröder, V. N. (1956). Role du metabolisme des geniteurs dans la génèse du sexe de la descendance. *USP Sovrem Biol.* **42**, 33–50.

Schull, W. J. (1958). Empirical risks of consanguinous marriages: sex ratio, malformation and viability. *Am. J. hum. Gen.* **10**, 294–343.

—— and Neel, J. V. (1958). Radiation and the sex ratio in man. *Science, N.Y.* **128**, 343.

Shah, G. V., Desai, R. B., and Sheth, A. R. (1976). The effect of prolactin on metabolism of human spermatozoa. *Fert. Steril.* **27**, 1292–1294.

—— and Sheth, A. R. (1979). Is prolactin involved in sperm capacitation? *Med. Hypoth.* **5**, 909–914.

Shastry, P. R., Hegde, U. C., and Rao, S. S. (1977). Use of Ficoll sodium metrizoate density gradient to separate human X- and Y-bearing spermatozoa. *Nature, Lond.* **269**, 58–60.

Sheth, A. R., Mugatwala, P. P., Shah, G. V., and Rao, S. S. (1975). Occurrence of prolactin in human semen. *Fert. Steril.* **26**, 905–907.

—— and Rao, S. S. (1962). Potassium levels in human semen with reference to sperm motility. *Experientia* **18**, 324–325.

—— Vaidya, R. A., and Raiker, R. S. (1976). Presence of prolactin in human cervical mucus. *Fert. Steril.* **27**, 397–398.

Shettles, L. B. (1961). Human spermatozoa types. *Gynaecologia* **152**, 154–162.

—— (1970). Factors influencing sex ratios. *Int. J. Gynec. Obstet.* **8**, 643–647.

—— (1976). Separation of X and Y spermatozoa. *J. Urol.* **140**, 462–464.

—— and Rorvik, D. M. (1985). *How to choose the sex of your baby – a complete update of the method best supported by the scientific evidence.* Angus and Robertson, London.

Shishito, S., Shirai, M., and Sasaki, M. (1975). Galvanic separation of X and Y bearing human spermatozoa. *Int. J. Fert.* **20**, 13–16.

Simpson, J. L. (1979). More than we ever wanted to know about sex – should we be afraid to ask? *New Engl. J. Med.* **300**, 1483–1484.

Slater, E. (1944). A demographic study of a psychopathic population. *Annls Eugen.* **12**, 121–137.

Smith, M. L. and Luqman, W. A. (1982). Prolactin in seminal fluid. *Arch. Androl.* **9**, 105–113.

Snyder, R. G. (1961). The sex ratio of offspring of pilots of high performance aircraft. *Hum. Biol.* **33**, 1–10.

Steeno, O., Adimoelja, A., and Steeno, J. (1975). Separation of X- and Y-bearing spermatozoa with the sephadex gel-filtration method. *Andrologia* **7**, 95–97.

Stolkowski, J. and Choukroun, J. (1981). Preconception of sex in man. *Israel J. Med. Sci.* **17**, 1061–1067.

—— (1977). Le controle du sexe par la methode alimentaire: regimes pour 'garcons' et regimes pour 'filles'. Modalites d'application: les 23 premiers resultats. *Rev. Fr. Endocrinol. Clin. Biol.* **18**, 95–105.

—— and Lorrain, J. (1980). Preconceptional selection of fetal sex. *Int. Gynaec. Obstet.* **18**, 440–443.

Teitelbaum, M. S. (1972). Factors associated with the sex ratio in human populations. In *The structure of human populations* (ed. G. A. Harrison and A. J. Boyce) pp. 90–109. Clarendon Press, Oxford.

Thomson, J. A. (1908). *Heredity.* John Murray, London.

Turpin, R. and Schutzenberger, M. P. (1948). Recherche statistique su la distribution du sexe à la naissance. *C. r. hebd. Séanc. Acad. Sci., Paris* **226**, 1845–1846.

Unterberger, F. (1930). Das problem der wilkurlichen Beeinflussing des Geschlechts beim Menschen. *Deut. mediz. Wochensch.* **56**, 304–307.

—— (1932). Geschlechtsbestimmung und Wasserstoffionen-konzentration (Sex determination and hydrogen ion concentration). *Deut. mediz. Wochensch.* **58**, 729–731.

Van Dujin, C. (1960). Nuclear structure of human spermatozoa. *Nature, Lond.* **188**, 916–918.

Vogel, F., Kruger, J., Kurth, M., and Schroder, T. M. (1966). On the absence of a relation between coffee-drinking in parents and the sex ratio of children. *Humangenetik* **2**, 119–132.

Vollman, R. F., James, W. H., Corson, S. L., and Rosner, F. (1979). Gender of infants conceived on different days of the menstrual cycle. *New Engl. J. Med.* **301**, 1125–1126.

Warnock, M. (1984). *Report on the committee of inquiry into human fertilisation and embryology.* Command 9314. Her Majesty's Stationary Office, London.

Wagner, G. and Levin, R. J. (1978). Oxygen tension of the vaginal surface during sexual stimulation in the human. *Fert. Steril.* **30**, 50–53.

—— —— (1980). Electrolytes in vaginal fluid during the menstrual cycle of coitally active and inactive women. *J. Reprod. Fert.* **60**, 17–27.

Whelan, E. M. (1974). Human sex ratio as a function of the timing of insemination within the menstrual cycle: A review. *Soc. Biol.* **21**, 379–384.

—— (1977). *Boy or girl? The sex selection technique that makes all others obsolete.* Bobbs-Merrill, Indianapolis.

Williamson, N. E. (1983). Parental sex preferences and sex selection. In *Sex selection of children.* (ed. N. G. Bennett) pp. 129–145. Academic Press, New York.

—— Lean, T. H., and Vengadasalam, D. (1978). Evaluation of an unsuccessful sex preselection clinic in Singapore. *J. biosoc. Science* **10**, 375–388.

Winston, S. (1931). The influence of social factors upon the sex ratio at birth. *Am. J. Sociol.* **37**, 1–21.

—— (1932). Birth control and the sex ratio at birth. *Am. J. Sociol.* **38**, 225–231.

Yanagimachi, R. (1984). Zona free hamster eggs. Their use in assessing fertilising capacity and examining chromosomes of human spermatozoa. *Gamete Res.* **10**, 187–232.

Zaneveld, L. J. D., Gould, K. G., Humphreys, W. J., and Williams, W. L. (1971). Scanning electron microscopy of mammalian spermatozoa. *J. reprod. Med.* **6**, 13–17.

Zavos, P. M. and Albertson, D. F. (1984). Prolactin release during sexual stimulation. *Infertility* **7**, 109–119.

6   Cancer of the prostate: endocrine factors

K. GRIFFITHS, P. DAVIES, C. L. EATON, M. E. HARPER, W. B. PEELING, A. O. TURKES, A. TURKES, D. W. WILSON AND C. G. PIERREPOINT

I   **Introduction**

II   **Endocrine factors in the treatment of prostatic cancer**
  1   The role of diethylstilboestrol (DES).
  2   Androgen receptor levels in prostatic cancer: prognostic value.
  3   Premalignancy and endocrine status.
  4   The clinical potential of other prognostic factors.
  5   Luteinizing hormone-releasing hormone (LH-RH) analogues and prostatic cancer.

III   **Regulation of prostatic growth and function**
  1   Plasma steroids and ageing.
  2   Adrenal steroids and the prostate.
  3   Metabolic activity of the prostate in relation to disease.
  4   Prolactin, growth hormone and the prostate.
  5   Hormonal status and carcinoma of the prostate.
  6   Prostate biochemistry: growth factors and oncogenes.

I   Introduction

Through many decades, the prostate gland has been the subject of considerable attention, not only from urologists and pathologists, but also endocrinologists and reproductive biologists. This interesting, but relatively inaccessible, gland would seem to have no obvious function, although one assumes that it must be concerned with some aspects of sperm viability. It is not essential for life. To many men, prostatic disease and its associated difficulties with micturition are merely signs of advancing years, a fact of life, and the clinician similarly accepts that the pathological changes in the prostate which can result in benign or malignant tumours in the gland are associated with the ageing process.

In England and Wales, cancer of the prostate gland is however the fourth most common cause of death from malignant disease in men, being responsible for about 4000 deaths each year (Office of Population Censuses and Surveys 1978; Alderson 1981). In the United States of America,

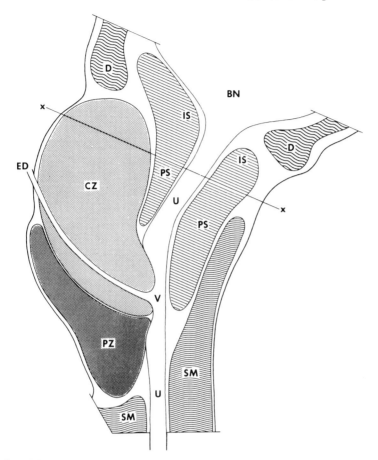

Fig. 6.1. Diagrammatic representation of the anatomy of the prostate with reference to the McNeal concept (McNeal, 1975). CZ, central zone; PZ, peripheral zone; V, verumontanum; U, urethra; ED, ejaculatory duct; PS, preprostatic sphincter; IS, internal sphincter; SM, external sphincter muscle; BN, bladder neck; D, detrusor muscle.

prostatic cancer is the second most common form of malignancy in men (Cutler and Young 1975) with 56,000 new cases and more than 18,000 deaths each year (Silverberg and Holleb 1975). The disease is certainly established as a feature of advancing years, rarely presenting under 50 years of age. Indeed it is often present, but asymptomatic, in patients dying of old age, or other causes. Unlike benign hypertrophy of the prostrate, which appears to develop from prostatic periurethral glands (McNeal 1970, 1972, 1975, 1979; Blacklock 1976) opening into the upper segment of the urethra (Fig. 6.1), prostatic cancer originates in the peripheral zone of the true prostate (McNeal 1975) and could become comparatively large before causing symptoms of urinary obstruction (Peeling and Griffiths 1978). Although observations from autopsies showed a high incidence of nodular hyperplasia associated with prostatic cancer (Sommers, 1957), which suggested that hyperplasia represented a premalignant condition (Armenian, Lilienfeld, Diamond and Bross 1974), there is now general support for the concept that benign and malignant diseases of the prostate are independent clinical conditions arising in different regions of the gland, probably with separate aetiologies. One might expect to find cancer and nodular hyperplasia together in the same prostate since hypertrophy of the gland is so common in older men. Incidental carcinoma (World Health Organization 1980), the small foci of cancer cells often found associated with clinically benign hyperplasia of the prostate (Labess 1952; Turner and Belt 1957; Harbitz and Haugen 1972), is quite common, just as there is a high incidence of latent, or autopsy-diagnosed, cancer without clinical manifestation of the disease (Scott, Mutchnik, Laskowski and Schmalhorst 1969; Lundberg and Berge 1970). A classical autopsy study demonstrated foci of carcinoma in the prostate glands of 30% of men by the age of 50 (Rich 1935).

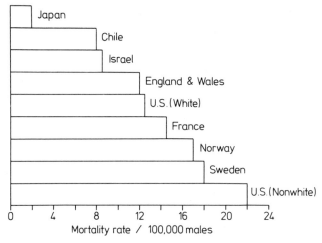

Fig. 6.2.   Mortality rates for cancer of the prostate, age-standardised to world population, 1964–1965 (Skeet, 1976).

Of interest, for example, is the fact that the incidence of prostatic carcinoma is influenced not only by age but also by race (Fig. 6.2). The incidence is high in both white Americans and northern Europeans, but not in the oriental races. The disease is more prevalent in North American black men, whereas it is comparatively rare in the black populations of West Africa, although prostatic cancer is not often diagnosed below the age of 50 (Huben, Mettlin, Natarajan, Smart, Pontes and Murphy 1982) and the mean life expectancy in many black African countries is not much beyond 40 years (Rotkin 1979). In contrast to this, however, the incidence of prostatic carcinoma is low in the Japanese male, whose mean life expectancy and socio-economic standards are comparable to those in North America and Northern Europe.

Environmental and socio-economic factors obviously influence aetiology since studies of migrating populations indicate that the mortality rate from cancer of the prostate gland, although low in Japan, increases to half that of the indigenous American population for those Japanese who move to the U.S.A. (Haenszel and Kurihara 1968; Wynder, Mabuchi and Whitmore 1971). Particularly noteworthy, however, is the fact that the small foci of latent carcinoma, found at autopsy, is just as common (Oota and Misu 1958) in Japanese men as in Caucasians of corresponding age. What initiates the growth and development of these areas of cells and the manifestation of clinical cancer remains to be ascertained, but studies on testosterone metabolism in Japanese men (Okamoto, Setoishi, Horiuchi, Mashimo, Moriji and Itoh 1971) were clearly in order.

The basis of all endocrine investigations concerning the prostate has been that the growth, maintenance and functional activity of the gland are largely dependent upon testosterone secreted by the testis. Although Starling originally introduced the term 'hormone' in 1905 in his Croonian lecture, it was a little earlier that the suggestion was made that the testes, like the thyroid, might secrete a chemical substance which maintained the male characteristics (Griffiths 1895). It is well accepted, however, that 200 years ago John Hunter (1786) drew attention to the relationship between testicular function and the prostate when he reported that castration was followed by a decrease in the size of the prostate. The work of Zondek and Aschheim (1927) subsequently suggested a pituitary control of the gonads when it was observed that precocious puberty could be induced in laboratory animals by administration of an anterior pituitary extract.

Our understanding of the endocrine factors that may be concerned in controlling prostatic growth and function has increased immensely during the ensuing years. Figure 6.3 illustrates some aspects of the relationship between various endocrine glands and the prostate. It shows the transfer of releasing factors from hypothalamus to the anterior pituitary with the consequent secretion of pituitary hormones that exercise control of steroid hormone synthesis and secretion by testis and adrenal. Androgenic hormones produced by the testis and adrenal can be taken up by the prostate, further metabolised, and become involved in the complex biochemical processes leading to the control of cell growth and function.

Furthermore, the cornerstone of our understanding of prostatic cancer is the series of investigations of Dr. Charles Huggins and his colleagues (Huggins and Clark 1940; Huggins and Hodges 1941; Huggins, Stevens and Hodges 1941) which established that prostatic cancer cells retained some degree of androgen-dependence. Although references to the use of orchidectomy for the treatment of enlarged prostates appeared towards the

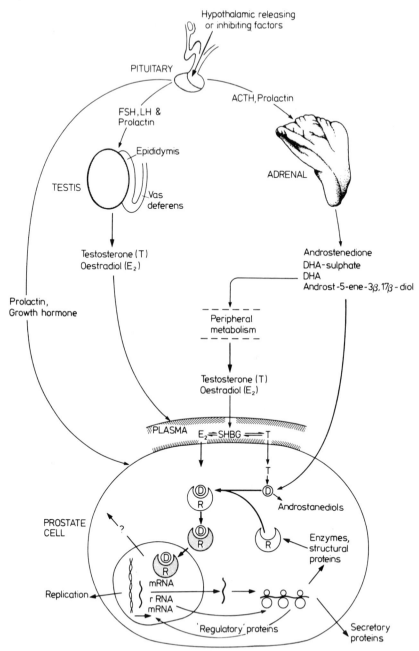

Fig. 6.3.   Schematic representation of the endocrine factors concerned with the regulation of prostatic function. DHA, dehydroepiandrostosterone; D, 5α-dihydrotestosterone; R, specific intracellular receptor; SHBG, sex hormone-binding globulin; mRNA, messenger RNA; rRNA, ribosomal RNA.

latter part of the 19th century (White 1895; Cabot 1896), it was the experiments of Huggins and his colleagues that provided the scientific basis for the acceptance of antiandrogen therapy, either by oestrogen administration or orchidectomy, for the management of patients with carcinoma of the prostate. Huggins, moreover, introduced the use of the newly synthesised oestrogen, diethylstilboestrol (DES) (Dodds, Goldberg, Lawson and Robinson 1938) for the treatment of prostate cancer and in some clinics today this synthetic oestrogen is still the preferred form of therapy. New approaches to therapy are being developed however, and it is probably appropriate to consider these at an early stage of this review.

II Endocrine factors in the treatment of prostatic cancer

1 THE ROLE OF DIETHYLSTILBOESTROL (DES)

Through the past three or four decades it has generally been accepted that DES provides an inexpensive, but relatively successful, form of oestrogen therapy in the management of carcinoma of the prostate, for which it has probably been the most widely used of drugs. Although there is evidence for a direct action of DES on both testis and prostate (Griffiths, Davies, Harper, Peeling and Pierrepoint 1979), DES would appear to exercise its therapeutic effect indirectly, by suppressing the pituitary gland, inhibiting the release of luteinizing hormone (LH) and, consequently, the secretion of testosterone by the testis.

The thromboembolic problems caused by the clinical use of DES have however elicited considerable controversy through the years, especially after the American VACURG (Veterans Administration Cooperative Urological Research Group) investigations (Mellinger 1964; Arduino 1967; Byar 1977), and there has been a great deal of concern about other side effects of DES administration, including feminisation, decreased libido, impotence, azoospermia, nausea and various gastric disturbances. These are, however, dose-related and are minimal if just 1 mg DES is given daily, although in the United Kingdom, it became accepted that 1 mg, three times a day (t.d.s.), provides a valuable form of primary treatment for advanced prostatic cancer. The rationale behind the use of DES-diphosphate (Honvan), in doses often exceeding 200 mg daily, was that the free, presumably active, DES was released directly in the prostate by the high concentration of prostatic phosphatases present in the tissue. There is little evidence that DES-diphosphate accumulates in the prostate or that DES is released in any considerable amount (Abramson and Miller 1982) although DES could well have a cytotoxic action within the gland. Endocrine studies from our laboratories (Griffiths, Peeling, Harper, Davies and Pierrepoint 1981) indicate that DES, 1 mg t.d.s. is just as effective as 200 mg Honvan in reducing the concentration of plasma testosterone to 'castrate' levels, whilst the elevation of plasma prolactin and growth hormone (GH) levels, observed after Honvan treatment, would suggest that it should not be used on a long-term basis (Griffiths et al. 1979).

Metastatic carcinoma of the prostate does, however, respond to primary endocrine therapy with approximately 70–80 per cent of patients experiencing symptomatic relief (Resnick and Grayhack 1975; Peeling and Griffiths 1986). Relapse generally occurs within 1–2 years (Whitmore 1973; Jordan, Blackard and Byar 1977) and the median survival is then 6 months. Disease progression obviously occurs because only a proportion of the neoplastic cells are hormone-responsive and the autonomous proliferation of the remainder eventually kills the patient.

Patients that relapse rarely respond to further endocrine therapy (Brendler 1973) although hypophysectomy appears to give worthwhile palliation of pain without affecting survival (Fitzpatrick, Gardiner, Williams, Riddle and O'Donoghue 1980). Generally, plasma testosterone levels of patients on long-term DES therapy do not show any change after administration of luteinizing hormone-releasing hormone (LH-RH) and the concentration of testosterone is similar to that of castrated men (Kemp, Read, Fahmy, Pike, Gaskell, Queen, Harper and Griffiths 1981). One of the six patients investigated, who had been taking DES for some months, had a comparatively high concentration of plasma testosterone and his testes responded after LH-RH administration (Fig. 6.4). Clearly, in this patient, prolonged daily treatment with 3 mg DES had failed to suppress the pituitary-testicular axis. Any lack of response to the treatment may not therefore have been due to the presence initially of a hormone-independent cancer, but presumably to too low a dosage of DES. Therefore, although it is now recognised that orchidectomy is generally of no value as secondary endocrine therapy following prolonged oestrogen administration (Stone, Hargreave and Chisholm 1980), it has to be accepted that in this particular patient in which the requisite depression of testicular function has been inadequate, the subsequent removal of his testes could well have been beneficial.

Although drugs other than DES have been used for the treatment of advanced prostatic cancer, none can be said to-date to have gained universal acceptance, and none is as inexpensive as DES. The long-acting polyoestradiol phosphate (Estradurin), injected intramuscularly, only weakly inhibits gonadotrophin secretion (Jönsson, Olsson, Luttrop, Cekan, Purvis and Diczfalusy 1975), whereas Premarin, a mixture of conjugated equine oestrogens, or ethinyloestradiol, are less effective than DES. Various progestational steroids have been used, including medroxyprogesterone acetate (Provera), hydroxyprogesterone caproate (Delalutin), cyproterone acetate (Androcur) and chlormadinone acetate and although some would consider the clinical results promising (Geller, Vazakas, Fruchtman, Newman, Nakao and Loh 1968) their effectiveness has not been established by rigorous randomised clinical trials and consequently these drugs cannot yet be said to have a major role in the management of prostatic cancer.

2   ANDROGEN RECEPTOR LEVELS IN PROSTATIC CANCER: PROGNOSTIC VALUE

Endocrine procedures for the treatment of prostatic cancer will result in disease regression or stabilisation in 70–80 per cent of patients. Treatment is palliative, the disease almost inevitably recurs and patients are not cured of cancer. Initial response followed by relapse clearly expresses the inexorable progression of the clones of hormone-independent cancer cells. Those who fail to respond initially to primary endocrine therapy and are the 'high-risk', bad prognosis group represent patients in whom a large proportion of cancer cells, if not all, are hormone-independent.

Just as oestrogen receptor analysis of breast tumours offers an extremely valuable parameter for predicting a response to endocrine therapy (Consensus Meeting 1980) and for assessing prognosis (Griffiths, Nicholson, Blamey and Elston 1986), so it was hoped that the knowledge of the receptor status of prostatic cancer would similarly identify the patient who would respond to endocrine treatment. A considerable amount of research has been directed to the characterisation and quantitation of the androgen receptors of prostatic tissue. An earlier review from the Tenovus Institute (Davies 1978) considered the technological difficulties of androgen receptor analysis of prostatic cancer and the problems associated with sex hormone binding globulin (SHBG) contamination, a protein thought by some (Cowan, Cowan and Grant 1977) to be a reservoir of C_{19}-steroids within prostatic cells. Overall, however, despite the amount of data that has now been generated, the clinical value of the androgen receptor status of a prostatic cancer, in the management of the patients, has still to be established and the following sections review the current situation.

Quantitation of cytosolic androgen receptors (AR) in human prostate cancer has proved inadequate as an index of tissue hormone-dependence (de Voogt and Dingjan 1978; Ekman 1982; Trachtenberg and Walsh 1982; Brendler, Isaacs, Follansbee and Walsh 1984; Gonor, Lakey and McBlain 1984) and increasingly, nuclear AR-content is being promoted as the more useful prognostic indicator (Ghanadian, Auf, Williams, Davies and Richards 1981; Trachtenberg and Walsh 1982; Brendler *et al.* 1984; Gonor *et al.* 1984; Fentie *et al.* 1986). Ghanadian and his colleagues (1981), and also Trachtenberg and Walsh (1982), have attempted to establish the critical concentrations of nuclear AR-content that relate to duration of response, or time to disease progression. The data from these groups tend to be a little equivocal. In a series of 32 patients, Ghanadian *et al.* (1981) found that 10 of 13 patients whose tumours contained nuclear AR concentrations of more than 500 fmol per mg DNA were still responsive to

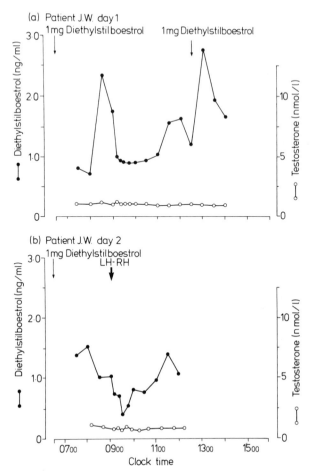

Fig. 6.4. Sequential plasma testosterone and diethylstilboestrol (DES) levels in patients (J.W. and F.C.) with carcinoma of the prostate on DES therapy (1 mg t.d.s.). Day 1 is represented by (a) and (c), Day 2 by (b) and (d). LH-RH (100 μg i.v.) was administered on day 2. (From Kemp *et al.* 1981).

endocrine therapy after two years. Trachtenberg and Walsh (1982) however established their critical concentration of extractable AR, with regard to duration of response, at 100 fmol per mg DNA; it is clear from the results however that there is a considerable overlap in receptor concentrations for the responders and non-responders.

Both these studies considered only those receptors which could be extracted from the nucleus using high-molarity salt. It has been established during recent years that a considerable 'population' of nuclear AR of rat ventral prostate is resistant to extraction by such high ionic strength solutions, or exhaustive nucleolysis (Barrack and Coffey 1980, 1982;

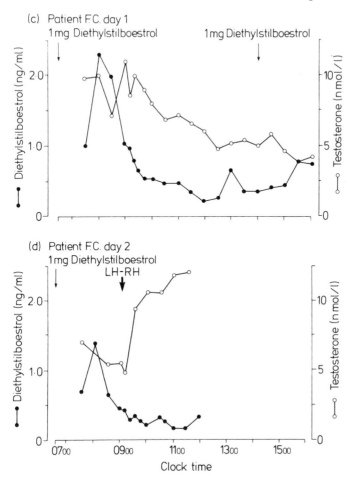

Fig. 6.4.   (cont.)

Davies et al., 1980; Davies, Thomas and Giles 1982; Davies, 1983a; Rennie, Bruchovsky and Cheng 1983). This significant proportion of AR has been located on the nuclear matrix of rat ventral-prostate cells (Barrack and Coffey 1980; Rennie *et al.* 1983) and in view of the multiple biological functions ascribed to the nuclear matrix (Barrack and Coffey 1982), including RNA and DNA synthesis, the steroid-receptor associated with it must be considered a biologically active component. Androgen receptors have subsequently been detected on the nuclear matrix prepared from human prostate samples (Donnelly, Lakey and McBlain 1984; Gonor *et al.* 1984) and these matrix, or salt-resistant, receptors have now been taken into consideration in the assessment of possible androgen-responsiveness of human prostate cancer (Barrack, Bujnovsky and Walsh 1983; Brendler *et al.* 1984; Gonor *et al.* 1984; Fentie, Lakey and McBlain 1986).

In a small study, Gonor *et al.* (1984) reported disease progression in only one patient with a matrix AR concentration greater than 200 fmol per g tissue, whereas no patient with a matrix-bound AR concentration of less than 100 fmol per g tissue showed evidence of disease regression. Concentrations between 100 and 200 fmol per g tissue were of little predictive value. Moreover, nuclear matrix-bound AR was not found to be superior to extractable AR as an index of hormone-dependence and in fact the concentration of total nuclear AR was found to be most useful in distinguishing between responders and nonresponders. Brendler *et al.* (1984) also reported that salt-extractable receptor was superior to nuclear nonextractable AR in predicting patient response to endocrine therapy. Furthermore in a subsequent re-analysis, McBlain and colleagues (Fentie *et al.* 1986) found that only nuclear AR concentrations (salt-extractable, matrix-bound or total) correlate well with patient survival, whilst cytosolic AR and total cellular AR levels did not correlate with survival; however, when patients were grouped as either receptor-poor, or receptor-rich, cytosolic, total cellular and nuclear salt-extractable AR populations could be used to identify groups of patients with significantly different survival time. Nuclear matrix and total nuclear AR were not significantly discriminatory. Although the real value of these studies is, however, compromised by the small numbers of patients involved and it is probably necessary to accumulate more data to determine with certainty the importance of studying subnuclear AR populations, the clinical value of total nuclear AR assays is becoming more evident.

The tendency to concentrate on nuclear AR assays fits contemporaneously with changing opinions on the molecular processes by which steroid hormones exercise their control of target cells. New enucleation methods (Welshons, Liebermann and Gorski 1984) and improved immunocytochemistry (King and Greene 1984) suggest that oestradiol receptors are found only in the nucleus of oestrogen-dependent cells. The non-cooperativity of steroid binding in intact cells, or to receptor immobilised on hydroxyapatite, compared to cooperativity in soluble systems, has been interpreted in favour of an insoluble receptor *in vivo* (Gorski, Welshons and Sakai 1984), although there are arguments against this new model system for steroid action (Szego and Pietras 1985). Re-modelling, or retaining the old concept, does not however alter the fact that hormonal agonists do effect gene expression and do this through some changes in certain aspects of receptor structure.

With the assumption that receptors for all steroid hormones are similarly distributed in the various target tissues, the old and new models are compared in Fig. 6.5, illustrated as an adaptation for androgen action. *In the*

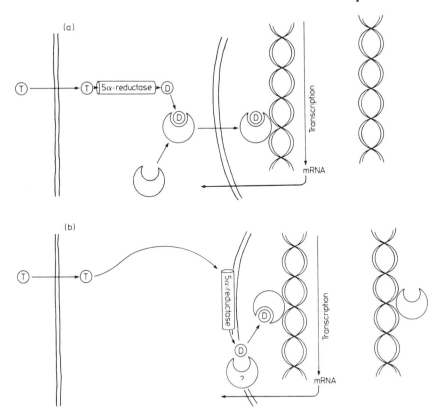

Fig. 6.5. Intracellular processes involved in androgen action. (a) accepted 'old' model showing translocation of the DHT-receptor complex into the nucleus. The 5α-reductase converting testosterone (T) to DHT (D) is illustrated as an extra-nuclear enzyme. (b) 'new' model, in which the receptor and the 5α-reductase are associated with the nuclear membrane.

'old' model (a), extracellular testosterone (T) diffuses into the cell, is metabolized by 5α-reductase to 5α-dihydrotestosterone (D), which binds to a cytoplasmic receptor of selective high affinity. This occupation of the receptor confers upon it increased affinity for nuclear components resulting in its translocation into the nucleus (the 'two-step' hypothesis). The interaction of the androgen-receptor complex with tissue-specific acceptor sites, which may be DNA, chromatin-associated proteins or matrix components, or distributed among all three, influences nuclear processes, represented most rapidly and significantly in the production of mRNA precursors. *In the remodelled scheme (b)*, 5α-reduction of T to D occurs in the region of the nucleus. Receptor molecules are constantly situated *in vivo* in the location of their action, but in an inactive structure. The entry of steroid into the nucleus and subsequent binding to the receptor causes a conformational change and regulatory interlocking with the transcriptional apparatus to increase RNA synthesis as in (a). The process may be facilitated by receptors located on the nuclear membrane. Cytoplasmic receptors *in vitro* could result from unoccupied or uncommitted receptors being more susceptible to leaching out in the buffer systems used, or from nuclear damage. Perhaps quickening the demise of the two-step transloca- tion mechanism for androgen action are observations of androgen interaction with intact nuclear envelopes prepared from rat ventral prostate tissue (Lefebvre *et al.*, 1985) and the exclusive nuclear localisation of 5α-reductase (Habib, Smith, Robinson and Chisholm 1985; Houston, Chisholm and Habib 1985).

It is to be hoped therefore that the current, new studies with nuclear receptor assays will now lead to a greater understanding of androgen receptor distribution in human prostate tissue both normal and abnormal. Certainly, both prostatic cancer and benign prostatic hypertrophy are conditions in which nuclear androgen receptor levels may differ, either quantitatively or in their localisation within the nucleus, from those found in the normal prostate (Barrack *et al.* 1983; Davies 1983b; Donnelly *et al.* 1984; Grimaldo and Meikle 1984; Benson, Utz, Holicky and Veneziale 1985; Kyprianou and Davies 1986). Nuclear preparations from benign prostatic adenoma have however been characterised by higher concentrations of extractable receptors (Grimaldo and Meikle 1984; Kyprianou and Davies 1986), equivalent concentrations of extractable and nonextractable recep- tors (Barrack *et al.* 1983), and by a higher concentration of matrix receptors (Donnelly *et al.* 1984). Furthermore, nuclear preparations from prostatic carcinoma have been reported to be enriched in extractable (Barrack *et al.* 1983; Brendler *et al.* 1984), but also in nonextractable AR (Davies 1983b; Gonor *et al.* 1984; Fentie *et al.* 1986).

Accumulation of AR does not predispose toward malignancy *per se*: indeed, 'accumulation' may be an artefactual function of the problems associated with the analytical procedures, discussed later. Re-allocation of functioning receptor could be a more significant feature. In our studies, nuclear preparations from benign prostatic adenomatous tissue were found to be characterised by a large variable pool of apparently uncommitted receptors, not associated with nuclear components when probed with micrococcal nuclease (Kyprianou and Davies 1986). These receptors were functional, but apparently not functioning, although interaction with hypersensitive fragile regulatory sites cannot be excluded. On the other hand, nuclear preparations from carcinomatous tissue were characterised by an increase in nuclease-resistant AR (Davies 1983b) under conditions which indicate an increase in acceptor sites. Nuclei from rat ventral prostate respond to short-term castration with an increase in the number of matrix acceptor sites (Davies, Thomas and Manning 1986), accounting for the rapid, excessive, reoccupation of these sites after androgen replacement

(Davies *et al.* 1982; Rennie, Bruchovsky and Cheng 1983). This may represent a restorative rescue-phenomenon which is regulated back to normal levels when equilibrium is re-established. A dys-regulation of such a system would clearly have drastic effects on the biochemistry of the tissue.

Androgen receptor assays now, as always, are bedevilled by non-uniform distribution of AR among areas of the same prostate (Van Aubel, Bolt-de Vries, Blankenstein, ten Kate and Schroder, 1985; Bowman, Barnes, Blacklock and Sullivan 1986), between different prostate tissue types (Sirett, Cowan, Janecska, Grant and Glen 1980; Bashirelahi, Young, Sidh, and Sanefuji 1980; Lahotnen, Bolton, Kontturi and Vihko 1983; Smith, Sutherland, Chisholm and Habib 1983; Kyprianou and Davies 1986) and possibly among different subnuclear compartments, as discussed earlier. These problems are intensified when only small samples of tissue are available, such as those obtained from needle biopsies. New methods are still forthcoming with which it is hoped that such difficulties can be overcome (Bowman, Barnes and Blacklock 1985), but the absence of a universally accepted, reproducible method for the assay of AR is undoubtedly the major contributory factor in producing such conflicting results. Most assays are influenced by the slow dissociation of androgen from receptor at low temperature, which means that exchange assays at $0°$ are incomplete unless allowed to proceed for exceedingly long periods of time. The instability of the receptor is also not helpful and means that exchange assays at elevated temperatures may be non-quantitative. These problems have recently been discussed in detail (Hechter, Mechaber, Zwick, Campfield, Eychenne, Baulieu and Robel 1983) and new approaches to AR-assay are still being proposed (Robel, Eychenne, Blondeau, Baulieu and Hechter 1985; Traish, Muller and Wotiz 1985). The observation that circulating autoantibodies to human androgen receptor are present in high titres in the sera of some patients with prostate diseases (Liao & Witte 1985) may provide a route to the rapid production of monoclonal antibodies against androgen receptors. Furthermore, reports that both human glucocorticoid receptor (Weinberger, Hollenberg, Rosenfeld, and Evans 1985) and human oestrogen receptor (Green, Walter, Kumar, Krust, Bornert, Argos and Chambon 1986) show homology with the v-erb A oncogene product suggest that these three genes are derived from a common primordial regulatory gene. The possibility of conserved sequences corresponding to domains important in the function of steroid receptors opens the door to investigation and purification of the androgen receptor using nucleic acid technologies.

It is well accepted that androgen receptors have a fundamental role in the androgenic control of prostatic function and if the analytical problems are solved, the determination of the receptor concentration of a prostate tumour should provide a valuable prognostic parameter. It is important, however, to consider the tissue sampling procedures. Transurethral resection (TUR) of prostatic tissue involves a hot cutting loop of a resectoscope which chars the tissue chippings. This cannot be good since androgen receptors are unstable at high temperature, although certain research groups believe that if the chippings are large, it is possible effectively to remove burnt tissue (Snochowski, Pousette, Ekman, Bression, Anderson, Hogberg and Gustafsson 1977; Ghanadian and Auf 1980). Geller and his colleagues have reported however that the frequency of the electric current used for resection does influence the receptor analysis (Albert, Geller and Nachtsheim 1982).

A study in this Institue (Kyprianou, Williams, Peeling, Davies and Griffiths 1986) compared nuclear receptor levels in tissue taken by cold punch resection with the concentration in tissue removed by electro-resection. Twenty patients with benign prostatic hypertrophy, with tumour mass assessed by ultrasonography to be greater than 70 g, were treated by retropubic prostatectomy. After preliminary cyto-urethroscopy, several samples of tissue were taken from the right lobe using a cold punch resectoscope. Using a hot loop resectoscope further biopsies were then taken transurethrally from regions in the right lobe, adjacent to those taken previously and the patient was then repositioned and a standard retropubic prostatectomy performed. Specimens were removed from both right and left lobes of the enucleated tissue.

Table 6.1 gives the data on the receptor content of the various tissue samples. Nuclear androgen receptors were detected in all cold punch samples and in enucleated tissue, with similar concentrations observed in tissue from the same gland. Comparison of levels in samples from left and right lobes taken after prostatectomy (n = 10) gave a correlation coefficient of 0.94, and comparison between cold punch-resected tissue and the remainder of the right lobe, provided a correlation coefficient of 0.87. Of the 20 specimens obtained by transurethral resection, only six contained detectable nuclear receptors and at significantly lower concentrations than those of cold punch resections or enucleated tissue. These six specimens were however the largest, approximate weight 500–800 mg, and the charred tissue was easily removed before processing.

The method of sampling clearly affects the analysis of the tissue receptor content and must therefore be taken into account as data continue to accumulate on AR-levels in relation to hormone responsiveness and prognosis. The need for larger TUR samples for such analysis has also been stressed in reports from other research groups (Habib, Smith, Robinson and Chisholm 1985; Bowman, Barnes and Blacklock 1985).

3 PREMALIGNANCY AND ENDOCRINE STATUS

It is interesting that the clinical behaviour of prostatic cancer does reflect androgen stimulation and in a large proportion of patients the disease

Table 6.1

Androgen receptor content ([AR] fmol per mg DNA: undetectable is signified by '0'; not done is signified by a blank in a column) and binding affinity (Kd, nmol per l) of nuclei prepared from specimens of benign prostatic hyperplasia prostate tissue obtained by transurethral resection (TUR), cold punch resection (CP) and retropubic prostatectomy (RPP: 'Right' and 'Left' refer to separate lobes).
(From Kyprianou et al. 1986).

Patients		TUR		CP		RPP Right		RPP Left	
Initials	Age (y)	[AR]	Kd	[AR]	Kd	[AR]	Kd	[AR]	Kd
(A.H.)	74	0		1190	14.1	911	5.0	915	3.8
(R.M.)	72	0		836[a]	7.5	968	7.6	1000	6.9
(L.H.)	72	0		1010	3.1	1040	9.4	807	2.8
(W.G.)	59	0		576[b]	4.8	458[b]	4.7	598[b]	3.7
(C.D.)	87	0		1320	9.3	1073	10.4	969	7.9
(H.B.)	78	0		1850	19.3	1491	4.4	1584	11.2
(I.W.)	61	648[c]	8.0	1180	3.5				
(L.B.)	72	359[c]	2.4	903	3.2	1116	9.3	1200	15.4
(A.C.)	70	0		2100	6.6				
(G.E.)	67	391[b,c]	12.1	561[b]	7.4	487[b]	14.5	496[b]	9.4
(J.T.)	58	436[c]	5.5	745	4.3				
(A.T.)	84	0		1010	2.3				
(R.R.)	59	299[c]	4.8	767	5.6	832	9.1	937	4.9
(R.H.)	67	0		374[b]	6.2				
(F.V.)	68	380[c]	10.4	898	6.1				
(F.G.)	77	0		514[a]	6.7				
(W.N.)	74	77[c]	1.7	969	4.7				
(J.A.)	61	0		561	12.7				
(A.L.)	76	0		813	12.2	697	1.1	796	13.4
(D.S.)	55	0		1030	5.7	708	6.0		

[a]Insufficient amount of tissue available for complete saturation analysis of [³H]-DHT binding to nuclear androgen receptor.
[b]Prostate samples were stored under liquid N₂(−170°C) for 3 to 4 weeks before being assayed.
[c]TUR specimens were obtained with 27F resectoscope loop and were relatively larger (500–800 mg) and comparatively free of charring.

responds well to treatment that effectively lowers plasma levels of testosterone, or is essentially anti-androgenic. The disease has not been reported in the prepubertally-castrated male, yet the endocrine factors concerned in the aetiology of the disease are little understood. Certainly there are few leads to indicate that hormones are concerned in the initiation of prostatic cancer and it may be that the role of the hormones in the pathogenesis of prostatic cancer may be permissive rather than inciting.

Attention has been directed to the investigations of McNeal on the pathogenesis of prostatic cancer (McNeal 1975). The central and peripheral zones of the prostate slowly but progressively atrophy with advancing years, although certain glands show a glandular morphology similar to that of young men. McNeal describes, however, an age-related increase in atypical hyperplasia, a diffuse or multifocal proliferation of ductal and epithelial tissue and he considered this active glandular tissue to be a premalignant change closely associated with prostatic carcinoma. Another retrospective study of the pathology of the prostate (Koppel, Heranze, and Shimkin 1967) also showed that these diffuse areas of hyperplasia were more often found in prostatic tissue from cancer patients than controls. It is not easy to assess the relationship between these observations and the well substantiated (Baba 1982) high incidence of small, clinically asymptomatic, latent carcinoma seen at post-mortem predominantly in the peripheral zone of the prostate. It is considered by some (Varkarakis, Castro, and Azzopardi 1970; Correa, Anderson, Gibbons, and Tate 1974) that these small foci of carcinoma rarely become clinically manifest and it is suggested (Hirst and Bergman 1954) that the latent period between initial development of cancer to the clinical condition is at least 20 years. We know nothing about the endocrine or biochemical conditions which might promote the change from 'inactive' latent carcinoma into the aggressive cancer. The time-course for such a process is difficult to ascertain and it is quite clear that extensive growth of the cancer in the peripheral zone with capsular involvement and distant spread of the disease is possible before the disease becomes symptomatic.

Possibly relevant, therefore, is that premalignancy and the foci of carcinoma are associated with an 'active' prostate with glandular morphology more generally present in the gland of a younger man and it is tempting to contemplate the possible correlation of this situation to high levels of plasma or tissue androgens which might promote this active state. Rotkin (1979) attributes carcinogenesis of the prostate gland to an 'overbalance of androgenic components', as described later.

The pattern is indeed interesting and analogous in many ways to that recently reported by our group from observations of breast tissue (Simpson, Mutch, Halberg, Griffiths and Wilson 1982). This study, concerned with breast cancer pathogenesis and the age frequency distribution of epitheliosis or epithelial hyperplasia, in 500 consecutive cancer mastectomies, was stimulated by the reports of Jensen and her colleagues (Wellings, Jensen, and Marcum 1975; Jensen, Rice, and Wellings 1976a, 1976b; Wellings, Jensen, and de Vault 1976) on the frequency and distribution, in breast tissue, of atypical lobules showing excess epithelium and corresponding to epitheliosis in routine histopathology. These 'hyperplastic' lobules were more commonly found in association with cancer in mammary tissue both ipsilateral and contralateral to the cancerous breast and suggests that cancer may develop, not in normal mammary epithelium, but in proliferating tissue. Noteworthy however are the statements of Sir Robert Muir (1941) in his classical work on the genesis of mammary cancer, where he reported that the stages of evolution of malignancy within ducts and acini cannot be followed and that malignancy arises without hyperplastic change. Many pathologists tend to agree with this (Gray 1979), but this study (Simpson *et al.* 1982) of 500 mastectomies showed a bimodal age-frequency distribution, with a high probability of epitheliosis in cancer mastectomies from women in their early 40's, low in the 50's and high in the elderly. When the results were compared to the normal age-frequency distribution (Sandison 1962), it was found that the cancerous breasts from patients aged 26–55 contain 2.6–9.5 times the expected frequency, whereas that seen in postmenopausal women were close to that normally expected (Table 6.2).

Table 6.2

Comparison between the observed frequency of epitheliosis in the non-neoplastic tissue of cancer mastectomies and the expected frequency as found in the mammary tissue of consecutive autopsies. (From Simpson et al. 1982).

| Age (y) | Percent breasts with epitheliosis | | |
	Cancer mastectomies	Autopsy* breasts	Observed 'expected'
26–30	33	8	4.1
31–35	28	7	4.0
36–40	57	6	9.5
41–45	59	27	2.6
46–50	55	14	3.9
51–55	53	17	3.1
56–60	17	23	0.7
61–65	32	27	1.2
66–70	29	27	1.1
71–75	29	27	1.1
76–80	35	35	1.0
81–85	50	33	1.5
86–90	50	No data	–
Note x̄	41	21	2.8

*Sandison, 1962.

Does breast cancer therefore initiate in these active, proliferative regions of epitheliosis, promoted by oestrogen stimulation, in a manner similar to that observed in prostatic tissue under the influence of androgenic steroids?

Does cancer of the prostate develop from proliferating epithelial cells stimulated by high levels of androgen? The foci of carcinomatous cells are well differentiated and presumably represent hormone-dependent cells. It is generally accepted that the more malignant the cancer, the less hormone-dependent and the less the degree of differentiation (Gleason 1977). It seems that only a proportion of these areas of latent cancer become aggressive, grow and metastasise and it is clearly important that we understand more about the biology of this process. Certainly the most devastating aspect of prostatic cancer, as with that of the breast, is the propensity to metastasise from its primary site before the disease is clinically manifest. Greater understanding is required of the process by which certain phenotypes develop in the primary tumour, then spread and metastasise, and of the biological factors that must influence the invasive nature of the cancer. At present there are few leads that relate hormonal factors with the initiation of neoplasia, with the process of metastasis, or to the aetiology of the disease (Griffiths et al. 1979) and there is, as yet, no real evidence that the hormonal status of a person, prior to cancer development, will assist in defining a high-risk group.

4 THE CLINICAL POTENTIAL OF OTHER PROGNOSTIC FACTORS

Therefore despite intensive investigation, we are still searching for evidence to indicate that carcinoma of the prostate is caused by excessive androgen stimulation. Although the disease is generally androgen-dependent, there is little data to suggest an endocrine abnormality associated with patients with prostatic cancer when compared to the hormonal status of clinically healthy men of similar age or in relation to disease progression (British Prostate Study Group 1979; Griffiths et al. 1979; 1981; Hoisaeter, Haukaas, Bakke, Hoiem, Segadal, and Thorsen 1982).

The problem with prostatic cancer is one of early diagnosis and it has to be stated that the principal means of diagnosis is still, as it was in the early part of the century, the rectal examination. If the endocrinology of men could be investigated when preneoplasia or early malignancy is established, then it might be possible to understand better some of the biological and hormonal factors that influence the promotion of the disease. The introduction of procedures for ultrasonic scanning of the prostate gland by rectal probes (Fig. 6.6) may however be of particular value in recognising early pathological changes in the tissue (Peeling, Griffiths and Evans 1986) and appropriate sampling techniques may allow the withdrawal of tissue for biochemical investigations.

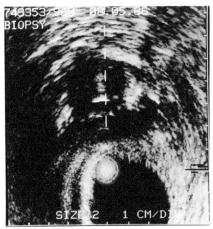 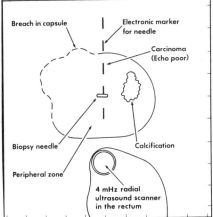

Fig. 6.6.   Transrectal ultrasound scan of the prostate. Areas of calcification can be readily seen. The echo-poor region identifies the cancer. Using ultrasonic scanning, biopsy needles can be inserted to remove tissue samples from well defined regions of the gland.

Of importance in this respect was the more detailed statistical analysis of data from the measurement of the plasma hormone concentrations and biological factors of patients presenting to the various clinics in the United Kingdom (British Prostate Study Group, 1979) which revealed potentially valuable correlations. Canonical variate analysis of pretreatment hormone levels indicated that an elevated plasma growth hormone concentration related to metastatic disease and furthermore, sequential sampling of plasma from a limited number of control subjects and patients tended to suggest that higher nocturnal levels of growth hormone were associated with the disease (Fig. 6.7). There was also evidence that pretreatment plasma levels of LH and testosterone were relevant to patient-survival after endocrine therapy (Harper, Pierrepoint, and Griffiths 1984), those patients dying within one year of diagnosis having higher LH levels and lower concentrations of testosterone in plasma. A similar report from Adlercreutz and his colleagues (Rannikko, Kairento, Karonen and Adlercreutz 1981) indicated that high testosterone/oestradiol or testos-terone/prolactin ratios characterised patients who would respond to endocrine therapy.

Using the Cox Proportional Hazards Regression Model (Cox 1972, 1975), a prognostic index was derived from data of 88 patients who presented with histologically proven metastatic carcinoma of the prostate (Wilson, Harper, Jensen, Richards, Peeling, Pierrepoint, and Griffiths 1985). Higher plasma testosterone concentrations (>4.0 ng per ml) or lower levels of growth hormone (<2.0 µU per ml) gave a better prognosis ($P < 0.01$), as illustrated in Fig. 6.8. Age was not shown to be a significant

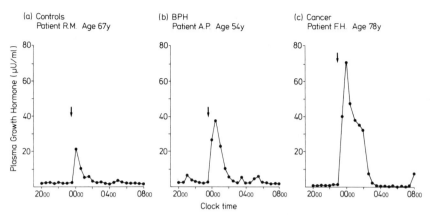

Fig. 6.7.   Growth hormone concentrations in serum samples collected sequentially through the night. Examples are shown for a control subject, a patient with benign prostatic hyperplasia and a patient with cancer. Arrows indicate when the subjects went to sleep.

prognostic factor except that the younger men tended to have the more aggressive disease as previously suggested (Wilson, Kemp, and Stein 1984).

The cumulative survival data as a function of time following diagnosis (Fig. 6.9) for patients grouped according to histological grade assessed by the Gleason procedure (Gleason 1977), indicated a better prognosis for grade 3 patients compared to grades 4 or 5. From the Cox regression analysis, four covariates assumed statistical significance; testosterone ($P \sim 0.001$), Gleason grade ($P \sim 0.03$), growth hormone ($P \sim 0.045$) and age ($P \sim 0.09$) and the values of the coefficients for these covariates are shown in Table 6.3 together with a derived Prognostic Index.

The value of the Index must of course be influenced by the analytical procedures for hormone analysis, the subjective assessment of grade, the factors affecting growth hormone secretion, activity-rest schedules of patients, sample collection and possibly many other parameters, but the prospect of predicting the high-risk patients with prostatic cancer is exciting. The concept is being re-assessed with new data emerging from a Phase III prostatic cancer trial (Turkes, Peeling, and Griffiths 1987) discussed later in this chapter, but the early recognition of aggressive hormone-independent cancer could well offer a rational means of introducing selected primary cytotoxic therapy, probably in association with accepted forms of endocrine treatment, for the management of those patients identified as a high risk group.

5   LUTEINIZING HORMONE-RELEASING HORMONE (LH-RH) ANALOGUES AND PROSTATIC CANCER

After many years, firstly offering orchidectomy, then DES, for the

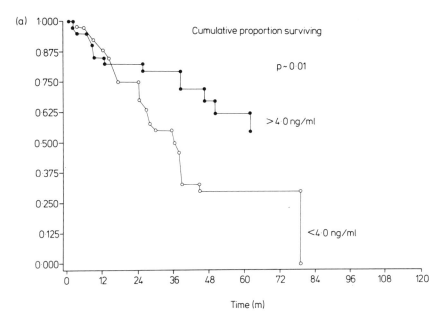

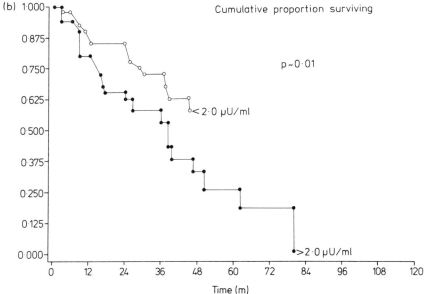

Fig. 6.8.   Survival curves (Kaplan and Meier, 1958) for prostatic cancer patients with metastases stratified according to (a) testosterone concentration and (b) growth hormone concentration in plasma. Statistical significance assessed by the Mantel-Cox test (Mantel, 1966).

Table 6.3

Values for the coefficients of covariates to be significant using the Cox proportional hazards model. Data were obtained from patients with histologically proven carcinoma of the prostate with metastases. T = testosterone, GH = growth hormone, GG = Gleason grade. (From Wilson et al. 1985).

Variable	Coeff.	S.E.	Coeff./S.E.
Age (months)	−0.0032	0.0019	−1.6994
T (ng/mL)	−0.4950	0.1549	−3.1946
GH (mU/L)	0.1224	0.0462	2.6504
GG (1 to 5)	0.6665	0.2381	2.7987

The prognostic index (PI) is given by
PI = −0.495 T + 0.67 GG + 0.12 GH −0.003 age

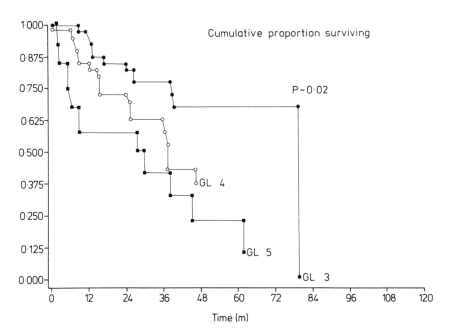

Fig. 6.9.   Kaplan-Meier survival curves for prostatic cancer patients with metastases stratified according to Gleason grade of their tumours. (From Wilson *et al*. 1985).

treatment of prostatic cancer, it is noteworthy that urologists in the United Kingdom have latterly tended to favour orchidectomy, or subcapsular orchidectomy, for primary first-line endocrine therapy. Both forms of treatment remove the testicular source of testosterone, responsible for the production of 95% of the daily output of the hormone (Baird, Uno, and

Melby 1969), and thereby decreases the level to approximately 2 nmol per l in plasma. Dexamethasone suppression and corticotrophin stimulation tests have indicated that the adrenal gland is the source of the 'residual' testosterone (Robinson and Thomas 1971; Griffiths, Harper, and Peeling 1976). Androstenedione, synthesised and secreted by the adrenal gland, is converted peripherally to testosterone. Although it has often been stated in the literature that patient relapse after primary endocrine therapy is related to increasing adrenal output of C_{19}-steroids, a major study of the British Prostate Study Group (1979) failed to show any association between such adrenal activity and disease progression. It has often been suggested that it might offer a more effective form of therapy if both testicular and adrenal androgen production were removed, although those critical of such an approach would indicate that neither adrenalectomy, nor hypophysectomy, produced any further response in the relapsed patient (Brendler 1973). Secondary surgical treatment was however generally undertaken on the very ill, debilitated patient and results might not be expected to be particularly encouraging. In later years the possibility of using, immediately post-orchidectomy, either an antiandrogen such as cyproterone acetate or flutamide to block androgenic action at the target organ, or aminoglutethimide or a synthetic glucocorticoid to suppress adrenal activity, as a form of a complete 'androgen blockade', has gained considerable interest. Cyproterone acetate and flutamide are administered in smaller doses sufficient to inhibit the action of the 'residual' adrenal source of testosterone, whereas aminoglutethimide inhibits the synthesis of C_{19}-steroids by the adrenal gland.

This approach forms the basis of current studies world-wide to assess the value of complete androgen withdrawal and the use of new forms of treatment involving LH-RH analogues rather than orchidectomy for primary therapy (Labrie, Dupont, Belanger, Lacoursiere, Raynaud, Husson, Gareau, Fazekas, Sandow, Monfette, Girard, Emond, and Houle 1983; Labrie, Dupont, and Belanger 1985).

Preliminary investigations from the Tenovus Institute indicated that atrophy of the gonads resulted from the administration of pharmacological doses of the LH-RH analogue, ICI 118630, (Zoladex, D-Ser(Bu^t)^{6}Azgly^{10}-LH-RH, I.C.I. Pharmaceuticals, Macclesfield, Cheshire, U.K.) to female rats (Maynard and Nicholson 1979; Nicholson, Walker, Davies, Turkes, Turkes, Dyas, Blamey, Williams, Robinson, and Griffiths 1984). Such studies, together with the similar, pioneering work of Schally and his colleagues with other LH-RH analogues (Auclair, Kelly, Labrie, Coy, and Schally 1977) directed attention to their potential value for the treatment of hormone-dependent breast and prostate cancer. Low doses mimic the physiological action of LH-RH, stimulating the secretion of gonadotrophins from the pituitary whereas the continuous administration of pharmacological amounts produces a paradoxical anti-gonadal effect. Clinical studies in various centres have now produced evidence of the tremendous potential of the LH-RH analogues for the treatment of advanced cancer of the prostate (Tolis, Ackman, Stellos, Mehta, Labrie, Fazekas, Comaru-Schally, and Schally 1982; Walker, Nicholson, Turkes, Turkes, Griffiths, Robinson, Crispin, and Dris 1983; Trachtenberg, 1983) and of premenopausal breast cancer (Nicholson *et al.* 1984).

Zoladex in solution and administered subcutaneously significantly reduced plasma levels of testosterone, LH and FSH in patients in a Phase I clinical study of our South Wales research group. These hormonal changes were associated with satisfactory clinical responses of patients and the treatment was without significant side-effects (Walker *et al.* 1983).

A slow-release (depot) formulation of Zoladex has now been developed allowing the analogue to be administered as a single monthly injection. The deca-peptide has been incorporated into a 50:50 lactide-glycolide co-polymer in the form of a small cylindrical rod which is injected into the subcutaneous tissue of the anterior abdominal wall (Fig. 6.10).

The depot formulation was recently tested in twenty two patients with metastatic carcinoma of the prostate who had not received any previous therapy (Walker, Turkes, Turkes, Zwink, Beacock, Buck, Peeling, and Griffiths 1984). Patients were randomised to one of three doses of depot Zoladex, 3.6, 1.8 or 0.9 mg administered every four weeks and corresponding to a daily release rate of 120, 60 or 30 µg respectively for 28 days. After Zoladex administration, serum concentrations of LH increased within the first 8 hours, demonstrating the short-term stimulatory effect, then subsequently fell by day 15, to pre-treatment levels or less. The concentration of LH remained slightly higher at day 22 in one patient (101/3), compared to the others. Complete pituitary desensitisation was observed (Fig. 6.11) following the second and third injection of Zoladex. The concentrations of testosterone in serum were also found to increase within five days after the first injection (Fig. 6.12) and then declined to castrate levels (< 2.0 nmol/L) after 15–22 days, with the higher dose of 3.6 mg being marginally more effective.

These preliminary data clearly demonstrated the efficacy of the slow-release depot formulation in producing a medical castration. The treatment now offers a very practical, innovative means of administering the LH-RH analogue which is clinically effective (Walker *et al.* 1984). A Phase III randomised clinical trial has recently completed recruitment, where the long-term effect of depot Zoladex has been compared to orchidectomy for the management of metastatic carcinoma of the prostate (Turkes *et al.* 1986). Figure 6.13(a) illustrates the changes in testosterone and LH through the first six months in patients treated with Zoladex and followed clinically using various 'tumour markers' such as prostatic acid phosphatase (PAP) (Fig. 6.13b). This trial was also of particular interest in indicating that the determination of prostate specific antigen (PSA) offers a better guide to

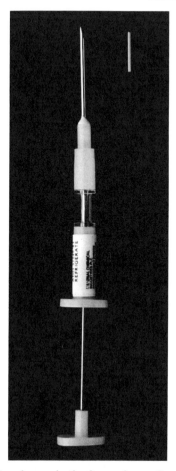

Fig. 6.10.   The LH-RH analogue, in the form of a small cylindrical rod is injected under local analgesia, through a 16-gauge needle into the subcutaneous tissue of the abdominal wall.

disease progression than PAP, Fig. 6.14 showing two examples of patient relapse, when PSA levels increased, but PAP concentrations remained low in one of the patients, rising in the other. There is currently a rapidly increasing interest in the use of LH-RH agonists to treat hormone-dependent cancer and many trials are underway to test the effectiveness of complete 'androgen blockade', using an LH-RH analogue together with antiandrogen or adrenal suppressive therapy.

Labrie and his colleagues (1983) have assessed an LH-RH analogue, Buserelin (Hoechst, HOE-766) together with Anandron (RU 23908 Roussel Uclaf, Paris, France), an antiandrogen that also blocks the 'residual' testosterone effect at the level of the target cell by interfering with the availability of androgen receptor (Raynaud, Bonne, Moguilewsky,

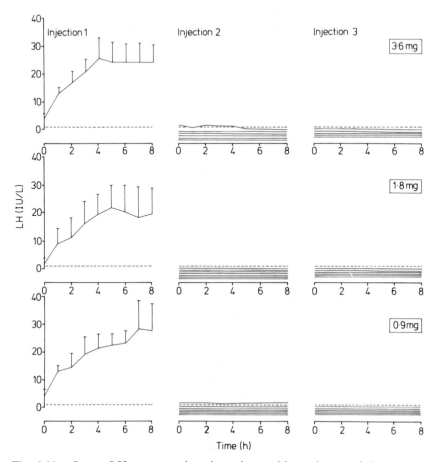

Fig. 6.11.   Serum LH concentrations in patients with carcinoma of the prostate following long-term treatment with depot Zoladex. (From Pierrepoint *et al.* 1985).

Lefebvre, Belanger, and Labrie 1984). Together with clinical colleagues associated with the British Prostate Study Group and with Schering, U.K., the Tenovus Institute is participating in a new randomised clinical trial of Zoladex and the antiandrogen, cyproterone acetate, against Zoladex alone, as part of this new and very interesting approach to the treatment of prostatic cancer.

It remains to be seen whether after longer-term objective assessment, there is justification for the early excitement created by Labrie's preliminary report (Labrie, Dupont, and Belanger 1984) of a virtually 100 per cent remission rate in 30 patients on combined LH-RH analogue and antiandrogen therapy, who had been followed for 18 months. Geller and his colleagues, using progestational agents such as megestrol acetate (Geller and Albert 1983) have long been advocates of attempts to develop drug regimens to

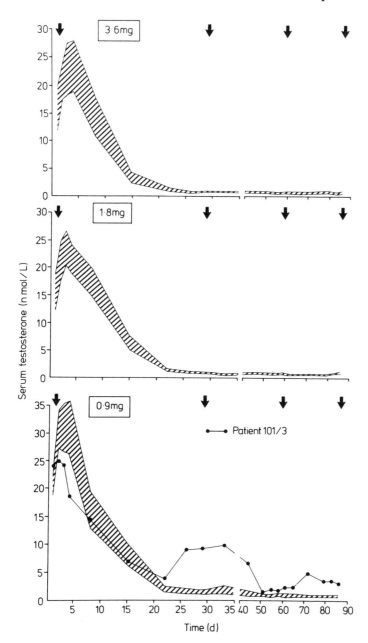

Fig. 6.12. Serum testosterone concentrations in patients with carcinoma of the prostate following long-term treatment with depot Zoladex. Arrows indicate when Zoladex was administered. Shaded area represents range of values ± S.E.M.. (From Pierrepoint *et al*. 1985).

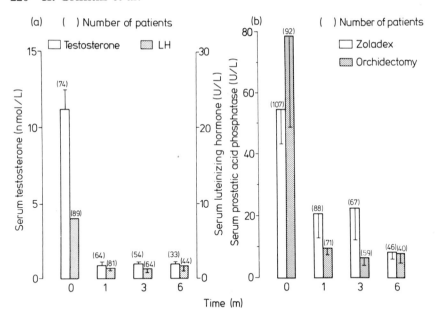

Fig. 6.13. (a) Testosterone and LH concentrations in samples of serum obtained from patients on long-term Zoladex depot treatment. (b) Comparison of serum prostatic acid phosphatase concentrations in patients participating in the Phase III, orchidectomy *versus* Zoladex depot trial. Data show mean ± S.E.M. for the number patients indicated in bracket. (Unpublished results).

remove the effects of both testicular and adrenal androgens. It is certainly important to ascertain whether more aggressive primary endocrine therapy to remove the effects of testicular and adrenal androgens will extend the time to relapse and increase survival time, and this new approach using LH-RH analogues must be recognised as a major advance in this field of cancer.

III   Regulation of prostatic growth and function

Some of the endocrine factors that may be concerned in controlling the growth and functional activity of the prostate are diagrammatically illustrated in Fig. 6.3. Although the clinical features of prostatic carcinoma indicate that the growth of the tumour is, to some degree, functionally dependent upon androgenic stimulation, the role of the C_{19}-steroids circulating in the blood, or synthesised in the prostate cells, in the aetiology of the disease remains obscure, as does the manner in which the various peptide hormones influence the biology of the gland. It is generally well accepted that endocrine factors probably have an important role in the aetiology of prostatic cancer, although case-control studies of patients with

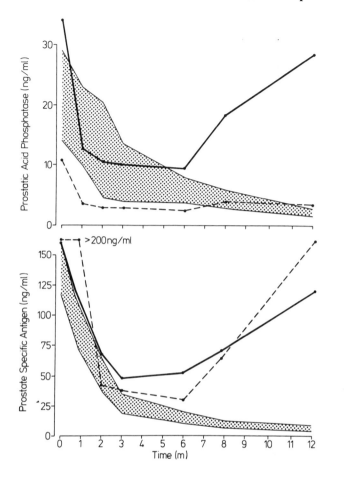

Fig. 6.14. Prostatic acid phosphatase (PAP) and prostatic specific antigen (PSA) levels in patients treated with Zoladex or by orchidectomy and in whom the disease is controlled (Hatched area). In two patients, disease progression became clinically evident and PSA levels reflected the change. PAP concentrations changed in one of the two patients. Shaded areas represents range of values ± S.E.M.. (Unpublished results).

the disease, and matched controls have not unequivocally demonstrated any significant hormonal abnormality (Griffiths *et al.* 1979). For example, as might be expected, various research groups have reported that the concentrations of testosterone in the plasma of patients with prostatic cancer are normal (Harper, Peeling, Cowley, Brownsey, Phillips, Groom, Fahmy, and Griffiths 1976; British Prostate Study Group, 1979; Bartsch, Horst, Becker, and Nehse, 1977), lower than expected (Zumoff, Levin, Strain,

Rosenfeld, O'Connor, Freed, Kream, Whitmore, Fukushima and Hellman 1982) or elevated (Ghanadian, Pinah, and O'Donoghue 1979; Drafta, Proca, Zamfir, Schindler, Neacsu, and Stroe 1982). Clearly, such case-control studies have limited value and it is important as stated previously, if we are to understand the aetiology of the disease, to direct investigations to the earlier periods of the male life, when the endocrine status might influence the early asymptomatic phase of the development of the cancer.

1   PLASMA STEROIDS AND AGEING

The principal circulating androgen is testosterone, of which approximately 95 per cent of a daily 6–7 mg producton rate originates in the Leydig cells of the testis (Baird et al. 1969; Lipsett 1970) and the plasma concentration of its free, non-protein bound form is generally accepted as the biologically active moiety and a reasonable indicator of androgenic status. The concentration of testosterone in the spermatic vein is of the order of 40–50 µg per 100 ml, approximately 70 times higher than the average concentration of 600 ng per 100 ml in the peripheral circulation.

Other C_{19}-steroids in plasma, androstenedione, DHA (dehydroepiandrosterone), DHA-sulphate and androst–5–ene–3β,17β–diol are predominantly of adrenal origin and the remainder of the plasma testosterone is produced peripherally from these adrenal C_{19}-steroids, but principally from androstenedione. The blood production rates of these adrenal C_{19}-steroids have been determined and are approximately 4.0, 8.0 and 25 mg per 24 hours for androstenedione, DHA and DHA-sulphate respectively, and 0.2–0.3 mg testosterone daily is formed from the peripheral metabolism of androstenedione.

It is now universally accepted that 5α-dihydrotestosterone (DHT) rather than testosterone is the active, intracellular androgenic hormone in the prostate gland (Bruchovsky and Wilson 1968a, 1968b; Anderson and Liao 1968). The conversion of testosterone to DHT by human prostatic preparations, in vitro, was originally reported following the now classical experiments of Farnsworth and Brown (1963), and it is well established that various other metabolites of testosterone, particularly the diols, 5α-androstane-3α,17β diol and 5α-androstane-3β,17β-diol, are synthesised by such tissue (Fig. 6.15). Investigations in the Tenovus Institute (Pike, Peeling, Harper, Pierrepoint and Griffiths 1970) also indicated that when [7α-³H]testosterone was infused into the cephalic vein of patients undergoing open prostatectomy for benign prostatic hypertrophy, the radioactive steroids isolated from the adenoma, enucleated 30 minutes later, showed a similar pattern of testosterone metabolism to that observed in vitro with DHT the principal metabolite. Voigt and his colleagues

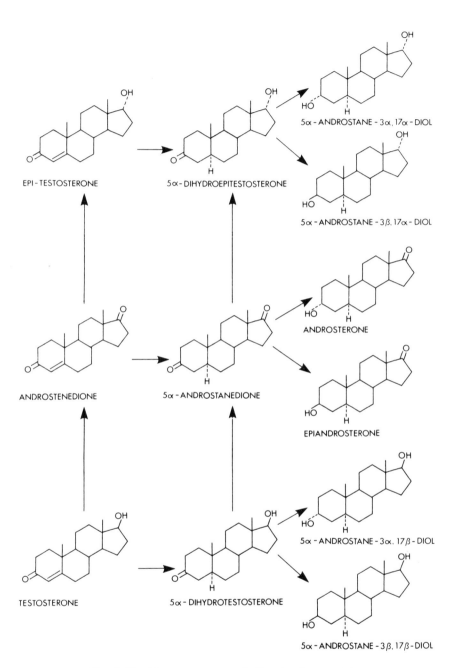

Fig. 6.15.   Testosterone metabolism by the prostate.

(Becker, Kaufman, Klosterhalfen, and Voigt 1972) produced similar data from their impressive series of investigations.

Therefore, although evidence is accumulating to support the concept that both benign and malignant growth of the prostate are mediated by androgens, it is interesting that these conditions become clinically manifest at a period of time in life when testicular function would be generally considered to be declining. Despite this, the plasma testosterone concentration in the adult male is maintained at a relatively constant level until the sixth decade, principally because of the decrease in metabolic clearance rate (Vermeulen, Rubens and Verdonck 1972). After this period of time, the concentration falls. These precise, excellent studies of Vermeulen and his collegues (Vermeulen, van Camp, Mattelaer, and De Sy 1979) clearly illustrated that certain changes in the androgen-oestrogen balance with ageing could well be implicated in the aetiology of prostatic disease. It was established that sex hormone-binding globulin (SHBG) retained a constant binding capacity until the fifth decade after which a gradual increase occurred until the age of 80, when its value was twice that of younger men. Vermeulen has reported that 57 per cent of the plasma testosterone is bound to SHBG, 40 per cent to albumin, 1 per cent to corticosteroid binding globulin (CBG) and the free non-protein bound fraction represents 2 per cent of the total testosterone concentration in younger males. This latter fraction decreased to 1.75 per cent in the fifth decade and to 1.25 per cent in the eighth when the biologically active, free moiety represents approximately only 20 per cent of the 'free level' of the younger man. Results from other groups generally support this concept (Pirke and Doerr 1975; Nieschlag, Kley, Wiegelmann, Solbach, and Kruskemper 1973) and it is interesting that the determination of the testosterone concentration in saliva also showed a significant decrease with age (Fig. 6.16) (Read, Harper, Peeling, and Griffiths 1981). Data from the Tenovus Institute, now confirmed by many other groups, have indicated that the levels of steroid in saliva relate closely not only to the concentration of the steroid in plasma, but to the free non-protein bound fraction (Fahmy, Read, Walker, and Griffiths 1982; Read, Fahmy, Walker, and Griffiths 1984) and the use of salivary steroid analysis offers a new, non-invasive, procedure, to study the endocrine status of subjects, using multiple sampling regimens that allow a more detailed assessment of rhythmic hormone changes with time.

Plasma oestradiol-17β concentration increases with age in the normal male (Pirke and Doerr 1975) probably explaining the eleveation in SHBG binding capacity in the older man. Peripheral conversion of androstenedione and testosterone to oestrone and oestradiol-17β is responsible for up to 90 per cent of the oestrogen production in the young healthy male (Siiteri and MacDonald 1973; Hemsell, Grodin, Brenner, Siiteri, and MacDonald 1974; MacDonald 1976), resulting in the synthesis of approximately 40 μg per day. The remainder is secreted by the testis (Kelch, Jenner, Weinstein,

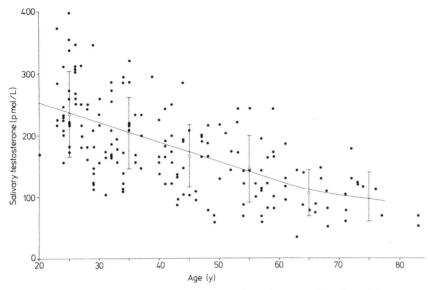

Fig. 6.16. Salivary testosterone concentrations in normal male subjects as a function of age. (From Read *et al*. 1981).

Kaplan, and Grumbach 1972; Longcope, Widrich and Sawin 1972; Weinstein, Kelch, Jenner, Kaplan and Grumbach 1974). The increased peripheral aromatisation of C_{19}-steroids in the elderly man, associated with elevated levels of SHBG, results in an increase in the ratio of free oestradiol-17β/free testosterone in plasma (Vermeulen 1976).

With advancing years, therefore, the 'ageing prostate' is subjected to a changing androgen/oestrogen balance, increasing levels of specific binding protein which must influence the transfer of steroid from plasma to tissue and the rate of accumulation of steroid within the cells, and a significant difference in the ratio of free biologically active testosterone and oestradiol-17β. There still remains the need for a greater understanding of the biochemical processes concerned in the transfer of plasma steroids into the target cells and the mechanisms by which the intracellular concentrations of steroids are controlled.

2 ADRENAL STEROIDS AND THE PROSTATE

Through the years, there has been considerable evidence from animal experimentation to indicate that adrenal steroids can influence the growth and function of the prostate gland (Tullner 1963; Tisell 1970, 1972; Muntzing 1971) and a great deal of interest is currently centred on the role of the adrenal androgens in promoting the growth of human prostatic cancer.

There is, however, little evidence from studies with the intact male rat that the adrenal C_{19}-steroids have any major biological effect in the prostate gland. Adrenalectomy does not affect the size of the prostate whilst after orchidectomy the gland and other accessory sex organs quickly atrophy. After castration there is no compensatory increase in adrenal androgen synthesis and secretion (Sanford, Paulson, Rohner, Drago, Santen and Bardin 1977; Parker, Lai, Wolk, Lifrak, Kim, Epstein, Hadley, and Miller 1984) with subsequent change in prostate weight, although ACTH administration does produce a significant increase in size of the prostate (Tullner 1963; Tisell 1970; Walsh and Gittes 1970). Tisell (1970) also reported that cortisone induced the growth and secretory activity of prostatic epithelium, an effect accentuated by oestradiol–17β. Adrenal hyperplasia and urinary retention was reported in mice with oestrogen-induced hyperplasia of the prostate (Fingerhut and Veenema, 1967). Subsequent adrenalectomy caused urine flow and the prostate gland then appeared histologically normal. Although oestrogens may have a direct growth-promoting effect on the prostate, the authors considered the adrenal activity, stimulated by the oestrogen-promoted release of ACTH, was a factor in the pathogenesis of prostatic hyperplasia. Certainly oestrogen administration results in adrenal hyperplasia (Danutra, Harper, Boyns, Cole, Brownsey and Griffiths 1973).

In men there is no evidence to implicate the adrenal gland in the pathogenesis of prostatic cancer, although experiments from these laboratories have indicated that DHT in human benign hypertrophic prostatic tissue can be formed from adrenal C_{19}-steroids (Harper, Pike, Peeling, and Griffiths 1974). The influence of adrenal androgens may only be significant, however, in the patient with prostatic cancer who has received primary endocrine therapy, be it castration, LH-RH analogue or oestrogen. There is considerable discussion, as already stated, regarding the question as to whether 'complete androgen blockade', removal of the testicular source of testosterone simultaneously with the suppression of the adrenal androgen effect, will lengthen the period to relapse and ultimately extend survival time. Earlier attempts (Geller and Albert 1983) both to suppress testicular activity and inhibit the action of testosterone at the level of the target organ by administration of megestrol acetate produced encouraging results. Labrie and his colleagues (1983) have claimed a 100 per cent response rate with patients treated simultaneously with an LH-RH agonist and an antiandrogen, although details as to how patients were referred to his clinics are not yet clear. There are many urologists who would refute the concept that all patients who present consecutively to a normal urology clinic will respond to endocrine therapy and as previously stated, in many centres in Europe and the USA, randomised clinical trials are being established to test this particular idea.

There is no doubt however that the 'residual' testosterone in plasma of castrated patients, or those treated with oestrogen or LH-RH analogues, is of adrenal origin (Cowley, Brownsey, Harper, Peeling, and Griffiths 1976;

Pierrepoint, Turkes, Walker, Harper, Wilson, Peeling, and Griffiths 1985), although relapse is rarely associated with elevated androgen levels (British Prostate Study Group 1979) and there is little evidence that a secondary rise in plasma testosterone levels relates to a disease progression. Whether the adrenal androgens do play a role in maintaining the growth of the hormone-dependent cancer cells is still being questioned. Although it is generally accepted that secondary endocrine therapy in the form of adrenalectomy is of little value in the management of prostatic cancer (Brendler 1973), data are still being presented to suggest that treatment with aminoglutethimide (Drago, Santen, Lipton, Worgel, Harvey, Boucher, Mann, and Rohner 1984; Murray and Pitt 1985) or megestrol acetate (Geller and Albert 1985a) may be valuable in certain patients who have relapsed. To many urologists, however, relapse expresses the inexorable progression of the androgen-independent cancer cells, probably clones of cells present in the primary tumour from the outset and relates to an average survival time after relapse, of only six months. Cytotoxic therapy or radiotherapy may be the only options available for the more successful management of these patients.

3 METABOLIC ACTIVITY OF THE PROSTATE IN RELATION TO DISEASE

Reference has already been made to the now classical experiments of Farnsworth and Brown (1963) demonstrating the synthesis, *in vitro*, of DHT from testosterone by preparations of human hypertrophic prostatic tissue. Subsequent investigations with human and animal prostatic tissue by various groups including our own (Ofner 1969; Farnsworth 1970; Harper *et al.* 1974) confirmed these original observations and established the pattern of metabolic activity illustrated in Fig. 6.15. It has often been considered that an abnormality in the steroid metabolic activity may be concerned in the pathogenesis of the diseased state and there is some evidence that metabolites of testosterone, other than DHT, may have specific roles to play within the target tissue. Interconversion of these steroids within prostatic cells may provide a regulatory process for the control of growth and function, with any consequent imbalance becoming responsible for glandular dysfunction. Studies have been directed to the establishment of the 'intracellular androgenic status' of prostatic tissue, normal, hyperplastic and malignant, in the hope that the analysis might help to elucidate the role the metabolites play in carcinogenesis. In this respect, the extensive investigations of Voigt, Krieg and their colleagues (Voigt, Horst, and Krieg 1975; Krieg, Bartsch, Herzer, Becker, and Voigt 1977; Krieg, Bartsch, Janssen, and Voigt 1979) have provided extremely valuable data, complementary to the results of many other groups involved in this work.

Following an early observation (Shimazaki, Kirihara, Yoshikazu, and Shida 1965) that the 5α-reductase activity of both prostatic carcinoma and hypertrophic tissue was less than that in normal tissue, subsequent reports tended to support the concept that malignant tissue had the least active reductase, and with comparatively inactive 3-hydroxysteroid-oxidoreductases in the malignant tissue, resulted in an accumulation of both testosterone and DHT. In normal prostatic tissue, the metabolic activity tends to promote the formation of the androstanediols and protects against the accumulation of testosterone and DHT (Krieg *et al.* 1979). In benign hypertrophic prostatic tissue, the activities of the 5α-reductase and 3-hydroxysteroid-oxidoreductases lead to the accumulation of DHT (Krieg *et al.* 1979; Bruchovsky and Lieskovsky 1979). Bruchovsky and Lieskovsky (1979) attempted to show some degree of linearity between 5α-reductase activity and age, but eventually inferred that the increased activity of the enzyme was a consequence of benign prostatic hypertrophy, not age.

The concentration of testosterone is therefore high in cancer tissue (Habib *et al.* 1976), and Siiteri and Wilson (1970) also showed the concentration of DHT in benign prostatic adenomas to be five times higher than in normal prostate tissue taken at autopsy, data subsequently confirmed by others (Voigt and Krieg 1978; Geller, Albert, Lopez, Geller, and Niwayama 1976). The elevation of DHT levels in the prostate glands of men over 60 years of age and the association between the size of the canine prostate and the endogenous DHT concentration provoked Siiteri and Wilson (1970) to suggest that this androgen was indeed responsible for the abnormal tissue growth. It should be noted, however, that from a study of the endogenous and plasma concentrations of C_{19}-steroids in BIO 87.2 hamsters, which develop prostatic hypertrophy with increasing age, the data suggested that the elevated levels of androgen in the benign tumour was a consequence of the growth, not the cause (Finney, Harper, Gaskell, and Griffiths 1980). Higher levels of C_{19}-steroids in tissue and blood and associated enzyme activity in the tumour were observed once the tumour had become established.

The tendency of the dog, like man, to develop an age-related hypertrophy of the prostate (Zuckerman and Groome 1937), but less commonly, neoplasia (Berg 1958), is interesting in relation to the report by Walsh and Wilson (1976) that administration of exogenous 5α-androstane-3α,17β-diol to castrated dogs induced benign prostatic hyperplasia, an effect enhanced by oestrogen. Despite a possible specific androgenic role for the androstanediol, it has generally been assumed that the induced growth was due to the conversion of the diol to DHT and Coffey and his collegues subsequently demonstrated (DeKlerk, Coffey, Ewing, McDermott, Reiner, Robinson, Scott, Strandberg, Talalay, Walsh, Wheaton, and Zirkin 1979) that DHT and oestrogen produced a similar response in castrated beagle dogs.

Previous discussion has directed attention to the fact that approximately 70–80 per cent of prostatic cancers contain clones of hormone-dependent cells and these tumours show some degree of response to primary anti-androgen therapy. Geller and Albert (1985b), on the basis that DHT mediates androgenic action and comprises 90% of the steroid in the nucleus of androgen-dependent prostate cells, determined the DHT content of prostatic cancer tissue from previously-untreated patients and of those who had relapsed after primary endocrine therapy. The DHT level was found to offer an effective guide in predicting response to therapy. The average disease-free interval of patients with concentrations of DHT > 2.5 ng per g was 24 months whereas the period was less than 10 months if the level of the hormone was less than 2.0 ng per g. Although anti-androgen therapy, castration or oestrogen treatment generally resulted in a DHT concentration which was less than 2.5 ng per g, certain patients had tissue levels greater than this value, suggesting inadequate dosage, poor patient compliance in taking DES, or an excessive adrenal contribution to the prostatic androgen concentration. It will be interesting to await further data on this potentially valuable prognostic indicator.

4   PROLACTIN, GROWTH HORMONE AND THE PROSTATE

Early experimental studies in animals showed that factors other than androgenic steroids, probably pituitary hormones, influenced the growth and function of the prostáte gland. Their actions were generally considered permissive in that they synergised with steroids to promote their biological effects, such that a more marked atrophy of the rat prostate was observed after hypophysectomy and castration than after castration alone (Lostroh and Li 1957). Huggins and Russell (1946) reported a similar effect in the dog. Administration of growth hormone (GH) or ACTH to hypophysectomised-castrated rats stimulated prostatic growth whereas LH, FSH and prolactin did not (Lostroh and Li 1957), although the simultaneous treatment of castrated rats with ACTH and prolactin produced greater prostatic growth than ACTH alone (Tullner 1963). Furthermore, an impaired uptake of testosterone into the ventral prostate has been reported in hypophysectomised rats (Laurence and Landau 1965). Growth hormone was generally found to complement the effects of prolactin and testosterone in regulating prostatic weight (Chase, Geschwind, and Bern 1957). There were many such reports, but it has to be accepted that the preparations of pituitary hormones available in those early days were in all probability not as pure as at present and some degree of uncertainty has remained about these early data.

Sufficient evidence has however accumulated to suggest that prolactin has a specific role in the control of prostatic function (Grayhack, Bunce, Kearns, and Scott 1955; Grayhack and Lebowitz 1967; Jacobi 1982). Studies *in vitro* indicate that the presence of prolactin increased the uptake of testosterone into prostatic tissue (Laurence and Landau 1965; Farnsworth 1970; Lasnitzki 1972; Lloyd, Thomas and Mawhinney 1973; Boyns, Griffiths, Pierrepoint, and Peeling 1975) and the uptake and distribution of zinc in the prostate is influenced by prolactin (Gunn, Gould, and Anderson 1965). Results from this Institute (Harper *et al.* 1976b) have however indicated that administration of CB154 (2-bromo-α-ergocryptine, Sandoz Ltd.), an inhibitor of prolactin secretion (Billeter and Fluckiger 1971; Fluckiger 1972), to male rats for periods of up to 30 days, failed to affect the weights of the accessory sex glands although there was an effect on the distribution of zinc in the prostate particularly in the lateral lobe of the gland. Recently, it was demonstrated (Holland and Lee 1980) that elevated plasma prolactin stimulates the growth of the lateral prostate of the rat. Specific binding sites for prolactin have been identified in prostatic tissue (Aragona and Freisen 1975a; Kledzik, Marshall, Campbell, Gelato, and Meites 1976), although the evidence is equivocal that the hormone stimulates adenyl cyclase (Golder, Boyns, Harper, and Griffiths 1972) and recently, immunocytochemical evidence was obtained for prolactin and prolactin binding sites in canine (El Etreby and Mahrous 1979) and human (Harper, Sibley, Peeling, and Griffiths 1981; Witorsch 1981) prostatic tissues.

It is clear that a greater appreciation of the effects of prolactin on prostatic growth could be provided by tissue culture techniques and investigations were directed to the influence of prolactin on human benign prostatic hyperplasia epithelial cell proliferation, a monolayer culture system which had previously been shown to be sensitive to both testosterone and DHT (Syms, Harper, Battersby, and Griffiths 1982). Prolactin was found to stimulate the growth of the epithelial cells in culture (Syms, Harper, and Griffiths 1985), a response which did not merely reflect increased androgen uptake, since the effect was observed with cells maintained in steroid-free medium. Synergistic effects of prolactin and androgen were observed on the growth of these cells, and the study complements others (Yamanka, Kirdani, Saroff, Murphy, and Sandberg 1977) that indicate a relationship between the roles of these hormones in regulating the biology of the prostate.

Prolactin also acts indirectly on the prostate by virtue of its synergistic effect with LH on testicular secretion of testosterone (Hafiez, Lloyd, and Bartke 1972) and may influence the synthesis and secretion of C_{19}-steroids by the adrenal, a topic discussed later. Furthermore the data derived from these various experimental approaches suggests that prolactin exercises a direct action on the prostate gland but whether the hormone is concerned in the development of prostatic disease remains to be elucidated. The effect on plasma prolactin of the increasing oestradiol/testosterone ratio in blood of ageing men requires further study, but the stimulatory action of oestrogen on prolactin release is well established (Meites, Lu, Wuttke, Welsch,

Nagasawa, and Quadri 1972; Boyns, Cole, Phillips, Hiller, Cameron, Griffiths, Shahmanesh, Feneley, and Hartog 1974). At present, there is little evidence that the levels of prolactin in patients with prostate cancer differ from normal (Aragona and Freisen 1975b; Harper et al. 1976a; Bartsch et al. 1977; Hammond, Kontturi, Maattala, Puuka, and Vihko 1977; Jacobi, Rathgen, and Altwein 1980a). High-dose oestrogen therapy with its associated increase in prolactin levels in plasma (Harper et al. 1976a) may thereby adversely influence the biochemistry of the prostate and might be contra-indicated. The use of bromocriptine for the management of patients with prostatic cancer has not been found to be of particular clinical value (Jacobi, Altwein, and Hohenfellner 1980b; Coune and Smith 1975), although treatment with L-dopa, which reduces the plasma prolactin concentration, tends to relieve bone pain (Sadoughi, Razvi, Busch, Ablin, and Guinan 1974).

Equally interesting is the role of growth hormone (GH) in the prostate. Immunocytochemical procedures (Sibley, Harper, Peeling, and Griffiths 1984) detected endogenous human GH, and also the binding of exogenously applied human GH, in tumour tissue from patients with benign prostatic hyperplasia or prostatic carcinoma. The binding of GH was primarily to the connective tissue in both benign and carcinomatous tissue, although some evidence was obtained for the identification of a binding site within the cytoplasm of epithelial cells. A similar type of study has been reported on canine prostate (El Etreby and Mahrous 1979). The functional significance of these binding sites in both epithelial and stromal elements obviously requires further investigation. For example, GH stimulates the synthesis and breakdown of collagen and the proliferation of fibroblasts (Deyl, Rosmus, and Adam 1976), and somatomedin is produced from fibroblasts in response to GH stimulation (Atkison, Weidman, Bhaumick, and Bala 1980; Clemmons, Underwood, and Van Wyk 1981). Furthermore GH binding sites have been characterised on mouse fibroblasts using iodinated GH (Murphy, Vrhovsek, and Lazarus 1983). The observation that patients with metastatic disease of the prostate have elevated plasma GH levels compared to patients without metastases (British Prostate Study Group 1979) might suggest a role for GH in the dissemination of the cancer cells.

The effect of high-dosage oestrogen therapy in increasing plasma prolactin concentrations and possibly thereby influencing adrenal function has long been a topic of discussion. Specific binding sites for prolactin have been reported in adrenal tissue (Posner, Kelly, Shiu, and Friesen 1974; Marshall, Kledzik, Gelato, Campbell, and Meites 1976) and it has been speculated that prolactin may act as an adrenal androgen-stimulating hormone. Although evidence for this is not convincing (Anderson 1980; Adams 1985), it is established that adrenal glucocorticoid and C_{19}-steroid synthesis and secretion are 'dissociated' in adrenarche, obesity, fasting and in the ageing process. Preliminary studies from this Institute (Boyns et al., 1972) produced some evidence of a prolactin influence on adrenal C_{19}-steroid synthesis, and since then many reports have appeared describing an association between hyperprolactin-aemia and increased concentrations of plasma DHA and DHA sulphate (Vermeulen, Suy, and Rubens 1977; Carter, Tyson, Warne, McNeilly, Faiman, and Fries 1977; Kandeel, Rudd, Butt, Logan-Edwards, and London 1978; Vermeulen and Ando 1978; Schiebinger, Chronsos, Cutler, and Loriaux 1986). Although O'Hare and his colleagues (1980) were unable to show an effect of ovine prolactin, with or without ACTH, on adrenal cortical cells in culture, others (Serio, Forti, Guisti, Bassi, Giannotti, Calabresi, Manvero, Armato, Fiorelli, and Pinchera 1980) were able to demonstrate an increased secretion of DHA sulphate in the presence of prolactin. Higuchi, Nawaka, Maki, Higashizima, Kato, and Ibayashi (1984) using monolayer cultures of human adrenal cells, showed that prolactin potentiated the stimulatory effect of ACTH on DHA and DHA sulphate secretion, but not on androstenedione or cortisol.

A synergistic effect of prolactin on adrenal androgen production would seem to be generally accepted and Adams (1985) has summarised the relevant data on the possible biochemical processes within the adrenal cortex that may be involved. Interesting relationships between diet and prolactin secretion are discussed, including the elevation of prolactin in plasma by high-fat diets (Chan, Didato, and Cohen 1975) and the depression of the nocturnal elevation in plasma prolactin by changing from an omnivorous to a vegetarian diet (Hill and Wynder 1976). The relationship between prolactin and the adrenal, and thereby the prostate and dietary factors may be relevant to the study of prostatic cancer. More detailed information is required on the biochemical processes concerned with the differentiation of the zona reticularis in adrenal gland, C_{19}-steroid production and adrenarche, and the influence of prolactin on these processes. The precise biological role of adrenal C_{19}-steroids is not yet established. They are metabolised to DHT in the prostate but conversion to the 'oestrogenic' androst-5-ene-3β,17β-diol (androstenediol) may also be relevant, since nutritional status may result in the higher levels of DHA-sulphate, DHA and androstenedione in the Oriental races (Adams 1985).

5   HORMONAL STATUS AND CARCINOMA OF THE PROSTATE

In the programmes directed to the search for endocrine or biochemical factors that might be concerned in the development of prostatic carcinoma, the determination of steroid and peptide hormone concentrations in plasma of patients with the disease have failed to identify any endocrine disturbance, or any differences from the asymptomatic normal male, that could lead to a greater understanding of disease aetiology. Probably the biochemical processes within the prostatic cells may induce increased responsiveness to normal hormone levels in plasma, and research should be directed more towards the elucidation of such intracellular changes.

The conditions which predispose towards the development of prostatic carcinoma may be imprinted on the gland at a much earlier stage in life than at the time when the disease becomes clinically manifest. This has been alluded to previously in this review and it is noteworthy that prostatic cancer does not occur in the pre-pubertally castrated man. The various endocrine factors involved in puberty may create a new cellular hormone milieu in which it becomes possible for cancer to develop. Moreover, Rotkin (1979), investigating epidemiological factors associated with carcinoma of the prostate, considered that early 'oncogene transformation' of the gland may occur as a consequence of the diverse physiological and secretory patterns involved with the onset of pubertal change. When factors such as the development of secondary sexual characteristics, sex-drive or libido were considered, Rotkin suggested that the men at risk of developing prostatic cancer had a 'strong overbalance of androgenic components'.

The processes concerned in the regulation of the maturation of the prostate gland and its growth to a particular size must be complex, requiring a fine balance of growth stimulatory factors including androgenic steroids, and also growth inhibitory factors. Minor changes in this balance, while not affecting glandular function, could allow the activation of cellular oncogenes, and the hormonal profiles concerned with puberty may be more informative than those associated with ageing in understanding prostatic cancer. Fewer patients with prostatic cancer developed gynaecomastia and obesity early in life compared to controls. It may be that certain environmental factors, or an influence endogenous to the prostate itself, may protect an individual in whom latent carcinomatous foci have developed. Rotkin suggests the endogenous oestrogens may have a protective role. In considering environmental factors, the level of phyto- oestrogens in the diet should also be considered and this Institute is currently developing selected ion monitoring mass spectrometry techniques (Gould, Turkes, and Gaskell 1986) in order to investigate this possibility.

In relation to these studies, those of Meikle and his colleagues (1985) are also of interest. A series of early reports (Woolf 1960; Krain 1973; Thiessen 1974; Cannon, Bishop, Skolnick, Hunt, Lyon, and Smart 1982) have suggested that a family history of prostatic cancer increases an individual's risk of developing the same condition. Meikle and his colleagues reported (1982a; 1982b; 1985) not only that familial factors have an important influence on the risk of developing cancer of the prostate, but that they also have an effect on the levels of plasma testosterone, with men at risk showing plasma concentrations in the lower range of normal. Studies from this

Institute (Harper *et al.* 1984; Pierrepoint *et al.* 1985) and from the laboratories of Adlercreutz and his colleagues (1981) have indicated a relationship between plasma testosterone levels of patients presenting with prostatic cancer and prognosis, those with low concentrations surviving only a short period of time. Equally significant, however, is the recent report from Finland (Haapiainen, Rannikko, Adlercreutz, and Alfthan 1986) of a high concentration of both the 'free' oestradiol and total oestradiol in the plasma of patients in whom the cancer has not spread beyond the capsule. The authors also suggest that endogenous oestrogens play an inhibitory role in controlling the growth and metastasising tendency of the tumour. Another interesting approach to the study of the prostate was that of Rose and his colleagues (1984) who have analysed the steroid content of prostatic fluid from patients with cancer and from normal control subjects. The composition of prostatic fluid, a secretory product of the gland and obtained by transrectal massage, could be expected to reflect the steroid metabolising capacity of the tissue. The analysis indicated a significantly higher concentration of oestradiol in the fluid of patients with prostatic cancer than in controls. It would seem reasonable to suggest that despite considerable research, little is known about the role of the oestrogens in the control of prostatic biochemistry and that more work is clearly necessary.

In the search for an aetiology for prostatic carcinoma and appreciating that only man and dog, throughout the animal kingdom, suffer from the condition, it would seem reasonable that a factor or factors common to both species might predispose them to this neoplastic change. It is certainly true that man and dog share the same habitat and, to a large extent, similar life styles, and it may eventually be shown that these two considerable influences may be instrumental in the development of prostatic cancer as well as of other neoplastic diseases. It has been argued (Bostock and Curtis 1984; Pierrepoint 1985) that, as the ageing process is faster in the dog than in man, this animal could possibly function as 'an indicator species' to herald similar human disease processes some 30 years in advance.

Be that as it may, such environmental factors producing the same disease in man as in his companion animals have yet to be identified. There is however an endogenous feature that appears to be peculiar solely to man and dog and which relates directly to the prostate gland. The vasa deferentia, originating from the caudae epididymidum, pass directly through the prostates of these two species to reach the urethra (Pierrepoint 1975). This does not occur in the other domestic animals in which they pass directly into the urethra.

These ducts are essential for the transport of sperm to the penis at the time of ejaculation and are thus uniquely positioned for the transfer of other genital products. It was hypothesised (Pierrepoint 1975) that such products may include androgens, of which of course, the testis is the main source, and that these may be carried directly to the prostate without recourse to the general circulation in which considerable dilution takes place.

The experimental approach adopted to test this hypothesis was simply in the first instance to examine the effect of bilateral vasectomy on the functional activity of the ventral prostate of the rat (Pierrepoint and Davies 1973).

The activity of the androgen-regulated prostatic nuclear enzyme, RNA polymerase, was evaluated in such operated animals compared with sham-operated controls. Weights of organs after varying post-operative periods and changes in DNA and RNA content were also assessed. The effect of ablative surgery was quite profound in reducing the levels of these parameters significantly below those of the control organs. Because of the inevitable suspicion that such surgical procedures could be affecting the testis and its appendages and thus reducing the circulating levels of androgenic hormones, a unilateral approach was adopted (Pierrepoint, Davies, and Wilson 1974) so that the ipsi- and contra-lateral (to the vasectomy) organs could be compared in the same animal. The two sets of organs obviously receive the same systemic arterial supply of androgens and the same animal thus acts as its own control. The organs, ipsilateral to the vasectomy showed significant reduction in the parameters examined compared with those contralaterally.

During surgery, it was noted that the vas deferens is accompanied by a large tightly-opposed vein, which was designated the deferential vein (Pierrepoint, Davies, Millington, and John 1975) and which, by the surgical procedure adopted at the time, was ligated concomitantly with the vasectomy. The question had to be asked, was it the severance of this vein or that of the vas, that was responsible for the androgen-ablative effect of the surgical procedure?

A further series of experiments, again simple in concept and procedure, were performed, including bilateral deferential venotomy, bilateral vasectomy, maintaining an intact vein, bilateral vasectomy/venotomy and sham-operated. The activities of the Mg^{++} and Mn^{++}-dependent species of RNA polymerase in the ventral prostate were subsequently assessed and found to be significantly reduced in each of the three groups subjected to ligative procedures. The Mg^{++}-dependent or ribosomal RNA polymerase enzyme was especially sensitive to ligation of the deferential vein (Pierrepoint et al., 1975; Pierrepoint, Davies, Lewis, and Moffat 1975). On further examination it was shown that the androgen content of deferential vein blood was of the order of that found in the testicular vein. This finding has been confirmed (Chapdelaine, Boulanger, Dupuy, Beau, and Roberts 1978) and extended to man in which a similar situation has also been shown to exist. The deferential vein drains, in the dog, variably into the venous effluent of the prostate (Dhabuwala and Pierrepoint 1977) and radio-opaque materal injected into this vessel will readily pass into prostatic tissue (Dhabuwala, Roberts, and Pierrepoint 1978). It has been further shown by per-rectal ultrasound that prostatic volume, especially that of the periurethral region, is significantly reduced in man following vasectomy (Jakobsen and Juul 1985).

The implications of these findings may be profound in that a high concentration of androgens is being diverted into the prostate gland in the two species that develop prostatic hyperplasia and carcinoma. It will, of course be of interest to note any differences in the incidence of these two conditions in those men that have been vasectomised compared with the remainder of the population.

6   PROSTATE BIOCHEMISTRY: GROWTH FACTORS AND ONCOGENES

Although the role of steroid hormones as modulators of normal prostatic differentiation has been extensively investigated the regulation of prostatic growth and function by factors derived from the organ's 'micro-environment' itself, is less well understood. Evidence for the existence of intra-glandular growth control mechanisms arises from several areas of research and should be considered in this review.

Studies of adult human prostatic histopathology have suggested that the changes in epithelial growth and functional differentiation associated with glandular atrophy and with hyperplasia are preceded by alterations in the composition of prostatic connective tissue or stroma (McNeal 1979). Prostatic hyperplasia and premalignant changes have been associated with a 'persistently youthful' glandular stroma (McNeal 1979) implying an influence of factors derived from the connective tissue. These observations, although circumstantial, acquire greater significance when related to the now well established phenomenon of stromal regulation of epithelial differentiation, demonstrated during the development of the prostate and of other glandular structures in the foetus (Kratochwil 1969: Cunha 1972; 1976; Cunha, Chung, Shannon, and Reese 1980). These studies suggest that organotypic epithelial differentiation is primarily controlled by an inductive influence associated with organ specific embryonic stroma and subsequent investigations have indicated that such factors can affect the activity of normal adult epithelial cells (Cunha, Lung, and Reese 1980) and of epithelial tumours (Fujii, Cunha, and Norman 1982).

Many studies have sought to identify and characterise diffusible factors, derived from connective tissue, that may be involved in the induction of specific patterns of epithelial differentiation, both in embryonic tissues and in adult organs. Clearly-defined mediators of this type of cellular interaction, however, have yet to be demonstrated. Early studies in the prostate suggested that differential patterns of androgen metabolism in epithelial and stromal elements and subsequent intercellular exchange of metabolites may be responsible for the inductive process (Cowan, Cowan, Grant, and Elder 1977). It is now clear, however, that prostatic epithelial and stromal cell populations are both capable of synthesising physiologically active metabolites from circulating androgens (Voigt *et al.* 1975; Eaton and Pierrepoint 1986). Although earlier studies have suggested that oestrogen sensitivity may be confined to prostatic stroma (Chaisiri and Pierrepoint 1980), more recent investigations (Eaton, Hamilton, Kenvyn, and Pierrepoint 1985) indicated that in the adult, both stromal and epithelial cell populations are potential steroid target tissues with similar patterns of steroid receptor content.

Until recently the lack of suitable model systems to investigate and characterise the activities of individual cell populations has limited progress in this field. Despite the increasing availability of culture systems (Eaton and Pierrepoint, 1982a) and the fact that the growth and functional responses of both stromal and epithelial cell populations to steroid and protein hormones can be separately characterised (Eaton and Pierrepoint 1982b; Eaton *et al.* 1985; Eaton 1984), diffusible factors acting between prostatic cell populations still remain to be demonstrated. It is also becoming increasingly apparent that, while diffusible paracrine inductors may exist within the prostate, other mechanisms, in particular those involving the physical association between cells, require further exploration. Studies to define culture conditions for the growth of epithelia *in vitro* with the maintenance of *in vivo* functional differentiation have clearly demonstrated a high degree of dependence of this cell type on attachment to specific types of supportive substrate (Kidwell, Wicha, Salomon, and Liotta 1980; Gospodarowicz, Delgado, and Vlodavsky 1980; Eaton and Pierrepoint 1982a). In particular substrates containing the various isotypes of collagen have not only allowed the maintenance of functional activity in cultured epithelial cells, but have also induced rapid, uniform growth patterns in these cell populations. Collagen and associated glycoproteins and glycosaminoglycans, are the major constituents of prostatic stroma and it has been suggested (Mariotti, Thornton, and Mawhinney 1981) that quantitative and possibly qualitative alterations in collagen content of glandular stromas may relate to ageing and to changing circulating levels of steroid hormones. The complexity of the extracellular matrix provides the potential for wide variations in epithelial attachment and growth control and such a process may be responsible for the alteration in epithelial activity observed during prostatic atrophy and hypertrophy.

In contrast, structural autonomy, with the maintenance of varying levels of prostatic differentiation, often in sites distal from the prostate, is a major feature of prostatic carcinoma. Studies in this Institute using a spontaneous canine prostatic adenocarcinoma, have demonstrated a high degree of prostatic differentiation and rapid growth in athymic mice xenografts, generated by subcutaneous injection of cloned sublines of this tumour. Both the parent cell line and the substrains are substrate-independent in respect of growth of morphological differentiation, *in vivo* and *in vitro*.

Fundamental differences in attachment requirements and consequent growth modulation clearly exist betwen epithelia derived from normal and benign and malignant tumour tissues. Studies with other model systems have suggested that normal cellular attachment is mediated by a specific class of cell surface glycoproteins, which include fibronectin and laminin (Grinnell and Field 1979; Terranova, Rohrbach, and Martin 1980). These glycoproteins display a high avidity for collagen binding (Hahn and Yamada 1979), are major components of cellular adhesion plaques and are specifically associated with elements of the cell cytoskeleton (Ali and Hynes 1977; Yamada and Olden 1978), a structure intimately involved with the process of cell division. Transformation *in vitro* is associated with down-regulation of attachment glycoprotein expression (Adams, Sobel, Howard, Olden, Yamada, de Crombrughe, and Pastan 1977), whereas addition of exogenous glycoprotein can modulate the growth and differentiation of tumour cells (Vlodavsky and Gospodarowicz 1981). Together, these studies re-emphasise the importance of the attachment process in cellular growth control and provide some indication of the altered attachment requirements observed in tumours. It seems likely that substrate-independence in transformed cells is achieved by an autonomous cytoskeletal organisation in which the presence of cell surface glycoproteins, with organised attachment plaque formation, is not a prerequisite for successful cell division. The modulating influences of normal attachment are therefore not applicable. Some of our current studies in this field are aimed at providing a more detailed understanding of the interrelationships between the cytomatrix and the attachment plaque in normal and transformed epithelium.

The prostate epithelial cell is therefore the focus of a barrage of regulatory stimuli. Not only is it a target, by virtue of its receptors, for androgens, oestrogens and progestins, but it may also be influenced by protein hormones, peptide growth factors and undefined factors secreted from the surrounding stroma. Elucidating the intracellular processes connecting the interaction of a growth-stimulating agent and its receptor, and the initiation of DNA synthesis, is a prerequisite to understanding cancer biology. This observation is self-evident to cancer investigators in view of the demonstration that the oncogenes carried by strongly transforming retroviruses are derived from normal cellular genes, called proto-oncogenes. Proto-oncogenes are not themselves oncogenic, but may become oncogenic through alterations in structure or expression (Marshall and Rigby 1984; Varmus 1985; Bishop 1985) and cellular oncogenes (c-*onc*) are genes essential for normal growth and differentiation (Bishop 1983; this review provides access to the terminology which has developed for oncogenes and their products). The interest in the latter has been enhanced by evidence of the relationship between growth factors and oncogenes (Stiles 1985; Goustin, Leof, Shipley, and Moses 1986). Thus, homologies exist between the c-*sis* product and the

B-chain of platelet-derived growth factor (PDGF) (Waterfield, Scrace, Whittle, Stroobant, Johnsson, Wasteson, Westermark, Heldin, Huang and Deuel 1983; Doolittle, Hunkapiller, Hood, Devare, Robbins, Aaronson, and Antoniades 1983), the c-*erb* B codes for the epidermal growth factor (EGF)-receptor (Downward, Yarden, Mayes, Scrace, Totty, Stockwell, Ulrich, Schlessinger, and Waterfield 1984) the c-*fms* product resembles the CSF (colony stimulating factor)-1 receptor (Sherr, Rettenmier, Sacca, Roussel, Look, and Stanley 1985) and that of v-*ros* is insulin receptor-related (Ullrich *et al.* 1985; Ebina *et al.* 1985). The *neu* oncogene, and probably the c-*erb* B-2 gene (Yamamoto, Ikawa, Akiyama, Semba, Nomura, Miyajima, Saito, and Togoshima 1986) repeatedly activated in rat neuro- and glioblastomas, is related to, but distinct from, the c-*erb* B gene encoding EGF-receptor (Schechter et al., 1984; 1985), and encodes an EGF receptor-related protein (Bargmann, Hung, and Weinberg 1986) for which the ligand is not yet known.

Several other oncogene products are similar to growth factor receptors in that both have transmembrane and tyrosine kinase domains (Hunter 1984). The product of the c-*ras* oncogene, p21, may be an obligatory intermediate in the transduction of the growth factor signal (Mulcahy, Smith, and Stacey 1985), and growth factors have been shown to increase transcription of c-*myc* and c-*fos* genes (Kelly, Cochran, Stiles, and Leder 1983; Greenberg and Ziff 1984; Muller, Bravo, Burckhardt, and Curran 1984; Kruijer, Cooper, Hunter, and Verma 1984; Stiles 1985). The c-*myc* and c-*fos* genes encode nuclear-located proteins (Muller *et al.* 1984; Alitalo, Ramsay, Bishop, Pfeiffer, Colby, and Levinson 1983; Eisenman, Tachibana, Abrams, and Hann 1985; Persson and Leder 1984) which may influence other genes necessary for cell proliferation. The ability of a cell to produce and respond to its own growth factors confers a particularly interesting and definite advantage (Sporn and Roberts 1985); the alternative to autocrine stimulation is paracrine regulation wherein growth stimulatory signals are passed between neighbouring cell populations.

Jacobs and associates (Jacobs, Pinka, and Lawson 1979; Jacobs and Lawson 1980) demonstrated the presence of a fibroblast growth-promoting factor (FGF) in crude extracts of human benign prostatic hypertrophy tissue, well-differentiated carcinoma, and postpubertal normal prostate. The factor from the benign adenoma, which did not compete with EGF for the latter's receptor (Story, Jacobs, and Lawson 1983), had a molecular weight $(M_r) > 67,000$ in low ionic strength and $< 33,000$ in high-salt buffer. A growth factor capable of stimulating DNA synthesis of BALB/3T3 (M_r 11,000–13,000) cells was purified by heparin-Sepharose chromatography from cytosol prepared from benign adenoma (Nishi, Matuo, Mugurama, Yoshitake, Nishikawa, and Wada 1985): it differed from bovine FGF. Simpson, Harrod, Eilon, Jacobs, and Mundy (1985) identified a mRNA fraction, approximately 1800 bases in length, from PC3 cells, which encoded an osteoblast-stimulating activity. As there is evidence for steroid-hormone regulation of growth factor content and activity (Byyny, Orth, Cohen, and Doyne 1974; Hosoi, Tanaka, and Ueha 1981; Gospodarwicz and Moran 1974; Baker, Barsh, Carney, and Cunningham 1978; Mukku and Stancel 1985; Dickson, Huff, Spencer, and Lippman 1986; Murphy, Sutherland, Stead, Murphy, and Lazarus 1986) the possibility of an androgen-regulated growth factor in human prostate is intriguing, especiallyu if its constitutive expression is essential to the maintenance of the transformed phenotype.

Rijnders, van der Korput, van Steenbrugge, Romijn, and Trapman (1985) studied the expression of cellular oncogenes in four different human prostatic carcinoma cell lines. Large amounts of Ha-*ras* and *myc* RNA were detected in all cell lines with smaller quantities of Ki-*ras* and N-*ras* mRNA. Some androgen-independent cell lines contained transcripts of *myb* and *fms*, while *sis* RNA was greatest in PC3 cells. The hormone-dependent cell line, PC82, contained significant amounts of *fos* mRNA. No correlation could be found between expression of *myc* and growth of PC82; Ha-*ras* expression was slightly sensitive to androgen withdrawal, but *fos* transcripts were drastically reduced after androgen withdrawal. Fleming, Hamel, MacDonald, Ramsey, Pettigrew, Johnston, Dodd, and Matusik (1986) reported higher levels of expression of c-*myc* in prostate adenocarcinoma than in benign hypertrophy. In our Institute, we have found transcripts of c-*myc*, c-Ha-*ras*, c-Ki-*ras*, c-*sis*, c-*fos*, c-*myb*, c-*erbA*, c-*erbB* and p53 in benign adenoma and/or carcinoma tissue. Expression of p53 was high in all samples: expression of c-*myc* was consistently elevated in carcinoma compared to benign hypertrophic tissue. Expression of c-Ha-*ras* increased in correlation with dedifferentiation. Only c-*fos* (positively) showed any correlation with androgen receptor content and it is very likely that the comparison of levels of expression of cellular oncogenes among tissue samples will be similarly affected by the complications of tissue hetero-geneity, as has been found for steroid receptor measurement.

Using a histochemical technique, Viola, Fromowitz, Oravez, Deb, Finkel, Lundy, Hand, Thor, and Schlom (1986) were unable to detect p21[ras] antigen in epithelial and stromal cells from either normal or benign hypertrophic prostate tissue, but p21[ras] was detected in two of six tumours with grade I carcinoma, four of six with grade II, and all of 17 with higher grades. Compared with carcinoembryonic antigen and prostate-specific antigen, p21[ras] was the only phenotypic marker that correlated with histological tumour grade.

Apart from providing data on their value as potential prognostic indicators, studies on cellular oncogenes and growth factors may lead to newer approaches to therapy. As stated above, the identification of a prostate growth factor, facultatively expressed under androgenic influence or constitutively expressed when autonomy is attained, could lead to a more general form of treatment, more curative than palliative. As a number of growth factors show some degree of tissue specificity, therapy based at this level could be more precisely targeted and less generalised than current chemotherapy. Also, the possibility that the autologous or unregulated production of a factor in the mitogenic cascade could arise from deletion or dysfunction of an antagonistic modulatory function (Sporn and Roberts 1985; Knudson* 1985) could offer even more therapeutic potential. Whatever the outcome, we are entering a most exciting phase which should extensively increase our understanding of the endocrine and biochemical control of prostatic growth and function.

References

Abramson, F. P. & Miller, H. C. Jr. (1982). Bioavailability, distribution and pharmacokinetics of diethylstilboestrol produced from stilphostrol. *J. Urol.*, **128**, 1336–1339.

Adams, J. B. (1985). Control of secretion and the function of C_{19}-Δ^5-steroids of the human adrenal gland. *Mol. Cell. Endocr.* **41**, 1–17.

Adams, S. L., Sobel, M. E., Howard, B. H., Olden, K., Yamada, K. M., de Crombrughe, B. and Pastan, I. (1977). Levels of translatable mRNAs for cell surface protein, collagen precursors and two membrane proteins are altered in Rous sarcoma virus transformed chick embryo fibroblasts. *Proc. natl. Acad. Sci. U.S.A.* **74**, 3399–3403.

Adlercreutz, H., Rannikko, S., Kairento, A-L., and Karonen, S-L. (1981). Hormonal pattern in prostatic cancer II. Correlation with primary response to endocrine treatment. *Acta endocr. Copenh.* **98**, 634–640.

Albert, J., Geller, J., and Nachtsheim, D. A. (1982). The type of current frequency used in transurethral resection of prostate affects the androgen receptor. *The Prostate* **3**, 221–224.

Alderson, M. R. (1981). Epidemiology in Prostate Cancer. In *Recent results in cancer research*, No. 78. (Ed W. Duncan) pp. 1–19. Springer-Verlag, Heidelberg.

Ali, I. V., and Hynes, R. O. (1977). Effects of cytochalasin B and colchicine on attachment of a major surface protein of fibroblasts. *Biochim. Biophys. Acta* **471**, 16–24.

Alitalo, K., Ramsay, G., Bishop, J. M., Pfeiffer, S. O., Colby, W. W., and Levinson, A. D. (1983). Identification of nuclear proteins encoded by viral and cellular *myc* oncogenes. *Nature, Lond.* **306**, 274–277.

Anderson, D. C. (1980). The adrenal androgen-stimulating hormone does not exist. *Lancet* ii, 454–456.

Anderson, K. M., and Liao, S. (1968). Selective retention of dihydrotestosterone by prostatic nuclei. *Nature Lond.* **219**, 277–279.

Aragona, C. and Freisen, H. (1975a). Specific prolactin binding sites in the prostate and testis of rats. *Endocrinology* **97**, 677–684.

—— —— (1975b). Prolactin and aging. In *Benign Prostatic Hyperplasia*. (Eds J. T. Grayhack, J. D. Wilson and M. J. Scherbenske) pp. 165–175. National Institutes of Health, Bethesda, Md.

Arduino, L. J. (1967). Veterans Administration Cooperative Urological Research Group. Carcinoma of the prostate. Treatment and comparisons. *J. Urol.* **98**, 516–522.

Armenian, H. K., Lilienfeld, A. M., Diamond, E. L., and Bross, J. D. (1974). Relation between benign prostatic hypertrophy and cancer of the prostate. A prospective and retrospective study. *Lancet* ii, 115–117.

Atkison, P. R., Weidman, E. R., Bhaumick, B., and Bala, R. M. (1980). Release of somatomedin-like activity by cultured WI-38 human fibroblasts. *Endocrinology* **106**, 2006–2012.

Auclair, C., Kelly, P. A., Labrie, F., Coy, D. H., and Schally, A. V. (1977). Inhibition of testicular LH/HCG receptor levels by treatment with a potent LH-RH agonist or HCG. *Biochem. Biophys. Res. Commun.* **76**, 855–862.

Baba, S. (1982). Epidemiology of cancer of the prostate: analysis of countries of high and low incidence. In *Prostate Cancer.* (Eds G. H. Jacobi and R. Hohenfellner) pp. 11–28. *International Perspectives in Urology* 3, Williams and Wilkins, Baltimore.

Baird, D. T., Uno, A., and Melby, J. C. (1969). Adrenal secretion of androgens and oestrogens. *J. Endocr.* **45**, 135–136.

Baker, J. B., Barsh, G. S., Carney, D. H., and Cunningham, D. D. (1978). Dexamethasone modulates binding and action of epidermal growth factor in serum-free cell culture. *Proc. natl. Acad. Sci. U.S.A.* **75**, 1882–1886.

Bargmann, C. L., Hung, M-C., and Weinberg, R. A. (1986). The *neu* oncogene encodes an epidermal growth factor receptor-related protein. *Nature, Lond* **319**, 226–230.

Barrack, E. R. and Coffey, D. S. (1980). The specific binding of estrogens and androgens to the nuclear matrix of sex hormone responsive tissues. *J. biol. Chem.* **255**, 7265–7275.

—— —— (1982). Biological properties of the nuclear matrix: steroid hormone binding. *Recent Prog. Horm. Res.* **38**, 133–189.

—— Bujnovsky, P., and Walsh, P. C. (1983). Subcellular distribution of androgen receptors in human normal, benign hyperplastic and malignant prostate tissue: characterization of nuclear salt-resistant receptors. *Cancer Res.* **43**, 1107–1116.

Bartsch, W., Horst, H. J., Becker, H., and Nehse, G. (1977). Sex hormone binding globulin capacity, testosterone, 5α-dihydrotestosterone, oestradiol and prolactin in plasma of patients with prostatic carcinoma under various types of hormonal treatment. *Acta endocr., Copenh.* **85**, 650–664.

Bashirelahi, N., Young, S. D., Sidh, S. M., and Sanefuji, H. (1980). Androgen, oestrogen and progestogen and their distribution in epithelial and stromal cells of human prostate. In *Steroid Receptors, Metabolism and Prostatic Cancer.* (Eds F. H. Schroder and H. J. de Voogt). pp. 240–255. Excerpta Medica, Amsterdam.

Becker, H., Kaufmann, J., Klosterhalfen, H., and Voigt, K. D. (1972). *In vivo* uptake and metabolism of ^{3}H-testosterone and ^{3}H-5α-dihydrotestosterone by human benign prostatic hypertrophy. *Acta endocr., Copenh.* **71**, 589–599.

Benson, R. C., Utz, D. C., Holicky, E., and Veneziale, C. M. (1985). Androgen receptor binding activity in human prostate cancer. *Cancer* **55**, 382–388.

Berg, O. A. (1958). Parenchymatous hypertrophy of the canine prostate gland. *Acta endocr., Copenh.* **27**, 140–154.

Billeter, E. and Fluckiger, E. (1971). Evidence for a luteolytic function of prolactin in the intact cyclic rat using 2-Br-α-ergocryptine (CB154). *Experientia, Basel* **27**, 464–465.

Bishop, J. M. (1983). Cellular oncogenes and retroviruses. *A. Rev. Biochem.* **52**, 301–354.

—— (1985). Viruses, genes and cancer. II. retroviruses and cancer genes. *Cancer* **55**, 2329–2333.

Blacklock, N. J. (1976). Surgical anatomy of the prostate. In *Scientific Foundation of Urology*, Vol. 2. (Eds D. I. Williams and G. D. Chisholm). pp. 113–125. Year Book Medical Publishing, Chicago.

Bostock, D. E. and Curtis, R. (1984). Comparison of canine oropharyngeal malignancy in various geographical locations. *Vet. Rec.* **114**, 341–342.

Boulanger, P., Desaulniers, M., Blean, G., Roberts, K. D., and Chapdelaine, A. (1983). Sex hormone concentrations in plasma from the canine deferential vein. *J. Endocr.* **96**, 223–228.

Bowman, S. P., Barnes, D. M., and Blacklock, N. J. (1985). A mini-column method for routine measurement of human prostatic androgen receptors. *J. steroid Biochem.* **23**, 421–430.

—— —— —— and Sullivan, P. J. (1986). Regional variation of cytosol androgen receptors throughout the diseased human prostate gland. *The Prostate* **8**, 167–180.

Boyns, A. R., Cole, E. N., Golder, M. P., Danutra, V., Harper, M. E., Brownsey, B. G., Cowley, T. H., Jones, G. E., and Griffiths, K. (1972). Prolactin studies with the prostate. In *Prolactin and Carcinogenesis.* (Eds A. R. Boyns and K. Griffiths). pp. 207–216. Alpha Omega Alpha, Cardiff.

—— —— Phillips, M. E. A., Hiller, S. G., Cameron, E. H. D., Griffiths, K., Shahmanesh, M., Feneley, R. C. L., and Hartog, M. (1974). Plasma prolactin, GH, LH, FSH, TSH and testosterone during treatment of prostatic carcinoma with oestrogens. *Eur. J. Cancer* **10**, 445–449.

—— Griffiths, K., Pierrepoint, C. G., and Peeling, W. B. (1975). Prolactin and the prostate. In *Normal and abnormal growth of the prostate.* (Ed. M. Goland) pp. 431–444. Charles C. Thomas, Springfield, Illinois.

Brendler, C. B., Isaacs, J. T., Follansbee, A. L., and Walsh, P. C. (1984). The use of multiple variables to predict response to endocrine therapy in carcinoma of the prostate. *J. Urol.* **131**, 694–700.

Brendler, H. (1973). Adrenalectomy and hypophysectomy for prostatic cancer. *Urology* **2**, 99–102.

British Prostate Study Group (1979). Evaluation of plasma hormone concentrations in relation to clinical staging in patients with prostatic cancer. *Brit. J. Urol.* **51**, 382–389.

Bruchovsky, N. and Wilson, J. D. (1968a). The intranuclear binding of testosterone and 5α-andostan-17β-ol-3-one by rat prostate. *J. biol. Chem.* **243**, 5953–5960.

—— —— (1986b). The conversion of testosterone to 5α-androstan-17β-ol-3-one by rat prostate *in vivo* and *in vitro*. *J. biol. Chem.* **243**, 2012–2021.

—— and Lieskovsky, G. (1979). Increased ratio of 5α-reductase: 3α(β)-hydroxysteroid dehydrogenase activities in hyperplastic human prostate. *J. Endocr.* **80**, 289–301.

Byar, D. P. (1977). VACURG studies on prostatic cancer and its treatment. In *Urological Pathology: The Prostate.* (Ed. M. Tannenbaum). pp241–267, New York, Lea and Febiger.

Byyny, R. L., Orth, D. N., Cohen, S., and Doyne, E. S. (1974). Epidermal growth factor; effects of androgens and adrenergic agents. *Endocrinology* **95**, 776–782.

Cabot, A. T. (1896). The question of castration for enlarged prostate. *Ann. Surg.* **24**, 265–309.

Cannon, L., Bishop, D. T., Skolnick, M., Hunt, S., Lyon, J. L., and Smart, C. R. (1982). Genetic epidemiology of prostate cancer in the Utah Mormon Geneology. *Cancer Surv.* **1**, 47–69.

Carter, N. J., Tyson, J. E., Warne, G. L., McNeilly, A. S., Faiman, C., and Fries, H. G. (1977). Adrenocortical function in hyperprolactinaemic women. *J. clin. Endocr. Metab.* **45**, 973–980.

Chaisiri, N. and Pierrepoint, C. G. (1980). Examination of the distribution of oestrogen receptor between stromal and epithelial compartments of the canine prostate. *The Prostate* **1**, 357–366.

Chan, P-C., Didato, F., and Cohen, L. A. (1975). High dietary fat, elevation of rat serum prolactin and mammary cancer. *Proc. Soc. exp. Biol. Med.* **149**, 133–135.

Chapdelaine, A., Boulanger, P., Dupuy, G. M., Beau, G., and Roberts, K. D. (1978). Sex steroids in the vas deferens and deferential vein: possible role in BPH. *60th Ann. Meeting Endocrine Soc.*, June, 1978. Williams & Wilkins, Baltimore, pp252.

Chase, M. D., Geschwind, I. I., and Bern, H. A. (1957). Synergistic role of prolactin in response of male sex accessories to androgen. *Proc. Soc. exp. Biol. Med.* **94**, 680–683.

Clemmons, D. R., Underwood, L. E., and Van Wyk, J. J. (1981). Hormonal control of immunoreactive somatomedin production by cultured human fibroblasts. *J. clin Invest.* **67**, 10–19.

Consensus Meeting on steroid receptors in breast cancer (1980) NIH Bethesda, USA. *Cancer* **46**, (No 12).

Correa, R. J., Anderson, R. G., Gibbons, R. P., and Tate, M. J. (1974). Latent carcinoma of the prostate—why the controversy? *J. Urol.* **111**, 644–646.

Coune, A., and Smith, P. (1975). Clinical trial of 2-bromo-α-ergocryptine (NSC-169774) in human prostatic cancer. *Cancer Chemother. Rep.* **59**, 209–213.

Cowan, R. A., Cowan, S. K., and Grant, J. K. (1977). Binding of methyltrienolone (R1881) to a progesterone receptor-like component of human prostate cytosol. *J. Endocr.* **74**, 281–289.

—— —— —— and Elder, H. Y. (1977). Biochemical investigations of separated epithelium and stroma from benign hyperplastic prostatic tissue. *J Endocr.* **74**, 111–120.

Cowley, T. H., Brownsey, B. G., Harper, M. E., Peeling, W. B., and Griffiths, K. (1976). The effect of ACTH on plasma testosterone and androstenedione concentrations in patients with prostatic carcinoma. *Acta endocr., Copenh.* **81**, 310–320.

Cox, D. R. (1972). Regression models and life tables. *J. Roy. Stat. Soc.* **B 34**, 187–220.

—— (1975). Partial likelihood. *Biometrika* **62**, 269–276.

Cunha, G. R. (1972). Tissue interactions between epithelium and mesenchyme of urogenital and integumental origin. *Anat. Rec.* **172**, 529–541.

—— (1976). Epithelial-stromal interactions in development of the urogenital tract. *Int. Rev. Cytol.* **47**, 137–194.

—— Chung, L. W. K., Shannon, J. M., and Reese, B. A. (1980). Stromal-epithelial interactions in sex differentiation. *Biol. Reprod.* **22**, 19–42.

—— Lung, B., and Reese, B. A. (1980). Glandular epithelium induction by embryonic mesenchyme in adult bladder epithelium of BALB/C mice. *Invest. Urol.* **17**, 302–304.

Cutler, S. J., and Young, J. L. (1975). Third national cancer survey: incidence data. *National Cancer Institute Monograph*, **41**.

Danutra, V., Harper, M. E., Boyns, A. R., Cole, E. N., Brownsey, B. G., and Griffiths, K. (1973). The effect of certain stilboestrol analogues on plasma prolactin and testosterone in the rat. *J. Endocr.* **57**, 207–215.

Davies, P. (1978). Receptors in the human prostate. In *Tumour makers, determination and clinical significance, Sixth Tenovus Workshop* (Eds K. Griffiths, C. G. Pierrepoint, and A. M. Neville). pp. 175–187. Alpha Omega Alpha, Cardiff.

—— Thomas, P., Borthwick, N. M., and Giles, M. G. (1980). Distribution of acceptor sites for androgen-receptor complexes between transcriptionally active and inactive fractions of rat ventral prostate chromatin. *J. Endocr.* **87**, 225–240.

—— —— and Giles, M. G. (1982). Responses to androgens of rat ventral prostate nuclear androgen-binding sites sensitive and resistant to micrococcal nuclease. *The Prostate* **3**, 439–457.

—— (1983a). Extraction of androgen-receptor complexes from regions of rat ventral prostate nuclei sensitive or resistant to nucleases. *J. Endocr.* **99**, 51–61.

—— (1983b). Androgen receptors in normal and malignant prostate. In *Advances in hormone receptors. Methodological and clinical aspects.* (Ed. M. A. Navarro Moreno). pp. 106–125. Jarpyo Editoros, Madrid.

—— Thomas, P., and Manning, D. L. (1986). Correlations between prostate chromatin structure and transcriptional activity and acceptor site distribution. *The Prostate* **8**, 151–166.

DeKlerk, D. P., Coffey, D. S., Ewing, L. L., McDermott, I. R., Reiner, W. G., Robinson, C. H., Scott, W. W., Strandberg, J. D., Talalay, P., Walsh, P. C., Wheaton, L. G., and Zirkin, B. R. (1979). A comparison of spontaneous and experimentally induced canine prostatic hyperplasia. *J. clin. Invest.* **64**, 842–849.

De Voogt, H. J., and Dingjan, P. (1978). Steroid receptors in human prostatic cancer: a preliminary evaluation. *Urol. Res.* **6**, 151–158.

Deyl, Z., Rosmus, J., and Adam, M. (1976). Pituitary and collagen. In *Hypothalamus, pituitary and aging* (Eds A. V. Everitt and J. A. Burgess). Charles C. Thomas, Illinois, pp. 171–192.

Dhabuwala, C. B., and Pierrepoint, C. G. (1977). Venous drainage and functional control of the canine prostate gland. *J. Endocr.* **75**, 105–108.

—— Roberts, E. E., and Pierrepoint, C. G. (1978). The radiographic demonstration of the dynamic transfer of radio-opaque material from the deferential vein to the prostate in the dog. *Invest. Urol.* **15**, 346–347.

Dickson, R. B., Huff, K. K., Spencer, E. M., and Lippman, M. E. (1986). Induction of epidermal growth factor-related polypeptides by 17β-estradiol in MCF-7 human breast cancer cells. *Endocrinology* **118**, 138–142.

Dodds, E. C., Goldberg, L., Lawson, W., and Robinson, R. (1938). Oestrogenic activity of certain synthetic compounds. *Nature, Lond.* **141**, 247–248.

Donnelly, B. J., Lakey, W. H., and McBlain, W. A. (1984). Androgen binding sites on nuclear matrix of normal and hyperplastic human prostate. *J. Urol.* **131**, 806–811.

Doolittle, R. F., Hunkapiller, M. W., Hood, L. E., Devare, S. G., Robbins, K. C., Aaronson, S. A., and Antoniades, H. N. (1983). Simian sarcoma virus *onc* gene, *v-sis*, is derived from the gene (or genes), encoding a platelet-derived growth factor. *Science (Washington)* **221**, 275–277.

Downward, J., Yarden, Y., Mayes, E., Scrace, G., Totty, N., Stockwell, P., Ulrich, A., Schlessinger, J., and Waterfield, M. D. (1984). Close similarity of epidermal growth factor receptor and v-erb-B oncogene protein sequences. *Nature, Lond.* **307**, 521–527.

Drafta, D., Proca, E., Zamfir, V., Schindler, A. C., Neacsu, E., and Stroe, E. (1982). Plasma steroids in benign prostatic hypertrophy and carcinoma of the prostate. *J. steroid. Biochem.* **17**, 689–693.

Drago, J. R., Santen, R. J., Lipton, A., Worgel, T. J., Harvey, H. A., Boucher, A., Mann, A., and Rohner, T. J. (1984). Clinical effects of aminoglutethimide, medical adrenalectomy, in treatment of 43 patients with advanced prostatic carcinoma. *Cancer* **53**, 1447–1450.

Eaton, C. L., and Pierrepoint, C. G. (1982a). Epithelial and fibroblastoid cell lines derived from the normal canine prostate: I. Separation and characterisation of epithelial and stromal components. *The Prostate* **3**, 277–290.

—— —— (1982b). Epithelial and fibroblastoid cell lines derived from the normal canine prostate: II. Cell proliferation in response to steroid hormones. *The Prostate* **3**, 493–506.

—— (1984). *Studies with isolated stromal and epithelial cells of the canine prostate gland.* PhD Thesis. University of Wales College of Medicine, Cardiff, Wales.

—— Hamilton, T. C., Kenvyn, K., and Pierrepoint, C. G. (1985). Studies of androgen and oestrogen binding in normal canine prostatic tissue and in epithelial and stromal cell lines derived from the canine prostate. *The Prostate* **7**, 377–388.

—— and Pierrepoint, C. G. (1986). Cell lines derived from the normal canine prostate. In *In Vitro Models for Cancer Research.* (Ed M. Webber). CRC Press, Boca Raton, (in press).

Ebina, Y., Ellis, L., Jarnagin, K., Edery, M., Graf, L., Clauser, E., Ou, J-H., Masiarz, F., Kan, Y. W., Goldfine, I. D., Roth, R. A., and Rutter, W. J. (1985). The human insulin receptor cDNA: the structural basis for hormone-activated transmembrane signalling. *Cell* **40**, 747–758.

Eisenman, R. N., Tachibana, C. Y., Abrams, H. D., and Hann, S. R. (1985). V-myc and c-myc-encoded proteins are associated with the nuclear matrix. *Mol. Cell. Biol.* **5**, 114–126.

Ekman, P. (1982). The application of steroid receptor assays in human prostate cancer research and clinical management. *Anticancer Res.* **2**, 163–176.

El Etreby, M. P., and Mahrous, A. T. (1979). Immunocytochemical technique for detection of prolactin (PRL) and growth hormone (GH) in hyperplastic and neoplastic lesions of dog prostate and mammary gland. *Histochemistry* **64**, 279–286.

Fahmy, D. R., Read, G. F., Walker, R. F., and Griffiths, K. (1982). Steroids in saliva for assessing endocrine function. *Endocr. Rev.* **3**, 367–395.

Farnsworth, W. E. and Brown, J. R. (1963). Metabolism of testosterone by the human prostate. *J. Amer. med. Ass.* **183**, 436–439.

—— (1970). The normal prostate and its endocrine control. In *Some aspects of the aetiology and biochemistry of prostatic cancer*, 3rd Tenovus Workshop (Eds K. Griffiths and C. G. Pierrepoint). pp 3–15. Alpha Omega Alpha, Cardiff.

Fentie, D. D., Lakey, W. H., and McBlain, W. A. (1986). Applicability of nuclear androgen receptor quantification to human prostate adenocarcinoma. *J. Urol.* **135**, 167–173.

Fingerhut, B., and Veenema, R. J. (1967). The effect of bilateral adrenalectomy on induced benign prostatic hyperplasia in mice. *J. Urol.* **97**, 508–517.

Finney, R. W., Harper, M. E., Gaskell, S. J., and Griffiths, K. (1980). Tissue androgen concentrations in golden hamsters with benign prostatic tumours. *J. Endocr.* **84**, 353–361.

Fitzpatrick, J. M., Gardiner, R. A., Williams, J. P., Riddle, P. R., and O'Donoghue, E. P. N. (1980). Pituitary ablation in the relief of pain in advanced prostatic carcinoma. *Br. J. Urol* **52**, 301–304.

Fleming, W. H., Hamel, A., MacDonald, R., Ramsey, E., Pettigrew, N. M., Johnston, B., Dodd, J. G., and Matusik, R. J. (1986). Expression of the c-myc proto-oncogene in human prostatic carcinoma and benign prostatic hyperplasia. *Cancer Res.* **46**, 1535–1538.

Fluckiger, E. (1972). Drugs and the control of prolactin secretion. In *Prolactin and Carcinogenesis*. Fourth Tenovus Workshop. (Eds A. R. Boyns and K. Griffiths) pp. 162–180. Alpha Omega Publ.

Fujii, H., Cunha, G. R., and Norman, J. T. (1982). The induction of adenocarcinomatous differentiation in neoplastic bladder epithelium by an embryonic prostatic inductor. *J. Urol.* **128**, 858–861.

Geller, J., Vazakas, G., Fruchtman, B., Newman, H., Nakao, K., and Loh, A. (1968). The effect of cyproterone acetate on adenocarcinoma of the prostate. *Surg. Gynec. Obst.* **127**, 748–758.

—— Albert, J., Lopez, D., Geller, S., and Niwayama, G. (1976). Comparison of androgen metabolites in benign prostatic hypertrophy (BPH) and normal prostate. *J. clin. Endocr. Metab.* **43**, 686–688.

—— and Albert, J. D. (1983). Comparison of various hormonal therapies for prostatic carcinoma. *Semin. Oncol.* **10**, 34–42.

—— —— (1985a). Adrenal androgen blockade in relapsed prostate cancer. *Eur. J. Cancer clin. Oncol* **21**, 1127–1131.

—— —— (1985b). DHT in prostatic cancer tissue—a guide to management and therapy. *The Prostate* **6**, 19–25.

Ghanadian, R., Pinah, C. M., and O'Donoghue, E. P. N. (1979). Serum testosterone and dihydrotestosterone in carcinoma of the prostate. *Br. J. Cancer* **39**, 696–699.

—— and Auf, G. (1980). Receptor proteins for androgens in BPH and carcinoma of the prostate. In *Steroid Receptors, Metabolism and Prostatic Cancer* (Eds F. H. Schroder & H. J. de Voogt) pp. 110–125. Excerpta Medica, Amsterdam.

—— —— Williams, G., Davies, A., and Richards, A. (1981). Predicting response of postate carcinoma to endocrine therapy. *Lancet* ii, 1418.

Gleason, D. F. (1977). Histological grading and clinical staging of prostatic carcinoma. In *Urologic Pathology: The Prostate.* (Ed. M. Tannenbaum). pp. 171–197. Lea & Febiger, Philadelphia.

Golder, M. P., Boyns, A. R., Harper, M. E., and Griffiths, K. (1972). An effect of prolactin on prostatic adenylate cyclase activity. *Biochem. J.* **128**, 725–727.

Gonor, S. E., Lakey, W. H., and McBlain, W. A. (1984). Relationships between concentrations of extractable and matrix-bound nuclear androgen-receptor and clinical response to endocrine therapy for prostatic carcinoma. *J. Urol.* **131**, 1196–1201.

Gorski, J., Welshons, W., and Sakai, D. (1984). Remodelling the estrogen receptor model. *Mol. Cell. Endocrinol.* **36**, 11–15.

Gospodarowicz, D., and Moran, J. (1974). Effect of fibroblast growth factor, insulin, dexamethasone and serum on the morphology of 3T3 cells. *Proc natn Acad Sci U.S.A.* **71**, 4648–4652.

—— Delgado, D., and Vlodavsky, I. (1980). Permissive effect of extracellular matrix on cell proliferation *in vitro. Proc natn Acad Sci U.S.A.* **77**, 4094–4098.

Gould, V. J., Turkes, A. O., and Gaskell, S. J. (1986). GC-MS analysis of salivary testosterone with reference to DES-treated prostatic cancer patients. *J. ster. Biochem.* (in press).

Goustin, A. S., Leof, E. B., Shipley, G. D., and Moses, H. L. (1986). Growth factors and cancer. *Cancer Res.* **46**, 1015–1029.

Gray, L. A. (1979). Are hyperplasias or atrophies associated with breast cancer? In *Breast Disease.* (Eds D. J. Marchant and I. Nyirjesy). pp. 42–51. Grune & Stratton, N. Y.

Grayhack, J. T., Bunce, P. L., Kearns, J. W., and Scott, W. W. (1955). Influence of the pituitary on prostatic response to androgen in the rat. *Bull. Johns Hopkins Hosp.* **96**, 154–163.

—— and Lebowitz, J. M. (1967). Effect of prolactin on citric acid of lateral lobe of prostate of Sprague-Dawley rat. *Invest. Urol.* **5**, 87–94.

Green, S., Walter, P., Kumar, V., Krust, A., Bornert, J-M., Argos, P., and Chambon, P. (1986). Human oestrogen receptor cDNA: sequence, expression and homology to v-erb-A. *Nature, Lond.* **320**, 134–139.

Greenberg, M. E., and Ziff, E. B. (1984). Stimulation of 3T3 cells induces transcription of the c-fos proto-oncogene. *Nature, Lond.* **311**, 433–438.

Griffiths, J. (1895). Three lectures upon the testes. *The Lancet*, March 30th, 791–799.

Griffiths, K., Harper, M. E., and Peeling, W. B. (1976). Hormone studies. In *Scientific Foundation of Urology* Vol. II (Eds D. I. Williams and G. D. Chisholm). pp. 354–361. Heinemann Medical Books, London.

—— Davies, P., Harper, M. E., Peeling, W. B., and Pierrepoint, C. G. (1979). The etiology and endocrinology of prostatic cancer. In *Endocrinology of Cancer* Vol 2 (Ed. D. Rose). pp. 1–55. CRC Press.

—— Peeling, W. B., Harper, M. E., Davies, P., and Pierrepoint, C. G. (1981). Prostatic Cancer: Rationale for Hormone Therapy. In *Hormonal Management of Endocrine-Related Cancer* (Ed. B. A. Stoll). pp. 131–147. Lloyd-Luke (Medical Books) Ltd.

—— Nicholson, R. I., Blamey, R. W., and Elston, C. W. (1986). Estrogen receptor determination: studies in relation to rapidly progressive carcinoma of the breast. In *Estrogen/antiestrogen action in breast cancer therapy* (Ed. Y. C. Jordan). pp. 325–340. University of Wisconsin Press.

Grimaldo, J. I., and Meikle, A. W. (1984). Increased levels of nuclear androgen receptors in hyperplastic prostates of aging men. *J. steroid Biochem.* **21**, 147–150.

Grinnell, F., and Field, M. K. (1979). Initial adhesion of human fibroblasts in serum-free medium: possible role of secreted fibronectin. *Cell* **18**, 117–128.

Gunn, S. A., Gould, T. C., and Anderson, W. A. D. (1965). The effect of growth hormone injections and prolactin preparations on the control of interstitial cell-stimulating hormone uptake. *J. Endocr.* **32**, 205–214.

Haapiainen, R., Rannikko, S., Adlercreutz, H., and Alfthan, O. (1986). Correlation of pretreatment plasma levels of estradiol and sex hormone-binding globulin-binding capacity with clinical stage and survival of patients with prostatic cancer. *The Prostate* **8**, 127–137.

Habib, F. K., Smith, T., Robinson, R., and Chisholm, G. D. (1985). Influence of surgical techniques on receptor levels and 5α-reductase activity of the human prostate gland. *The Prostate* **7**, 287–292.

—— Lee, I. R., Stitch, S. R., and Smith, P. H. (1976). Androgen levels in the plasma and protastic tissues of patients with benign hypertrophy and carcinoma of the prostate. *J. Endocr.* **71**, 99–107.

Haenzel, W., and Kurihara, M. (1968). Studies of Japanese migrants. I. Mortality from cancer and other diseases among Japanese in the United States. *J. natn. Cancer Inst.* **40**, 43–68.

Hafiez, A. A., Lloyd, C. W., and Bartke, A. (1972). The role of prolactin in the regulation of the testis function: the effects of prolactin and luteinising hormone on the plasma levels of testosterone and androstenedione in hypophysectomized rats. *J. Endocr.* **52**, 327–332.

Hahn, L-H. E., and Yamada, K. M. (1979). Identification and isolation of a collagen binding fragment of the adhesive glycoprotein fibronectin. *Proc. natn. Acad. Sci. U.S.A.* **76**, 1160–1163.

Hammond, G. L., Kontturi, M., Maattala, P., Puukka, M., and Vihko, R. (1977). Serum FSH, LH and prolactin in normal males and patients with prostatic disease. *Clin. Endocrin.* **7**, 129–135.

Harbitz, T. B., and Haugen, O. A. (1972). Histology of the prostate in elderly men. *Acta pathol. microbiol. scand.* **80**, 756–768.

Harper, M. E., Pike, A., Peeling, W. B., and Griffiths, K. (1974). Steroids of adrenal origin metabolised by human prostatic tissue *in vivo* and *in vitro. J. Endocr.* **60**, 117–125.

—— Peeling, W. B., Cowley, T., Brownsey, B. G., Phillips, M. E. A., Groom, G., Fahmy, D. R., and Griffiths, K. (1976a). Plasma steroid and protein hormone concentrations in patients with prostatic carcinoma before and during oestrogen therapy. *Acta endocr., Copenh.* **81**, 409–429.

—— Danutra, V., Chandler, J. A., and Griffiths, K. (1976b). The effect of 2-bromo-α-ergocryptine (CB154) administration on the hormone levels, organ weights, prostatic morphology and zinc concentration in the male rat. *Acta endocr., Copenh.* **83**, 211–224.

—— Sibley, P. E. C., Peeling, W. B., and Griffiths, K. (1981). The immunocytochemical detection of growth hormone and prolactin in human prostatic tissues. In *The prostatic cell: structure and function, part B*. (Eds G. P. Murphy, A. A. Sandberg, and J. P. Karr). pp. 115–128. Alan R. Liss, New York.

—— Pierrepoint, C. G., and Griffiths, K. (1984). Carcinoma of the prostate: relationship of pretreatment hormone levels to survival. *Eur. J. Cancer clin. Oncol.* **20**, 477–482.

Hechter, O., Mechaber, D., Zwick, A., Campfield, A., Eychenne, B., Baulieu, E-E., and Robel, P. (1983). Optimal radioligand exchange conditions for measurement of occupied androgen receptor sites in rat ventral prostate. *Archs Biochem. Biophys.* **224**, 49–68.

Hemsell, D. L., Grodin, J. M., Brenner, P. F., Siiteri, P. K., and MacDonald, P. C. (1974). Plasma precursors of estrogen. II. Correlation of the extent of conversion of plasma androstenedione to estrogen with age. *J. clin. Endocr. Metab.* **38**, 476–479.

Higuchi, K., Nawaka, H., Maki, T., Higashizima, M., Kato, K-I., and Ibayashi, H. (1984). Prolactin has a direct effect on adrenal androgen secretion. *J. clin. Endocr. Metab.* **59**, 714–718.

Hill, P., and Wynder, F. (1976). Diet and prolactin release. *Lancet* ii, 806–807.

Hirst, A. E. Jr., and Bergman, R. T. (1954). Carcinoma of the prostate in men 80 or more years old. *Cancer (Philadelphia)* 7, 136–141.

Hoisaeter, P. A., Haukaas, S., Bakke, A., Hoiem, L., Segadal, E., and Thorsen, T. (1982). Blood hormone levels related to stages and grades of prostatic cancer. *The Prostate* 3, 375–381.

Holland, J. M., and Lee, C. (1980). Effects of pituitary grafts on testosterone stimulated growth of rat prostate. *Biol. Reprod.* 22, 351–355.

Hosoi, K., Tanaka, I., and Ueha, T. (1981). Induction of epidermal growth factor by tri-iodo-L-thyronine in the submandibular glands of mice with testicular feminization. *J. Biochem.* **90**, 267–272.

Houston, B., Chisholm, G. D., and Habib, F. K. (1985). Evidence that human prostatic 5α-reductase is located exclusively in the nucleus. *Febs letters* **185**, 231–234.

Huben, R., Mettlin, C., Natarajan, N., Smart, C. R., Pontes, E., and Murphy, G. P. (1982). Carcinoma of prostate in men less than 50 years old. *Urology* **20**, 585–588.

Huggins, C., and Clark, P. J. (1940). Quantitative studies of prostatic secretion. II. The effect of castration and of estrogen injection on the normal and on the hyperplastic prostatic glands of dogs. *J. exp. Med.* **72**, 747–762.

—— and Hodges, C. V. (1941). Studies on prostatic cancer. The effect of castration, of estrogen and of androgen injection on serum phosphatases in metastatic carcinoma of the prostate. *Cancer Res.* **1**, 293–297.

—— Stevens, R. E. Jr., and Hodges, C. V. (1941). Studies on prostatic cancer. II. The effects of castration on advanced carcinoma of the prostate gland. *Arch. Surg.* **43**, 209–223.

—— and Russell, P. S. (1946). Quantitative effects of hypophysectomy on testis and prostate of dogs. *Endocrinology* **39**, 1–7.

Hunter, J. (1786). Observations on certain parts of the animal oeconomy, 1st ed., Bibliotheca Osteriana. pp. 38–39. London.

Hunter, T. (1984). The proteins of oncogenes. *Sci. Am.* **251**, 70–79.

250 K. Griffiths *et al.*

Jakobsen, H., and Juul, N. (1985). Influence of vasectomy on the volume of the non-hyperplastic prostate in men. *Int. J. Androl.* **8**, 13–20.

Jacobi, G. H., Rathgen, R. H., and Altwein, J. E. (1980a). Serum prolactin and tumors of the prostate: unchanged basal levels and lack of correlation to serum testosterone. *J. endocrin. Invest.* **3**, 15–19.

—— Altwein, J. E., and Hohenfellner, R. (1980b). Adjunct bromocriptine treatment as palliation for prostate cancer: experimental and clinical evaluation. *Scand. J. Urol. Nephrol. Suppl* **55**, 107–112.

—— (1982). Experimental rationale for the investigation of antiprolactins as palliative treatment for prostatic cancer. In *Prostate Cancer, Int Prospectives in Urology*, Vol 3 (Eds G. H. Jacobi and R. Hohenfellner). pp. 419–431. Williams & Wilkins, Baltimore.

Jacobs, S. C., Pinka, D., and Lawson, R. K. (1979). Prostatic osteoblastic factor. *Invest. Urol.* **17**, 195–199.

—— and Lawson, R. K. (1980). Mitogenic factor in human prostate extracts. *Urology* **16**, 488–493.

Jensen, H. M., Rice, J. D., and Wellings, S. R. (1976a). Preneoplastic lesions in the human breast. *Science, N.Y.* **191**, 295–297.

—— —— —— (1976b). Breast lobules. *Science, N.Y.* **193**, 919.

Jönsson, G., Olsson, A. M., Luttrop, W., Cekan, Z., Purvis, K., and Diczfalusy, E. (1975). Treatment of prostatic carcinoma with various types of oestrogen derivatives. In *Vitamins and Hormones* Vol. 33 (Eds P. L. Munson, E. Diczfalusy, J. Glover, and R. E. Olsen). pp. 351–376. Academic Press, London.

Jordan, W. B. Jr., Blackard, C. E., and Byar, D. P. (1977). Reconsideration of orchiectomy in the treatment of advanced prostatic carcinoma. *Southern med. J.* **70**, 1411–1413.

Kandeel, F. R., Rudd, B. T., Butt, W. R., Logan-Edwards, R., and London, D. R. (1978). Androgen and cortisol responses to ACTH stimulation in women with hyperprolactinaemia. *Clin. Endocrinol.* **9**, 123–130.

Kaplan, E. L. and Meier, P. (1958). Nonparametric estimation from incomplete observations. *J Amer. statist. Assoc.* **53**, 457–481.

Kelch, R. P., Jenner, M. R., Weinstein, R., Kaplan, S. L., and Grumbach, M. M. (1972). Estradiol and testosterone secretion by human, simian and canine testes, in males with hypogonadism and in male pseudohermaphrodites with the feminizing testes syndrome. *J. clin. Invest.* **51**, 824–830.

Kelly, K., Cochran, B. N., Stiles, C. D., and Leder, P. (1983). Cell-specific regulation of the c-myc gene by lymphocyte mitogens and platelet-derived growth factor. *Cell* **35**, 603–610.

Kemp, H. A., Read, G. F., Fahmy, D. R., Pike, A. W., Gaskell, S. J., Queen, K., Harper, M. E., and Griffiths, K. (1981). Measurement of diethylstilboestrol in plasma from patients with cancer of the prostate. *Cancer Res.* **41**, 4693–4697.

Kidwell, W. R., Wicha, M. S., Salomon, D., and Liotta, L. A. (1980). In *Control mechanisms in animal cells* (Ed. L. Jimenez de Asua). pp. 333–340 Raven Press, London.

King, W. J., and Greene, G. L. (1984). Monoclonal antibodies localize oestrogen receptor in the nuclei of target cells. *Nature, Lond.* **307**, 745–747.

Kledzik, G. S., Marshall, S., Campbell, G. A., Gelato, M., and Meites, J. (1976). Effects of castration, testosterone, estradiol and prolactin on specific prolactin-binding activity in ventral prostate of male rats. *Endocrinology* **98**, 373–379.

Knudson, A. G. (1985). Hereditary cancer, oncogenes and antioncogenes. *Cancer Res* **45**, 1437–1443.

Koppel, M., Heranze, D. R., and Shimkin, M. B. (1967). Characteristics of patients with prostatic carcinoma: a control case study on 83 autopsy pairs. *J. Urol.* **98**, 229–233.

Krain, L. S. (1973). Epidemiologic variables in prostatic cancer. *Geriatrics* **28**, 93–98.

Kratochwil, K. (1969). Organ specificity in mesenchymal induction demonstrated in the embryonic development of the mammary gland of the mouse. *Devl. Biol.* **20**, 46–71.

Krieg, M., Bartsch, W., Herzer, S., Becker, H., and Voigt, K. D. (1977). Quantification of androgen binding, androgen tissue levels, and sex hormone-binding globulin in prostate, muscle and plasma of patients with benign prostatic hyperplasia. *Acta endocri. Copenh.* **86**, 200–215.

—— —— Janssen, W., and Voigt, K. D. (1979). A comparative study of binding, metabolism and endogenous levels of androgens in normal, hyperplastic and carcinomatous human prostate. In *Prostate Cancer* (Eds D. S. Coffey & J. T. Isaacs). pp. 93–111. UICC Tech. Workshop No: 48.

Kruijer, W., Cooper, J. A., Hunter, T. & Verma, I. M. (1984). Platelet-derived growth factor induces rapid but transient expression of the c-fos gene and protein. *Nature, Lond.* **312**, 711–716.

Kyprianou, N. and Davies, P. (1986). Association states of androgen receptors in nuclei of human benign hypertrophic prostate. *The Prostate* **8**, 363–380.

—— Williams, H., Peeling, W. B., Davies, P., and Griffiths, K. (1986). Evaluation of biopsy techniques for androgen receptor assay in human prostatic tissue. *Br. J. Urol.* **58**, 41–44.

Labess, M. (1952). Occult carcinoma in clinically benign hypertrophy of the prostate: a pathological and clinical study. *J. Urol.* **68**, 893–896.

Labrie, F., Dupont, A., Belanger, A., Lacoursiere, Y., Raynaud, J. -P., Husson, J. M., Gareau, J., Fazekas, A. T. A., Sandow, J., Monfette, G., Girard, J. G., Emond, J., and Houle, J. G. (1983). New approach in the treatment of prostatic cancer. Complete instead of partial withdrawal of androgens. *The Prostate* **4**, 579–594.

—— —— —— (1984). Spectacular response to combined antihormonal treatment in advanced prostate cancer. In Endocrinology, International Congress Series 655 (Eds F. Labrie and L. Proulx). pp. 450–453. Excerpta Medica.

—— —— —— (1985). Complete androgen blockade for the treatment of prostate cancer. In *Important advances in oncology* (Eds V. T. De Vita, S. Hellman and S. A. Rosenberg). Lippincott Co., Philadelphia, pp. 193–217.

Lahotnen, R., Bolton, N. J., Kontturi, M., and Vihko, R. (1983). Nuclear androgen receptors in the epithelium and stroma of human benign prostatic hypertrophic glands. *The Prostate* **4**, 129–139.

Lasnitzki, I. (1972). The effect of prolactin on the prostate gland in organ culture. In *Prolactin and Carcinogenesis, Fourth Tenovus Workshop* (Eds A. R. Boyns and K. Griffiths). pp. 200–206. Alpha Omega Publ.

Laurence, D. M. and Landau, R. L. (1965). Impaired ventral prostate affinity for testosterone in hypophysectomised rats. *Endocrinology* **73**, 1119–1125.

Lefebvre, Y. A., Golsteyn, E. J., and Michiel, T. L. (1985). Androgen interactions with intact nuclear envelopes from the rat ventral prostate. *J. steroid Biochem.* **23**, 107–113.

Liao, S., and Witte, D. (1985). Autoimmune anti-androgen-receptor anitbodies in human serum. *Proc. natn. Acad. Sci U.S.A.* **82**, 8345–8348.

Lipsett, M. B. (1970). Steroid secretion by the human testis. In *The Human Testis* (Eds E. Rosenberg and C. A. Paulsen). pp. 407–421. Plenum Press, New York.

Lloyd, J. W., Thomas, J. A., and Mawhinney, M. G. (1973). A difference in the *in vitro* accumulation and metabolism of testosterone-1,2-^{3}H by the rat prostate gland following incubation with ovine or bovine prolactin. *Steroids* 22, 473–483.

Longcope, C., Widrich, W., and Sawin, C. T. (1972). The secretion of estrone and estradiol-17β by human testis. *Steroids* 20 439–448.

Lostroh, A. J., and Li, C. H. (1957). Stimulation of the sex accessories of hypophysectomised male rat by non-gonadotrophin hormones of the pituitary gland. *Acta endocr. Copenh.* 25, 1–16.

Lundberg, S., and Berge, T. (1970). Prostatic carcinoma: an autopsy study. *Scand. J. Urol. Nephrol.* 4, 93–97.

MacDonald, P. C. (1976). Origin of estrogen in men. In *Benign prostatic hyperplasia.* (Eds J. T. Grayhack, J. D. Wilson and M. J. Saherbenske). pp. 191–192. U.S. Government Printing Office, Washington.

Mantel, N. (1966). Evaluation of survival data and two new rank order statistics arising in its consideration. *Cancer Chemother. Rep.* 50, 163–170.

Mariotti, A., Thornton, M., and Mawhinney, M. G. (1981). Actions of androgens and oestrogen on collagen levels in male accessory sex organs. *Endocrinology* 109, 837–843.

Marshall, S., Kledzik, G. S., Gelato, M., Campbell, G. A., and Meites, J. (1976). Effects of estrogen and testosterone on specific prolactin binding in the kidney and adrenal of rats. *Steroids* 27, 187–195.

Marshall, C. J., and Rigby, P. W. J. (1984). Viral and cellular genes involved in oncogenesis. *Cancer Surveys* 3, 183–214.

Maynard, P. V., and Nicholson, R. I. (1979). Effect of high doses of a series of new luteinizing hormone releasing hormone analogues in intact female rats. *Br. J. Cancer* 39, 274–279.

McNeal, J. E. (1970). Age related changes in prostatic epithelium associated with carcinoma. In *Some Aspects of the Aetiology and Biochemistry of Prostatic Cancer*, 3rd Tenovus Workshop (Eds K. Griffiths and C. G. Pierrepoint). pp. 23–32. Alpha Omega Publ, Cardiff.

—— (1972). The prostate and prostatic urethra, a morphologic synthesis. *J. Urol.* 107, 1008–1016.

—— (1975). Structure and pathology of the prostate. In *Normal and Abnormal Growth of the Prostate* (Ed. M. Goland). pp. 55–65. Charles C. Thomas, Springfield, Illinois.

—— (1979). New morphological findings relevant to the origin and evolution of carcinoma of the prostate and BPH. In *Prostate Cancer* (Eds D. S. Coffey, and J. T. Isaacs), pp. 24–37. UICC Technical Report Series, Vol 48, Geneva, UICC.

Meikle, A. W. and Stanish, W. M. (1982a). Familial prostatic cancer risk and low testosterone. *J. clin. Endocr. Metab.* 54, 1104–1108.

—— —— Taylor, N., Edwards, C. Q., and Bishop, C. T. (1982b). Familial effects on plasma sex-steroid content in men: testosterone, estradiol and sex hormone-binding globulin. *Metabolism* 31, 6–9.

—— Smith, J. A., and West, D. W. (1985). Familial factors affecting prostatic cancer risk and plasma sex-steroid levels. *The Prostate* 6, 121–128.

Meites, J., Lu, K. H., Wuttke, W., Welsch, C. W., Nagasawa, H., and Quadri, S. S. (1972). Recent studies on functions and control of prolactin secretion in rats. *Recent Prog. Horm. Res.* 28, 471–526.

Mellinger, G. T. (1964). Veterans Administration Cooperative Urological Research Group. Carcinoma of the prostate: A continuing co-operative study. *J. Urol.* 91, 590–594.

Muir, R. (1941). The evolution of carcinoma of the mamma. *J. Pathol. Bact.* 11, 155–172.

Mulcahy, L. S., Smith, M. R., and Stacey, D. W. (1985). Requirement for *ras* proto-oncogene function during serum-stimulated growth of NIH 3T3 cells. *Nature, Lond.* **313**, 241–243.

Muller, R., Bravo, R., Burckhardt, J., and Curran, T. (1984). Induction of *c-fos* gene and protein by growth factors precedes activation of *c-myc. Nature, Lond.* **312**, 716–720.

Mukku, V. R., and Stancel, G. M. (1985). Regulation of epidermal growth factor receptor by estrogen. *J. biol. Chem.* **260**, 9820–9824.

Muntzing, J. (1971). The androgenic action of adrenal implants in the ventral prostate of adult, castrated and oestrogen treated rats. *Acta pharmacol. toxicol.* **30**, 203–207.

Murphy, L. J., Sutherland, R. L., Stead, B., Murphy, L. C., and Lazarus, L. (1986). Progestin regulation of epidermal growth factor receptor in human mammary carcinoma cells. *Cancer Res.* **46**, 728–734.

—— Vrhovsek, E., and Lazarus, L. (1983). Characterization of specific growth hormone binding sites in mouse fibroblasts. *Endocrinology* **113**, 750–757.

Murray, R., and Pitt, P. (1985). Treatment of advanced prostatic cancer, resistant to conventional therapy, with aminoglutethimide. *Eur. J. Cancer clin. Oncol.* **21**, 453–458.

Nicholson, R. I., Walker, K. J., and Maynard, P. V. (1980). Anti-tumour potential of a new luteinising hormone releasing hormone analogue ICI 118630. In *Breast Cancer: Experimental and Clinical Aspects*. (Eds H. T. Mourisden and T. Palshof). pp. 296–299. Pergamon Press, Oxford/New York.

—— —— Davies, P., Turkes, A., Turkes, A. O., Dyas, J., Blamey, R. W., Williams, M., Robinson, M. R. G., and Griffiths, K. (1984) Use and mechanism of action of the LH-RH agonist ICI 118630 in the therapy of hormone-sensitive breast and prostate cancer. In *Progress in Cancer Research and Therapy* Vol. 31 (Eds F. Bresciani, R. J. B. King, M. E. Lippman, M. Namer, and J. P. Raynaud). pp. 519–532. Raven Press, N.Y.

Nieschlag, E., KJey, H. K., Wiegelmann, W., Solbach, H. G., and Kruskemper, H. L. (1973). Lebensalter und endokrine funktion der testes des erwachsenen Mannes. *Dtsch. med. Wochenschr.* **98**, 1281–1284.

Nishi, N., Matuo, Y., Mugurama, Y., Yoshitake, Y., Nishikawa, K., and Wada, F. (1985). A human prostatic growth factor (hPGF): partial purification and characterization. *Biochem. Biophys. Res. Commun.* **132**, 1103–1109.

Office of Population Census and Surveys (1978). *Mortality Statistics 1976*. Series DH2, No:3, HMSO, London.

Ofner, P. (1969). Effects and metabolism of hormones in normal and neoplastic prostate tissue. *Vitams Horm.* **26**, 237–291.

O'Hare, M. J., Nice, E. C., and Neville, A. M. (1980). Regulation of androgen secretion and sulfoconjugation in the adult human adrenal cortex: studies with primary monolayer cell cultures. In *Adrenal Androgens* (Eds A. R. Genazzani, H. J. J. Thijssen and P. K. Siiteri). pp. 7–25. Raven Press, N.Y.

Okamoto, M., Setoishi, C., Horiuchi, Y., Mashimo, K., Moriji, K., and Itoh, S. (1971). Urinary excretion of testosterone and epitestosterone and plasma levels of LH and testosterone in the Japanese and Ainu. *J. clin Endocrinol Metab* **32**, 673–674.

Oota, K., and Misu, Y. (1958). A study on latent carcinoma of the prostate in Japanese. *GANN* **49**, 283.

Parker, L., Lai, M., Wolk, F., Lifrak, E., Kim, S., Epstein, L., Hadley, D., and Miller, J. (1984). Orchiectomy does not selectively increase adrenal androgen concentrations. *J. clin. Endocr. Metab.* **59**, 547–550.

Peeling, W. B., and Griffiths, K. (1978). Prostatic Cancer. In *Surgical Review*, Vol 1

(Eds J. Lumley and J. Craven) pp. 303–327. Pitman Medical.

—— Griffiths, G. J., and Evans, K. T. (1986). Clinical staging of prostatic cancer. In *The Prostate* (Eds J. P. Blandy and B. Lytton). pp. 121–146. Butterworths.

—— Griffiths, K. (1986). Endocrine treatment of prostatic cancer. In *The Prostate* (Eds J. P. Blandy and B. Lytton). pp. 188–207. Butterworths.

Persson, H., and Leder, P. (1984). Nuclear localization and DNA binding properties of a protein expressed by human *c-myc* oncogene. *Science N.Y.* **225**, 718–721.

Pierrepoint, C. G. (1975). Does hormone transfer along the vasa deferentia contribute to the control of prostatic function and could it be a factor in the etiology of prostatic hyperplasia? In *Normal and Abnormal Growth of the Prostate* (Ed. M. Goland). pp. 517–529. Charles C. Thomas, Springfield, Illinois.

—— and Davies, P. (1973). The effect of vasectomy on the activity of prostatic RNA polymerase in rats. *J. Reprod. Fert.* **35**, 149–152.

—— —— Wilson, D. W. (1974). The role of the epididymis and ductus deferens in the direct and unilateral control of the prostate and seminal vesicles of the rat. *J. Reprod. Fert.* **41**, 413–423.

—— —— Millington, D., and John, B. (1975). Evidence that the deferential vein acts as a local transport system for androgen in rat and dog. *J. Reprod. Fert.* **43**, 293–303.

—— —— Lewis, M. H., and Moffat, D. B. (1975). Examination of the hypothesis that a direct control system exists for the prostate and seminal vesicles. *J. Reprod. Fert.* **44**, 395–409.

—— (1985). Possible benefits to veterinary medicine of considering the dog as a model for human cancer. *J. small anim. Pract.* **26**, 43–47.

—— Turkes, A. O., Walker, K. J., Harper, M. E., Wilson, D. W., Peeling, W. B., and Griffiths, K. (1985). Endocrine factors in the treatment of prostatic cancer. In *EORTC Genitourinary Group Monograph 2, Part A: Therapeutic Principles in Metastatic Prostatic Cancer*. pp. 51–72. Alan R. Liss, New York.

Pike, A., Peeling, W. B., Harper, M. E., Pierrepoint, C. G., and Griffiths, K. (1970). Testosterone metabolism *in vivo* by human prostatic tissue. *Biochem. J.* **120**, 443–445.

Pirke, K. M., and Doerr, P. (1975). Age related changes in free plasma testosterone, dihydrotestosterone, and oestradiol. *Acta endocr. Copenh.* **89**, 171–178.

Posner, B. I., Kelly, P. A., Shiu, R. P. C., and Friesen, H. G. (1974). Studies of insulin, growth hormone and prolactin binding: tissue distribution, species variation and characterisation. *Endocrinology* **95**, 521–531.

Rannikko, S., Kairento, A-L., Karonen, S-L., and Adlercreutz, H. (1981). Hormonal patterns in prostatic cancer 1. Correlation with local extent of tumour, presence of metastases and grade of differentiation. Acta endocr. Copenh. **98**, 625–633.

Raynaud, J. P., Bonne, C., Moguilewsky, M., Lefebvre, F. A., Belanger, A., and Labrie, F. (1984). The pure anti-androgen RU 23908 (Anandron), a candidate of choice for the combined antihormonal treatment of prostatic cancer: A review. *The Prostate* **5**, 299–311.

Read, G. F., Harper, M. E., Peeling, W. B., and Griffiths, K. (1981). Changes in male salivary testosterone concentration with age. *Int. J. Androl.* **4**, 623–627.

—— Fahmy, D. R., Walker, R. F., and Griffiths, K. (Eds.) (1984). Immunoassay of Steroids in Saliva. *Proc. Ninth Tenovus Workshop*, Alpha Omega Publ, Cardiff.

Rennie, P. S., Bruchovsky, N., and Cheng, H. (1983). Isolation of 3S androgen receptors from salt-resistant fractions and nuclear matrices of prostatic nuclei after mild trypsin digestion. *J. biol Chem* **258**, 7623–7630.

Resnick, M. I., and Grayhack, J. T. (1975). Treatment of Stage IV carcinoma of the prostate. *Urol. Clin. N Amer* **2**, 141–161.

Rich, A. R. (1935). On frequency of occurrence of occult carcinoma of prostate. *J. Urol.* **33**, 215–223.

Rijnders, A. W. M., van der Korput, J. A. G. M., van Steenbrugge, G. J., Romijn, J. C., and Trapman, J. (1985). Expression of cellular oncogenes in human prostatic carcinoma cell lines. *Biochem. Biophys. Res. Commun.* **132**, 548–554.

Robel, P., Eychenne, B., Blondeau, J-P., Baulieu, E-E., and Hechter, O. (1985). Sex steroid receptors in normal and hyperplastic human prostate. *The Prostate* **6**, 255–267.

Robinson, M. R. G., and Thomas, B. S. (1971). Effect of hormonal therapy on plasma testosterone levels in prostatic carcinoma. *Br. med. J.* **4**, 391–394.

Rose, D. P., Laakso, K., Sotarauta, M., and Wynder, E. (1984). Hormone levels in prostatic fluid from healthy Finns and prostate cancer patients. *Eur. J. Cancer. clin. Oncol.* **20**, 1317–1319.

Rotkin, I. D. (1979). Epidemiological factors associated with prostatic cancer. In *Prostatic Cancer, UICC Technical Report Series* Vol 48 (Eds D. S. Coffey and J. T. Isaacs) pp. 56–80. Geneva.

Sadoughi, N., Razvi, M., Busch, I., Ablin, R., and Guinan, P. (1974). Cancer of the prostate, relief of bone pain with levodopa. *Urology* **4**, 107–108.

Sandison, A. T. (1962). An autopsy study of the adult human breast. *Natn cancer Inst. Monogr.* **8**, 1–145.

Sanford, E. J., Paulson, D. F., Rohner, T. J., Drago, J. R., Santen, R. J., and Bardin, C. W. (1977). The effects of castration on adrenal testosterone secretion in men with prostatic cancer. *J. Urol.* **118**, 1019–1021.

Schechter, A. L., Stern, D. F., Vaidyanathan, L., Decker, S. J., Drebin, J. A., Greene, M. I., and Weinberg, R. A. (1984). The *neu* oncogene: an *erb-B*-related gene encoding a 185,000-M$_r$ tumour antigen. *Nature, Lond.* **312**, 513–516.

—— Hung, M-C., Vaidyanathan, L., Weinberg, R. A., Yang-Feng, T. L., Francke, U., Ullrich, A., and Coussens, L. (1985). The *neu* gene: an erb B-homologous gene distinct from and unlinked to the gene encoding the EGF receptor. *Science, N.Y.* **229**, 976–978.

Schiebinger, R. J., Chronsos, G. P., Cutler, G. B., and Loriaux, D. L. (1986). The effect of serum prolactin on plasma adrenal androgens on the production and metabolic clearance rate of dehydroepiandorsterone sulphate in normal and hyperprolactinaemic subjects. *J. clin. Endocr. Metab.* **62**, 202–209.

Scott, R., Jr., Mutchnik, D. L., Laskowski, T. Z., and Schmalhorst, W. R. (1969). Carcinoma of the prostate in elderly men: incidence, growth characteristics and clinical significance. *J. Urol.* **101**, 602–607.

Serio, M., Forti, G., Guisti, G., Bassi, F., Giannotti, P., Calabresi, E., Manvero, F., Armato, U., Fiorelli, G., and Pinchera, A. (1980). *In vivo* and *in vitro* effects of prolactin on adrenal androgen secretion. In *Adrenal Androgens* (Eds A. R. Genazzani, J. H. H. Thijssen and P. K. Siiteri). pp. 71–81. Raven Press.

Sherr, C. J., Rettenmier, C. W., Sacca, R., Roussel, M. F., Look, A. T., and Stanley, E. R. (1985). The *c-fms* proto-oncogene product is related to the receptor for the mononuclear phagocyte growth factor, CSF-1. *Cell* **41**, 665–676.

Shimazaki, J., Kirihara, H., Yoshikazu, I., and Shida, K. (1965). Testosterone metabolism in the prostate; formation of androstan-17βol-3-one and androst-4-ene-3β,17-dione and inhibitory effects of natural and synthetic estrogens. *Gunma J. med. Sci.* **14**, 313–325.

Sibley, P. E. C., Harper, M. E., Peeling, W. B., and Griffiths, K. (1984). Growth hormone and prostatic tumours: localization using a monoclonal human growth hormone antibody. *J. Endocr.* **103**, 311–315.

Siiteri, P. K. and Wilson, J. D. (1970). Dihydrotestosterone in prostatic hypertrophy. I. The formation and content of dihydrotestosterone in the hypertrophic prostate of man. *J. clin. Invest.* **49**, 1737–1745.

—— and MacDonald, P. C. (1973). Role of extraglandular estrogen in human endocrinology. In *Handbook of Physiology, section 7. Endocrinology* Vol II (Eds S. R. Geiger, R. O. Greep, and E. B. Astwood) pp. 615–629. Williams & Wilkins, Baltimore.

Silverberg, E., and Holleb, A. I. (1975). Major trends in cancer: 25-year survey. *Cancer Journal for Clinicians* **25**, 2–8.

Simpson, H. W., Mutch, F., Halberg, F., Griffiths, K., and Wilson, D. W. (1982). Bimodal age-frequency distribution of epitheliosis in cancer mastectomies. *Cancer* **50**, 2417–2422.

Simpson, E., Harrod, J., Eilon, G., Jacobs, J. W., and Mundy, G. R. (1985). Identification of a messenger ribonucleic acid fraction in human prostatic cancer cells coding for a novel osteoblast-stimulating factor. *Endocrinology* **117**, 1615–1620.

Sirett, D. A. N., Cowan, S. K., Janecska, A. E., Grant, J. K., and Glen, E. S. (1980). Prostatic tissue distribution of 17β-hydroxy-5α-androstan-3-one and of androgen receptors on benign hyperplasia. *J. steroid Biochem.* **13**, 723–728.

Skeet, R. G. (1976). Epidemiology of urogenital tumours. In *Scientific Foundation of Urology*, Vol II (Eds D. I. Williams & G. D. Chisholm). pp. 199–211. Heinemann, London.

Smith, T., Sutherland, F., Chisholm, G. D., and Habib, F. K. (1983) Factors affecting the reproducibility of androgen receptor determinations in human prostate. *Clin. chim Acta* **131**, 129–141.

Snochowski, M., Pousette, A., Ekman, P., Bression, D., Anderson, L., Hogberg, B., and Gustafsson, J. A. (1977). Characterisation and measurement of the androgen receptor in human benign prostatic hyperplasia and prostatic carcinoma. *J. clin. Endocr. Metab.* **45**, 920–930.

Sommers, S. C. (1957). Endocrine changes with prostatic carcinoma. *Cancer (Philadelphia)* **10**, 345–358.

Sporn, M. B. and Roberts, A. B. (1985). Autocrine growth factors and cancer. *Nature, Lond.* **313**, 745–747.

Stone, A. R., Hargreave, T. B., and Chisholm, G. D. (1980). The diagnosis of oestrogen escape and the role of secondary orchidectomy in prostatic cancer. *Br. J. Urol.* **52**, 535–538.

Stiles, C. D. (1985). The biological role of oncogenes-insights from platelet-derived growth factor. *Cancer Res.* **45**, 5215–5218.

Story, M. T., Jacobs, S. C., and Lawson, R. K. (1983). Epidermal growth factor is not the major growth-promoting agent in extracts of prostatic tissue. *J. Urol.* **130**, 175–179.

—— —— —— (1984). Partial purification of a prostatic growth factor. *J. Urol.* **132**, 1212–1215.

Syms, A. J., Harper, M. E., Battersby, S., and Griffiths, K. (1982). Proliferation of human prostatic epithelial cells in culture: Aspects of identification. *J. Urol.* **127**, 361–367.

—— —— and Griffiths, K. (1985). The effect of prolactin on human BPH epithelial cell proliferation. *The Prostate* **6**, 145–153.

Szego, C. M., and Pietras, R. J. (1985). Subcellular distribution of oestrogen receptors (Matters, arising, with replies from G. L. Greene, W. J. King and J. Gorski). *Nature, Lond.* **317**, 88–89.

Terranova, V. P., Rohrbach, D. H., and Martin, G. R. (1980). Role of laminin in the attachment of PAM 212 (epithelial) cells to basement membrane collagen. *Cell* **22**, 719–726.

Thiessen, E. (1974). Concerning a familial association between breast cancer and both prostatic and uterine malignancies. *Cancer* **34**, 1102–1107.

Tisell, L-E. (1970). Effect of cortisone on the growth of the ventral prostate, the dorsolateral prostate, the coagulating glands and the seminal vesicles in castrated, adrenalectomised and castrated non-adrenalectomised rats. *Acta endocr. Copenh.* **64**, 637–655.

Tisell, L-E. (1972). Adrenal effect on the growth of the ventral and dorso-lateral prostate in castrated rats injected with oestradiol. *Acta endocr. Copenh.* **71**, 191–204.

Tolis, G., Ackman, D., Stellos, A., Mehta, A., Labrie, F., Fazekas, A. T. A., Comaru-Schally, A. M., and Schally, A. V. (1982). Tumour growth inhibition in patients with luteinizing hormone-releasing hormone agonist. *Proc. natn. Acad. Sci. U.S.A.* **79**, 1658–1662.

Trachtenberg, J. and Walsh, P. C. (1982). Correlation of prostatic nuclear androgen receptor content with duration of response and survival following hormonal therapy in advanced prostatic cancer. *J. Urol.* **127**, 466–471.

—— (1983). The treatment of metastatic prostatic cancer with a potent luteinizing hormone releasing hormone analogue. *J. Urol.* **129**, 1149–1152.

Traish, A. M., Muller, R. E., and Wotiz, H. H. (1985). A new exchange procedure for the quantitation of prostatic androgen receptor complexes formed in vivo. *J. steroid Biochem.* **23**, 405–413.

Tullner, W. W. (1963). Hormonal factors in the adrenal-dependent growth of the rat ventral prostate. *Natn. Cancer Inst. Monogr.* **12**, 211–223.

Turkes, A. O., Peeling, W. B., and Griffiths, K. (1987). Treatment of patients with advanced cancer of the prostate: phase III trial, Zoladex against castration. *J. steroid Biochem.* **26**, in press.

Turner, R. D., and Belt, E. (1957). A study of 229 consecutive cases of total perineal prostatectomies for cancer of the prostate. *J. Urol.* **77**, 82–77.

Ulrich, A., Bell, J. R., Chen, E. Y., Herrera, R., Petruzzelli, L. M., Dull, T. J., Gray, A., Coussens, L., Liao, Y-C., Tsubokawa, M., Mason, A., Seeburg, P. H., Grunfeld, C., Rosen, O. M., and Ramachandran, J. (1985). Human insulin receptor and its relationship to the tyrosine kinase family of oncogenes. *Nature, Lond.* **313**, 756–761.

Van Aubel, O. G. J. M., Bolt-de Vries, J., Blankenstein, M. A., ten Kate, F. J. W., and Schroder, F. H. (1985). Nuclear androgen receptor content in biopsy specimens from histologically normal, hyperplastic and cancerous human prostate tissue. *The Prostate* **6**, 185–194.

Varkarakis, M., Castro, J. E., and Azzopardi, J. G. (1970). Prognosis of stage 1 carcinoma of the prostate. *Proc. R. Soc. Med.* **63**, 91–93.

Varmus, H. E. (1985). Viruses, genes, and cancer. I. The discovery of cellular oncogenes and their role in neoplasia. *Cancer* **10**, 2324–2328.

Vermeulen, A. (1976). Testicular hormonal secretion and aging in males. In *Benign prostatic hyperplasia* (Eds J. T. Grayhack, J. D. Wilson, and M. J. Saherbenske) pp. 177–182. DHEW Publication No (NIH) 76–1113.

—— and Ando, S. (1978). Prolactin and adrenal androgen secretion. *Clin. Endocrinol.* **8**, 295–303.

—— Rubens, R., and Verdonck, L. (1972). Testosterone secretion and metabolism in male senescence. *J. clin. Endocr. Metab.* **34**, 730–735.

—— Suy, E., and Rubens, R. (1977). Effect of prolactin on plasma DHEA(S) levels. *J. clin. Endocr. Metab.* **44**, 1222–1225.

—— Van Camp, A., Mattelaer, J., and De Sy, W. (1979). Hormonal factors related to abnormal growth of the prostate. In *Prostate Cancer* (Eds D. S. Coffey & J. T. Issacs). UICC Technical Workshop Series, Vol 48. pp. 81–92. UICC, Geneva.

Viola, M. V., Fromowitz, F., Oravez, S., Deb, S., Finkel, G., Lundy, J., Hand, P., Thor, A., and Schlom, J. (1986). Expression of *ras* oncogene p21 in prostate cancer. *New Engl. J. Med.* **314**, 133–137.

Vlodavsky, I., and Gospodarowicz, D. (1981). Respective involvement of laminin and fibronectin in the adhesion of human carcinoma and sarcoma cells. *Nature, Lond.* **289**, 304–306.

Voigt, K. D., Horst, H. J., and Krieg, M. (1975). Androgen metabolism in patients with benign prostatic hypertrophy. *Vitams Horm.* **33**, 417–436.

—— and Krieg, M. (1978). Biochemical endocrinology of prostatic tumours. In 'Current Topics in Experimental Endocrinology, Vol III' (Eds L. Martini and V. H. T. James) pp. 177–184. Academic Press, N.Y.

Walker, K. J., Nicholson, R. I., Turkes, A. O., Turkes, A., Griffiths, K., Robinson, M., Crispin, Z., and Dris, S. (1983). Therapeutic potential of the LHRH agonist, ICI 118630 in the treatment of advanced prostatic carcinoma. *Lancet* ii, 413–416.

—— Turkes, A. O., Turkes, A., Zwink, R., Beacock, C., Buck, A. C., Peeling, W. B., and Griffiths, K. (1984). Treatment of patients with advanced cancer of the prostate using a slow-release (depot) formulation of the LHRH agonist ICI 118630 (Zoladex). *J. Endocr.* **103**: R1-R4.

Walsh, P. C., and Gittes, R. F. (1970). Inhibition of extratesticular stimuli to prostatic growth in the castrate rat by antiandrogens. *Endocrinology* **87**, 624–627.

—— and Wilson, J. D. (1976). The induction of prostatic hypertrophy in the dog with androstanediol. *J. clin. Invest.* **57** 1093–1097.

Waterfield, M. D., Scrace, G. T., Whittle, N., Stroobant, P., Johnsson, A., Wasteson, A., Westermark, B., Heldin, C. H., Huang, J. S., and Deuel, T. F. (1983). Platelet-derived growth factor is structurally related to the putative transforming protein p28 *sis* of simian sarcoma virus. *Nature, Lond.* **304**, 35–39.

Weinberger, C., Hollenberg, S. M., Rosenfeld, M. G., and Evans, R. M. (1985). Domain structure of human glucocorticoid receptor and its relationship to the v-erb A oncogene product. *Nature, Lond.* **318**, 670–672.

Weinstein, R. L., Kelch, R. P., Jenner, M. R., Kaplan, S. L., and Grumbach, M. M. (1974). Secretion of unconjugated androgens and estrogens by the normal and abnormal human testis before and after human chorionic gonadotrophin. J. clin. Invest. **53**, 1–6.

Wellings, S. R., Jensen, H. M., and Marcum, R. G. (1975). Atlas of subgross pathology of the human breast with special reference to possible precancerous lesions. *J. natn. Cancer Inst.* **55**, 231–273.

—— —— and de Vault, M. R. (1976). Persistent and atypical lobules in the human breast may be precancerous. *Experientia* **32**, 1463–1465.

Welshons, W. V., Liebermann, M. E., and Gorski, J. (1984). Nuclear localization of unoccupied oestrogen receptors. *Nature, Lond.* **307**, 747–749.

White, J. W. (1895). The results of double castration in hypertrophy of the prostate. *Ann. Surg.* **22**, 1–80.

Whitmore, W. F. (1973). The natural history of prostatic cancer. *Cancer* **32**, 1104–1112.

Wilson, J. M., Kemp, L. W., and Stein, G. J. (1984). Cancer of the prostate. Do younger men have a poorer survival rate? *Br. J. Urol.* **56**, 391–396.

Wilson, D. W., Harper, M. E., Jensen, H., Richards, G., Peeling, W. B., Pierrpoint, C. G., and Griffiths, K. (1985). A prognostic index for the clinical management of patients with advanced prostatic cancer. *The Prostate* **7**, 131–141.

Witorsch, R. J. (1981). Visualization of prolactin binding sites in prostate tissue. In *The prostatic cell: structure and function*, part B. (Eds G. P. Murphy, A. A. Sandberg & J. P. Karr). pp. 89–113. Alan R. Liss, New York.

Woolf, C. M. (1960). An investigation of the familial aspects of carcinoma of the prostate. *Cancer* **13**, 739–744.

World Health Organization (1980). Histological Typing of Prostate Tumours. *International Histological Classification of Tumours No. 22*. Geneva: World Health Organization.

Wynder, E. L., Mabuchi, K., and Whitmore, W. F. (1971). Epidemiology of cancer of the prostate. *Cancer (Philadelphia)* **28**, 344–360.

Yamada, K. M. and Olden, K. (1978). Fibronectin—adhesive glycoproteins of cell surface and blood. *Nature, Lond.* **275**, 179–184.

Yamamoto, T., Ikawa, S., Akiyama, T., Semba, K., Nomura, N., Miyajima, N., Saito, T., and Togoshima, K. (1986). Similarity of protein encoded by the human c-erb-B-2 gene to epidermal growth factor receptor. *Nature, Lond.* **230**, 231–234.

Yamanka, H., Kirdani, R. Y., Saroff, J., Murphy, G. P., and Sandberg, A. A. (1977). Hormonal control of zinc uptake and binding in the rat dorsolateral prostate. *Invest. Urol.* **14**, 492–495.

Zondek, B. and Aschheim, S. (1927). Das Hormon des Hypophysenvorderlappens. *Klin-Wochenschr.* **6**, 348–352.

Zuckerman, S. and Groome, J. R. (1937). The aetiology of benign enlargement of the prostate in the dog. *J. Pathol. Bact.* **44**, 113–124.

Zumoff, B., Levin, J., Strain, G. W., Rosenfeld, R. S., O'Connor, J., Freed, S. Z., Kream, J., Whitmore, W. S., Fukushima, D. K., and Hellman, L. (1982). Abnormal levels of plasma hormones in men with prostate cancer: evidence toward a 'time-disease' theory. *The Prostate* **3**, 579–588.

7　Endogenous opioid peptides and human reproduction

TREVOR A. HOWLETT AND LESLEY H. REES

I　Introduction

Since mankind has used and abused opiate alkaloids for many centuries, opiate effects on human reproductive function were recognized long before the isolation, over 10 years ago, of endogenous peptides with opioid activity (Hughes, Smith, Kosterlitz, Fothergill, Morgan, and Morris 1975). Since this discovery however research has expanded greatly and a vast body of literature now documents the physiology and pathology of opioid modula-

Table 7.1
Major endogenous opioid peptides: Amino acid sequences

Pro enkephalin derived

Met Enkephalin	Tyr-Gly-Gly-Phe-Met$_5$
Leu Enkephalin	Tyr-Gly-Gly-Phe-Leu$_5$
Extended ⎫	⎰ Tyr-Gly-Gly-Phe-Met-Arg-Phe$_7$
met-enkephalins ⎭	⎱ Tyr-Gly-Gly-Phe-Met-Arg-Gly-Leu$_8$
Peptide E	Tyr-Gly-Gly-Phe-Met-Arg-Arg...............
	.... Tyr-Gly-Gly-Phe-Leu$_{25}$

Pro dynorphin derived

Leu-enkephalin	Tyr-Gly-Gly-Phe-Leu$_5$
Dynorphin	Tyr-Gly-Gly-Phe-Leu-Arg-Arg-Ile$_8$-Arg-Pro-Lys-Leu-Lys-Trp-Asp-Asn-Gln$_{17}$
Rimorphin	Tyr-Gly-Gly-Phe-Leu-Arg-Arg-Gln-Phe-Lys-Val-Val-Thr$_{13}$
α-Neo Endorphin	Tyr-Gly-Gly-Phe-Leu-Arg-Lys-Tyr-Pro-Lys$_{10}$
β-Neo Endorphin	Tyr-Gly-Gly-Phe-Leu-Arg-Lys-Tyr-Pro$_9$

Pro opiomelancortin derived

β-Endorphin	Tyr-Gly-Gly-Phe-Met .... Gln$_{31}$

tion of hormonal and other aspects of reproduction. This review will concentrate on the human data but will refer to animal studies if important differences occur or if human data are unavailable. We will, by convention (Hughes and Kosterlitz 1983), use the term *opiate* to refer to plant-derived or synthetic alkaloids such as morphine and *opioid* to refer to endogenous peptides. Alternative generic terms such as 'endorphins' or 'enkephalins' have been used but may cause confusion since they also refer to specific peptide sequences.

II   Opioid peptides, precursors and receptors

The structures of the major opioid peptides are shown in Table 7.1. Met-enkephalin and leu-enkephalin were the original opioids isolated from porcine brain (Hughes *et al.* 1975); they share the opioid-active N-terminal sequence Tyr-Gly-Gly-Phe and differ only in their C-terminal residue methionine and leucine respectively. All opioid peptides since discovered share the sequence of either met- or leu-enkephalin at their N-terminal but possess C-terminal extensions of variable length which may confer greater potency (Goldstein, Tachibana, Lowney, Hunkapiller, and Hood 1980), increased stability against degradation (Corbett, Paterson, McKnight, Magnan, and Kosterlitz 1982) and different receptor specificity (see below). Analysis of the DNA sequence in recent years has allowed the prediction of the amino-acid sequence of 3 precursor molecules (Rossier 1982) (Fig. 7.1).

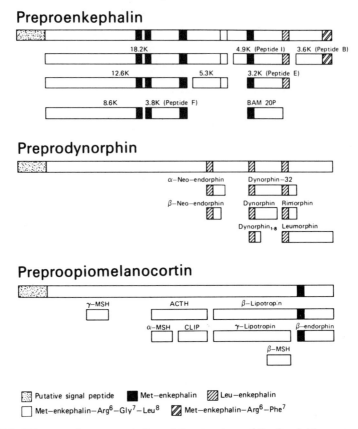

Fig. 7.1. Diagramatic representation of the structure of the 3 opioid precursors and related opioids.

Pro-opiomelanocortin (POMC) is the precursor for ACTH, LPH and also β-endorphin but does not appear to be further processed to met-enkephalin. Pro-enkephalin is the major biological source of met-enkephalin and its extended forms (met-enkephalin-Arg-Phe, met-enkephalin Arg-Gly-Leu and Peptide E) and one source of leu-enkephalin. Pro-dynorphin (pro-enkephalin B) contains 3 copies of leu-enkephalin and its extended forms dynorphin, rimorphin (dynorphin B), leumorphin and α-neo-endorphin. The three precursors are of similar molecular size and also contain a number of sequence homologies, suggesting that all may have originally derived from a primordial opioid gene.

A variety of types of opioid receptors have been described. Behavioural studies in dogs suggested that opiates could be divided into μ-, κ- or σ-agonists (Martin, Eades, Thompson, Huppler and Gilbert 1976) and studies with opioid peptides using bioactivity in peripheral tissue preparations (Wuster, Schulz, and Herz 1981) or binding studies of radioactively labelled ligands in brain tissue (Herz 1984) confirmed the existence of μ, δ, κ and possibly ε-receptors. The μ-receptor is the original morphine receptor and is highly sensitive to the opiate antagonist naloxone; β-endorphin is probably the most active endogenous ligand (although this is also active at other receptors). The enkephalins interact with the δ-receptor and pro-dynorphin-derived peptides (other than leu-enkephalin) are largely selective for κ-receptors. Both δ and κ receptors are relatively resistant to naloxone (by about an order of magnitude) (Paterson, Robson, and Kosterlitz 1983) but are blocked *in vivo* by high doses; the κ-receptor is probably the more sensitive. Finally specific ε-receptors for β-endorphin have been characterized pharmacologically (Wuster *et al.* 1981). The σ-receptor is not now considered to be an opioid receptor but may have its own endogenous ligand.

Since receptor-specific opioid antagonists are not yet available the receptor or opioid subtype involved in a given physiological process is inferred, in man, from the dose of naloxone required to block (or produce) that response. Low doses of naloxone (under 1 mg total dose) suggests μ- or ε-receptor involvement (and thus β-endorphin as the opioid agonist involved) and higher doses (up to 10 or 20 mg) suggests δ- or κ-receptors (and thus pro-enkephalin or pro-dynorphin-derived opioids). Some caution is necessary however since at very high concentration naloxone may exhibit non-opioid actions (Sawynok, Pinsky, and Labella 1979).

III Localization of opioid peptides and receptors

Opioids are localized throughout the central nervous system in specific neurone tracts and, as expected, appear to be separated on the basis of the known precursors (Cuello 1983). It is of particular interest for this review that all opioids are concentrated in the hypothalamus, pituitary, peri-aqueductal gray matter (PAG) and spinal cord. Hypothalamic β-endorphin-ergic neurones originate largely in the arcuate nucleus from where they project to the median eminence as well as other parts of the hypothalamus and brain. Enkephalinergic neurones exist in several hypothalamic nuclei whilst pro-dynorphin derived peptides are mostly concentrated in the supraoptic and paraventricular nuclei, although some are also present in the arcuate nucleus and posterior hypothalamus (Watson, Khachaturian, Coy, Taylor, and Akil 1982). μ- and κ-receptors are present in hypothalamus in roughly equal numbers but δ-receptors account for less than 10 per cent of opioid binding (Herz 1984).

The pituitary contains large amounts of β-endorphin in the anterior pituitary corticotroph cells and pro-dynorphin derived opioids in the posterior pituitary vasopressin neurones (Watson *et al.* 1982). Met-en-kephalin is present in both anterior pituitary somatotrophs (Weber, Voigt, and Martin 1978) and posterior pituitary oxytocin neurones (Martin and

Voigt 1981). The role of PAG and spinal opioids in pain perception is considered below.

Outside the CNS opioids are concentrated in the adrenal medulla and other parts of the sympathetic nervous system, and in neurones in the gut. Smaller quantitites are also described in the testis (see below) and pancreas (Bruni, Watkins, and Yen 1979). β-endorphin (McLoughlin, Lowry, Ratter, Besser, and Rees 1980) and met-enkephalin (Clement-Jones, Lowry, Rees, and Besser 1980a, b) both circulate in human plasma and their role is considered in greater detail below.

IV   Opioids and neuroendocrine regulation

Opiates and opioids have been demonstrated to have effects on every pituitary hormone, both in animals (Meites, Bruni, VanVugt, and Smith 1979) and in man (Grossman 1983; Morley 1981). Only effects which have relevance to human reproduction will be considered here.

1   PROLACTIN

Administration of both morphine and a variety of other opiate alkaloids (Tolis, Hickey, and Guyda 1975; Delitala, Grossman, and Besser 1983), the enkephalin analogue DAMME (FK33–824) (Stubbs, Delitala, Jones, Jeffcoate, Edwards, Ratter, Besser, Bloom, and Alberti 1978; VonGraf-fenreid, del Pozo, Roubicek, Krebs, Poldinger, Burmeister, and Kerp 1978; Demura, Suda, Wakabayashi, Yoshimura, Jibiki, Odagiri, Demura, and Shizume 1981; Grossman, Stubbs, Gaillard, Delitala, Rees, and Besser 1981) and β-endorphin (Reid, Hoff, Yen, and Li 1981; Reid, Quigley, and Yen 1983) all cause prompt release of prolactin in man. In the rat β-endorphin is a more potent secretagogue than met-enkephalin (DuPont, Barden, Cusan, Merand, Labrie, and Vaudry 1980) or dynorphin (VanVugt, Sylvester, Aylsworth, and Meites 1981) and intra-ventricular administration of antisera for β-endorphin lowers both basal and stress-induced prolactin secretion (Ragavan and Frantz 1981), sugges-ting that β-endorphin is indeed the opioid involved. This is compatible with the naloxone-sensitivity of opioid stimulation in man (Stubbs *et al.* 1978) which suggests involvement of μ- or ε-selective opioids. The opioid effect is blocked by administration of dopamine agonists and potentiated by dopamine antagonists (Delitala, Grossman, and Besser 1981) suggesting that opioids act by inhibition of release of dopamine, the major prolactin inhibitory factor, from the median eminence. This concept is supported by direct experimental evidence in the rat where opioids decrease hypothal-amic dopamine turnover (Van Loon, Ho, and Kim 1980) and release (Gudelsky and Porter 1979; Haskins, Gudelsky, Moss, and Porter 1981; Wilkes and Yen 1980).

Most studies have found no direct effect of opiates or opioid peptides on prolactin secretion by isolated pituitary cells *in vitro*. Recently however it has been suggested that two more recently discovered opioids, α neo-endorphin (Matsushita, Kato, Shimatsu, Katakami, Fujino, Matsuo, and Imura 1982) and leumorphin (Tojo, Kato, Onta, Matsushita, Shimatsu, Kabayama, Inoue, Yanaihara, and Imura 1985) may have a direct stimulatory effect on pituitary cells.

In the rat naloxone lowers basal, stress-induced and suckling-induced prolactin secretion, suggesting an important physiological role for opioids in prolactin secretion (Bruni, VanVugt, Marshall, and Meites 1977; Shaar, Frederickson, Dininger, and Jackson 1977; VanVugt, Bruni, and Meites 1978). In man, in contrast, most studies have failed to demonstrate an effect of naloxone, even in high dosage, on basal or stess-induced release of prolactin in normal subjects (Martin, Tolis, Wood, and Guyda 1979; Morley, Baranetsky, Wingert, Carlson, Hershman, Melmed, Levin, Jamison, Weitzman, Chang, and Varner 1980; Quigley, Sheehan, Casper and Yen 1980*a*; Spiler, and Molitch 1980; Wakabayashi, Demura, Miki, Ohmura, Miyoshi, and Shizume 1980; Grossman *et al.* 1981), or on elevated prolactin levels in the puerperium (Martin *et al.* 1979; Grossman, West, Williams, Evans, Rees, and Besser 1982), or in patients with prolactinomas (Blankstein, Reyes, Winter, and Faiman 1979; Quigley *et al.* 1980; Tolis, Jukier, Wiesen, and Krieger 1982). A minority of studies has however reported inhibition of basal or stess-related prolactin release (Rubin, Swezey, and Blascke 1979; Pontiroli, Baio, Stella, Crescenti, and Girardi 1982; Saltiel, Passa, Kuhn, Fiet, and Canivet 1982). High doses of naloxone (15 mg) may abolish the exercise-induced release of prolactin in highly trained male athletes (Fig. 7.2) (Moretti, Fabbri, Gnessi, Cappa, Calzolari, Fraioli, Grossman, and Besser 1983), but other workers, generally using different exercise regimes and untrained subjects, have found no change or even a slight increase in prolactin secretion (Mayer, Wessel, and Koberling 1980; Spiler and Molitch 1980; Grossman, Bouloux, Price, Drury, Lam, Turner, Thomas, Besser, and Sutton 1984).

Recent work has suggested another, inhibitory opioid control of prolactin release. Thus a naloxone infusion (1.6 mg per h) may stimulate pulsatile prolactin release in women during the late follicular and mid luteal phases of the menstrual cycle (Cetel, Quigley, and Yen 1985) and in women on the oral contraceptive (Casper, Bhanot, and Wilkinson 1984), but not in early follicular or late luteal phase, or in hypogonadal women. This pulsatile release of prolactin is synchronous with that of LH (Fig. 7.3) suggesting that a common mechanism, presumably involving LHRH, mediates the release of both LH and prolactin under these circumstances. Indeed LHRH may stimulate prolactin release at least in postmenopausal women (Casper and Yen 1981) and *in vitro* evidence suggests that this might involve paracrine gonadotroph-lactotroph interactions (Denef and Andries 1983).

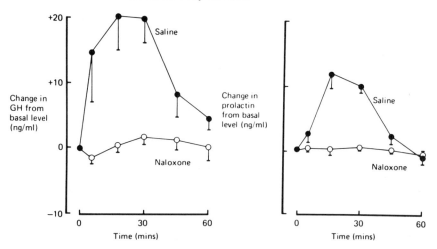

Fig. 7.2. Prolactin and growth hormone response to severe exercise in highly trained male athletes on a control day (●–●) compared to pretreatment with naloxone (○–○). Both responses are inhibited by naloxone. (From Moretti *et al.* 1983, with permission).

PRL RESPONSE TO NALOXONE

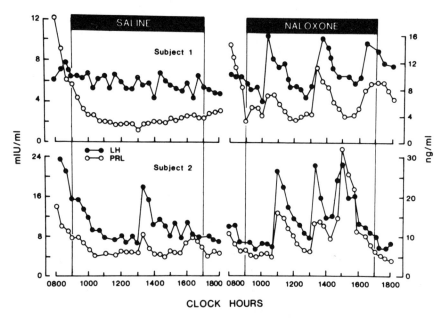

Fig. 7.3. Increase in the pulsatility of prolactin secretion during a naloxone infusion. Prolactin pulses appear to coincide with those of LH. (From Cetel *et al.* 1985, with permission).

Overall however, in spite of the existence of stimulatory and possibly inhibitory opioidergic mechanisms for control of prolactin secretion, no clear physiological or pathological role has emerged.

2   GONADOTROPHINS

In contrast to prolactin the predominent opioid modulation of gonado-trophin secretion is inhibitory. Chronic opiate addiction has been long recognized to cause hypogonadism: loss of libido and impotence associated with low serum testosterone in the male and amenorrhoea and infertility in the female (Morley 1981). Morphine and other opiates decrease serum LH (Delitala, Grossman, and Besser 1983), as do a met-enkephalin analogogue (Stubbs *et al.* 1978; Von Grafftenreid *et al.* 1978) and β-endorphin (Reid *et al.* 1981a); inhibition of FSH is less consistently noted, perhaps due to its longer half life in the circulation.

Naloxone acutely increases serum LH, and possibly FSH, in both sexes, indicating tonic inhibitory opioid control of gonadotrophin secretion (Morley *et al.* 1980; Quigley and Yen 1980; Grossman, Moult, Gaillard, Delitala, Toft, Rees, and Besser 1981; Lightman, Jacobs, Maguire, McGarrick, and Jeffcoate 1981; Moult, Grossman, Evans, Rees, and Besser 1981; Ropert, Quigley, and Yen 1981; Ellingboe, Veldhuis, Mendelson, Kuehnle, Mello, and Holbrook 1982). Studies of LH pulsatility indicate an increase in both frequency and amplitude of LH pulses with naloxone in both sexes (Quigley and Yen 1980; Ellingboe *et al.* 1981; Grossman *et al.* 1981; Moult *et al.* 1981; Ropert *et al.* 1981). In females this effect is most marked in the late follicular and particularly mid-luteal phase of the menstrual cycle and opioids may thus mediate the slowing of LH pulsatility seen at these times. In support of this concept in experimental animals, both hypothalamic (Hulse, Coleman, Copolou, and Clements 1984) and hypophyseal portal blood (Wehrenberg, Wardlaw, Frantz, and Ferin 1982) concentrations of β-endorphin are maximal after ovulation.

Relatively low doses of naloxone are required to modulate LH release, suggesting primarily involvement of μ- or ε-receptors and thus β-endorphin (Grossman, Moult, Cunnah, and Besser 1986). However in the immature rat intrahypothalamic injections of antisera to both β-endorphin and dynorphin (but not met-enkephalin) raise LH (Schulz, Wilhelm, Pirke, Gramsch, and Herz 1981), suggesting that more than one opioidergic pathway might be involved. Opioid effects occur at hypothalamic level and involve modulation of LHRH release. Thus opioids have no effect on the LH response to the LHRH test (Grossman *et al.* 1981), LHRH antagonists block naloxone stimulation in the rat (Blank and Roberts 1982), naloxone stimulates LHRH release from both human (Rasmussen, Liu, Wolf, and Yen 1983) and rat (Drouva, Epelbaum, Tapia-Arachcibia, Laplante, and Kordon 1981) hypothalami *in vitro* and morphine decreases hypothalamic

LHRH content (Kalra and Simpkins 1981). In the rat there is ample evidence that opioids act via stimulatory adrenergic pathways (Kalra and Simpkins 1981; VanVugt, Aylsworth, Sylvester, Leung, and Meites 1981; Kalra and Crowley 1982; Dyer, Mansfield, Corbet, and Dean 1985). However in man neither adrenergic (Veldhuis, Rogol, Williams, and Johnson 1983) nor dopaminergic (Delitala et al. 1981) mechanisms appear to be involved.

In man, as in the rat, opioidergic mechanism appear to be closely connected with the feedback of gonadal steroids on LHRH secretion. Thus, whilst naloxone responsiveness has been reported in postmenopausal women (Lightman et al. 1981), most studies agree that naloxone is unable to elevate further serum LH levels in the postmenopausal or oophorectomized female, and that naloxone sensitivity is restored after oestrogen and/or progesterone replacement (Reid, Quigley, and Yen 1983; Melis, Paoletti, Gambacciani, Mais, and Fioreti 1984; Casper and Alapin-Rubillovitz 1985; Shoupe, Montz, and Lobo 1985). Furthermore, the time of maximum naloxone sensitivity in the normal menstrual cycle is in the luteal phase when endogenous oestrogen and progesterone are at their higher levels (Quigley and Yen 1980), and in the ovariectomized monkey hypophyseal portal plasma concentrations of β-endorphin are low but rise after gonadal steroid replacement (Wardlaw, Wehrenberg, Ferin, Antunes, and Frantz 1982). Gonadal steroid feedback on LHRH may thus be mediated via opioid pathways but this is by no means certain since evidence as to whether opioids can inhibit LH secretion in postmenopausal women is contradictory (Grossman et al. 1981; Reid et al. 1983). It is therefore possible, as an alternative, that the presence of gonadal steroids might be necessary for the activation of a separate opioid pathway which inhibits LH release.

Changes in hypothalamic opioid activity have been implicated in a number of pathological conditions in which gonadotrophin secretion is reduced. Patients with oligomenorrhoea or amenorrhoea secondary to hyperprolactinaemia have normal mean gonadotrophin levels but only infrequent LH pulses of large amplitude (Moult, Rees, and Besser 1982). Infusion of naloxone in such patients promptly restores normal LH pulsatility, at least in the short term (Fig. 7.4) (Quigley et al. 1980; Lightman et al. 1981; Grossman, Moult, McIntyre, Evans, Silverstone, Rees, and Besser 1982), without alteration in the serum prolactin. This suggests that prolactin inhibits LHRH secretion via opioidergic mechanisms. Similar LH responses to naloxone have been reported in patients with so-called hypothalamic amenorrhoea (Quigley, Sheenan, Casper and Yen 1980b) and in the amenorrhoeic athletes (McArthur, Bullen, Betlins, Pagane, Badger, and Klibanski 1980). The therapeutic use of orally active opiate antagonists in such conditions remains an exciting future possibility.

In contrast to the rat, where fasting impairs LH secretion by activating an inhibitory opioid pathway (Dyer et al. 1985), the majority of patients with anorexia nervosa show no gonadotrophin response to naloxone (Grossman et

Serum LH (●) and FSH (○) in a patient with hyperprolactinaemia
(A) and one with Anorexia Nervosa (B) infused with high–dose
Naloxone

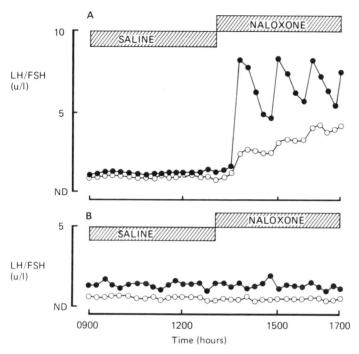

Fig. 7.4. Restoration of LH pulsatility by naloxone infusion in a patient with hyperprolactinaemia. In contrast no response is seen in anorexia nervosa. (From Grossman *et al.* 1982, with permission).

al. 1982; Baranowska, Rozbicka, Jeske, and Abdel-Fattah 1984), and those patients who do respond may have another pre-existing cause for amenorrhoea (Baranawska *et al.* 1984). Similarly patients with Kallman's syndrome and idiopathic hypopituitarism show no response to naloxone (Veldhuis, Kulin, Warner, and Santner 1982). The absence of response of LH to naloxone has been reported in polycystic ovary syndrome, and thus by implication a pathophysiological role for decreased hypothalamic opioid activity in the LH hypersecretion of this condition (Cumming, Reid, Quigley, Rebar, and Yen 1984). Other workers have however reported normal responses (Lightman *et al.* 1981). Finally it has been suggested that opioids might play a role in the pre-menstrual syndrome (Reid 1983).

Changes in the activity of hypothalamic inhibitory opioid pathways are unlikely to initiate puberty in man since responsiveness to naloxone is absent before puberty and does not develop until near maturity (Veldhuis, Kulin, Warner, and Satner 1982; Fraioli, Cappa, Fabbri, Gnesi, Moretti,

Borrelli, and Isidori 1984). In the rat in contrast removal of opioid inhibition may initiate puberty (Bhanot and Wilkinson 1983).

β-endorphin is present in the testis (see below), and in the rat intratesticular injection of naloxone has been shown to lower testosterone secretion, implying a local modulating role (Gerendai, Shaha, Thau, and Bardin 1984). Whether such a mechanism exists in man is unknown, but certainly the overall effect of 24 h intravenous infusion of naloxone is to raise serum testosterone via LH secretion (Delitala, Giusti, Mazzochi, Granzierra, Tarditi, and Giordano 1983), so that a major physiological role seems unlikely.

3   OXYTOCIN

For obvious ethical reasons no data exist as to the role of opioids in the control of oxytocin secretion in man. In the rat, however, opioids appear to inhibit oxytocin release *in vivo* in response to suckling and other stimuli (Clarke, Wood, Merrick, and Lincoln 1979; Holdar, Hoffman, and Zimmerman 1982; Van Wimersma Greidanus and Ten Haaf 1984; Bicknell 1985), and naloxone potentiates secretion of oxytocin from electrically stimulated rat neurohypophysis *in vitro* (Bicknell and Leng 1982). Met-enkephalin is colocalized with oxytocin in the magnocellular neurones projecting to the posterior pituitary (Martin and Voigt 1981) and may thus be coreleased into plasma; opioid inhibition of oxytocin release might thus represent ultra-short loop feedback by this met-enkephalin at pituitary level, particularly in the *in vitro* investigations (Bicknell and Leng 1982). Recent studies in the rat have suggested that central inhibitory opioid control of oxytocin secretion may regulate 'spacing' of successive births during parturition (Leng, Mansfield, Bicknell, Dean, Ingram, Marsh, Yates, and Dyer 1985) and mediate the interruption of parturition by environmental stresses (Leng, Bicknell, Mansfield, and Dyer 1986).

If similar inhibitory opioid control occurs in man then clearly administration of opiate drugs for pain relief during labour might inhibit release of oxytocin. This possibility has however not been explored.

4   OTHER ENDOCRINE EFFECTS

Opioids are able to modulate release of most other anterior and posterior pituitary hormones (Morley 1981; Grossman 1983), but since they are of little relevance to reproduction these will be considered only briefly.

ACTH secretion is tonically inhibited by an opioid pathway. This inhibition is independent of the circadian rhythm (Grossman, Gaillard, McCartney, Rees, and Besser 1982) and naloxone-insensitive δ- or κ-receptors are involved (Grossman *et al.* 1986). Opioids probably inhibit release of hypothalamic corticotrophin-releasing factor via an intervening noradrenergic pathway (Grossman and Besser 1982).

Opioids stimulate release of growth hormone and TSH, and inhibit release of vasopressin. It is uncertain however if this is of any physiological significance.

Opioids inhibit adrenal medullary catecholamine release in response to a variety of stressful stimuli (Grossman and Bouloux 1984; Bouloux, Grossman, Lytras, and Besser 1985), although it is as yet unclear whether this involves a central or peripheral mechanism. Opioids may also modulate pancreatic insulin secretion although results are conflicting since naloxone (Morley *et al*. 1980) and β-endorphin (Reid and Yen 1981) are both reported to stimulate insulin release *in vivo*; *in vitro* however, opioids clearly stimulate insulin secretion (Green, Perrin, Penman, Yaseen, Ray, and Havell 1983).

V   Opioids in plasma

1   MEASUREMENT OF OPIOID PEPTIDES IN HUMAN PLASMA

Changes in levels of endogenous opioid peptides in human plasma and CSF have been examined under a variety of physiological and pathological circumstances. Before considering these studies in detail, however, a brief note on the methodology of these measurements is necessary, since results obtained need to be interpreted in the light of the methods used.

Characterization and measurement of opioid peptides in plasma are subject to a number of important methodological problems.

i   Lability

The half life of opioid peptides in plasma is very short, often of the order of a few minutes (Reid *et al*. 1981*a*; Clements-Jones *et al*. 1980*a*; Howlett, Walker, Besser, and Rees 1984). Great care therefore needs to be taken in the collection of samples for RIA, and they must generally be rapidly centrifuged at 4°C and then frozen, often in the presence of enzyme inhibitors or after acidification (Clement-Jones *et al*. 1980*b*), in order to avoid loss of peptide. Furthermore the transfer of plasma peptidases to the RIA incubation mixture may result in enzymatic degradation of both standard peptides and the radioactive tracer peptide, thereby causing an apparent immunoreactivity which is an artefact, since the degraded tracer may no longer bind to the antiserum, thus giving the impression of displacement by larger amounts of unlabelled peptide. This effect is a particular problem in assays of unextracted plasma, and is part of the rationale for extraction of any labile peptide from plasma, using a variety of techniques, prior to RIA, This type of enzymatic artefact can however also occur in extracted RIA's (Howlett *et al*. 1984).

ii   Sensitivity

Basal levels of opioids in the plasma are low, generally in the picomolar range. Hence bioassays of opioid activity or radioreceptor assays measuring binding to opioid receptors, which might otherwise yield the most useful information about the biological activity of opioids in the plasma, cannot be employed. Even using sensitive radioimmunoassays basal levels of plasma opioids are often close to the limits of detection so that caution is necessary in interpretation of reported small changes within the normal range. Extraction of opioids from plasma prior to assay can provide a concentrating step and is a further reason for the use of such techniques.

iii   Specificity

All opioid peptides share at least the opioid active sequence Tyr-Gly-Gly-Phe at their N-terminal and many show much more extensive homologies. Specificity is therefore a major problem in the assay of these peptides and a variety of techniques has been devised to overcome this problem: Met- and leu-enkephalin differ in structure by only their C-terminal residue and thus many assays showed considerable cross reactivity between the 2 peptides. This problem was however overcome by the development of assays to oxidized met-enkephalin, with the subsequent oxidation of all samples prior to assay (Clement-Jones *et al.* 1980*a*, *b*). With the discovery of the other major families of opioid peptides and their precursors, a bewildering array of opioids, fragments and larger peptides have been described, all of which may cross react to different extents in various assays, whilst having different receptor specificity and therefore presumably different biological functions. Thus, for example, the opioid dynorphin has been reported to occur in various tissues in a large number of molecular forms varying widely in length from 8 to 32 residues (Fischli, Goldstein, Hunkapiller, and Hood, 1982; Lemaire, Chouinard, Denis, and Morris 1982; Weber, Evans, and Barchas 1982; Suda, Tozawa, Tachibana, Demura, Shizume, Sasaki, Mouri, and Miura 1983; Howlett, Patience, Besser, and Rees 1986). Clearly such closely related peptides cannot be distinguished by simple radioimmuno-assay and ultimately some chromatographic technique, in particular high pressure liquid chromatography (HPLC), must be used in order to characterize precisely opioid peptides and their physiological changes in plasma and elsewhere.

β-endorphin illustrates additional problems of specificity since its entire sequence is contained at the C-terminal of its immediate precursor β lipotrophin (β-LPH) which does not act as an opiate. Most assays for β-endorphin thus show considerable, often 100 per cent, crossreactivity with β-LPH, necessitating an initial chromatographic step, such as gel-filtration, prior to assay if only β-endorphin is to be measured (Nakao, Nakai, Oki,

Horii, and Imura 1978; Wardlaw and Frantz 1979; McLoughlin, Lowry, Ratter, Besser, and Rees 1980). Furthermore a high proportion of pituitary β-endorphin is N-acetylated, rendering the peptide inactive at opioid receptors but still detectable in most radioimmunoassays (Smyth, Massey, Zakarian, and Finnie 1979), and this can only be distinguished by techniques such as HPLC.

Simple measurements of immunoreactive opioids in plasma thus cannot be taken to indicate the presence of a particular molecular form, or indeed even of an opiate-active peptide, unless combined with more sophisticated chromatographic techniques. Unfortunately in many studies, including some mentioned below, this has not been achieved.

iv Precursors

Several studies have suggested that large molecular weight opioid precursors may be secreted into plasma, often in higher concentrations than the smaller peptides such as met-enkephalin (Smith, Grossman, Gaillard, Clement-Jones, Ratter, Mallinson, Lowry, Besser, and Rees 1981; Evans, Medbak, Hinds, Tomlin, Varley, and Rees 1984). Such precursors, being inactive at opioid receptors but less labile in the circulation, clearly might fulfil a hormonal role if on arrival at a distant site they were further processed to form active opioid peptides. These precursors are, however, generally not detected in RIA unless the assays are combined with chromatography and with controlled tryptic digestion to release smaller immunoreactive peptides.

v Artefact

As mentioned above enzymatic artefact has in the past resulted in erroneous identification of immunoreactive opioids in plasma and elsewhere (Howlett *et al*. 1984). Other less obvious problems can however also cause artefacts mimicking changes in opioid levels. For example the presence of contrast medium in CSF, used in many experimental studies, has recently been shown to increase apparent levels of immunoreactive β-endorphin by over 100 per cent via an alteration of assay binding (Fang, Fessler, Rachlin, and Brown 1984).

In spite of these problems a number of opioid peptides have been clearly characterized in human plasma.

2 β-ENDORPHIN

β-endorphin is cosecreted with ACTH from the corticotrophs of the anterior pituitary (Guillemin, Vargo, Rossier, Minick, Ling, Rivier, Vale, and Bloom 1977) and circulates in human plasma in concentrations comparable to ACTH (Nakao *et al*. 1978; Wardlaw and Frantz, 1979; McLoughlin *et al*.

1980). Levels of plasma β-endorphin change in parallel with those of ACTH under most circumstances studied: thus β-endorphin has the same circadian rhythm as ACTH (Dent, Guillemault, Albert, Posner, Cox, and Goldstein 1981; Shanks, Clement-Jones, Linsell, Mullen, Rees, and Besser 1981), increases, as does ACTH, after insulin-induced hypoglycaemia (Nakao, Nakai, Jingami, Oki, Fukuata, and Imura 1979; Smith *et al.* 1981), or following the injection of corticotrophin-releasing factor, CRF (McLoughlin, Tomlin, Grossman, Lytras, Coy, Besser, and Rees 1984), and is suppressed by administration of dexamethasone (Smith *et al.* 1981). β-endorphin levels are elevated in states of ACTH hypersecretion, such as Cushing's disease, ectopic ACTH production or Addison's disease (Nakai, Nakao, Oki, Imura, and Li 1978; Suda, Liotta, and Krieger 1978; Smith *et al.* 1981), and are undetectable after hypophysectomy. Changes in plasma β-endorphin during stress, exercise and pregnancy are considered in greater detail below.

3   MET-ENKEPHALIN

Immunoreactive met-enkephalin, coeluting with the synthetic pentapeptide on chromatography, has been clearly demonstrated to circulate in human plasma (Clement-Jones *et al.* 1980*a,b*). Its source is uncertain since, although a gradient of secretion into the adrenal vein can be demonstrated in both man and animals, suggesting secretion by the adrenal medulla (Clement-Jones *et al.* 1980*a*; Medbak *et al.* 1983), total adrenalectomy does not result in a lowering of plasma levels (Smith *et al.* 1981). Possible alternative sources for plasma met-enkephalin include sympathetic nerve terminals and the gut. Unlike β-endorphin, the level of met-enkephalin in plasma does not show a circadian rhythm (Shanks *et al.* 1981), nor does it show a response to dexamethasone or insulin-induced hypoglycaemia (Smith *et al.* 1981), or indeed to most other stimuli studied. More recent experiments have suggested that large molecular weight met-enkephalin precursors may be released during stress in animals (Medbak, Mason, and Rees 1985). Very high levels of plasma met-enkephalin are however found in patients with renal failure, including large amounts of high molecular weight precursors (Smith *et al.* 1981), although the mechanism of this elevation of met-enkephalin in unknown. Smaller but still significant increases of plasma met-enkephalin are seen after oral administration of ethanol in patients or volunteers taking the oral hypoglycaemic drug chlorpropamide (Medbak, Wass, Clement-Jones, Cooke, Bowcock, Cudworth, and Rees 1981), although it is unclear if this is related to the 'flush' experienced after alcohol by some patients on such treatment. Changes in plasma met-enkephalin during exercise, shock and acupuncture are considered below.

4 LEU-ENKEPHALIN

Leu-enkephalin has been reported to circulate in human plasma and to rise following insulin-induced hypoglycaemia (Ryder and Eng 1981). We have however been unable to confirm this in our laboratory (Clement-Jones and Rees unpublished observations).

5 DYNORPHIN

In early studies large quantities of immunoreactive dynorphin were reported in the plasma of stressed rats (Millan, Tsang, Przewlocki, Hollt, and Herz 1981). This 'dynorphin' was however subsequently shown to be an artefact produced by enzymatic tracer degradation in assay (Howlett *et al.* 1984; Spampinato and Goldstein 1984). Very small quantities of a dynorphin-like peptide, or a larger precursor, may circulate in human plasma (Boarder, Erdelyi, and Barchas 1982; Howlett *et al.* 1984), but as yet nothing is known of their physiology.

6 'STRESS' AND 'SHOCK'

Since β-endorphin is coreleased with ACTH it is hardly surprising that plasma levels of β-endorphin have been demonstrated to rise under a wide variety of 'stressful' situations including insulin-induced hypoglycaemia (Nakao *et al.* 1979; Smith *et al.* 1981), surgical stress (Dubois, Pickar, Cohen, Roth, MacNamara, and Bunney 1981), obstetric delivery (see below) and septic shock (Evans *et al.* 1984). This release of β-endorphin into plasma has been suggested to be the physiological basis of the 'stress analgesia' noted both in experimental animals and, more familiarly, in injured sportsmen on the sports field or soldiers in battle. However such a direct relationship seems unlikely since β-endorphin is unable to penetrate the blood brain barrier and peak levels are generally barely high enough to stimulate any known opioid receptors. In addition detailed chromotography shows that much of this immunoreactive β-endorphin may not be authentic, opioid-active, β-endorphin (L. McLoughlin, unpublished observations).

Despite the fact that met-enkephalin is costored with catecholamines in the adrenal medulla (Viveros, Diliberto, Hazum, and Chang 1979) and has been shown to be coreleased with catecholamines *in vitro* (Viveros *et al.* 1979; Corder, Mason, Perrett, Lowry, Clement-Jones, Linton, Besser, and Rees 1982), no change in plasma met-enkephalin has been observed under most conditions of 'stress'. One important exception however is during the stress of septic shock, where release of met-enkephalin, in addition to β-endorphin, has been demonstrated both in experiments on dogs (Evans *et al.* 1984) and in a small study of burned patients with septicaemia (Elliot,

Everitt, Gault, Quaba, Hackett, Howlett, Tomlin, and Rees 1985). In the animal study definite secretion of both met-enkephalin and a large molecular weight precursor into the adrenal vein were noted. This release of opioids has added to the evidence for a role of brain, or even plasma, opioids in the pathophysiology of septic shock: opioids may modulate cardiovascular function both at local level in the heart, at the brain stem cardiovascular centre, or indirectly by inhibiting adrenal medullary release of catecholamines (Bouloux *et al.* 1985), and could therefore mediate the hypotension associated with shock. Further evidence comes from the reported efficacy of the opioid antagonist naloxone to reverse hypotension in such circumstances, both in extensive experimental animal research into shock (Holaday 1983) and in rather more anecdotal clinical case reports (Dirksen, Otten, Wood, Verbaan, Haalebos, Verdoun, and Nijhuis 1980; Lenz, Druml, Gassner, Hruby, Kleinberger, and Laggner 1981; Peters, Johnson, Friedman, and Mitch 1981; Swinburn and Phelan 1982). It should be noted however that this effect of naloxone appears to be only temporary, and therefore cannot yet be recommended for routine clinical usage.

7   EXERCISE AND TRAINING

Several groups have reported the release of β-endorphin during exercise in man (Fraioli, Moretti, Paolucci, Alicicco, Crescenzi, and Fortunio 1980; Bortz, Angwin, Mefford, Boarder, Noyce, and Barchas 1981; Carr, Bullen, Skrinar, Arnold, Rosenblatt, Martin, and McArthur 1981; Colt, Wardlaw, and Frantz, 1981; Farrell, Gates, Maksud, and Morgan 1981; Howlett, Tomlin, Ngahfoong, Rees, Bullen, Skrinar, and McArthur 1984); heat and dehydration during exercise may potentiate this release (Kelso, Herbert, Gwazdauskas, Goss, and Hess 1984).

Release of met-enkephalin has also been reported in some, but not all, untrained women during treadmill exercise (Howlett *et al.* 1984), but not in another study using a bicycle ergometer in men (Grossman *et al.* 1984); it remains unclear whether this difference is one of sex, training, intensity or type of exercise.

Training may modulate this release of plasma opioids. Potentiation of β-endorphin release by training was found in one study (Carr *et al.* 1981), but could not be confirmed in another (Howlett *et al.* 1984). The release of met-enkephalin was practically abolished during an intensive training programme in one study (Howlett *et al.* 1984).

Competitive women athletes frequently suffer from disturbances of the menstrual cycle, including delayed menarche (Warren 1980), secondary amenorrhoea (Feicht, Johnson, Martin, Sparkes, and Wagner 1978) and inadequate luteal phase (Shangold, Freeman, Thysen, and Gatz 1979), the incidence of which is directly related to amount of training (Feicht *et al.* 1978). Since it appears that the abnormalities of LH pulsatility in such women can be corrected by naloxone (McArthur *et al.* 1980; see above) it is suggested that changes in endogenous opioids may mediate these abnormalities of the menstrual cycle (and indeed the less well documented hypogonadism of male athletes). Once again it is unclear whether changes in plasma opioids might be involved since, although the median eminence is outside the blood-brain barrier and thus, at least theoretically, accessible to peptides, peak levels of opioids are probably too low to bring about the observed effects. It is perhaps more likely that local changes in hypothalamic opioid activity are responsible. Other, non-opioid, gonadotrophin inhibitors including prolactin (Moretti *et al.* 1983) and melatonin (Carr, Reppert, Bullen, Skrinar, Beitins, Arnold, Rosenblatt, Martin, and McArthur 1981) are released during exercise and may also play a role.

8 PREGNANCY AND LABOUR

Several of groups of workers have confirmed that plasma β-endorphin rises steadily during pregnancy, reaching a peak at term (Csontos, Rust, Hollt, Mahr, and Teschemacher 1979; Genezzani, Facchinetti, and Parrini 1981; Browning, Butt, Lynch, and Shakespear 1983; Hoffman, Abboud, Haase, Hung, and Goebelmann 1984; Newnham, Tomlin, Ratter, Bourne, and Rees 1984), in parallel with the previously-reported changes in plasma ACTH (Rees, Burke, Chard, Evans, and Letchworth 1975). β-endorphin levels in the first trimester of pregnancy are lower than those in non-pregnant women, in spite of elevation of plasma β-LPH at that time (Genazzani ′*et al.* 1981; Browning *et al.* 1983; Hoffman *et al.* 1984), suggesting differential changes in either POMC processing or in peptide degradation during pregnancy. Whether the increased β-endorphin at term is of pituitary or placental origin and indeed its relationship, if any, to increased circulating CRF41 levels in pregnancy (Sasaki, Liotta, Luckey, Margioris, Suda, and Krieger 1984) is not yet determined. Further elevation of plasma β-endorphin occurs during labour, reaching a peak (which is often extremely high) at delivery, and falling again rapidly to reach non-pregnant levels within 24 hours (Fletcher, Thomas, and Hill 1980; Genazzani *et al.* 1981; Goland, Wardlaw, Stark, and Frantz 1981; Jouppila, Jouppila, Karlqvist, Kaukoranta, Leppaluotto, and Vuolteenaho 1983; Hoffman *et al.* 1984; Hewnham, Dennett, Ferron, Tomlin, Legg, Bourne, and Rees 1984) (Fig. 7.5). This release of β-endorphin is presumably in response to the stress of labour, and is paralleled by a similar rise in plasma cortisol (Maltau, Eielson, and Stokke 1979). Release is substantially reduced by epidural anaesthesia (Browning *et al.* 1983*b*; Hoffman *et al.* 1984), or by intrathecal morphine (Abboud, Goebelsmann, Raya, Hoffman, DeSousa, Brizgys, Kotelko, and Shnider 1984). An inverse relationship between plasma β-endorphin and perception of pain during labour has been reported by some workers (Newnham *et al.* 1984) but not by others (Fletcher *et al.* 1980;

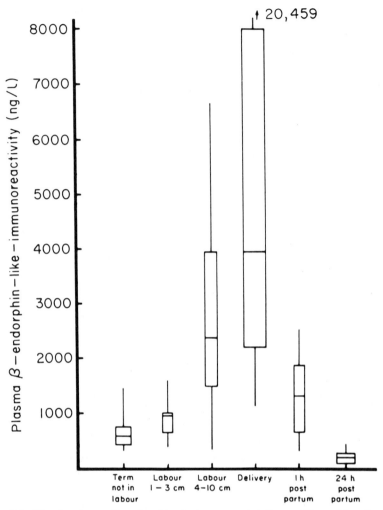

Fig. 7.5. Rise in plasma β-endorphin during normal delivery. (From Newnham *et al.* 1984, with permission).

Kimball, Chang, and Chapman 1984). A relationship has also been noted between plasma β-endorphin and the incidence of post-partum 'blues' (Kimball *et al.* 1984; Newnham *et al.* 1984).

β-endorphin is also present in umbilical cord blood. Levels do not correlate with maternal levels and are higher in umbilical artery than vein, suggesting fetal secretion. They are significantly increased in hypoxic or acidotic neonates (Browning *et al.* 1983*b*; Smith, Redman, Howlett, Rees, Anand, Harkness, Cotes, and Dawes 1986; Wardlaw, Stark, Baxi, and Frantz 1979). Less dramatic changes in human fetal heart rate or in periods of apnoea in the fetal sheep do not however correlate with changes in either plasma β-endorphin or met-enkephalin (Hofmeyr, Bamford, Howlett, and Parkes 1984; Smith *et al.* 1986).

VI Opioids in reproductive tract and placenta

The mouse vas deferens is one of the classical tissues used for opiate bioassay and it is thus perhaps not surprising that a variety of opioids have been found in the male reproductive tract of several mammalian species. β-endorphin, met- and leu-enkephalin have all been isolated in relatively high concentrations from human seminal plasma and spermatozoa, and are also found in mammalian Leydig cells, epididymis, vas deferens, seminal vesicles and prostate (Sharp and Pekary 1981; Sastry, Janson, Owens, and Tayeb 1982; Tsong, Philips, Halmi, Liotta, Margioris, Bardin, and Krieger 1982). Testicular levels of β-endorphin are not altered by hypophysectomy suggesting local synthesis (Tsong *et al.* 1982), and appear to increase with stimulation by gonadotrophin (Shaha, Liotta, Kreiger, and Bardin 1984). The role of opioids found in the reproductive tract is unknown.

The human placenta is also rich in a variety of opioids including β-endorphin (Nakai, Nakao, Oki, and Imura 1978), met- and leu-enkephalin (Tan and Yu 1981) and dynorphin (Lemaire, Valette, Chouinard, Dupuis, Day, Porthe, and Cros 1983). In the case of β-endorphin, synthesis of β-endorphin, β-LPH and ACTH, as well as a number of higher molecular weight precursors, has been demonstrated *in vitro* by pulse-chase experiments (Liotta and Krieger 1981). It is unclear whether the placenta contributes to the elevated levels of these peptides in plasma during pregnancy or what role, if any, opioids play in the regulation of other placental secretions, although there is some preliminary evidence that κ-opioids may potentiate hCG secretion (Valette, Tafani, Porthe, Pontonnier, and Cros 1983).

VII Other opioid functions

1 NOCICEPTION

Opioid peptides are involved in pain perception at several levels in the central nervous system. The primary afferent pain pathways, containing substance P as neurotransmitter, terminate in the dorsal horn of the spinal cord, synapsing with ascending neurones of the spinothalamic tract. This area of the cord is rich in both the enkephalins and dynorphin in neurones which can be demonstrated by immunohistochemistry to be in close relationship with those containing substance P. Furthermore opiates inhibit release of substance P from isolated rat trigeminal nucleus *in vitro* (Jessell and Iverson 1977). Opioidergic neurones thus seem to modulate the afferent pain signal at spinal cord level and may represent the biochemical basis of the 'pain gate' of Melzack and Wall (1965).

The periaqueductal gray matter (PAG) is rich in opioids and opioid receptors (Cuello 1983) and also intimately involved in pain perception, being the origin of descending modulatory pathways to the spinal cord. Electrical stimulation of the PAG in patients with chronic pain results in effective analgesia which is antagonized by naloxone (Richardson and Akil 1977). Such stimulation has also been reported to cause release of both met-enkephalin (Akil, Richardson, Hughes, and Barchas 1978) and β-endorphin (Akil, Richardson, Barchas, and Li 1978; Hosobuchi, Rossier, Bloom, and Guillemin 1979) into CSF; some caution is however necessary in the interpretation of these latter results as it has recently been shown that the apparent release of β-endorphin may in fact be an artefact caused by use of contrast medium (Fessler, Brown, Rachlin, and Mullan 1984).

The ancient art of acupuncture analgesia has also recently been shown to be associated with the release of opioid peptides into CSF. Pioneer work in China first demonstrated the release of a transferable, analgesic factor into CSF during acupuncture (Research Group for Acupuncture Analgesia, 1974) and subsequent work in our laboratory has confirmed the release of met-enkephalin during high-frequency auricular electroacupuncture for heroin withdrawal (Clement-Jones, McLoughlin, Lowry, Besser, Rees, and Wen 1979), and β-endorphin during low-frequency body electroacupuncture for the relief of pain (Clement-Jones, McLoughlin, Tomlin, Besser, Rees, and Wen 1980). The opioid released depends on the type of acupuncture given rather than the condition being treated (Price, McLoughlin, Till, Fitzpatrick, Wen, Besser, and Rees 1984). Dynorphin is also probably involved in the acupuncture response since intrathecal administration of anti-dynorphin antiserum blocks the effect of acupuncture in animals (Han and Xie 1984). Elevation of plasma met-enkephalin has also been reported after traditional acupuncture (Kiser *et al.* 1983), and minor changes in plasma β-endorphin after analgesic transcutaneous nerve stimulation (Facchinetti, Sandrini, Petraglia, Alfonsi, Nappi, and Genazzani 1984). It is also of some interest that placebo analgesia is blocked by naloxone (Levine, Gordon, and Fields 1978).

Does this knowledge of the role of opioids in pain perception increase our clinical ability to provide effective, safe, analgesia, particularly during labour, when the classical side effects of opiate analgesia, particularly respiratory depression, are least desired? Unfortunately little has yet filtered through to clinical practice although a few potentially useful aspects will be mentioned.

β-endorphin does not pass the blood brain barrier. However administered intrathecally it provides prolonged and effective obstetric analgesia (Oyama, Matsuki, Taneichi, Ling, and Guillemin 1980), although the route of administration and cost mean that this is unlikely to be a practical analgesic procedure. Dynorphin and synthetic κ-selective opiates are potent analgesics acting at spinal cord level without unwanted central effects (Piercey, Varner, and Schroeder 1982), so they might find clinical use in the future. Finally a novel approach has been the administration of the enkephalinase inhibitor thiorphan. This produces analgesia to some pain stimuli in animals which can be reversed by naloxone, but has not yet been used in man (Roques, Fournie-Zaluski, Soroca, Lecomte, Malfroy, Llorens, and Schwarzt 1980).

2 IMMUNE REGULATION

Heroin addicts have been long known to suffer from a variety of complications due to infection (Louria, Hensle, and Rose 1967) and to have abnormalities of immune response (Brown, Stimmel, Taub, Kochwa, and Rosenfield 1974). Opioid receptors have now been demonstrated on lymphocytes, granulocytes and monocytes and on complement (Hazum, Chang, and Cuatrecasas 1979; Lopker, Abood, Hoss, and Lionetti 1980; Schweigerer, Bhakdi, and Teschemacher 1982) and an increasing number of specific opioid effects on the immune system have recently been reported (Weber and Pert 1984), including suppression of natural killer cell activity (Shavit, Lewis, Terman, Gale, and Liebeskind 1984) and stimulation of mast cell degranulation (Casale, Bowman, and Kaliner 1984). It thus seems likely that opioids may play an important role in immune regulation and the obvious possibility arises that placental opioids may be involved in the relative immunosuppression of pregnancy. This aspect has as yet received little attention.

VIII   Conclusions

In this chapter we have attempted to summarize the current knowledge of the role of endogenous opioid peptides with particular reference to reproductive biology. It will be seen that many effects of opioid peptides have been described but that our knowledge of their physiological significance is at best incomplete and often totally absent. As the growth of publications continues to accelerate, with a corresponding increase in the list of endogenous opioids and related peptide neurotransmitters, interpretation of the significance of opioid peptides is thus likely to become even more complex. Nevertheless opioids do appear to have important pathophysiological roles, particularly in neuroendocrine regulation.

IX   Glossary

COELUTE — to elute on a chromatography system in the same position as another substance (generally a synthetic standard peptide)

CORELEASED — (of 2 or more substances) released at the same time, and generally from the same cell, to a given stimulus

COSTORED — stored in the same cell, and often same secretory granule

POTENTIATION (POTENTIATE) — an increase in the physiological response to a given stimulus in the presence of another factor which may or may not cause the same response itself, alone. The response to both factors is greater than the sum of the individual responses

References

Abboud, T. K., Goebelsmann, U., Raya, J., Hoffman, D. I., De Sousa, B., Brizgys, R. V., Kotelko, D. M., and Shnider, S. M. (1984). Effects of intrathecal morphine during labor on maternal plasma β-endorphin level. *Am. J. Obstet. Gynec.* **149**, 709–10.

Akil, H., Richardson, D. E., Barchas, J. D., and Li, C. H. (1978). Appearance of β-endorphin-like immunoreactivity in human ventricular cerebospinal fluid upon analgesic electrical stimulation. *Proc. natn. Acad. Sci., U.S.A.* **75**, 5170–5172.

—— —— Hughes, J., and Barchas, J. D. (1981). Enkephalin-like material elevated in ventricular cerebrospinal fluid of pain patients after analgetic focal stimulation. *Science, N.Y.* **201**, 463–465.

Baranowska, B., Rozbicka, G., Jeske, W., and Abdel-Fattah, M. H. (1984). The role of endogenous opiates in the mechanism of inhibited luteinizing hormone (LH) secretion in women with anorexia nervosa: The effect of naloxone on LH, follicle-stimulating hormone, prolactin and β-endorphin secretion. *J. clin. Endocr. Metab.* **59**, 412–416.

Bhanot, R. and Wilkinson, M. (1983). Opiatergic control of gonadotrophin secretion during puberty in the rat: A neurochemical basis for the hypothalamic 'gonadostat'. *Endocrinology* **113**, 596–603.

Bicknell, R. J. (1985). Endogenous opioid peptides and hypothalamic neuroendocrine neurones. *J. Endocr.* **107**, 437–446.

—— and Leng, G. (1982). Endogenous opiates regulate oxytocin but not vasopressin release from the neurohypophysis. *Nature, Lond.* **298**, 161–162.

Blank, M. S. and Roberts, D. L. (1982). Antagonist of gonadotrophin-releasing hormone blocks naloxone-induced elevations in serum luteinizing hormone. *Neuroendocrinology* **35**, 309–312.

Blankstein, J., Reyes, F., Winter, J., and Faiman, C. (1979). Failure of naloxone to alter growth hormone and prolactin levels in acromegalic and in hyperprolactinaemic patients. *Clin. Endocr.* **11**, 474–479.

Boarder, M. R., Erdelyi, E., and Barchas, J. D. (1982). Opioid peptides in human plasma: evidence for multiple forms. *J. clin. Endocr. Metab.* **54**, 715–720.

Bortz, W. M., Angwin, P., Mefford, I. N., Boarder, M. R., Noyce, N., and Barchas, J. D. (1981). Catecholamines, dopamine and endorphin levels during extreme exercise. *New Engl. J. Med.* **305**, 466–7.

Bouloux, P. M. G., Grossman, A., Lytras, N., and Besser, G. M. (1985). Evidence for the participation of endogenous opioids in the sympathoadrenal response to hypoglycaemia in man. *Clin. Endocr.* **22**, 49–56.

Brown, S. M., Stimmel, B., Taub, R. N., Kochwa, S., and Rosenfield, R. E. (1974). Immunologic dysfunction in heroin addicts. *Arch. int. Med.* **134**, 1001–6.

Browning, A. J. F., Butt, W. R., Lynch, S. S., and Shakespear, R. A. (1983a). Maternal plasma concentration of β-lipotrophin, β-endorphin and γ-lipotrophin throughout pregnancy. *Br. J. Obstet. Gynaec.* **90**, 1147–1151.

―― ―― ―― ―― and Crawford, J. S. (1983*b*). Maternal and cord plasma concentrations of β-lipotrophin, β-endorphin and γ-lipotrophin at delivery: effect of analgesia. *Br. J. Obstet. Gynaec.* **90**, 1152–6.

Bruni, J. F., VanVugt, D., Marshall, S., and Meites, J. (1977). Effects of naloxone morphine and methionine enkephalin on serum prolactin, luteinizing hormone, follicle stimulating hormone, thyroid stimulating hormone and growth hormone. *Life Sci.* **21**, 461–466.

―― Watkins, W. B., and Yen, S. S. C. (1979). β-endorphin in the human pancreas. *J. clin. Endocr. Metab.* **49**, 649–51.

Carr, D. B., Bullen, B. A., Skrinar, G. S., Arnold, M. A., Rosenblatt, M., Beitins, I. Z., Martin, J. B., and McArthur, J. W. (1981). Physical conditioning facilities the exercise-induced secretion of β-endorphin and β-lipotropin in women. *New. Engl. J. Med.* **305**, 560–563.

―― Reppert, S. M., Bullen, B., Skrinar, G., Beitins, I., Arnold, M., Rosenblatt, M., Martin, J. B., and McArthur, J. W. (1981). Plasma melatonin increases during exercise in women. *J. clin. Endocr. Metab.* **53**, 224–5.

Casale, T. B., Bowman, S., and Kaliner, M. (1984). Induction of human cutaneous mast cell degranulation by opiates and endogenous opioid peptides: evidence for opiate and non-opiate receptor participation. *J. All. clin. Immun.* **73**, 775–81.

Casper, R. F. and Alapin-Rubillovitz, S. (1985). Progestins increase endogenous opioid peptide activity in postmenopausal women. *J. clin. Endocr. Metab.* **60**, 34–36.

―― Bhanot, R., and Wilkinson, M. (1984). Prolonged elevation of hypothalamic opioid peptide activity in women taking oral contraceptives. *J. clin. Endocr. Metab.* **58**, 582–584.

―― and Yen, S. S. C. (1981). Simultaneous pulsatile release of prolactin and luteinizing hormone induced by luteinizing hormone-releasing factor. *J. clin. Endocr. Metab.* **52**, 934–936.

Cetel, N. S., Quigley, M. E., and Yen, S. S. C. (1985). Naloxone-induced prolactin secretion in women: evidence against a direct prolactin stimulatory effect of endogenous opioids. *J. clin. Endocr. Metab.* **60**, 191–196.

Clarke, G., Wood, P., Merrick, L., and Lincoln, D. W. (1979). Opiate inhibition of peptide release from the neurohumoral terminal of hypothalamic neurones. *Nature, Lond.* **282**, 746–748.

Clement-Jones, V., Lowry, P. J., Rees, L. H., and Besser, G. M. (1980*a*). Met-enkephalin circulates in human plasma. *Nature, Lond.* **283**, 295–297.

―― ―― ―― ―― (1980*b*). Development of a specific extracted radioimmunoassay for met-enkephalin in human plasma and cerebrospinal fluid. *J. Endocr.* **86**, 231–243.

―― McLoughlin, L., Lowry, P. J., Besser, G. M., Rees, L. H., and Wen, H. L. (1979). Acupuncture in heroin addicts – changes in met-enkephalin and β-endorphin in blood and cerebrospinal fluid. *Lancet* **ii**, 380–382.

―― ―― Tomlin, S., Besser, G. M., Rees, L. H., and Wen, H. L. (1980). Increased β-endorphin but not met enkephalin levels in human cerebrospinal fluid after acupuncture for recurrent pain. *Lancet* **ii**, 946–948.

Colt, E. W. D., Wardlaw, S. L., and Frantz, A. G. (1981). The effect of running on plasma β-endorphin. *Life Sci.* **28**, 1637–1640.

Corbett, A. D., Paterson, S. J., McKnight, A. T., Magnan, J., and Kosterlitz, H. W. (1982). Dynorphin (1–8) and dynorphin (1–9) are ligands for the kappa-subtype of opiate receptor. *Nature, Lond.* **229**, 79–81.

Corder, R., Mason, D. F. J., Perrett, D., Lowry, P., Clement-Jones, V., Linton, E. A., Besser, G. M., and Rees, L. H. (1982). Simultaneous release of neurotensin, somatostatin, enkephalins and catecholamines from perfused cat adrenal glands. *Neuropeptides* **3**, 9–17.

Csontos, K., Rust, M., Hollt, V., Mahr, W., Kromer, W., and Teschemacher, H. J. (1979). Elevated plasma β-endorphin levels in pregnant women and their neonates. *Life Sci.* **25**, 835–844.

Cuello, A. C. (1983). Central distribution of opioid peptides. *Br. med. Bull.* **39**, 11–16.

Cumming, D. C., Reid, R. L., Quigley, M. E., Rebar, R. W., and Yen, S. S. C. (1984). Evidence for decreased endogenous dopamine and opioid inhibitory influences on LH secretion in polycystic ovary syndrome. *Clin. Endocr.* **20**, 643–648.

Delitala, G., Giusti, M., Mazzocchi, G., Granziera, L., Tarditi, W., and Giordano, G. (1983). Participation of endogenous opiates in regulation of the hypothalamic-pituitary-testicular axis in normal men. *J. clin. Endocr. Metab.* **57**, 1277–1281.

—— Grossman, A., and Besser, G. M. (1981). Changes in pituitary hormone levels induced by met-enkephalin in man – the role of dopamine. *Life Sci.* **29**, 1537–1544.

—— —— —— (1983). Differential effects of opiate peptides and alkaloids on anterior pituitary hormone secretion. *Neuroendocrinology* **37**, 275–279.

Demura, R., Suda, T., Wakabayashi, I., Yoshimura, M., Jibiki, K., Odagiri, E., Demura, H., and Shizume, K. (1981). Plasma pituitary hormone responses to the synthetic enkephalin analog (FK33–824) in normal subjects and patients with pituitary diseases. *J. clin. Endocr. Metab.* **52**, 263–266.

Denef, C. and Andries, M. (1983). Evidence for paracrine interaction between gonadotrophs and lactotrophs in pituitary cell aggregates. *Endocrinology* **112**, 813–822.

Dent, R. R. M., Guillemault, C., Albert, L. H., Posner, B. I., Cox, B. M., and Goldstein, A. (1981). Diurnal rhythm of plasma immunoreactive β-endorphin and its relationship to sleep stages and plasma rhythms of cortisol and prolactin. *J. clin. Endocr. Metab.* **52**, 942–947.

Dirksen, R., Otten, M. H., Wood, G. J., Verbaan, C. J., Haalebos, M. M. P., Verdouw, P. V., and Nijhuis, G. M. M. (1980). Naloxone in shock. *Lancet* **ii**, 1360–1.

Drouva, S. V., Epelbaum, J., Tapia-Arancibia, L., Laplante, E., and Kordon, C. (1981). Opiate receptors modulate LHRH and SRIF release from mediobasal hypothalamic neurones. *Neuroendocrinology* **32**, 163–167.

Dubois, M., Pickar, D., Cohen, M. R., Roth, Y. F., Macnamara, T., Bunney, W. E. (1981). Surgical stress in humans is accompanied by an increase in plasma β-endorphin immunoreactivity. *Life Sci.* **29**, 1249–1254.

Dupont, A., Barden, N., Cusan, L., Merand, Y., Labrie, F., and Vaudry, H. (1980). β-endorphin and met-enkephalins: their distribution, modulation by estrogens and haloperidol, and role in neuroendocrine control. *Fedn. Proc.* **39**, 2544–2550.

Dyer, R. G., Mansfield, S., Corbet, H., and Dean, A. D. P. (1985). Fasting impairs LH secretion in female rats by activating an inhibitory opioid pathway. *J. Endocr.* **105**, 91–97.

Ellinboe, J., Veldhuis, J. D., Mendelson, J. H., Kuehnle, J. C., Mello, N. K., and Holbrook, P. G. (1982). Effect of endogenous opioid blockade on the amplitude and frequency of pulsatile luteinizing hormone secretion in normal men. *J. clin. Endocr. Metab.* **54**, 854–857.

Elliot, D., Everitt, A. S., Gault, D., Quaba, A. A., Hackett, M. E. J., Howlett, T. A., Tomlin, S., and Rees, L. H. (1985). The role of endorphins in septicaemia shock: a pilot study in burned patients. *Burns* **11**, 387–392.

Evans, S. F., Medbak, S., Hinds, C. J., Tomlin, S. J., Varley, J. G., and Rees, L. H. (1984). Plasma levels and biochemical characterisation of circulating met-enkephalin in canine endotoxin shock. *Life Sci.* **34**, 1481–1486.

Facchinetti, F., Sandrini, G., Petraglia, F., Alfonsi, E., Nappi, G., and Genazzani, A. R. (1984). Concomitant increase in nociceptive flexion reflex threshold and plasma opioids following transcutaneous nerve stimulation. *Pain* **19**, 295–303.

Fang, V. S., Fessler, R. G., Rachlin, J. R., and Brown, F. D. (1984). Effect of contrast media on radioimmunoassay of β-endorphin in cerebrospinal fluid. *Clin. Chem.* **30**, 311–314.

Farrell, P. A., Gates, W. K., Maksud, M. G., and Morgan, W. P. (1982). Increases in plasma β-endorphin/β-lipotrophin immunoreactivity after treadmill running in humans. *J. appl. Physiol.* **52**, 1245–1249.

Fessler, R. G., Brown, F. D., Rachlin, J. R., and Mullan, S. (1984). Elevated β-endorphin in cerebrospinal fluid after electrical brain stimulation: artifact of contrast infusion? *Science, N.Y.* **224**, 1017–1019.

Fiecht, C. B., Johnson, T. S., Martin, B. J., Sparkes, K. E., and Wagner, W. W. (1978). Secondary amenorrhoea in athletes. *Lancet* ii, 1145–1146.

Fischli, W., Goldstein, A., Hunkapiller, M. W., and Hood, L. E. (1982). Two big dynorphins from porcine pituitary. *Life Sci.* **31**, 1769–1772.

Fletcher, J. E., Thomas, T. A., and Hill, R. G. (1980). β-endorphin and parturition. *Lancet* ii, 310.

Fraiol, F., Cappa, M., Fabbri, A., Gnessi, L., Moretti, C., Borrelli, P., and Isidori, A. (1984). Lack of endogenous opioid inhibitory time in early puberty. *Clin. Endocr.* **20**, 299–305.

—— Moretti, C., Paolucci, D., Alicicco, E., Crescenzi, F., and Fortunio, G. (1980). Physical exercise stimulates marked concomitant increase of β-endorphin and adrenocorticotrophin (ACTH) in peripheral blood in man. *Experientia* **36**, 987–989.

Gambert, S. R., Garthwaite, T. L., Pontzer, C. H., Cook, E. E., Tristani, F. E., Duthie, E. H., Martinson, D. R., Hargen, T. C., and McCarty, D. J. (1981). Running elevates plasma β-endorphin immunoreactivity and ACTH in untrained human subjects. *Proc. Soc. exp. Biol. Med.* **168**, 1–4.

Genazzani, A. R., Facchinetti, F., and Parrini, D. (1981). β-lipotrophin and β-endorphin plasma levels during pregnancy. *Clin. Endocr.* **14**, 409–418.

Gerendai, I., Shaha, C., Thau, R., and Bardin, C. W. (1984). Do testicular opiates regulate Leydig cell function? *Endocrinology* **115**, 1645–1647.

Goland, R. S., Wardlaw, S. L., Stark, R. I., and Frantz, A. G. (1981). Human plasma β-endorphin during pregnancy, labour and delivery. *J. clin. Endocr. Metab.* **52**, 74–78.

Goldstein, A., Tachibana, S., Lowney, L. I., Hunkapiller, M., and Hood, L. (1980). Dynorphin (1–13), an extraordinarily potent opioid peptide. *Proc. natn. Acad. Sci. U.S.A.* **76**, 6666–6670.

Green, I. C., Perrin, D., Penman, E., Yaseen, A., Ray, K., and Howell, S. L. (1983). Effect of dynorphin on insulin and somatostatin secretion, calcium uptake and c-AMP levels in isolated rat islets of Langerhans. *Diabetes* **32**, 685–690.

Grossman, A. (1983). Brain opiates and neuroendocrine function. *Clinics Endocr. Metab.* **12**, 725–746.

—— and Besser, G. M. (1982). Opiates control ACTH through a noradrenergic mechanism. *Clin. Endocr.* **17**, 287–290.

—— Bouloux, P. M. G. (1984). The involvement of opioid peptides in the catecholamine response to stress in man. In *Opioid peptides in the periphery* (ed. Fraiolo F. *et al.*) pp. 169–177. Elsevier.

—— —— Price, P., Drury, P. L., Lam, K. S. L., Turner, T., Thomas, J., Besser, G. M., and Sutton, J. (1984). The role of opioid peptides in the hormonal responses to acute exercise in man. *Clin. Sci.* **67**, 483–491.

—— Gaillard, R. C., McCartney, P., Rees, L. H., and Besser, G. M. (1982). Opiate modulation of the pituitary-adrenal axis: effects of stress and circadian rhythm. *Clin. Endocr.* **17**, 279–286.

—— Moult, P. J. A., Cunnah, D., and Besser, G. M. (1986). Different opioid mechanisms are involved in the modulation of ACTH and gonadotrophin release in man. *Neuroendocrinology* **42**, 357–60.

—— —— Gaillard, R. C., Delitalia, G., Toff, W. D., Rees, L. H., and Besser, G. M. (1981). The opioid control of LH and FSH release: effects of a met-enkephalin analogue and naloxone. *Clin. Endocr.* **14**, 41–47.

—— —— McIntyre, H., Evans, J., Silverstone, T., Rees, L. H., and Besser, G. M. (1982). Opiate mediation of amenorrhoea in hyperprolactinaemia and in weight-loss related amenorrhoea. *Clin. Endocr.* **17**, 379–388.

—— Stubbs, W. A., Gaillard, R. C., Delitala, G., Rees, L. H., and Besser, G. M. (1981). Studies of the opiate control of prolactin, GH and TSH. *Clin. Endocr.* **14**, 381–386.

—— West, S., Williams, J., Evans, J., Rees, L. H., and Besser, G. M. (1982). The role of opiate peptides in the control of prolactin in the puerperium, and TSH in primary hypothyroidism. *Clin. Endocr.* **16**, 317–320.

Gudelsky, G. A. and Porter, J. C. (1979). Morphine and opioid peptide-induced inhibition of the release of dopamine from tuberoinfundibular neurons. *Life Sci.* **25**, 1697–1702.

Guillemin, R., Vargo, T., Rossier, J., Minick, S., Ling, N., Rivier, C., Vale, W., and Bloom, F. (1977). β-endorphin and adrenocorticotrophin are secreted concomitantly by the pituitary gland. *Science, N.Y.* **197**, 1367–1369.

Haldar, J., Hoffman, D. L., and Zimmerman, E. A. (1982). Morphine beta-endorphin and [D Ala$_2$] met-enkephalin inhibit oxytocin release by acetylcholine and suckling. *Peptides* **3**, 663–668.

Han, J-S. and Xie, G-X. (1984). Dynorphin: important mediator for electroacupuncture analgesia in the spinal cord of the rabbit. *Pain* **18**, 367–376.

Haskins, J. T., Gudelsky, G. A., Moss, R. L., and Porter, J. C. (1981). Iontophoresis of morphine into the arcuate nucleus: effects on dopamine concentrations in hypophysial portal plasma and serum prolactin concentrations. *Endocrinology* **108**, 767–771.

Hazum, E., Chang, K-J., and Cuatrecazes, P. (1979). Specific non-opiate receptors for β-endorphin. *Science, N.Y.* **205**, 1033–1035.

Herz, A. (1984). Multiple opioid receptors. In *Opioid modulation of endocrine function.* (ed. G. Delitala *et al.*) pp. 11–19. Raven Press.

Hoffman, D. I., Abboud, T. K., Haase, H. R., Hung, T. T., and Goebelsmann, V. (1984). Plasma β-endorphin concentrations prior to and during pregnancy, in labor and after delivery. *Am. J. Obstet. Gynec.* **150**, 492–496.

Hofmeyr, J., Bamford, A. S., Howlett, T. A., Parkes, M. J. (1985). Methionine enkephalin and the arrest of foetal breathing. In *Physiological development of the fetus and the newborn* (ed. C. T. Jones, and P. W. Nathanielsz). pp. 653–657. Academic Press.

Holaday, J. W. (1983). Cardiovascular effects of endogenous opioid system. *A. Rev. Pharmac. Tox.* **23**, 541–594.

Hosobuchi, Y., Rossier, J., Bloom, F. E., and Guillemin, R. (1979). Stimulus of human periaqueductal gray for pain relief increases immunoreactive β-endorphin in ventricular fluid. *Science, N. Y.* **203**, 279–281.

Howlett, T. A., Patience, R., Besser, G. M., and Rees, L. H. (1986). Novel forms of immunoreactive dynorphin in human phaeochromocytomas. *J. Endocr.* **108**, *Suppl* Abstr. 196.

—— Tomlin, S., Ngahfoong, L., Rees, L. H., Bullen, B. A., Skrinar, G., and McArthur, J. (1984). Release of β-endorphin and met-enkephalin during exercise in normal women: Response to training. *Br. med. J.* **288**, 1950–1952.

—— Walker, J., Besser, G. M., and Rees, L. H. (1984). 'Dynorphin' in plasma: Enzymatic artifact and authentic immunoreactivity. *Regulatory Peptides* **8**, 131–140.

Hughes, J. and Kosterlitz, H. W. (1983). Opioid peptides: Introduction. *Br. med. Bull.* **39**, 1–13.

—— Smith, T. W., Kosterlitz, H. W., Fothergill, L. A., Morgan, B. A., and Morris, H. R. (1975). Identification of two related pentapeptides from the brain with potent opiate agonist activity. *Nature, Lond.* **258**, 577–579.

Hulse, G. K., Coleman, G. J., Copolou, D. L., and Clements, J. A. (1984). Relationship between endogenous opioids and the oestrous cycle in the rat. *J. Endocr.* **100**, 271–275.

Jessell, T. M. and Iversen, L. L. (1977). Opiate analgesics inhibit substance P release from rat trigeminal nucleus. *Nature, Lond.* **268**, 549–551.

Jouppila, R., Jouppila, P., Karqvist, K., Kaukoranta, P., Leppaluoto, J., and Vuolteenaho, O. (1983). Maternal and umbilical venous plasma immunoreactive β-endorphin levels during labour with and without epidural analgesia. *Am. J. Obstet. Gynec.* **147**, 799–802.

Kalra, S. P. and Crowley, W. R. (1982). Epinephrine synthesis inhibitors block naloxone-induced LH release. *Endocrinology* **111**, 1403–1405.

—— and Simpkins, J. W. (1981). Evidence for noradrenergic mediation of opioid effects on luteinizing hormone secretion. *Endocrinology* **109**, 776–782.

Kelson, T. B., Herbert, W. G., Gwazdauskas, F. C., Goss, F. L., and Hess, J. L. (1984). Exercise-thermoregulatory stress and increased plasma β-endorphin/β-lipotropin in humans. *J. appl. Physiol.* **57**, 444–449.

Kimball, C. D., Chang, C. M., and Chapman, M. B. (1984). Endogenous opioid peptides in intrapartum uterine blood. *Am. J. Obstet. Gynec.* **149**, 79–82.

—— —— Huang, S. M., and Houck, J. C. (1981). Immunoreactive endorphin peptides and prolactin in umbilical vein and maternal blood. *Am. J. Obstet. Gynec.* **140**, 157–164.

Kiser, R. S., Khatami, M., Gatchel, R. J., Huang, X-Y., Bhatia, K., and Albhuler, K. Z. (1983). Acupuncture relief of chronic pain syndrome correlation with increased plasma met-enkephalin concentration. *Lancet* **ii**, 1394–6.

Lemaire, S., Chouinard, L., Denis, D., Panico, M., and Morris, H. R. (1982). Mass spectrometric identification of various molecular forms of dynorphin in bovine adrenal medulla. *Biochem. Biophys. Res. Commun.* **108**, 51–58.

Lemaire, S., Valette, A., Chouinard, L., Dupuis, N., Day, R., Porthe, G., and Gos, J. (1983). Purification and identification of multiple forms of dynorphin in human placenta. *Neuropeptides* **3**, 181–191.

Leng, G., Mansfield, S., Bicknell, R. J., Dean, A. D. P., Ingram, C. D., Marsh, M. I. C., Yates, J. O., and Dyer, R. G. (1985). Central opioids: a possible role in parturition? *J. Endocr.* **106**, 219–224.

—— Bicknell, R. J., Mansfield, S., and Dyer, R. G. (1986). Oxytocin, opioids and parturition. *J. Endocrinol.* **108** *Suppl* Abstract 19.

Lenz, K., Druml, W., Gassner, A., Hruby, K., Kleinberger, G., and Laggner, A. (1981). Naloxone in shock. *Lancet* **i**, 834.

Levine, J. D., Gordon, N. C., and Fields, H. L. (1978). The mechanisms of placebo analgesia. *Lancet* **ii**, 654–657.

Lightman, S. L., Jacobs, H. S., Maguire, A. K., McGarrick, G., and Jeffcoate, S. L. (1981). Constancy of opioid control of luteinizing hormone in different pathophysiological states. *J. clin. Endocr. Metab.* **52**, 1260–1263.

Liotta, A. S. and Kreiger, D. T. (1981). In vitro biosynthesis and comparative post-translational processing of immunoreactive precursor corticotropin/β-endorphin by human placental and pituitary cells. *Endocrinology* **106**, 1504–1511.

Lopker, A., Abood, L. G., Hoss, W., and Lionetti, F. (1980). Stereoselective muscarinic acetylcholine and opiate receptors in human phagocytic leukocytes. *Biochem. Pharmac.* **29**, 1361–1365.

Louria, D. B., Henslet, T., and Rose, J. (1967). The major medical complications of heroin addiction. *Annls. Int. Med.* **67**, 1–22.

Maltau, J. M., Eielson, O. V., and Stokke, K. T. (1979). Effect of stress during labor on the concentration of cortisol and estriol in maternal plasma. *Am. J. Obstet. Gynec.* **134**, 681–684.

Martin, J. B., Tolis, G., Wood, I., and Guyda, H., (1979). Failure of naloxone to influence physiological growth hormone and prolactin secretion. *Brain Res.* **168**, 210–215.

Martin, R. and Voigt, K. H. (1981). Enkephalins coexist with oxytocin and vasopressin in nerve terminal of rat neurohypophysis. *Nature, Lond*, **289**, 502–504.

Martin, W. R., Eades, C. G., Thompson, J. A., Huppler, R. E., and Gilbert, P. E. (1976). The effects of morphine-and nalorphine-like drugs in the non-dependent and morphine dependent chronic spinal dog. *J. Pharmac. exp. Ther.* **197**, 517–532.

Matsushita, N., Kato, Y., Shimatsu, A., Katakami, H., Fujino, M., Matsuo, H., and Imura, H. (1982). Stimulation of prolactin secretion in the rat by α-neo-endorphin, β-neo-endorphin and dynorphin. *Biochem. Biophys. Res. Commun.* **107**, 735–741.

Mayer, G., Wessel, J., and Kobberling, J. (1980). Failure of naloxone to alter exercise-induced growth hormone and prolactin release in normal men. *Clin. Endocr.* **13**, 413–416.

McArthur, J. W., Bullen, B. A., Betting, I. Z., Pagane, M., Badger, T. M., and Klibanski, A. (1980). Hypothalamic amenorrhoea in runners of normal body distribution. *Endocr. Res. Commun.* **7**, 13–25.

McLoughlin, L., Lowry, P. J., Ratter, S., Besser, G. M., and Rees, L. H. (1980). β-endorphin and β-MSH in human plasma. *Clin. Endocr.* **12**, 287–292.

—— Tomlin, S., Grossman, A., Lytras, N., Coy, D., Besser, G. M., and Rees, L. H. (1984). CRF-41 stimulates the release of β-lipotrophin and β-endorphin in normal human subjects. *Neuroendocrinology* **38**, 282–284.

Medbak, S., Wass, J. A. H., Clement-Jones, V., Cooke, E. D., Bowcock, S. A., Cudworth, A. G., and Rees, L. H. (1981). Chlorpropamide alcohol flush and circulating met-enkephalin: a positive link. *Br. med. J.* **283**, 973–939.

—— Mason, D. F. J., and Rees, L. H. (1985). Plasma met-enkephalin responses to insulin-induced hypoglycaemia in dogs. *J. Endocr.* **107**, *Suppl.* 65.

Meites, J., Bruni, J. F., VanVugt, D. A., and Smith, A. F. (1979). Relation of endogenous opioid peptides and morphine to neuroendocrine function. *Life Sci.* **24**, 1325–1336.

Melis, G. B., Paoletti, A. M., Gambacciani, M., Mais, V., and Fioretti, P. (1984). Evidence that estrogens inhibit LH secretion through opioids in postmenopausal women using naloxone. *Neuroendocrinology* **39**, 60–63.

Melzack, R. and Wall, P. D. (1965). Pain mechanisms: a new theory. *Science, N.Y.* **150**, 971–979.

Millan, M. J., Tsang, Y. F., Prezwlocki, R., Hollt, V., and Herz, A. (1981). The influence of foot-shock stress upon brain pituitary and spinal cord pools off immunoreactive dynorphin in rats. *Neurosci. Lett.* **24**, 75–79.

Moretti, C., Fabbri, A., Gnessi, L., Cappa, M., Calzolari, A., Fraioli, F.,

Grossman, A., and Besser, G. M. (1983). Naloxone inhibits exercise-induced release of PRL and GH in athletes. *Clin. Endocr.* **18**, 135–138.

Morley, J. E. (1981). The endocrinology of the opiates and opioid peptides. *Metabolism* **30**, 195–209.

—— Baranetsky, N. G., Wingert, T. D., Carlson, H. E., Hersham, J. M., Melmed, S., Levin, S. R., Jamison, K. R., Weitzman, R., Chang, R. J., and Varner, A. A. (1980). Endocrine effect of naloxone-induced opiate receptor blockade. *J. clin. Endocr. Metab.* **50**, 251–257.

Moult, P. J. A., Rees, L. H., and Besser, G. M. (1982). Pulsatile gonadotrophin secretion in hyperprolactinaemic amenorrhoea and the response to bromocriptine therapy. *Clin. Endocr.* **16**, 153–162.

—— Grossman, A., Evans, J. M., Rees, L. H., and Besser, G. M. (1981). The effect of naloxone on pulsatile gonadotrophin release in normal subjects. *Clin. Endocr.* **14**, 321–324.

Nakai, Y., Nakao, K., Oki, S., and Imura, H. (1978). Presence of immunoreactive β-lipotrophin and β-endorphin in human placenta. *Life Sci.* **23**, 2013–2018.

—— —— —— —— and Li, C. H. (1978). Presence of immunoreactive β-endorphin in plasma of patients with Nelson's syndrome and Addison's disease. *Life Sci.* **23**, 2293–2298.

Nakao, K., Nakai, Y., Jingami, H., Oki, S., Fukata, J., and Imura, H. (1979). Substantial rise of plasma β-endorphin levels after insulin-induced hypoglycaemia in human subjects. *J. clin. Endocr. Metab.* **49**, 838–841.

—— Oki, S., Horii, K., and Imura, H. (1978). Presence of immunoreactive β-EP in normal human plasma. *J. clin. Invest.* **62**, 1395–8.

Newnham, J. P., Tomlin, S., Ratter, S. J., Bourne, G. L., and Rees, L. H. (1983). Endogenous opioid peptides in pregnancy. *Br. J. Obstet. Gynaec.* **90**, 535–538.

—— Dennett, P. M., Ferron, S. A., Tomlin, S., Legg, C., Bourne, G. L., and Rees, L. H. (1984). A study of the relationship between circulating β-endorphin-like immunoreactivity and post partum 'blues'. *Clin. Endocr.* **20**, 169–177.

Oyama, T., Matsuki, A., Taneichi, T., Ling, N., and Guillemin, R. (1980). β-endorphin in obstetric analgesia. *Am. J. Obstet. Gynec.* **137**, 613–616.

Paterson, S. J., Robson, L. E., and Kosterlitz, H. W. (1983). Classification of opioid receptors. *Br. med. Bull.* **39**, 31–36.

Peters, W. P., Johnson, M. W., Friedman, P. A., and Mitch, W. E. (1981). Pressor effects of naloxone in septic shock. *Lancet* **i**, 529–532.

Piercey, M. F., Varner, K., and Schroeder, L. A. (1982). Analgesic activity of intraspinally administered dynorphin and ethylketocyclazocine. *Eur. J. Pharmac.* **80**, 283–284.

Pontiroli, A. E., Baio, G., Stella, L., Crescenti, A., and Girardi, A. M. (1982). Effects of naloxone on prolactin, luteinizing hormone and cortisol responses to surgical stress in humans. *J. clin. Endocr. Metab.* **55**, 378–380.

Price, P., McLaughlin, L., Till, J., Fitzpatrick, A., Wen, H. L., Besser, G. M., and Rees, L. H. (1984). The frequency of electroacupuncture determines the nature of endogenous opioid release. In *Central and Peripheral Endorphins: Basic and Clinical Aspects* (ed. E. E. Muller and A. R. Genazzani) pp. 157–160. Raven Press, New York.

Quigley, M. E., Sheehan, K. L., Casper, R. F., and Yen, S. S. C. (1980*a*). Evidence for increased opioid inhibition of luteinizing hormone secretion in hyperprolactinaemic patients with pituitary microadenoma. *J. clin. Endocr. Metab.* **50**, 427–430.

—— —— —— —— (1980*b*). Evidence for increased dopaminergic and opioid activity in patients with hypothalamic hypogonadotrophic amenorrhea. *J. clin. Endocr. Metab.* **50**, 949–954.

290   Trevor A. Howlett and Lesley H. Rees

—— and Yen, S. S. C. (1980). The role of endogenous opiates on LH secretion during the menstrual cycle. *J. clin. Endocr. Metab.* **51**, 179–181.

Ragavan, V. V. and Frantz, A. G. (1981). Opioid regulation of prolactin secretion: evidence for a specific role of β-endorphin. *Endocrinology* **109**, 1769–1771.

Rasmussen, D. D., Liu, J. H., Wolf, P. L., Yen, S. S. C. (1983). Endogenous opioid regulation of gonadotrophin-releasing hormone release from the human fetal hypothalamus in vitro. *J. clin. Endocr. Metab.* **57**, 881–884.

Rees, L. H., Burke, C. W., Chard, T., Evans, S. W., and Letchworth, A. T. (1985). Possible placental origin of ACTH in normal human pregnancy. *Nature, Lond.* **254**, 620–622.

Reid, R. L. (1983). Endogenous opioid activity and the premenstrual syndrome. *Lancet* **ii**, 786.

—— Hoff, J. D., Yen, S. S. C., and Li, C. H. (1981). Effects of exogenous βh-endorphin on pituitary hormone secretion and its disappearance rate in normal human subjects. *J. clin. Endocr. Metab.* **52**, 1179–1184.

—— Quigley, M. E., and Yen, S. S. C. (1983). The disappearance of opioidergic regulation of gonadotropin secretion in postmenopausal women. *J. clin. Endocr. Metab.* **57**, 1107–1110.

—— and Yen, S. S. C. (1981). β-endorphin stimulates the secretion of insulin and glucagon in humans. *J. clin. Endocr. Metab.* **52**, 592–4.

Research Group for Acupuncture Anagesia (1974). The role of some neurotransmitters of brain in finger acupuncture analgesia. *Sci. Sin.* **17**, 112–130.

Richardson, D. E. and Akil, H. (1977). Pain reduction by electrical brain stimulation in man: Chromic self administration in the periventricular gray matter. *J. Neurosurg.* **47**, 184–194.

Ropert, J. F., Quigley, M. E., and Yen, S. S. C. (1981). Endogenous opiates modulate pulsatile luteinizing hormone release in humans. *J. clin. Endocr. Metab.* **52**, 583–585.

Roques, B. P., Fournie-Zaluski, M. C., Soroca, E., Lecomte, J. M., Malfroy, B., Llorens, C., and Schwartz, J. C. (1980). The enkephalinase inhibitor thiorphan shows antinociceptive activity in mice. *Nature, Lond.* **288**, 286–288.

Rossier, J. (1982). Opioid peptides have found their roots. *Nature, Lond.* **298**, 221–222.

Rubin, P., Swezey, S., and Blaschke, T. (1979). Naloxone lowers plasma prolactin in man. *Lancet* **i**, 1293.

Ryder, S. W. and Eng, J. (1981). Radioimmunoassay of leucine-enkephalin-like substance in human and canine plasma. *J. clin. Endocr. Metab.* **52**, 367–369.

Saltiel, H., Passa, P. H., Kuhn, J. M., Fiet, J., and Canivet, J. (1982). Mediation par les peptides opioides de la résponse antehypophysaire à l'hypoglycemie insulinique. *Nouv. Press. Med.* **11**, 847–850.

Sasaki, A., Liotta, A. S., Luckey, M. M., Margioris, A. N., Suda, T., and Krieger, D. T. (1984). Immunoreactive corticotropin-releasing factor is present in human maternal plasma during the third trimester of pregnancy. *J. clin. Endocr. Metab.* **59**, 812–814.

Sastry, B. V. R., Janson, V. E., Owen, L. K., and Tayeb, O. S. (1982). Enkephalin and substance P-like immunoreactivities of mammalian sperm and accessory sex glands. *Biochem. Pharmac.* **31**, 3519–3522.

Sawynok, J., Pinsky, C., and Labella, F. S. (1979). Minireview on the specificity of naloxone as an opiate antagonist. *Life Sci.* **25**, 1621–1632.

Schulz, R., Wilhelm, A., Pirke, K. M., Gramsch, C., and Herz, A. (1981). β-endorphin and dynorphin control serum luteinizing hormone level in immature female rats. *Nature, Lond.* **294**, 757–759.

Schweigerer, L., Bhakdi, S., and Teschmacher, H. (1982). Specific non-opiate binding sites for human β-endorphin on the terminal complex of human complement. *Nature, Lond.* **296**, 572–574.

Shaar, C. J., Frederickson, C. A., Dininger, N. B., and Jackson, L. (1977). Enkephalin analogues and naloxone modulate the release of growth hormone and prolactin: Evidence for regulation by an endogenous opioid peptide in brain. *Life Sci.* **21**, 853–860.

Shaha, C., Liotta, A. S., Krieger, D. T., and Bardin, C. W. (1984). The ontogeny of immunoreactive β-endorphin in fetal, neonatal and pubertal testes from mouse and hampster. *Endocrinology* **114**, 1584–1591.

Shangold, M., Freeman, R., Thysen, B., and Gatz, M. (1979). The relationship between long distance running, plasma progesterone and luteal phase length. *Fert. Steril.* **31**, 130–3.

Shanks, M. F., Clement-Jones, V., Linsell, C. J., Mullen, P. E., Rees, L. H., and Besser, G. M. (1981). A study of the 24-hour profiles of plasma met-enkephalin in man. *Brain Res.* **212**, 403–409.

Sharp, B. and Pekary, A. E. (1981). β-endorphin (61–91) and other β-endorphin immunoreactive peptides in human semen. *J. clin. Endocr. Metab.* **52**, 586–588.

Shavit, Y., Lewis, J. W., Terman, G. W., Gale, R. P., and Liebeskind, J. C. (1984). Opioid peptides mediate the suppressive effect of stress on natural killer cell cytotoxicity. *Science, N.Y.* **223**, 188–190.

Shoupe, D., Montz, F. J., and Lobo, R. A. (1985). The effects of estrogen and progestin on endogenous opioid activity in oophorectomized women. *J. clin. Endocr. Metab.* **30**, 178–183.

Smith, J. H., Redman, C. W. G., Howlett, T. A., Rees, L. H., Anand, K., Harkness, R. A., Cotex, P. L., and Davies, G. S. (1987). Relationship between antenatal heart rate variation and the respiratory, metabolic and β-endorphin environment at delivery. Submitted to *Br. J. Obstet. Gynaec.*

Smith, R., Grossman, A., Gaillard, R., Clement-Jones, V., Ratter, S., Mallinson, J., Lowry, P. J., Besser, G. M., and Rees, L. H. (1981). Studies on circulating met-enkephalin and β-endorphin: normal subjects and patients with renal and adrenal disease. *Clin. Endocr.* **15**, 291–300.

Smyth, D. G., Massey, D. E., Zakarian, S., and Finnie, M. D. A. (1979). Endorphins are stored in biologically active and inactive forms: isolation of α-N-acetyl peptides. *Nature, Lond.* **279**, 252–254.

Spampinato, S. and Goldstein, A. (1984). Immunoreactive dynorphin in rat tissues and plasma. *Neuropeptides* **3**, 193–212.

Spiler, I. J. and Molitch, M. E. (1980). Lack of modulation of pituitary hormone stress response by neural pathways involving opiate receptors. *J. clin. Endocr. Metab.* **50**, 516–520.

Stubbs, W. A., Delitala, G., Jones, A., Jeffcoate, W. J., Edwards, C. R. W., Ratter, S. J., Besser, G. M., Bloom, S. R., and Alberti, K. G. M. M. (1978). Hormonal and metabolic responses to an enkephalin analogue in normal man. *Lancet* **ii**, 1225–1227.

Suda, T., Tozawa, F., Tachibana, S., Demura, H., Shizume, K., Sasaki, A., Mouri, T., and Miura, Y. (1983). Multiple forms of immunoreactive dynorphin in human pituitary and phaeochromocytoma. *Life Sci.* **32**, 865–870.

—— Liotta, A. S., and Krieger, D. T. (1978). β-endorphin is not detectable in plasma from normal human subjects. *Science, N.Y.* **202**, 221–223.

Swinburn, W. R. and Phelen, P. (1982). Response to Naloxone in septic shock. *Lancet* **i**, 167.

Tan, L. T. and Yu, P. H. (1981). *De novo* biosynthesis of enkephalins and their homologues in the human placenta. *Biochem. Biophys. Res. Commun.* **98**, 752–760.

Tojo, K., Kato, Y., Ohta, H., Matsushita, N., Shimatsu, A., Kabayama, Y., Inoue, T., Yanihara, N., and Imura, H. (1985). Stimulation by leumorphin of prolactin secretion from the pituitary in rats. *Endocrinology* **117**, 1169–1174.

Tolis, G., Hickey, J., and Guyda, H. (1975). Effects of morphine on serum growth hormone, cortisol, prolactin and thyroid stimulating hormone in man. *J. clin. Endocr. Metab.* **41**, 797–800.

—— Jukier, L., Wiesen, M., and Krieger, D. T. (1982). Effect of naloxone on pituitary hypersecretory syndromes. *J. clin. Endocr. Metab.* **54**, 780–784.

Tsong, S. D., Philips, D., Halmi, N., Liotta, A. S., Margioris, A., Bardin, C. W., and Krieger, D. T. (1982). Adrenocorticotrophin and β-endorphin-related peptides are present in multiple sites in the reproductive tract of the male rat. *Endocrinology* **110**, 2204–2206.

Valette, A., Tafani, M., Porthe, G., Pontonnier, G., and Cros, J. (1983). Placental kappa binding site: interaction with dynorphin and its possible implication in hCG secretion. *Life Sci.* **33**, *Suppl.* **I**. 523–526.

Van Loon, G. R., Ho, D., and Kim, C. (1980). β-endorphin induced decrease in hypothalamic dopamine turnover. *Endocrinology* **106**, 76–80.

Van Vugt, D. A., Aylsworth, C. F., Sylvester, P. W., Leung, F. C., and Meites, J. (1981). Evidence for hypothalamic noradrenergic involvement in naloxone-induced stimulation of luteinizing hormone release. *Neuroendocrinology* **33**, 261–264.

—— Bruni, J. F., and Meites, J. (1978). Naloxone inhibition of stress-induced increase in prolactin secretion. *Life Sci.* **22**, 85–90.

—— Sylvester, P. W., Aylsworth, C. F., and Meites, J. (1981). Comparison of acute effects of dynorphin and β-endorphin on prolactin release in the rat. *Endocrinology* **109**, 2017–2018.

Van Wimersma Greidanus, T. B. and ten Haaf, J. A. (1984). Opioids and the posterior pituitary. In *Opioid modulation of endocrine function.* (ed. G. Delitala *et al.*) pp. 125–136. Raven Press, New York.

Veldhuis, J. D., Kulin, H. E., Warner, B. A., and Santner, S. J. (1982). Responsiveness of gonadotrophin secretion to infusion of an opiate-receptor antagonist in hypogonadotrophic individuals. *J. clin. Endocr. Metab.* **55**, 649–653.

—— Rogol, A. D., Williams, F. A., and Johnson, M. L. (1983). Do α-Adrenergic mechanisms regulate spontaneous or opiate-modulated pulsatile luteinizing hormone secretion in man? *J. clin. Endocr. Metab.* **57**, 1292–1296.

Viveros, O. H., Diliberto, E. J., Hazum, O., and Chang, K. J. (1979). Opiate-like materials in the adrenal medulla: evidence for storage and secretion with catecholamines. *Molec. Pharmac.* **16**, 1101–1108.

Von Graffenreid, B., Del Pozo, E., Roubicek, J., Krebs, E., Poldinger, W., Burmeister, P., and Kerp. P. (1978). Effects of the synthetic enkephalin analogue FK 33–824 in man. *Nature, Lond.* **272**, 729–730.

Wakabayashi, I., Demura, R., Miki, N., Ohmura, E., Miyoshi, H., and Shimuze, K. (1980). Failure of naloxone to influence plasma growth hormone, prolactin, and cortisol secretions induced by insulin hypoglycaemia. *J. clin. Endocr. Metab.* **50**, 597–599.

Wardlaw, S. L. and Frantz, A. G. (1979). Measurement of β-endorphin in human plasma. *J. clin. Endocr. Metab.* **48**, 176–180.

—— Stark, R. I., Baxi, L., and Frantz, A. G. (1979). Plasma β-endorphin and β-lipotrophin in the human foetus at delivery: correlation with arterial PH and PCO_2. *J. clin. Endocr. Metab.* **49**, 888–891.

—— Wehrenberg, W. B., Ferin, M., Antunes, J. L., and Frantz. A. G. (1982). Effect of sex steroids on β-endorphin in hypophyseal portal blood. *J. clin. Endocr. Metab.* **55**, 877–881.

Warren, M. P. (1980). The effects of exercise on pubertal progression and reproductive function in girls. *J. clin. Endocr. Metab.* **51**, 1150–1157.

Watson, S. J., Khachaturian, H., Coy, D., Taylor, L., and Akil, H., (1982). Dynorphin is located throughout the CNS and is often localised with α-neo-endorphin. *Life Sci.* **31**, 1773–1776.

Weber, E., Evans, C. J., and Barchas, J. D. (1982). Predominance of the amino-terminal octapeptide fragment of dynorphin in rat brain regions. *Nature, Lond.* **299**, 77–79.

—— Voigt, K. H., and Martin, R. (1978). Pituitary somatotrophs contain met-enkephalin-like immunoreactivity.. *Proc. natn. Acad. Sci. U.S.A.* **75**, 6134–6138.

Weber, R. J. and Pert, C. B. (1984). Opiatergic modulation of the immune system. In *Central and peripheral endorphins: Basic and clinical aspects* (ed. E. E. Muller and A. R. Genazzani). 35–42. Raven Press, New York.

Wehrenberg, W. B., Wardlaw, S. L., Frantz, A. G., and Ferin, M. (1982). β-endorphin in hypophyseal portal blood – variations throughout the menstrual cycle. *Endocrinology* **111**, 879–881.

Wilkes, M. M. and Yen, S. S. C. (1980). Reduction by β-endorphin of efflux of dopamine and dopac from superfused medial basal hypothalamus. *Life Sci.* **27**, 1387–1391.

Wuster, M., Schulz, R., and Herz, A. (1981). Multiple opiate receptors in peripheral tissue preparations. *Biochem. Pharmac.* **30**, 1883–1887.

8   Sperm – zona pellucida binding activity

SUZANNE FOURNIER-DELPECH AND MICHEL COUROT

I   Introduction

Fertilization is a fundamental process in sexually reproducing species whereby the bringing together of the male and female gametes, the oocyte and the spermatozoon, results in the fusion of the male and female genomes. Fertilization is not a chance event since it is species specific and is only possible under certain strict conditions. These conditions, related to the physiology of the two gametes and to the environment, must be satisfied to allow normal fertilization and the subsequent events leading to the development of the new embryo.

In mammals, fertilization occurs within the female genital tract where both the male and female gametes have to undergo the final stages of maturation before being able to fertilize. Then, in order to fuse, the gametes must firstly recognize and secondly adhere to each other before penetration of the oocyte by a spermatozoon can occur. In this review, we shall briefly discuss the events that take place when the two gametes are brought together, recognize and adhere to each other. We shall then go on to discuss the physiological ability of sperm to bind to the zona pellucida through yet unknown molecules, later on referred to as sperm 'receptors', and acquisition of these receptors in the epididymis.

II   The oocyte and its cellular barriers to fertilization

At the time of ovulation, most mammalian eggs are naturally protected by several surrounding layers. The outside layer, the cumulus oophorus, is simply a mass of cells made up of the remaining granulosa cells of the follicle. The cells can disperse spontaneously after a given period of time, the length of which depends on the species and chemical composition of the milieu, either *in vivo* or *in vitro* (Dudkiewicz 1984). While the cumulus oophorus as such does not constitute a strict barrier to the access of spermatozoa, cumulus cells may interfere with spermatozoa by enhancement of motility (Bradley and Garbers 1983), selection of spermatozoa (Krzanowska and Lorenc 1983; Sundström 1984), phagocytosis (Pijnenborg, Gordts, Ongkowidjojo, and Brosens 1985) and/or induction of the acrosome reaction (Cummins and Yanagimachi 1982; Yanagimachi and Phillips 1984).

The second layer is the zona pellucida, a thick, acellular component with an asymmetric organization (Phillips and Shalgi 1980; Aitken and Richardson 1981; Aitken, Rudak, Richardson, Dor, Djahanbahkch and Templeton 1981; Ahuja and Blowell 1983). This layer is made up of glycoproteins (Bleil and Wassarman 1980, 1980a; Gwatkin, Anderson, and Williams 1980; Ahuja and Tzartos 1981; Shimizu, Ito, and Dean 1982; Dunbar 1983; Shimizu, Tsuji, and Dean 1983) to which spermatozoa must bind before passing through it to reach the plasma membrane of the oocyte. The zona pellucida is, therefore, a major barrier to fertilization. Present around eggs of all mammals, it is an efficient system allowing the selection of spermatozoa based on criteria related to their functional efficiency such as capacitation (Fraser 1983, 1984; Gould, Overstreet, Yanagimachi, Yanagimachi, Katz, and Hanson 1983), hyperactivated motility (Bedford 1983) and morphological normality (Krzanowska and Lorenc 1983; Sundström 1984). Selection of spermatozoa also occurs mainly through species specificity (Austin and Braden 1956; Yanagimachi 1977, 1981) via reciprocal recognition molecules present on the zona pellucida (Aitken *et al.* 1981; Aitken, Holme, Richardson, and Hulme 1982; Drell and Dunbar 1984; Isojima, Koyama, Hasegawa, Tsunoda, and Hanada 1984; Sacco, Subramanian, and Yurewicz 1984; East, Gulyas, and Dean 1985; Koyama, Hasegawa, Tsuji, and Isojima 1985; Lee, Wong, Teh, and Nishizawa 1985; Mori,

Kamada, Yamano, Kinoshita, Kano, and Mori 1985). Recognition molecules are also present on the plasma membrane of spermatozoa (see p. 307) suggesting the presence of receptors on sperm for the zona (Hartmann and Gwatkin 1971; Hartmann, Gwatkin, and Hutchinson 1972; Bedford 1977; Peterson, Russell, Bundman, and Freund 1980, 1981) which have been discussed in recent reviews (Schmell and Gulyas 1980; Hartmann 1983). The presence of such receptors has been demonstrated using chemical probes as well as polyclonal and monoclonal antibodies raised against proteins of the sperm plasma membrane (Peterson, Robl, Dziuk, and Russell 1982; Shur and Hall 1982, 1982a; Florman, Bechtol, and Wassarman 1984; Huang and Yanagimachi 1984; Lambert 1984; Fournier- Delpech, Courot, and Dubois 1985; Peterson, Henry, Hunt, Saxena, and Russell 1985; Saling, Irons, and Waibel 1985; Saling and Lakoski 1985).

The final layer is the plasma membrane, or vitelline membrane of the oocyte which also interacts with the plasma membrane of spermatozoa at the time of fertilization. This interaction has been demonstrated by showing the migration of sperm antigens on the oocyte plasma membrane at the time of fertilization (Gaunt 1983) and by the finding of changes in the process of spermatozoal penetration when male gametes are pretreated with antibodies against their plasma membrane (Saling, Raines, and O'Rand 1983; Smith, Peterson, and Russell 1983; Hoffman and Curtis 1984; O'Rand and Irons 1984; Saling *et al.* 1985). This also raises the possibility of the presence of oocyte plasma membrane receptors on spermatozoa (Shalgi and Phillips 1980).

In the present paper we review the physiology of sperm receptors for the zona pellucida.

III   The interactions between the spermatozoon and the zona pellucida

1   SPERMATOZOAL BINDING TO THE ZONA PELLUCIDA

The early contact interactions between mammalian spermatozoa and oocytes require the sperm heads to be bound to the zona pellucida after they have either passed between the cells of the cumulus oophorus where present, or have dispersed them. Binding results are normally expressed as the percentage of eggs with spermatozoa bound to the zona surface and/or the average number of spermatozoa bound to each egg.

The gamete surface interactions involve the existence of adhesive structures on the two types of cells which react strongly and specifically and are saturable. The characteristics of these 'sperm receptors' for the zona have been investigated qualitatively (Shalgi and Phillips 1980) and quantitatively (Schmell and Gulyas 1980; Florman and Storey 1981, 1982; Singer, Lambert, Overstreet, Hanson, and Yanagimachi 1985). Furthermore, these receptors have been localized on the plasma membrane covering the acrosome, further referred to as 'periacrosomal' plasma membrane. As shown *in vitro*, cumulus-free eggs and isolated zona pellucida rapidly bind the heads of capacitated spermatozoa taken from the cauda epididymis, the sperm tail remaining free (Hartmann and Gwatkin 1971). The nature of the sperm zona binding of capacitated spermatozoa *in vitro* has been investigated by gentle washing in a micropipette. This has shown the existence of two sorts of spermatozoa, 'bound' and 'attached': bound spermatozoa are those which remain fixed to the zona pellucida after such washing, while attached spermatozoa are separated from it. These attached spermatozoa are considered to possess only a low affinity for the zona pellucida (Gwatkin 1977; Schmell and Gulyas 1980; Hartmann 1983).

Uncapacitated spermatozoa are also able to bind to the zona pellucida (boar: Peterson, Russell, and Hunt 1984; mouse: Saling 1982; rabbit: Swenson and Dunbar 1982). Surprisingly, except in the case of humans and primates, non-capacitated spermatozoa display a strong affinity for the zona of heterospecific as well as homospecific oocytes (Bedford 1977). This is in contrast to capacitated spermatozoa which bind only homospecific zona pellucida (Yanagimachi 1977). At present, little is known of the molecular basis of these differences as well as the basis of the sperm zona binding relevant to the fertilization process *in vivo* (Harper 1970; Moore and Bedford 1978) or *in vitro* (Gwatkin 1977; Rogers 1978). Only a limited number of characteristics of the sperm-zona pellucida binding interactions have been ascertained.

2 ZONA BINDING ABILITY, A SPERM FUNCTION DISTINCT FROM MOTILITY

Since the motility of spermatozoa may influence the interactions between the spermatozoa and oocytes by increasing the chances of spermatozoa binding, experiments have been carried out wherein spermatozoa were reversibly immobilized. Such immobilization was achieved by either lanthanum or low temperature treatment. When exposed *in vitro* to oocytes, immobilized spermatozoa were shown to adhere readily to the surface of the zona pellucida (mouse: Saling 1982; rat: Orgebin-Crist and Fournier-Delpech 1982; ram: Fournier-Delpech, Hamamah, Colas, and Courot 1983; Hamamah 1986; Table 8.1). However, in all cases, the immobilizing treatment resulted in a decreased number of bound spermatozoa (Peterson *et al*. 1980; Heffner and Storey 1982; Saling 1982).

3 LOCALIZATION OF ZONA PELLUCIDA BINDING STRUCT-URES ON SPERMATOZOA

At least three possibilities have been proposed for the binding of spermatozoa to the zona pellucida. Firstly, prior to the acrosome reaction,

Table 8.1

Binding of rat (cauda epididymal) or ram (ejaculated) spermatozoa to the zona pellucida of rat or sheep oocytes

Species	Temperature of incubation	Motility	% of oocytes with spermatozoa bound to the zona pellucida (No of eggs)
Rat	34° C	intense	100 (17)
	0–4° C	0	100 (11)
Sheep	34° C	intense	100 (8)
	0–4° C	0	100 (19)

From Orgebin-Crist and Fournier-Delpech 1982 and Hamamah 1986, by permission.

spermatozoa have been seen to bind by the external surface of the periacrosomal plasma membrane. This has been reported in the mouse (Florman and Storey 1982; Bleil and Wassarman 1983; Storey, Lee, Muller, Ward, and Wirtshafter 1984), the sheep (Crozet and Dumont 1984) and the cow (Crozet 1984). This implies that spermatozoa undergo the acrosome reaction only after binding to the egg (Gwatkin and Williams 1977; Saling, Sowinski, and Storey 1979; Saling and Storey 1979; Cherr, Lambert, Meizel, and Katz 1986). Furthermore, in the mouse it has been shown that proteins of the zona pellucida are able to initiate the acrosome reaction (Bleil and Wassarmann 1983).

The second possibility is that spermatozoa undergoing the acrosome reaction attach to the zona pellucida by the vesiculated acrosomal plasma membranes. This has been observed in the hamster where spermatozoa reaching the zona pellucida possess a fenestrated membrane. The vesicles fix the spermatozoa to the zona and the sperm then pass through the acrosomal ghost (Yanagimachi and Phillips 1984). This process may be important for binding the fertilizing spermatozoa with a particular orientation, since this determines the angle of their passage through the zona pellucida (Barros, Jedlicki, Bize, and Aguirre 1984). The presence of acrosomal ghosts at the surface of ewe and cow oocytes could indicate that these remnants are lost after the acrosome-reacted spermatozoa penetrate the egg (Crozet 1984; Crozet and Dumont 1984; Yanagimachi and Phillips 1984).

Thirdly, acrosome-reacted spermatozoa have been shown to be bound by the inner acrosomal membrane in the guinea pig. In this species, only spermatozoa whose acrosomal content is lost were able to attach to the zonal surface (Huang, Fleming, and Yanagimachi 1981; Huang and Yanagimachi 1984, 1985). Under experimental conditions in the rabbit, acrosome-reacted spermatozoa recovered from the perivitelline space of previously fertilized ova also are able to bind to the zona pellucida (Kuzan, Fleming, and Seidel 1985).

Each of these three possibilities has received ultrastructural support (Moore and Bedford 1983; Talbot 1985) and been illustrated by the different status of the envelopes covering the acrosome of capacitated spermatozoa (Meizel 1985).

The different possibilities of sperm zona binding could be due to species differences or to the fact that all three exist but are successive steps of the sperm zona interactions (Phillips, Shalgi, and Dekel 1985). The latter seems likely in some species where both intact spermatozoa and acrosome-reacted spermatozoa have been seen associated with the zona pellucida, for example in humans (Overstreet and Hembree 1976; Sathanantan, Trounson, Wood, and Leeton 1982; Barros *et al.* 1984; Singer *et al.* 1985), sheep (Fournier-Delpech, Crozet, and Courot 1986) and mice (Saling, Sowinski, and Storey 1979).

4 SPECIES-SPECIFICITY OF SPERM ZONA INTERACTIONS

Under natural conditions, species specificity of interactions between the male and female gamete is well accepted. However, with development of *in vitro* techniques, this specificity appears somewhat equivocal, since non-capacitated spermatozoa of several species are capable of binding strongly to the zona pellucida of heterospecific as well as homospecific oocytes (Yanagimachi 1977, 1981; Schmell and Gulyas 1980; Hartmann 1983). Marked exceptions are the spermatozoa of humans and other primates (Bedford 1977). However, these differences may simply be due to the different origins of the spermatozoa: in the studies involving rodents, spermatozoa were taken from the epididymis, whereas in primate (including human) studies ejaculated sperm were used. It is quite possible that secretions from the accessory glands which are components of the seminal plasma in some way contribute to the homospecificity of binding of the gametes. Moreover, it should be noted that the acrosome-intact spermatozoa which bind to heterospecific zona pellucida are nevertheless unable to undergo the acrosome reaction at zona contact (ram: Fournier- Delpech, Courtens, Pisselet, Delaleu, and Courot 1982; Jedlicki and Barros 1985).

As a result of molecular studies, explanations for these differences are now beginning to appear. Proteins responsible for the interactions between the sperm head and the zona pellucida have been characterized in several species including the human (Moore and Hartman 1984; Peterson *et al.* 1984, 1985; O'Rand, Matthews, Welch, and Fisher 1985; Saling and Lakoski 1985) and proteins from the zona pellucida, having affinity for receptors on sperm, have been isolated (Bleil and Wassarman 1983; East, Mattison, and Dean 1984; East *et al.* 1985). Antibodies against antigenic determinants of the zona pellucida have been assayed in homologous and heterologous systems, providing evidence that some common antigenic determinants of the zona pellucida are shared by different species. One such determinant is shared by the pig and the rabbit (Drell and Dunbar, 1984; Drell, Wood, Bundman, and Dunbar 1984), another by the human and the pig (Koyama *et al.* 1985; Mori *et al.* 1985) and three by the pig, human, hamster, rat and mouse (Isahakia and Alexander 1984; Isojima *et al.* 1984). Only monoclonal antibodies against the species-specific determinant are capable of blocking the binding of homologous spermatozoa to the zona pellucida. This is in agreement with the concept which postulates that species-specific antigens of the zona pellucida support binding of homologous sperm to the zona pellucida (Peterson *et al.* 1980).

Common antigens to the zona pellucida of different species could explain the heterospecific sperm-zona binding. These antigens are structurally different from the species-specific epitopes responsible for homologous sperm-zona binding. In fact monoclonal antibodies directed against foreign zonae pellucidae do not block homologous sperm-zona binding and zona penetration. This has been demonstrated in the hamster using monoclonal antibodies prepared against antigens of the pig zona pellucida (Isojima *et al.* 1984). This explains why fertilization is not inhibited in rats following immunization with heterologous zona pellucida such as those of the pig (Drell *et al.* 1984), since the heterologous antibodies that are formed are not active. The inhibition, when it does occur, of homologous sperm-zona pellucida binding by antibodies directed against heterospecific zona pellucida, could be explained by the experimental conditions, or a second antibody inducing steric hindrance of the receptor activity (Koyama *et al.* 1985).

We consider it to be of primary importance to elucidate the molecular mechanisms involved in the homospecific and heterospecific binding of spermatozoa to the zona pellucida. This will help to clarify the steps prior to fertilization such as activation of bound spermatozoa (that is, hyperactivated motility, acrosomal reaction) by components of the zona pellucida (Bleil and Wassarman 1980; Florman and Storey 1982; Bleil and Wassarman 1983; Storey *et al.* 1984).

IV   Acquisition of oocyte binding capacity by spermatozoa

1   EPIDIDYMAL ACQUISITION BY UNCAPACITATED SPERMA-TOZOA

The ontogenesis of the capacity of spermatozoa to interact with the zona pellucida has been investigated *in vitro*. These investigations have been carried out on several mammalian species by studying the ability of uncapacitated spermatozoa, taken at different levels of the male genital tract, to bind to the zona pellucida.

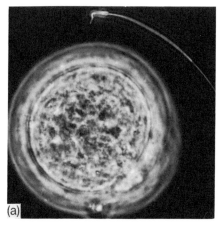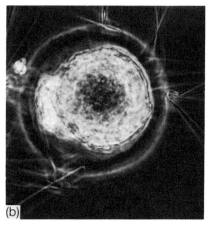

Fig. 8.1. Acquisition of zona pellucida binding capacity by testicular (a) and cauda epididymal (b) spermatozoa. Rat egg exposed to spermatozoa in appropriate medium (Toyoda and Chang 1974) 30 min at 37°C, then transferred to fresh medium. Notice that testicular spermatozoa are not, and epididymal spermatozoa are, bound to the egg. (Adapted from Orgebin-Crist and Fournier-Delpech 1982).

   In the epididymis, spermatozoa develop their ability to recognize and bind to the zona pellucida. This was first demonstrated in rodents, whereby rat testicular sperm failed to adhere to the zona pellucida while spermatozoa from the cauda epididymis bound readily (Orgebin-Crist and Fournier-Delpech 1982; Yanagimachi, Kamiguchi, Sugarawa and Mikamo, 1983; Fig. 8.1). As for ejaculated spermatozoa (see p. 297), these differences were not solely dependent on the motility of spermatozoa since similar results were obtained in experiments performed at 4°C where motility was abolished. Under these conditions, spermatozoa from the cauda epididymis were still able to bind to the zona pellucida while testicular spermatozoa remained unable to do so. These data suggested that the inability of testicular spermatozoa to interact with the zona pellucida was intrinsic and that the zona adhesiveness of the cauda epididymal spermatozoa was acquired in the epididymis. These results are consistent with data, collected *in vivo*, on the fertilizing capacity of rat spermatozoa, which is zero in testicular sperm but greater than 50 per cent in the cauda epididymal sperm (Dyson and Orgebin-Crist 1973; Orgebin-Crist and Olson 1984).
   Evidence has also been obtained in mice for the acquisition of sperm-zona pellucida binding sites in the epididymis. The ability of spermatozoa, recovered from the caput, corpus and cauda epididymis, to bind to the zona pellucida *in vitro* has been examined (Saling 1982). To avoid problems resulting from differences in motility, the spermatozoa were exposed to lanthanum under conditions which are not deleterious to binding to zona pellucida. The binding ability of the spermatozoa, as measured by the number of sperm cells bound to eggs, increased along the epididymis (Fig. 8.2).

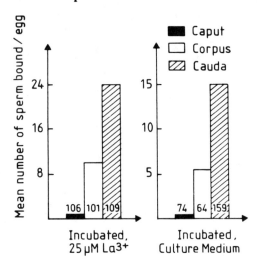

Fig. 8.2. Acquisition of zona pellucida binding capacity by mouse spermatozoa independent of motility. Incubation of spermatozoa with Lanthanum (La^{3+}) which inhibited motility did not change their relative capacity to bind zona pellucida: compare spermatozoa from caput, corpus or cauda epididymis with and without La^{3+} treatment. (Saling 1982).

Similar investigations performed in other mammalian species also show that spermatozoa gradually acquire the ability to bind to the zona pellucida in the body of the epididymis (ram: Fournier-Delpech *et al.* 1982, 1983; Hamamah 1983; boar: Dacheux and Paquignon 1983; Peterson *et al.* 1984; rat: Fournier-Delpech, Hamamah, Tananis-Anthony, Courot, and Orgebin-Crist 1984*a*; hamster: Cuasnicu, Gonzalez-Echeverria, Piazza, and Blaquier 1984).

Finally, in ejaculated spermatozoa, the zona-binding ability is strictly species-specific due to the disappearance of the affinity for heterologous zona pellucida (Table 8.2). These events indicate a molecular differentiation of the spermatozoa zona pellucida receptors by factors present in seminal plasma originating from the accessory glands.

2   ACQUISITION OF ZONA PELLUCIDA BINDING ABILITY BY CAPACITATED EPIDIDYMAL AND EJACULATED SPERMATOZOA

Under normal conditions, fertilization occurs only after spermatozoa have undergone capacitation (Yanagimachi 1977; Bedford 1983; Hartmann 1983; Chang 1984). The first studies with capacitated spermatozoa on the acquisition in the epididymis of their affinity for the egg have been done under heterospecific conditions with hamster vitelline membrane and

Table 8.2

Binding of ram cauda epididymal or ejaculated spermatozoa to the zona pellucida of rat or sheep oocytes. (Fournier-Delpech, unpublished observations).

Origin of *ram* spermatozoa	% of oocytes with spermatozoa bound to the ewe or rat zona pellucida	
	Sheep oocytes	*Rat* oocytes
Cauda epididymis	95–100 (20)	95–100 (200)
Ejaculate	95–100 (21)	< 10 (120)

Temperature of incubation: 34° C
(n) Number of eggs

human and marmoset spermatozoa capacitated *in vitro* (Hinrichsen and Blaquier 1980; Moore 1981*a*; Moore, Hartman, and Pryor 1983).

Homospecific experiments have been done in the sheep: intratubal ova were recovered after intrauterine insemination of spermatozoa originating from the different regions of the epididymis (Fournier-Delpech *et al.* 1986). Following the maturation by passage along the whole length of the epididymis an increase in the number of ova with spermatozoa bound to the zona pellucida has been observed, as well as an increase in the number of spermatozoa bound by the egg and the development of the surface of the periacrosomal membrane adhering to the zona surface. With spermatozoa from the head of the epididymis, the binding occurred only to a small portion of the apical periacrosomal membrane whereas with those from the body of the epididymis, the adhesive sperm membrane was clearly larger and vesiculated; in the tail the entire area of the periacrosomal membrane adhered to the zona pellucida (Fig. 8.3).

Experiments using ram testicular, epididymal and ejaculated spermatozoa recovered after intrauterine incubation lead to the same conclusion: once capacitation has been allowed to occur it can be recognized that those sperm which came from the cauda region of the epididymis had a higher incidence of binding to the zona than those which were derived from the head region. This increase in binding is in parallel with the fertilizing capacity (Fig. 8.4) as previously observed *in vitro* (but differently) for uncapacitated spermatozoa (Fig. 8.4). Thus the sperm-zona receptors react to the capacitation according to their degree of maturation.

V   Endocrine and molecular basis of interactions between the spermatozoa and the zona pellucida

The zona binding structures that develop on the surface of the sperm head in

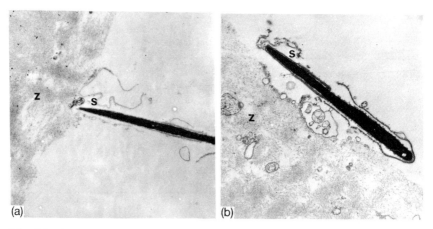

Fig. 8.3. *In vivo* sperm binding to the zona pellucida of sheep ova: a) Spermatozoa from caput epididymis: binding by the tip of periacrosomal plasma membrane; b) Spermatozoa from cauda epididymis: binding by the entire periacrosomal plasma membrane. z = zona pellucida, s = spermatozoa. (Adapted from Fournier-Delpech, Crozet, and Courot 1986, by permission).

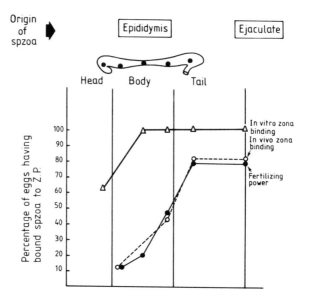

Fig. 8.4. Zona binding ability and fertilizing capacity of epididymal and ejaculated spermatozoa (sheep). (From Fournier-Delpech *et al.* 1983, 1986, by permission). △—△ = non-capacitated spermatozoa, ○—○ and ●—● = capacitated spermatozoa.

the epididymis are regulated by the endocrine activity of the testis.

1   ANDROGENIC CONTROL OF THE SPERM ZONA BINDING CAPACITIES

The development of the zona binding ability during epididymal transit is concomitant with the development of fertilizing ability under androgen regulation.

The zona binding ability of spermatozoa from the cauda epididymis is progressively lost between 3 days and 10 days after castration in the rat. Testosterone treatment maintains zona binding ability at control levels even 10 days after castration (Fig. 8.5) (Fournier-Delpech *et al.* 1984*a*).

In spermatozoa from castrated rams, whether treated with testosterone implants or not, it has been shown that the *in vitro* zona binding capacity is closely related to the level of testosterone in peripheral blood (Fig. 8.6) (Fournier-Delpech, Hamamah, Courot, and Kuntz 1984). In an *in vitro* culture system developed for the epididymal tubule, addition of androgens increased the zona binding ability of the spermatozoa from the proximal corpus as compared to the control cultures (Cuasnicu *et al.* 1984).

It seems likely that the zona binding structures are synthesized or activated under androgenic control during maturation of sperm as they pass through the epididymis. Their ontogenesis is relevant both to epididymal and to intrinsic spermatozoal factors.

2   EPIDIDYMAL ACQUISITION OF SPERM ZONA BINDING CAPACITIES

i   Epididymal factors

The increase in binding to the zona pellucida following exposure to 5αDHT of cultured spermatozoa from the proximal part of the corpus epididymis is lost when an inhibitor of protein synthesis, cycloheximide, is added to the culture dishes. Thus, protein synthesis is required for the androgenic effect (Cuasnicu *et al.* 1984; Gonzalez-Echeverria, Cuasnicu, Piazza, Pineiro, and Blaquier 1984). It is known that proteins secreted under androgenic control are incorporated into spermatozoa (rat: Fournier-Delpech, Bayard, and Boulard 1973; Killian and Amann 1973; Lea, Petrusz, and French 1978; Faye, Duguet, Mazzuca, and Bayard 1980; Kohane, Gonzalez-Echeverria, Pineiro, and Blaquier 1980; Lea and French 1981; Gonzalez-Echeverria, Cuasnicu, and Blaquier 1982; Jones and Brown 1982; Brooks 1983; Brooks and Tyver 1984; Jones, Brown, Von Glos, and Gaunt 1985; Hamster and rabbit: Moore 1980, 1981; Djakiew and Jones 1983; Klinefelter and Hamilton 1985).

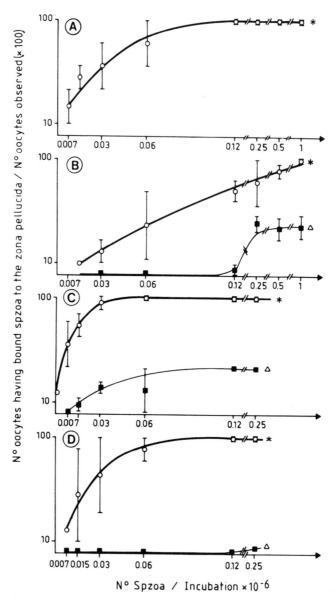

Fig. 8.5. Zona-binding ability of spermatozoa from the distal cauda epididymis from normal and castrated rats with (○) and without (■) testosterone treatment as a function of number of spermatozoa incubated: A) intact rats; B) rats castrated for 3 days; C) rats castrated for 7 days; D) rats castrated for 10 days. (From Fournier-Delpech *et al.* 1984*a*, by permission).

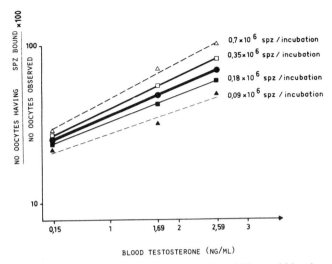

Fig. 8.6. Correlation between spermatozoa binding ability and blood testosterone in the ram. (Fournier-Delpech *et al.* 1984, by permission).

Some of these proteins play a key role in the sperm-zona binding ability: this has been demonstrated *in vitro* in rodents by exposing immature spermatozoa to an androgen dependent epididymal glycoprotein (rat: Orgebin-Crist and Fournier-Delpech 1982; hamster: Cuasnicu, Gonzalez-Echeverria, Piazza, Cameo, and Blaquier 1984*a*). Moreover, the presence of immunoglobulins against such epididymal proteins diminishes *in vitro* or *in vivo* the capacity of the spermatozoa to interact with the zona pellucida (Munoz and Metz 1978; Moore 1980, 1981; Mettler, Czuppon, and Tinneberg 1983; Cuasnicu, Gonzalez-Echeverria, Piazza, Cameo, and Blaquier 1984*b*; Fournier-Delpech *et al.* 1985).

These proteins secreted by the epididymis are strong candidates for a role in the sperm-zona recognition process: the glucosidase activity present in the epididymal plasma (Mann 1964) could modify the sperm surface, or alternatively it may be concerned with incorporating saccharide chains into the sperm plasma membrane to form new antigens of epididymal origin. Saccharides are known to be involved in intercellular adhesion (Nicolson, Usui, Yanagimachi, Yanagimachi, and Smith 1977; Yanagimachi 1977, 1981; Koehler 1981; Shur 1984; Sylvester, Skinner, and Griswold 1984).

ii   Intrinsic factors concerned with the epididymal acquisition of sperm zona binding capacities

Mammalian spermatozoa are thought to possess species-specific receptors for the zona pellucida that mediate the binding of sperm to the surface of the zona pellucida.

Indirect evidence for their existence has been obtained from the observed inhibitory effects on sperm attachment and fertilization after treatment of the gametes with antisera against antigens of zona pellucida both *in vitro* (Dunbar and Shiver 1976; Tsunoda and Chang 1976) and *in vivo* (Gwatkin and Williams 1977; Gwatkin, Williams, and Carlo 1977; Tsunoda and Chang 1978; Sacco 1981). The mechanism involved appears to be the inhibition of sperm binding to recognition molecules on the zona pellucida (Aitken and Richardson 1981; East *et al.* 1985; Koyama *et al.* 1985). Whether antibodies against zona pellucida inhibit the binding of spermatozoa through interaction with postulated specific receptors of the spermatozoa or by blockage by cross linking with structurally unrelated sites of the zona surface (Dunbar and Shiver 1976) remains an open question.

More direct evidence for the existence of receptors on spermatozoa for the zona pellucida has been provided by extraction from murine zona pellucida of proteins possessing sperm binding capacity. Among these proteins, the major one (ZP3) is active by virtue of its O-linked saccharide residues (Florman *et al.* 1984; Florman and Wassarman 1985). It is a sulfated glycoprotein of known molecular weight (80 KD), and is responsible both for the binding of spermatozoa with intact acrosomes to the zona pellucida and for induction of the acrosome reaction (Bleil and Wassarman 1980, 1980*a*, 1983). This system is thought to be general since a glycoprotein of different molecular weight but with a similar activity has been extracted from the pig zona pellucida. Boar spermatozoa pretreated with an antibody against this protein lost their ability to bind to the pig zona pellucida (Sacco *et al.* 1984).

The existence of specific receptors on spermatozoa for the zona pellucida, postulated from the inhibitory effects of antibodies against sperm plasma membrane on sperm eggs interactions (Peterson *et al.* 1982; Saling *et al.* 1983; Moore and Hartmann 1984; O'Rand and Irons 1984; Lee *et al.* 1985), has been confirmed by the extraction from mature spermatozoa of several antigenic surface components responsible of their binding to the zona pellucida. Their characteristics (for example, molecular weight), differ from species to species (Hartmann and Hutchinson 1981; O'Rand *et al.* 1985; Peterson *et al.* 1985; Saling and Lakoski 1985; Sullivan and Bleau 1985).

The components that enhance sperm-zona binding are acquired by the plasma membrane of uncapacitated sperm during passage through the epididymis. They are composed of some of the proteins characterized on the epididymal spermatozoa, and which differ from those on the testicular spermatozoa, as has been shown in several species such as bulls, rams, rats, rabbits, mice and boars (Killian and Amann 1973; Olson and Hamilton 1978; Oliphant and Singhas 1979; Voglmayr, Fairbanks, Jackowitz, and Colella 1980; Feutcher, Vernon, and Eddy 1981; Jones, Pholpramool, Setchell, and Brown 1981; Olson and Danzo 1981; Olson and Orgebin-Crist 1981; Hammerstedt, Hay, and Amann 1982; Brown, Von Glos, and Jones

1983; Wolf, Sokoloski, Dandekar, and Bechtol 1983; Russell, Peterson, Hunt, and Strack 1984; Orgebin-Crist and Olson 1984; Eddy, Vernon, Muller, Hahnel, and Fenderson 1985; Ellis, Hartmann, and Moore 1985; Jones *et al.* 1985; Peterson, Hunt, and Henry 1986; Saxena, Russell, Saxena, and Peterson 1986).

In view of the testicular origin of spermatozoal antigens which play a role in zona pellucida penetration (O'Rand, Irons, and Porter 1984), it could be postulated that the receptors on sperm for the zona might in part appear as proreceptors in the testis and then differentiate structurally and chemically in the epididymis under androgen dependent factors such as proteins whose synthesis is testosterone dependent. It seems likely that the receptors for zona pellucida which mature on the periacrosomal sperm plasma membrane in the epididymis exist on uncapacitated spermatozoa and bind non-specifically to the heterologous or homologous zona pellucida. The strict species-specificity is observed only with ejaculated spermatozoa before capacitation (see p. 303).

How such receptors recognize the zona pellucida remains a matter for conjecture. Galactosyltransferases are known to exist on the sperm membrane and have been postulated to be the receptors for the zona since they bind to the N-acetyl-glucosamine residues of the zona pellucida (Shur and Hall 1982, 1982*a*; Shur 1984). Such N-acetyl-glucosamine molecules could be part of the O-linked oligosaccharide chains which are the active binding sites of the protein ZP3 (Florman *et al.* 1984; Florman and Wassarman 1985). Like sperm membrane glycosyltransferase, galactosyltransferase extracted from bovine milk has also been shown to be active in sperm-zona binding (Lopez, Bayna, Litoff, Shaper, Shaper, and Shur 1985); it could represent receptors acquired by spermatozoa in the epididymis which are not strictly species and tissue specific. The binding capacity may be inhibited by the trypsin-like activity of the sperm membrane during sperm-zona interaction (Saling 1981). This releases small peptides from the zona pellucida, thus preventing the sperm-zona penetration (Hartmann and Hutchinson 1981).

This trypsin-like activity appears to be carried on the plasma membrane of mature capacitated spermatozoa and not by the zona pellucida. It enables spermatozoa to interact with the zona pellucida, but not with the egg plasma membrane (Saling 1981). Moreover, modulation of the activity of receptors on sperm appears to result from the sequential release of peptides during the sperm-zona pellucida interaction (Hartmann and Hutchinson 1981).

The enzymatic processes of the sperm-zona interaction also involve sperm sulphatases (Ahuja and Gilburt 1985).

These various processes suggest a programmed activity of receptors for the zona on the mature capacitated sperm and their transformation during the sperm-zona binding. The study of the differentiation and activity of such molecules during the post-testicular maturation of spermatozoa certainly represents a stimulating challenge.

An alternative possibility for the sperm-zona pellucida recognition and binding process relates to sperm surface carbohydrates (Ahuja 1982, 1985; Huang, Ohzu, and Yanagimachi 1982). These are associated with the saccharide chains which determine the species specificity of haptens of lectin-like zona proteins such as ZP3 (Lambert 1984). L-fucose and fucoidins have been shown to be part of the sperm-zona recognition signal in several mammals including humans (Huang, Ohzu, and Yanagimachi 1982; De Santis, Pinto, Cotelli, Rosati, Monroy, and D'Alessio 1983; Huang and Yanagimachi 1984; Lambert and Van Lee 1984; Ahuja 1985).

The postulated role of carbohydrate mediation in the sperm-zona interactions appears to be consistent with the changes known to affect the composition and topography of sperm membrane glycoproteins when spermatozoa pass through the epididymis (Fournier-Delpech and Courot 1980; Vierula and Rajaniemi 1981; Dacheux and Voglmayr 1983; Fournier-Delpech, Hamamah, Delaleu, Courtens, and Pisselet 1983*a*; Gaunt 1983; Gaunt, Brown, and Jones 1983; Arya and Vanha-Pertulla 1984; Fournier-Delpech, Hamamah, Courot and Pisselet 1984*b*). These changes occur in parallel with the development of the ability of sperm to bind to the zona pellucida. Carbohydrate moieties synthesized by spermatozoa (Ahuja 1985) raise the question of the differentiation of the saccharidic hapten in the epididymis; it could be under the control of the local environment, for example by the action of androgen-dependent epididymal secretory proteins with glycosyltransferase activities, on the spermatozoal carbo-hydrate moieties (Brown *et al.* 1983).

The mechanism of differentiation, under androgenic control, of the sperm-zona pellucida receptors in the epididymis requires further investiga-tion: the sperm receptors for zona appear to be complicated structures with different functional parts. Some are intrinsic to the structure of spermatozoa and capable of differentiation like the polysaccharidic chains. Others depend on epididymal factors such as specific androgen-dependent proteins.

VI   Conclusion

In this review, we have shown that the spermatozoa interact with eggs via components which are present both on the zona pellucida and the periacrosomal sperm plasma membrane. The 'receptors' of the spermato-zoa develop during epididymal transit in mammals. This implies a maturational process leading to the appearance of adhesive molecules on the sperm plasma membrane similar to those known in the echinoderms (Rossignol and Lennarz 1983; Vacquier 1983). In spite of differences in the sequence of events involving the appearance and the localization of the sperm receptors, this may be a general process of sperm/egg recognition common to both invertebrates and mammals.

The species specificity of the sperm 'receptors' for the egg in mammals has been found mainly on the periacrosomal sperm plasma membrane. In most species, except the guinea pig, the acrosome seems to remain unreacted until the time of gamete binding.

In sea urchins the first contact of spermatozoa with the egg is at the stage of the jelly surrounding the oocyte. Under appropriate ionic conditions, the jelly induces the acrosome reaction, a series of processes involving the assembly of actin filaments, the extrusion of the acrosomal vesicle and the exposure of an adhesive protein, the bindin, on the inner acrosomal membrane. This mediates the species specific binding of the sperm to the vitelline layer of the egg (Rossignol and Lennarz 1983; Vacquier 1983; Rossignol, Earles, Decker, Lennarz 1984; Shapiro 1984).

In mammals, species-specific receptors also exist on the periacrosomal sperm plasma membrane as seen in mice, boars, humans and rams. Their binding by homologous components of the zona pellucida seems to be associated with the initiation of the acrosome reaction.

The difference between invertebrates (for example, sea urchin), where homospecific receptors are found on the inner acrosomal membrane, and mammals, where they are found in the periacrosomal plasma membrane, may be related to the conditions of fertilization, that is, either in the external milieu or in the female genital tract. The presence of homospecific sites on the inner acrosomal membrane of mammals, as seen for example in the guinea pig, is probably related to a sort of primitive homospecific receptor, and may be a function of the large size of the acrosome in this species (Fawcett and Hollenberg 1963).

The homologous sperm zona receptors are either not differentiated, or only poorly differentiated at time of spermiation in mammals, in contrast to the sea urchin. They develop in the epididymis under endocrine testicular androgen control, thus ensuring fertilizing power (Orgebin-Crist, Danzo, and Davies, 1975).

The nature of the epididymal control of the differentiation of receptors for zona pellucida implies changes of the plasma membrane: 1) Oligosaccharide changes brought about by the activity of saccharide isomerases or transferases permit the binding of spermatozoa to zona pellucida (see p. 309); 2) Polypeptide changes inducing enzymatic activity (for example, galactosyltransferase) which allows binding of membrane components of spermatozoa to terminal residues in the zona pellucida; 3) Trypsin-like activity related to the sperm-zona pellucida adhesion.

Further research is needed on the identification and localization of binding sites in mammalian spermatozoa, the details of which are still too obscure. At present, the basic processes of their molecular maturation are still unknown. Likewise it is unclear how the sperm-zona pellucida complex induces the acrosome reaction.

In addition to their scientific interest, these questions are of fundamental importance for the understanding of the control of fertility.

Acknowledgements

The authors are indebted to Nicole Crozet, Marie-Claire Levasseur and Catherine Kuntz for critical discussion of the manuscript, together with Claudine Pisselet for checking the list of references, Josephine Arendt, Robert J. Kilgour and J. R. Clarke for improvement of the English.

References

Ahuja, K. K. (1982). Fertilization studies in the hamster: the role of cell-surface carbohydrates. *Expl. Cell. Res.* **140**, 353–362.
—— (1985). Inhibitors of glycoprotein biosynthesis block fertilization in the hamster. *Gamete Res.* **11**, 179–189.
—— and Blowell, G. P. (1983). Probable asymmetry in the organization of components of the hamster zona pellucida. *J. Reprod. Fert.* **69**, 49–55.
—— and Gilburt, D. J. (1985). Involvement of sperm sulphatases in early sperm. Zona interactions in the hamster. *J. Cell Sci.* **78**, 247–261.
—— and Tzartos, S. J. (1981). Investigation of sperm receptors in the hamster zona pellucida by using univalent (Fab) antibodies to hamster ovary. *J. Reprod. Fert.* **61**, 257–264.
Aitken, R. J., Holme, E., Richardson, D. W., and Hulme, M. (1982). Properties of intact and univalent (Fab) antibodies raised against isolated, solubilized, mouse zonae pellucidae. *J. Reprod. Fert.* **66**, 327–334.
—— and Richardson, D. W. (1981). Measurement of the sperm binding capacity of the mouse zona pellucida and its use in the estimation by anti-zona antibody titres. *J. Reprod. Fert.* **63**, 295–307.
—— Rudak, E. A., Richardson, D. W., Dor, J., Djahanbahkch, O., and Templeton, A. A. (1981). The influence of anti-zona and anti-sperm antibodies on sperm-egg interactions. *J. Reprod. Fert.* **62**, 597–606.
Arya, M. and Vanha-Perttula, T. (1984). Distribution of lectin binding in rat testis and epididymis. *Andrologia* **16**, 495–508.
Austin, C. R. and Braden, A. W. H. (1956). Early reactions of the rodent egg to spermatozoa penetration. *J. exp. Biol.* **33**, 358–365.
Barros, C., Jedlicki, A., Bize, I., and Aguirre, E. (1984). Relationship between the length of sperm preincubation and zona penetration in the golden hamster: a scanning Electron Microscopy study. *Gamete Res.* **9**, 31–43.
Bedford, J. M. (1977). Sperm-egg interaction: the specificity of human spermatozoa. *Anat. Rec.* **188**, 477–488.
—— (1983). Significance of the need for sperm capacitation before fertilization in eutherian mammals. *Biol. Reprod.* **28**, 108–120.
Bleil, J. D. and Wassarman, P. M. (1980). Mammalian sperm-egg interaction: identification of a glycoprotein in the mouse egg zonae pellucidae possessing receptor activity for sperm. *Cell* **20**, 873–882.
—— —— (1980a). Structure and function of the zona pellucida: identification and characterization of the proteins of the mouse oocyte's zona pellucida. *Devl. Biol.* **76**, 185–202.
—— —— (1983). Sperm-egg interactions in the mouse: sequence of events and induction of the acrosome reaction by a zona pellucida glycoprotein. *Devl. Biol.* **95**, 317–324.

Bradley, M. P. and Garbers, D. L. (1983). The stimulation of bovine caudal epididymal sperm forward motility by bovine cumulus-egg complexes *in vitro*. *Biochem. Biophys. Res. Commun.* **115**, 777–787.

Brooks, D. E. (1983). Selective binding of specific rat epididymal secretory proteins to spermatozoa and erythrocytes. *Gamete Res.* **7**, 367–376.

—— and Tyver, K. (1984). Analysis of surface proteins of rat spermatozoa during epididymal transit and identification of antigens common to spermatozoa rete-testis fluid and cauda epididymal plasma. *J. Reprod. Fert.* **71**, 249–257.

Brown, C. R., Von Glos, K. I., and Jones, R. (1983). Changes in plasma membrane glycoproteins of rat spermatozoa during maturation in the epididymis. *J. Cell Biol.* **96**, 256–264.

Chang, M. C. (1984). The meaning of sperm capacitation. *J. Androl.* **5**, 45–50.

Cherr, G. N., Lambert, H., Meizel, S., and Katz, D. F. (1986). *In vitro* studies of the golden hamster sperm acrosome reaction: completion on the zona pellucida and induction by homologous soluble zonae pellucidae. *Devl. Biol.* **114**, 119–131.

Crozet, N. (1984). Ultrastructural aspects of *in vivo* fertilization in the cow. *Gamete Res.* **10**, 241–251.

—— and Dumont, M. (1984). The site of the acrosome reaction during *in vivo* penetration of the sheep oocyte. *Gamete Res.* **10**, 97–105.

Cuasnicu, P. S., Gonzalez-Echeverria, F., Piazza, A., and Blaquier, J. A. (1984). Addition of androgens to cultured hamster epididymis increases zona recognition by immature spermatozoa. *J. Reprod. Fert.* **70**, 541–547.

—— —— —— Piñeiro, L., and Blaquier, J. A. (1984*a*). Epididymal proteins mimic the androgenic effect on zona pellucida recognition by immature hamster spermatozoa. *J. Reprod. Fert.* **71**, 427–431.

—— —— Piazza, A. D., Cameo, M. S. and Blaquier, J. A. (1984*b*). Antibodies against epididymal glycoproteins block fertilizing ability in rat. *J. Reprod. Fert.* **72**, 467–471.

Cummins, J. M. and Yanagimachi, R. (1982). Sperm-egg ratios and the site of the acrosome reaction during *in vivo* fertilization in the hamster. *Gamete Res.* **5**, 239–256.

Dacheux, J. L. and Paquignon, M. (1983). Influence of initiation of forward motility on the fertilizing ability of immature boar spermatozoa in *in vivo* homologous and *in vitro* heterologous systems in insemination. In *The sperm cell* (ed. J. André) pp. 99–102. M. Nijhoff, The Hague.

—— and Voglmayr, J. K. (1983). Sequence of sperm cell surface differentiation and its relationship to exogenous fluid proteins in the ram epididymis. *Biol. Reprod.* **29**, 1033–1046.

De Santis, R., Pinto, M. R., Cotelli, F., Rosati, F., Monroy, A., and D'Alessio, G. (1983). A fucosyl glycoprotein component with sperm receptor and sperm-activating activities from the vitelline coat of *Ciona intestinalis* eggs. *Expl. Cell Res.* **148**, 508–513.

Djakiew, D. and Jones, R. C. (1983). Sperm maturation, fluid transport, and secretion and absorption of protein in the epididymis of the echidna *(Tachyglossus aculeatus)*. *J. Reprod. Fert.* **68**, 445–456.

Drell, D. W. and Dunbar, B. S. (1984). Monoclonal antibodies to rabbit and pig zonae pellucidae distinguish species-specific and shared antigenic determinants. *Biol. Reprod.* **30**, 445–457.

—— Wood, D. M., Bundman, D., and Dunbar, B. S. (1984). Immunological comparison of antibodies to porcine zonae pellucidae in rats and rabbits. *Biol. Reprod.* **30**, 435–444.

Dudkiewicz, A. B. (1984). Purification of boar acrosomal aryl sulfatase A and possible role in the penetration of cumulus cells. *Biol. Reprod.* **30**, 1005–1014.

Dunbar, B. S. (1983). Morphological, biochemical and immunochemical characterization of the mammalian zona pellucida. In *Mechanism and control of animal fertilization* (ed. J. F. Hartmann) pp. 139–175. Academic Press, New York.

—— and Shiver, C. A. (1976). Immunological aspects of sperm receptors on the zona pellucida of mammalian eggs. *Immunol. Commun.* **5**, 375–385.

Dyson, A. L. M. B. and Orgebin-Crist, M. C. (1973). Effect of hypophysectomy, castration and androgen replacement upon the fertilizing ability of rat epididymal spermatozoa. *Endocrinology* **93**, 391–402.

East, I. J., Gulyas, B. J., and Dean, J. (1985). Monoclonal antibodies to the murine zona pellucida protein with sperm receptor activity: effects on fertilization and early development. *Devl. Biol.* **109**, 268–273.

—— Mattison, D. R., and Dean, J. (1984). Monoclonal antibodies to the major protein of the murine zona pellucida: effects on fertilization and early development. *Devl. Biol.* **104**, 49–56.

Eddy, E. M., Vernon, R. B., Muller, C. H., Hahnel, A. C., and Fenderson, B. A. (1985). Immunodissection of sperm surface modifications during epididymal maturation. *Am. J. Anat.* **174**, 225–237.

Ellis, D. H., Hartman, T. D., and Moore, H. D. M. (1985). Maturation and function of the hamster spermatozoon probed with monoclonal antibodies. *J. reprod. Immunol.* **7**, 299–314.

Fawcett, D. W. and Hollenberg, R. D. (1963). Changes in the acrosome of guinea pig spermatozoa during passage through the epididymis. *Z. Zellforsch.* **60**, 276–292.

Faye, J. C., Duguet, L., Mazzuca, M., and Bayard, F. (1980). Purification, radioimmunoassay and immunohistochemical localization of a glycoprotein produced by the rat epididymis. *Biol. Reprod.* **23**, 423–432.

Feuchter, F. A., Vernon, R. B., and Eddy, E. M. (1981). Analysis of the sperm surface with monoclonal antibodies: topographically restricted antigens appearing in the epididymis. *Biol. Reprod.* **24**, 1099–1110.

Florman, H. M., Bechtol, K. B., and Wassarman, P. M. (1984). Enzymatic dissection of the functions of the mouse egg's receptor for sperm. *Devl. Biol.* **106**, 243–255.

—— and Storey, B. T. (1981). Inhibition of *in vitro* fertilization of mice eggs: 3-quinuclidinyl benzylate specifically blocks penetration of zonae pellucidae by mouse spermatozoa. *J. exp. Zool.* **216**, 159–167.

—— —— (1982). Mouse gamete interactions: the zona pellucida is the site of the acrosome reaction leading to fertilization *in vitro*. *Devl. Biol.* **91**, 121–130.

—— and Wassarman, P. M. (1985). O-linked oligosaccharides of mouse ZP_3 account for its sperm receptor activity. *Cell* **41**, 313–324.

Fournier-Delpech, S., Bayard, F., and Boulard, C. (1973). Étude d'une protéine spécifique du sperme épididymaire chez le rat: hormonodépendance, relation avec l'acide sialique. *C.r. Séance. Soc. Biol.* **163**, 1989–1996.

—— and Courot, M. (1980). Glycoproteins of ram sperm plasma membrane. Relationship of protein having affinity for Con A to epididymal maturation. *Biochem. Biophys. Res. Commun.* **96**, 756–761.

—— and Dubois, M. P. (1985). Decreased fertility and motility of spermatozoa from rats isoimmunized with a prealbumin epididymal specific protein. *Int. J. Androl.* **6**, 246–150.

—— Courtens, J. L., Pisselet, C., Delaleu, B., and Courot, M. (1982). Acquisition of zona binding by ram spermatozoa during epididymal passage, as revealed by interaction with rat oocytes. *Gamete Res.* **5**, 403–408.

—— Crozet, N., and Courot, M. (1986). Electron microscopic study of the *in vivo* zona pellucida binding ability of epididymal ram spermatozoa. *Gamete Res.* **14**, 225–234.

—— Hamamah, S., Colas, G., and Courot, M. (1983). Acquisition of zona binding structures by ram spermatozoa during epididymal passage. In *The sperm cell* (ed. J. André) pp. 103–110. M. Nijhoff, The Hague.

—— —— and Kuntz, C. (1984). Androgenic control of zona binding capacity or ram spermatozoa. In *The male in farm animals reproduction*. (ed. M. Courot) pp. 103–107. M. Nijhoff, Kluwer Acad. Pub., Dordrecht.

—— —— Courot, M., and Pisselet, C. (1984*b*). Organization of glycoproteins of sperm-egg recognition system into the epididymis. In *Role of carbohydrates in cell recognition, Endogenous lectins*. Abstr. 2. (ed. M. Monsigny) Biology of the cell **51**, 49a.

—— —— Delaleu, B., Courtens, J. L., and Pisselet, C., (1983*a*). E. M. localization of sugars having affinity for lectins on the ram spermatozoa. In *The sperm cell* (ed. J. André) pp. 163–166. M. Nijhoff, The Hague.

—— —— Tananis-Anthony, C., Courot, M., and Orgebin-Crist, M. C. (1984*a*). Hormonal regulation of zona binding ability and fertilizing ability of rat epididymal spermatozoa. *Gamete Res.* **9**, 21–30.

Fraser, L. R. (1983). Mouse sperm capacitation assessed by kinetics and morphology of fertilization *in vitro*. *J. Reprod. Fert.* **69**, 419–428.

—— (1984). Mechanisms controlling mammalian fertilization. In *Oxford reviews of reproductive biology*, Vol. 6 (ed. J. R. Clarke) pp. 174–225. Clarendon Press, Oxford.

Gaunt, S. J. (1983). Spreading of a sperm surface antigen within the plasma membrane of the egg after fertilization in the rat. *J. Embryol. exp. Morphol.* **75**, 259–270.

—— Brown, C. R., and Jones, R. (1983). Identification of mobile and fixed antigens on the plasma membrane of rat spermatozoa using monoclonal antibodies. *Expl. Cell. Res.* **144**, 275–284.

Gonzalez-Echeverria, F. M., Cuasnicu, P. S., and Blaquier, J. A. (1982). Identification of androgen-dependent epididymal glycoprotein in the hamster epididymis and their association with spermatozoa. *J. Reprod. Fert.* **64**, 1–7.

—— —— Piazza, A., Piñeiro, L., and Blaquier, J. A. (1984). Addition of an androgen-free epididymal protein extract increases the ability of immature hamster spermatozoa to fertilize *in vivo* and *in vitro*. *J. Reprod. Fert.* **71** 433–437.

Gould, J. E., Overstreet, J. W., Yanagimachi, H., Yanagimachi, R., Katz, D. F., and Hanson, F. W. (1983). What functions of the sperm cell are measured by *in vitro* fertilization of zona free eggs? *Fert. Steril.* **40**, 344–352.

Gwatkin, R. B. L. (1977). *Fertilization mechanisms in man and mammals*. Plenum Press, New York.

—— Andersen, O. F., and Williams, D. T. (1980). Large scale isolation of bovine, and pig zonae pellucidae: chemical, immunological, and receptor properties. *Gamete Res.* **3**, 217–231.

—— and Williams, D. T. (1977). Receptor activity for the hamster and mouse solubilized zona pellucida before and after the zona reaction. *J. Reprod. Fert.* **49**, 55–59.

—— —— and Carlo, D. J. (1977). Immunization of mice with heat solubilized hamster zonae: production of anti-zona antibody and inhibition of fertility. *Fert. Steril.* **28**, 871–877.

Hamamah, S. (1983). *Evolution de caractéristiques membranaires des spermatozoïdes en relation avec le développement dans l'épididyme de leur capacité à reconnaître la zone pellucide. Etude chez le bélier.* Thèse de 3ème cycle, Univ. Montpellier.

—— (1986). *Evolution de caractéristiques membranaires des spermatozoïdes de bélier au cours de la maturation et de la capacitation. Sites de reconnaissance de la zone pellucide; régulation endocrinienne, étude de glycoprotéines..* Thèse Doc. ès Sci., Univ. Montpellier.

Hammerstedt, R. H., Hay, S. R., and Amann, R. P. (1982). Modification of ram sperm membranes during epididymal transit. *Biol. Reprod.* **27**, 745–754.

Harper, M. J. K. (1970). Factors influencing sperm penetration of rabbit egg *in vivo. J. exp. Zool.* **173**, 47–62.

Hartmann, J. F. (1983). Mammalian fertilization: gamete surface interactions *in vitro.* In *Mechanism and control of animal fertilization* (ed. J. F. Hartmann) pp. 325–364. Academic Press, New York.

—— and Gwatkin, R. B. L. (1971). Alteration of sites on the mammalian sperm surface following capacitation. *Nature, Lond.* **234**, 479–481.

—— —— and Hutchinson, C. F. (1972). Early contact interactions between mammalian gametes *in vitro*: Evidence that the vitellus influences adherence between sperm and zona pellucida. *Proc. natn. Acad. Sci. U.S.A.* **69**, 2767–2769.

—— and Hutchinson, C. F. (1981). Modulation of fertilization *in vitro* by peptides released during hamster sperm zona pellucida interaction. *Proc. natn. Acad. Sci. U.S.A.* **78**, 1690–1694.

Heffner, L. J. and Storey, B. T. (1982). Cold lability of mouse sperm binding to zona pellucida. *J. exp. Zool.* **219**, 155–161.

Hinrichsen, M. J. and Blaquier, J. A. (1980). Evidence for supporting the existence of sperm maturation in the human epididymis. *J. Reprod. Fert.* **60**, 291–294.

Hoffman, M. L. and Curtis, G. L. (1984). Prevention of monkey sperm penetration of zona free hamster ova by sperm antibody obtained from cynomolgus monkey. *Fert. Steril.* **42**, 101–111.

Huang, T. T. F., Fleming, A. D., and Yanagimachi, R. (1981). Only acrosome reacted spermatozoa can bind to and penetrate zona pellucida: a study using the guinea pig. *J. exp. Zool.* **217**, 287–290.

—— Ohzu, E., and Yanagimachi, R. (1982). Evidence suggesting that L-Fucose is part of a recognition signal for sperm-zona pellucida attachment in mammals. *Gamete Res.* **5**, 355–361.

—— and Yanagimachi, R. (1984). Fucoidin inhibits attachment of guinea pig spermatozoa to the zona pellucida through binding to the inner acrosomal membrane and equatorial domains. *Expl. Cell Res.* **153**, 363–373.

—— —— (1985). Inner acrosomal membrane of mammalian spermatozoa: its properties and possible functions in fertilization. *Am. J. Anat.* **174**, 249–268.

Isahakia, M. and Alexander, N. J. (1984). Interspecies cross-reactivity of monoclonal antibodies directed against human sperm antigens. *Biol. Reprod.* **30**, 1015–1026.

Isojima, S., Koyama, K., Hasegawa, A., Tsunoda, Y., and Hanada, A. (1984). Monoclonal antibodies to porcine zona pellucida antigen and their inhibitory effects on fertilization. *J. reprod. Immunol.* **6**, 77–87.

Jedlicki, A. and Barros, C. (1985). Scanning electron microscopy study of *in vitro* prepenetration gamete interactions. *Gamete Res.* **11**, 121–131.

Jones, R. and Brown, C. R. (1982). Association of epididymal secretory proteins showing lactalbumin like activity with the plasma membrane of rat spermatozoa. *Biochem. J.* **206**, 161–164.

—— —— Von Glos, K. I., and Gaunt, S. J. (1985). Development of a maturation antigen on the plasma membrane of rat spermatozoa in the epididymis and its fate during fertilization. *Expl. Cell Res.* **156**, 31–44.

—— Pholpramool, C., Setchell, B. P., and Brown, C. R. (1981). Labelling of membrane glycoproteins on rat spermatozoa collected from different regions of the epididymis. *Biochem. J.* **200**, 457–460.

Killian, G. J. and Amann, R. P. (1973). Immunoelectrophoretic characterization of fluid and sperm entering and leaving the bovine epididymis. *Biol. Reprod.* **9**, 439–449.

Klinefelter, G. R. and Hamilton, D. W. (1985). Synthesis and secretion of proteins by perifused caput epididymal tubules and association of secreted proteins with spermatozoa. *Biol. Reprod.* **33**, 1017–1027.

Koehler, J. K. (1981). Lectins as probe of the spermatozoon surface. *Arch. Androl.* **6**, 197–217.

Kohane, A. C., Gonzalez-Echeverria, F. M. C., Pineiro, L., and Blaquier, J. A. (1980). Interaction of proteins of epididymal origin with spermatozoa. *Biol. Reprod.* **23**, 737–742.

Koyama, K., Hasegawa, A., Tsuji, Y., and Isojima, S. (1985). Production and characterization of monoclonal antibodies to cross-reactive antigens of human and porcine zonae pellucidae. *J. Reprod. Immunol.* **7**, 187–198.

Krzanowska, H. and Lorenc, E. (1983). Influence of egg investments on *in vitro* penetration of mouse eggs by misshapen spermatozoa. *J. Reprod. Fert.* **68**, 57–62.

Kuzan, F. B., Fleming, A. D., and Seidel, G. E. Jr. (1985). Successful fertilization *in vitro* of fresh intact oocytes by perivitelline (acrosome-reacted) spermatozoa of the rabbit. *Fert. Steril.* **41**, 766–770.

Lambert, H. (1984). Role of sperm-surface glycoproteins in gamete recognition in two mouse species. *J. Reprod. Fert.* **70**, 281–284.

—— and Van Le, A. (1984). Possible involvement of a sialylated component of the sperm plasma membrane in sperm-zona interaction in the mouse. *Gamete Res.* **10**, 153–163.

Lea, O. A. and French, F. S. (1981). Characterization of an acidic glycoprotein secreted by the principal cells of the rat epididymis. *Biochim. Biophys. Acta.* **668**, 370–376.

—— Petrusz, P., and French, F. S. (1978). Purification and localization of acidic epididymal glycoprotein (AEG): a sperm coating protein secreted by the rat epididymis. *Int. J. Androl.* Suppl. **2**, 592–607.

Lee, C. Y., Wong, E., Teh, C. Z., and Nishizawa, Y. (1985). Generation of mouse oocyte monoclonal isoantibodies on *in vitro* fertilization. *J. reprod. Immunol.* **7**, 3–13.

Lopez, L. C., Bayna, E. M., Litoff, D., Shaper, N. L., Shaper, J. H., and Shur, B. D. (1985). Receptor function of mouse sperm surface galactosyltransferase during fertilization. *J. Cell Biol.* **101**, 1501–1510.

Mann, T. (1964). *The biochemistry of semen and of the male reproductive tract.* (ed. T. Mann). J. Wiley, New York.

Meizel, S. (1985). Molecules that initiate or help stimulate the acrosome reaction by their interactions with the mammalian sperm surface. *Am. J. Anat.* **174**, 285–302.

Mettler, L., Czuppon, A. B., and Tinneberg, H. R. (1983). Immunization with spermatozoal peptide antigens resulting in immuno-suppression of fertility rates in female rats. *Andrologia* **15**, 670–675.

Moore, H. D. M. (1980). Localization of specific glycoproteins secreted by the rabbit and hamster epididymis. *Biol. Reprod.* **22**, 705–718.

—— (1981). Glycoprotein secretions of the epididymis in the rabbit and hamster: localization on epididymal spermatozoa and the effect of specific antibodies on fertilization *in vivo*. *J. exp. Zool.* **215**, 77–85.

—— (1981a). An assessment of the fertilizing ability of spermatozoa in the epididymis of the marmoset monkey *(Callithrix jacchus)*. *Int. J. Androl.* **4**, 321–330.

318   Suzanne Fournier-Delpech and Michel Courot

—— and Bedford, J. M. (1978). An *in vivo* analysis of factors influencing fertilization of hamster eggs. *Biol. Reprod.* **19**, 879–885.

—— —— (1983). The interaction of mammalian gametes in the female. In *Mechanism and control of animal fertilization* (ed. J. F. Hartmann) pp. 453–497. Academic Press, New York.

—— Hartman, T. D., and Pryor, J. P. (1983). Development of the oocyte penetrating capacity of spermatozoa in the human epididymis. *Int. J. Androl.* **6**, 310–318.

—— —— (1984). Localization by monoclonal antibodies of various surface antigens of hamster spermatozoa and the effect of antibody on fertilization *in vitro. J. Reprod. Fert.* **70**, 175–183.

Mori, T., Kamada, M., Yamano, S., Kinoshita, T., Kano, K., and Mori, T. (1985). Production of monoclonal antibodies to porcine zona pellucida and their inhibition of sperm penetration through human pellucida *in vitro. J. Reprod. Immunol.* **8**, 1–11.

Munoz, M. G. and Metz, C. B. (1978). Infertility in female rabbits isoimmunized with subcellular sperm fractions. *Biol. Reprod.* **18**, 669–678.

Nicolson, G. L., Usui, N., Yanagimachi, R., Yanagimachi, H., and Smith, J R. (1977). Lectin-binding sites on the plasma membranes of rabbit spermatozoa. Changes in surface receptors during epididymal maturation and after ejaculation. *J. Cell Biol.* **74**, 950–962.

Oliphant, G. and Singhas, C. A. (1979). Iodination of rabbit sperm plasma membrane: relationship of specific surface proteins to epididymal function and sperm capacitation. *Biol. Reprod.* **21**, 937–944.

Olson, G. E. and Danzo, B. J. (1981). Surface changes in rat spermatozoa during epididymal transit. *Biol. Reprod.* **24**, 431–443.

—— and Hamilton, D. W. (1978). Characterization of the surface glycoproteins of rat spermatozoa. *Biol. Reprod.* **19**, 26–35.

—— and Orgebin-Crist, M. C. (1981). Surface changes during epididymal transit. In *The cell biology of the testis* (ed. C. W. Bardin and R. J. Sherins) pp. 372–391. Ann. N.Y. Acad. Sci.

O'Rand, M. G. and Irons, G. P. (1984). Monoclonal antibodies to rabbit sperm auto-antigens. II. Inhibition of human sperm penetration of zona free hamster eggs. *Biol. Reprod.* **30**, 731–736.

—— —— and Porter, J. P. (1984). Monoclonal antibodies to rabbit sperm auto-antigens. I. Inhibition of *in vitro* fertilization and localization on the egg. *Biol. Reprod.* **30**, 721–729.

—— Matthews, J. E., Welch, J. E., and Fisher, S. J. (1985). Identification of zona binding proteins of rabbit, pig, human and mouse spermatozoa on nitrocellulose blots. *J. exp. Zool.* **235**, 423–428.

Orgebin-Crist, M. C., Danzo, B. J., and Davies, J. (1975). Endocrine control of the development and maintenance of sperm fertilizing ability in the epididymis. In *Handbook of physiology*, sect. 7, *Endocrinology* 5, (ed. D. W. Hamilton and R. O. Greep), pp. 319–336. American Physiol. Soc. Bethesda.

—— and Fournier-Delpech, S. (1982). Sperm-egg interaction. Evidence for maturational changes during epididymal transit. *J. Androl.* **3**, 429–433.

—— and Olson, G. L. (1984). Epididymal sperm maturation. In *The male in farm animal reproduction* (ed. M. Courot) pp. 80–102. M. Nijhoff, Kluwer Acad. Pub., Dordrecht.

Overstreet, J. W. and Hembree, W. C. (1976). Penetration of the zona pellucida in non-living human oocytes by human spermatozoa *in vitro. Fert. Steril.* **27**, 815–831.

Peterson, R. N., Henry, L., Hunt, W., Saxena, N., and Russell, L. D. (1985). Further characterization of boar sperm plasma membrane proteins with affinity for the porcine zona pellucida. *Gamete Res.* **12**, 91–100.

—— Hunt, W. P., and Henry, L. H. (1986). Interaction of boar spermatozoa with porcine oocytes: increase in proteins with high affinity for the zona pellucida during epididymal transit. *Gamete Res.* **14**, 57–64.

—— Robl, J. M., Dziuk, P. J., and Russell, L. D. (1982). The effects of anti-sperm plasma membrane antibodies on sperm-egg binding, penetration and fertilization in the pig. *J. exp. Zool.* **223**, 79–81.

—— Russell, L. D., Bundman, D., and Freund, M. (1980). Sperm-egg interaction: evidence for boar sperm plasma membrane receptors for porcine zona pellucida. *Science, N.Y.* **207**, 73–74.

—— —— —— Conway, M. and Freund, M. (1981). The interaction of living boar sperm and sperm plasma membrane vesicles with the porcine zona pellucida. *Devl. Biol.* **84**, 144–156.

—— —— and Hunt, N. P. (1984). Evidence for specific binding of uncapacitated boar spermatozoa to porcine zonae pellucidae *in vitro. J. exp. Zool.* **231**, 137–147.

Phillips, D. M. and Shalgi, R. M. (1980). Surface properties of the zona pellucida. *J. exp. Zool.* **213**, 1–8.

—— —— and Dekel, N. (1985). Mammalian fertilization as seen with the scanning electron microscope. *Am. J. Anat.* **174**, 357–372.

Pijnenborg, R., Gordts, S., Ongkowidjojo, R., and Brosens, I. (1985). Sperm phagocytosis by corona cells in a human *in vitro* fertilization system. *Ann. N.Y. Acad. Sci.* **442**, 310–317.

Rogers, B. J. (1978). Mammalian sperm capacitation and fertilization *in vitro.* A critique of methodology. *Gamete Res.* **1**, 165–223.

Rossignol, D. P., Earles, B. J., Decker, G. L., and Lennarz, W. J. (1984). Characterization of the sperm receptor on the surface of eggs of *Strongylocentrotus purpuratus. Devl. Biol.* **104**, 308–321.

—— and Lennarz, W. J. (1983). The molecular basis of sperm-egg interaction in the sea urchin. *Ciba Fdn. Symp.* **98**, 268–296.

Russell, L. D., Peterson, R. N., Russell, T. A., and Hunt, W. (1983). Electrophoretic map of boar sperm plasma membrane polypeptide and localization and fractionation of specific polypeptide subclasses. *Biol. Reprod.* **28**, 393–413.

—— —— Hunt, W., and Strack, L. E. (1984). Post-testicular surface modification and contribution of reproductive tract fluid to the surface polypeptide composition of boar spermatozoa. *Biol. Reprod.* **30**, 959–978.

Sacco, A. G. (1981). Immunocontraception: consideration of the zona pellucida as a target antigen. *Obstet. Gynecol. Ann.* **10**, 1–26.

—— Subramanian, M. G., and Yurewicz, E. C. (1984). Association of sperm receptor activity with a purified pig zona antigen (PPZA). *J. reprod. Immunol.* **6**, 89–103.

Saling, P. M. (1981). Involvement of trypsin-like activity in binding of mouse spermatozoa to zonae pellucidae. *Proc. natn. Acad. Sci. U.S.A.* **78**, 6231–6235.

—— (1982). Development of the ability to bind to zonae pellucidae during epididymal maturation: reversible immobilization of mouse spermatozoa by lanthanum. *Biol. Reprod.* **26**, 429–436.

—— Irons, G., and Waibel, R. (1985). Mouse sperm antigens that participate in fertilization. I. Inhibition of sperm fusion with egg plasma membrane using monoclonal antibodies. *Biol. Reprod.* **33**, 515–526.

—— and Lakoski, K. A. (1985). Mouse sperm antigens that participate in fertilization. II. Inhibition of sperm penetration through the zona pellucida using monoclonal antibodies. *Biol. Reprod.* **33**, 527–536.

—— Raines, L. M., and O'Rand, M. G. (1983). Monoclonal antibody against mouse sperm blocks a specific event in the fertilization process. *J. exp. Zool.* **227**, 481–486.

—— Sowinski, J., and Storey, B. T. (1979). An ultrastructural study of epididymal mouse spermatozoa binding to the zonae pellucidae *in vitro*: sequential relationship to the acrosome reaction. *J. exp. Zool.* **209**, 97–104.

—— and Storey, B. T. (1979). Mouse gamete interactions during fertilization *in vitro*: chlortetracycline as a fluorescent probe for mouse sperm acrosome reaction. *J. Cell Biol.* **83**, 544–555.

Sathanantan, A. H., Trounson, A. O., Wood, C., and Leeton, J. F. (1982). Ultrastructural observations on the penetration of human sperm into the zona pellucida of the human. *J. Androl.* **3**, 356–364.

Saxena, N. K., Russell, L. D., Saxena, N., and Peterson, R. N. (1986). Immunofluorescence antigen localization on boar sperm plasma membranes: monoclonal antibodies reveal apparent new domains and apparent redistribution of surface antigens during sperm maturation and at ejaculation. *Anat. Rec.* **214**, 238–252.

Schmell, E. D. and Gulyas, B. J. (1980). Mammalian sperm-egg recognition and binding *in vitro*. I. Specificity of sperm interactions with live and fixed eggs in homologous and heterologous insemination of hamster, mouse and guinea pig oocytes. *Biol. Reprod.* **23**, 1075–1085.

Shalgi, R. and Phillips, D. M. (1980). Mechanics of *in vitro* fertilization in the hamster. *Biol. Reprod.* **23**, 433–444.

Shapiro, B. M. (1984). Molecular aspects of sperm-egg fusion. In *Cell fusion* (Ciba Fdn. Symp. 103) pp. 86–99. Pitman Press, London.

Shimizu, S., Ito, M., and Dean, J. (1982). Glycoproteins of mouse zona pellucida: analysis of their reactivity to lectins. *Biochem. Biophys. Res. Commun.* **109**, 449–454.

—— Tsuji, M., and Dean, J. (1983). *In vitro* biosynthesis of three sulfated glycoproteins of murine zonae pellucidae by oocytes grown in follicle culture. *J. biol. Chem.* **258**, 5858–5863.

Shur, B. D. (1984). The receptor function of galactosyltransferase during cellular interactions. *Mol. Cell. Biochem.* **61**, 143–158.

—— and Hall, N. G. (1982). Sperm surface galactosyltransferase activities during *in vitro* capacitation. *J. Cell Biol.* **95**, 567–573.

—— —— (1982a). A role for mouse sperm surface galactosyltransferase in sperm binding to the egg zona pellucida. *J. Cell Biol.* **95**, 574–579.

Singer, S. L., Lambert, H., Overstreet, J. W., Hanson, F. W., and Yanagimachi, R. (1985). The kinetics of human sperm binding to the human zona pellucida and zona-free hamster oocyte *in vitro*. *Gamete Res.* **12**, 29–39.

Smith, M., Peterson, R. N., and Russell, L. D. (1983). Penetration of zona-free hamster eggs by boar sperm treated with the ionophore A 23187 and inhibition of penetration by antiplasma membrane antibodies. *J. exp. Zool.* **225**, 157–160.

Storey, B. T., Lee, M. A., Muller, L., Ward, C. R., and Wirtshafter, D. G. (1984). Binding of mouse spermatozoa to the zonae pellucidae of mouse eggs in cumulus: evidence that the acrosomes remain substantially intact. *Biol. Reprod.* **31**, 1119–1128.

Sullivan, R. and Bleau, G. (1985). Interaction of isolated components from mammalian sperm and egg. *Gamete Res.* **12**, 101–116.

Sundström, P. (1984). Interaction between human gametes *in vitro*: a scanning electron microscopic study. *Arch. Androl.* **13**, 77–85.

Swenson, C. E. and Dunbar, B. S. (1982). Specificity of sperm-zona interaction. *J. exp. Zool.* **219**, 97–104.

Sylvester, S. R., Skinner, M. K., and Griswold, M. D. (1984). A sulfated glycoprotein synthesized by Sertoli cells and by epididymal cells is a component of the sperm membrane. *Biol. Reprod.* **31**, 1087–1101.

Talbot, P. (1985). Sperm penetration through oocyte investments in mammals. *Am. J. Anat.* **174**, 331–346.

Toyoda, Y. and Chang, M. C. (1974). Fertilization of rat eggs *in vitro* by epididymal spermatozoa and the development of eggs following transfer. *J. Reprod. Fert.* **36**, 9–22.

Tsunoda, Y. and Chang, M. C. (1976). Effect of anti-rat ovary antiserum on the fertilization of rat, mouse and hamster eggs *in vivo* and *in vitro*. *Biol. Reprod.* **14**, 354–361.

—— —— (1978). Effect of antisera against eggs and zonae pellucidae on fertilization and development of mouse eggs *in vivo* and in culture. *J. Reprod. Fert.* **54**, 223–237.

Vacquier, V. D. (1983). Purification of sea urchin sperm bindin by DEAE-cellulose chromatography. *Anal. Biochem.* **129**, 497–501.

Vierula, M. and Rajaniemi, H. (1981). Changes in surface protein structure of bull spermatozoa during epididymal maturation. *Int. J. Androl.* **4**, 314–320.

Voglmayr, J. K., Fairbanks, G., Jackowitz, M. A., and Colella, J. R. (1980). Post-testicular developmental changes in the ram sperm cell surface and their relationship to luminal fluid proteins of the reproductive tract. *Biol. Reprod.* **22**, 655–669.

Wolf, D. P., Sokoloski, J. E., Dandekar, P., and Bechtol, K. B. (1983). Characterization of human sperm surface antigens with monoclonal antibodies. *Biol. Reprod.* **29**, 713–723.

Yanagimachi, R. (1977). Specificity of sperm-egg interaction. In *Immunology of gametes* (ed. M. Edidin and M. H. Johnson) pp. 225–295. Cambridge University Press.

—— (1981). Mechanisms of fertilization in mammals. In *Fertilization and embryonic development* (ed. L. Mastroianni and J. D. Biggers) pp. 81–112. Plenum Press, New York.

—— Kamiguchi, Y., Sugawara, S., and Mikamo, K. (1983). Gametes and fertilization in the Chinese hamster. *Gamete Res.* **8**, 97–117.

—— and Phillips, D. M. (1984). The status of acrosomal caps of hamster spermatozoa immediately before fertilization *in vivo*. *Gamete Res.* **9**, 1–19.

9   Fertilization: Motility, the Cytoskeleton and the Nuclear Architecture

GERALD SCHATTEN AND HEIDE SCHATTEN

I Introduction

1 THE GOALS OF FERTILIZATION

Fertilization bridges the discontinuity in generations and represents, in one sense, an end point in the process of reproduction as well as the start of development. Fertilization is considered successful if one, and only one, sperm nucleus unites with the egg nucleus within an activated egg cytoplasm. In addition the fertilized egg must also be primed to prepare for its next challenges: the cell divisions leading to embryogenesis.

2 THE REQUIREMENTS FOR STRUCTURAL ALTERATIONS AND MOTILITY DURING FERTILIZATION

For fertilization to be accomplished, several architectural changes in the nuclei and cytoplasm must occur, and several movements must take place. Sperm are typically motile, and their swimming movement is usually required for the sperm to traverse the fluid medium to the egg surface. In some, but certainly not all, sperm, the acrosome reaction involves a reorganisation of the sperm's cytoskeleton. In many sperm from marine invertebrates, an explosive burst of actin assembly produces a bundle of microfilaments forming the elongated acrosomal process. It is this extension from the sperm's cytoskeleton that establishes the initial contact between the sperm and the egg surfaces.

The egg also effects several motions as well as numerous alterations in cytoplasmic and nuclear structure. The physical incorporation of the sperm and the formation of the fertilization or incorporation cone are changes in cell shape mediated by the egg cortex. In addition the eggs of most animals, including mammals, are actually oocytes and the events of fertilization overlap with the completion of meiosis; the elicitation of polar bodies, also mediated by the egg cortex, would then be included in the structural changes at fertilization. Once discharged into the cytoplasm, the sperm and egg nuclei (male and female pronuclei, respectively) must unite or become apposed, typically at the egg centre, and these nuclear migrations are the result of the elaboration of the egg cytoskeleton. While these events have direct implications in joining the parental genomes, there are also likely to be several structural alterations required for egg activation and the progression through the first cell cycle. These sorts of changes include microvillar formation and elongations as well as the assembly and disassembly of cytoplasmic asters in preparation for mitosis.

Evidence is accumulating that the architecture of the sperm and egg nuclei is quite modified and perhaps considerably reduced. Because a goal of fertilization is syngamy, the sperm and egg nuclei must be structurally primed to permit pronuclear fusion or chromosome intermixing at mitosis. This process will first require male pronucleus formation from the sperm nucleus. In those systems inseminated as oocytes in which pronuclear fusion does not occur, as in mammals, the formal process of fertilization will overlap with both meiosis and mitosis. Therefore the changes in chromatin structure permitting the maternal meiotic chromosomes to decondense into the female pronucleus and the second polar body nucleus, as well as the later condensation of the male and female pronuclei into the mitotic chromosomes, will also be considered.

3   OVERVIEW OF THE MOLECULAR BASES OF THE CYTO-SKELETON, KARYOSKELETON, AND MOTILITY MECHANISMS

Because this article will concentrate on the structural and motility events during fertilization, it may be useful to provide a brief overview of the current state of knowledge regarding intracellular architecture and models accounting for internal motion. These research areas are under very active investigation and many novel discoveries can be anticipated, even before the publication of this volume.

The cytoskeleton, found in all eukaryotic cells, is composed primarily of three fibrous elements, along with dozens of accessory proteins. Microfilaments are 7 nm filaments of indefinite length composed of polymers of the protein actin (reviewed by Clarke and Spudich 1977; Pollard 1981; Stossel 1984). Microtubules are 25 nm in diameter and are polymers of the dimeric protein tubulin (reviewed by Dustin 1978). Intermediate filaments, intermediate in size at 10 nm to microfilaments and microtubules, are the most heterogeneous class of cytoskeletal elements with several different sorts of proteins found in various tissues and in different organisms (reviewed by Lazarides 1980 and Franke, Schmid, Wellsteed, Grund, Gigi, and Geiger 1983). In addition to these classes of fibres, there are numerous accessory proteins, which serve several roles such as bundling the filaments together, capping their disassembling or assembling ends to regulate length, severing them in the middle to aid depolymerisation, and providing ATPase activities for motility. These proteins are named in various creative fashions and fall under the categories of 'actin-binding proteins' (ABPs), 'microtubule-associated proteins' (MAPs) and 'intermediate filament associated proteins' (Goldman, Goldman, and Yang 1985).

The cytoskeleton is responsible in animal cells for the maintenance of and all changes in cell shape and for effecting motility. Microfilament-mediated motility is familiar in muscle contraction, where the thin filaments are microfilaments and myosin serves as the ATPase, and during cytokinesis, when a contractile ring of microfilaments physically divides the cell in two. Microfilaments are responsible for maintaining cell shape in most cells, including the biconcave disc shape of erythrocytes and the bundled microfilaments forming 'stress fibers' in many cultured cells.

Microtubule-mediated motility is well understood in the case of ciliary, flagellar, and sperm tail movement, in which the ATPase dynein creates a lateral motion of one microtubule doublet along the next. This motion is translated into the characteristic bending behaviour by other proteins. Microtubules are also probably responsible for chromosome alignment and separation during mitosis and meiosis. Recently a new microtubule-associated ATPase kinesin (Schnapp, Vale, Sheetz, and Reese 1985) has been characterised, associated with axonal microtubules and microtubules in the mitotic apparatus.

Nuclear architecture involves the nuclear envelope and associated fibrous lamina-nuclear pore complex proteins (reviewed by Franke, Scheer, Krohne, and Jarasch 1981) and the less well defined interior nuclear matrix (reviewed by Berezney 1979). The behaviour of the lamins, discussed in Section VII, has been well explored by Gerace, Blobel and coworkers (Gerace, Blum, and Blobel 1978; Gerace and Blobel 1980, 1982) and recent progress has been fruitful in developing *in vitro* systems permitting nucleus formation and breakdown (Burke and Gerace 1986; Lohka and Maller 1985; Miake-Lye and Kirschner 1985).

4 SCOPE OF REVIEW

The aim of this review article is to consider the structural organisation of the egg cytoskeletal and nuclear architectural changes during fertilization. Because much of the evidence regarding fertilization has been derived from investigations on invertebrate and lower vertebrate systems, this information will typically be reviewed first in each subsection. Then the state of the knowledge regarding mammalian fertilization will be considered, with the inclusion of some of the remaining questions. The aim of this article is two-fold: to kindle further interest in the application of cellular and molecular structural investigations for solving fundamental problems in reproductive biology; and to highlight the importance of studying cells during reproduction in order to generate a fuller appreciation for cell and molecular biology.

II   Sperm incorporation

1   THE ACROSOME-REACTED SPERM

The first morphologically apparent event during fertilization is the elaboration of the acrosomal process of the sperm, and the first response of the egg occurs during sperm incorporation. In many marine invertebrates, sperm dramatically assemble or unwind a microfilament-containing process during the acrosome reaction. In some systems, like that of the sea cucumber *Thyone* the acrosomal process can be as long as the sperm tail, and it can extend to this length within 30 seconds (Tilney and Inoue 1981). In sea urchins (Fig. 9.1), the acrosomal process is typically less than 5 µm long. The formation of the acrosomal process has become a model in which to explore actin assembly and its ionic regulation (reviewed by Schackman, Eddy, and Shapiro 1978; Tilney and Tilney 1984). In sperm of other marine organisms, like the horseshoe crab *Limulus*, the microfilaments of the acrosomal process are already assembled in the mature sperm. The acrosome reaction then involves the unwinding of the previously assembled process rather than the *de novo* assembly of the acrosomal actin. Because the actin isolated from the periacrosomal cap of the sperm head is held in a non-filamentous state, and because it will not spontaneously assemble *in vitro*, this sort of actin has been referred to as 'pro-filamentous actin' or 'profilactin' (reviewed by Tilney and Tilney 1984).

2   THE QUESTION OF ACTIN POLYMERIZATION DURING THE ACROSOME REACTION IN MAMMALIAN SPERM

Although the dramatic assembly of actin in marine invertebrate sperm undergoing the acrosome reaction is serving as a model for the exploration of the actin behaviour, questions remain about the acrosome reaction in mammalian sperm. The acrosome reaction in mammalian sperm does not involve any similarly dramatic process. In addition, the site at which the sperm fuses its plasma membrane with the egg's differs from that in most fertilization systems in which fertilization is external. In Figure 9.1c of a sea urchin egg at fertilization, it is apparent that the sperm are held perpendicular to the egg surface, and the successful one will fuse initially at the apex of the sperm head, the tip of the extended acrosomal process. In contrast, as depicted in Figure 9.7 during mouse sperm incorporation, fusion of the mammalian sperm will occur along the equatorial region of the sperm head. This contrast raises questions about differences in the requirements for microfilaments in the sperm during the acrosome reaction.

Because mammalian sperm fuse with the egg along the equatorial region of the sperm head, investigations to detect and localise actin in this area have been performed. The rationale for these studies includes the question of whether actin need be present along the cortical plasma membrane faces of both the sperm and the egg surface to permit these membranous regions to be fusigenic. Some, but not all, mammalian sperm (Clarke and Yanagimachi 1978; Clarke, Clarke, and Wilson 1982; Flaherty, Breed, and Sarafis 1983; Halenda, Primakoff, and Myles 1984) have been reported to contain actin, and questions remain about the requirements for microfilaments during the acrosome reaction.

Because these experiments seem to indicate that some, but not all, mammalian sperm may contain polymerised actin at the fusigenic site, we have posed the question of whether actin assembly in the sperm is required for sperm incorporation. Recently Kashman, Groweiss, and Schmueli (1980) have isolated and characterised latrunculins, novel marine toxins from the Red Sea sponge *Latrunculia magnifica*, and Spector, Shochet, Kashman, and Groweiss (1983) have demonstrated that latrunculins will disrupt microfilament organisation in cultured cells at concentrations that do not affect the rate of actin polymerisation *in vitro*, suggesting that latrunculins may represent a new experimental drug with which to investigate microfilament organisation and behaviour.

To demonstrate that sea urchin sperm require the assembled microfilaments found after the acrosome reaction for sperm incorporation, insemination was performed with varying concentrations of sea urchin sperm induced to undergo the acrosome reaction in the presence of 2.6 µM latrunculin A. As shown in Table 9.1, latrunculin dramatically affects the sea urchin sperm's ability to fertilize sea urchin eggs.

Mammalian sperm, however, may not require actin assembly for successful sperm incorporation. As also demonstrated in Table 9.1, mouse sperm treated with an identical concentration of latrunculin are relatively unaffected in their ability to be incorporated into mouse oocytes during fertilization *in vitro*. This result appears to indicate that microfilaments in the mouse sperm may not be required for sperm incorporation and raises questions about the universality of actin assembly during the acrosome reaction.

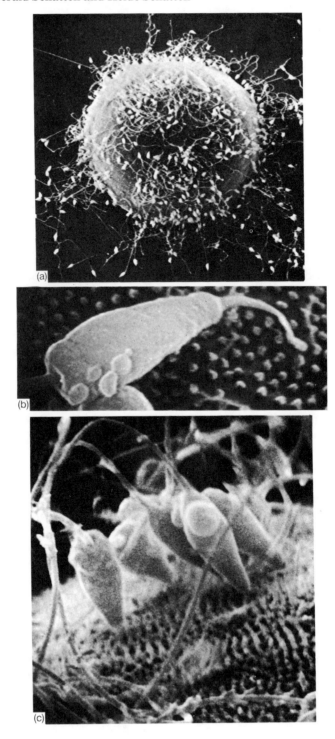

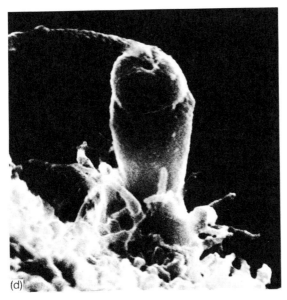

Fig. 9.1. Scanning electron microscopy of fertilization. (a) is a sea urchin egg at an early stage of insemination. In (b) elongation of the sperm acrosomal process is apparent extending from the apex of the spermhead. Sperm in this system typically bind perpendicular to the egg surface (c) unlike the case during mammalian fertilization. In (d) eruption of the egg surface around the successful sperm is shown. Microvilli elongate around the entering sperm to swell into the fertilization cone. (Reprinted, with permission, from Schatten and Schatten, 1980 and Schatten and Mazia, 1976).

3   EGG ACTIN POLYMERIZATION DURING SPERM INCOPORAT-ION

Microfilaments are active throughout sea urchin fertilization (reviewed by Vacquier 1981) including during the formation of the acrosomal process (Tilney, Hatano, Ishikawa, and Mooseker 1973), sperm incorporation (Longo 1980; Schatten and Schatten 1980, 1981; Tilney and Jaffe 1980; Cline, Schatten, Balczon, and Schatten 1983), microvillar elongation (Eddy and Shapiro 1976; Schroeder 1978) and cytokinesis (reviewed by Schroeder 1981). Changes in the state and organisation of actin in the sea urchin egg cortex have been studied by several investigators (Burgess and Schroeder 1977; Spudich and Spudich 1979; Begg, Rebhun, and Hyatt 1982; Otto, Kane, and Bryan 1980).

After the sperm has penetrated through the extracellular egg coatings, fusion between the sperm and egg plasma membranes may take place. Though the mechanisms of membrane fusion are only poorly understood, in some systems a reversal in the membrane potential of the egg from inside negative to positive results in a fast and temporary block to polyspermy

Table 9.1

Effects of latrunculin on the ability of sperm to fertilize sea urchin and mouse eggs

	% of fertilization	
	Control	Latrunculin
Sea Urchin Fertilization		
Low sperm concentration		
(ca. 3 sperm/egg)		
at 5 min. post-insemination	58.0% ± 16.4%	11.9% ± 1.5%
at 10 min. post-insemination	80.2% ± 8.9%	18.6% ± 2.8%
High sperm concentration		
(ca. 50 sperm/egg)		
at 5 min. post-insemination	98.0% ± 0.2%	28.7% ± 6.4%
at 10 min. post-insemination	98.7% ± 0.5%	35.3% ± 2.8%
Mouse Fertilization		
10^5 sperm/oocyte	67.8% ± 26.4%	58.4% ± 15.9%

Successful fertilization is assayed in sea urchin eggs by the elevation of the fertilization envelope and the centring of both pronuclei. Successful fertilization in mouse oocytes is determined by the introduction of the sperm tail into the egg and the deondensation of a male pronucleus. (From Schatten *et al.* 1986*a*).

(reviewed by Jaffe 1976). This electrical block to polyspermy, triggered at the time of fusion, does not seem to occur during mammalian fertilization.

Microfilaments assemble at the site of sperm-egg fusion, and this regional event is required for sperm incorporation in many systems. In sea urchins, as depicted in Figure 9.1, the egg surface erupts to form the 'fertilization cone,' which surrounds and eventually engulfs the successful sperm. A fluorescence-conjugated mushroom toxin, rhodamine-phalloidin, which specifically binds to polymerised actin, has recently been used to detect actin assembly during sperm incoporation (Cline and Schatten 1986). As depicted in Figure 9.2, there is a dramatic recruitment of rhodamine-phalloidin binding in and around the site of sperm incorporation. This result indicates that fusion initiates at first a regional accumulation of microfilaments at the incorporation site. It is also of interest that Lynn and Chambers (1985) have demonstrated the importance of the change in membrane potential for sperm incorporation, leading to the theory that membrane potential could be involved directly or indirectly in these structural events.

Inseminations performed with eggs treated with microfilament inhibitors demonstrated that microfilament assembly in the egg is required for sperm incorporation. Cytochalasins (Gould-Somero, Holland, and Paul 1977; Longo 1978, 1980; Schatten and Schatten 1980, 1981) and latrunculin (Schatten, Schatten, Spector, Cline, Paweletz, Simerly, and Petzelt 1986a) will prevent the incorporation of the sperm into the egg and the formation of the fertilization cone (Table 9.2). In addition these drugs will prevent

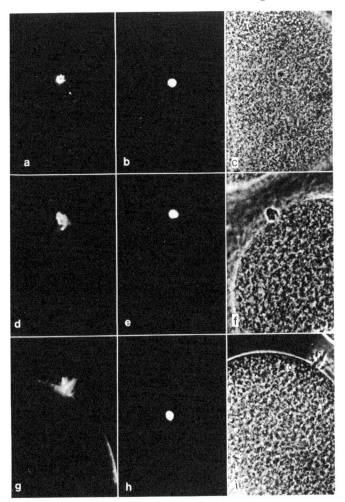

Fig. 9.2. Growth of fertilization cones in *Lytechinus*: (a) Thirty seconds after insemination the cone is detected as a bright fluorescent ring in the immediate vicinity of sperm contact; (b) Hoechst image of sperm DNA at the egg periphery 30 seconds after insemination; (c) Phase contrast of *a* and *b*; (d) By 60 seconds after insemination the cone is erect and has a petal-like appearance; (e) Hoechst image of sperm DNA as the sperm moves through the cone; (f) Phase contrast image of *d* and *e*; (g) At 7 minutes post insemination the erect cone is near its maximum size and has a "spike-like" appearance; (h) Hoechst image of sperm DNA when the sperm head had moved through the cone into the egg cortex. (i) Phase contrast image of *g* and *h*. X 1500. (Reprinted, with permission, from Cline and Schatten, 1986).

another microfilament-mediated event during fertilization, the elongation of the egg microvilli, as shown in Fig. 9.3.

4   MECHANISMS ACCOUNTING FOR MICROFILAMENT ACTIVITY DURING SPERM INCORPORATION

Microfilaments, like microtubules, are polarised fibres with opposing ends preferential for assembly and disassembly. Indeed this polarity is reflected in the possible direction for contraction and can be traced in the case of microfilaments if the fibres are labelled with a myosin fragment containing the ATPase activity: heavy meromyosin or 'S1' decoration. By analogy with skeletal muscle contraction, in which the heavy meromyosin 'arrowhead' complexes point in the direction of contraction, the direction of possible contraction of the microfilaments in the sperm acrosomal process and in the egg microvilli comprising the fertilization cone can be determined. Tilney and Jaffe (1980) have demonstrated that the polarity of both sets of microfilaments are opposite to that predicted for an acto-myosin sliding model. However, myosins have been isolated from egg cortices (Mabuchi 1974), leaving the possibility open that egg myosin participates in sperm incorporation.

Another manner in which assembling actin may permit sperm incoporation involves a different actin-binding protein, spectrin. Spectrin, so-called because it is isolated from plasma membrane 'ghosts' prepared by removing the haemoglobulin from red blood cells, has recently been isolated from a variety of non-erythrocyte cells (reviewed by Glenney and Glenney 1983). The spectrinesque protein from brain has been well studied by Willard and colleagues (Cheney, Hirokawa, Levine, and Willard 1983; Hirokawa, Cheney, and Willard 1983; Levine and Willard, 1981, 1983) and is referred to as fodrin. As shown in Figure 9.4, fodrin can be detected in the sperm acrosome and at the egg surface, with a marked accumulation in the fertilisation cone (Schatten, Cheney, Balczon, Cline, Simerly, Willard, and Schatten 1986b). Because a role of spectrin is to create a viscous gel with actin oligomers, a model for sperm incorporation, shown is Figure 9.5f, may be that spectrin creates a region of increased viscosity around the successful sperm so that it is not detached when the fertilization envelope elevates. This passive model predicts that another system of intracellular motility, like the formation of the sperm aster described in Section III, accounts for the discharge of the sperm from the egg cortex into the cytoplasm.

It is also of interest that non-erythrocyte spectrin has been found by several investigators in mammalian gametes, zygotes and embryos (Sobel 1983, Sobel, and Alliegro 1985; Sobel, Opas, and Kalnins 1985; Lehtonen *et al.* 1985, Schatten, Cheney, Balczon, Cline, Simerly, Willard, and Schatten 1986b). Although it appears that different types of spectrin-like proteins are found in different regions at various developmental stages, the precise role

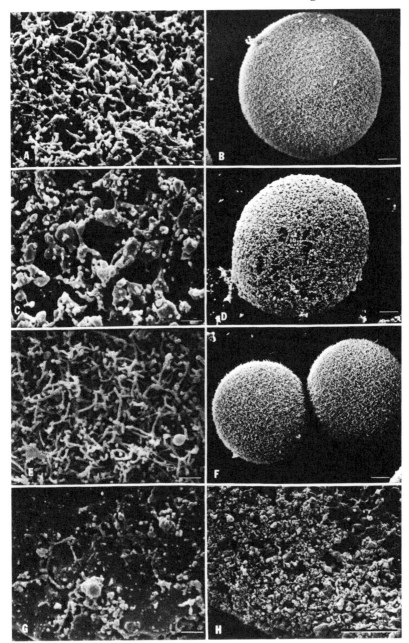

Fig. 9.3. Elongation of microvilli. Unlike control sea urchin eggs at 30 minutes (A,B) and 75 minutes (E,F) post-insemination, latrunculin-treated (2.6 μM) eggs are unable to undergo the normal elongation of microvilli at thirty minutes post-insemination (C,D), nor are they able to cleave by 75 minutes (G,H). The aberrant surface is convoluted with short stubby projections. Bars, A,C,E,G: 1 μm.; B,D,F,H,: 10 μm. (Reprinted, with permission, from Schatten *et al.*, 1986*a*).

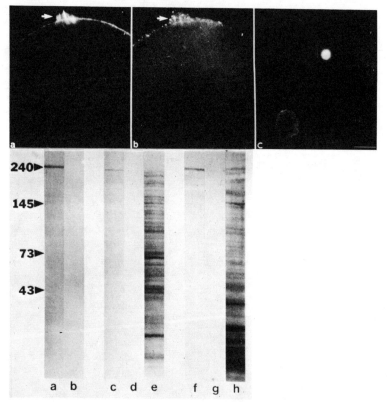

Fig. 9.4. Colocalization of fodrin and actin in a permeabilized fertilized sea urchin egg. At 10 minutes post-insemination, actin fluorescence (a) and fodrin immunofluorescence (b) microscopy demonstrates that these proteins are restricted to the egg cortex. There is a recruitment of actin and possibly fodrin in the fertilization cone (arrows). The incorporated sperm has decondensed to form the male pronucleus in c; the female pronucleus, at a different focal plane, is inset at the bottom of c. Triple labelled eggs stained for actin (a), fodrin (b) and pronuclear DNA (c). Bar = 10 μm. *Underneath*, Western blotting with antifodrin: (a) and (b), the purified 240 Kd subunit of mammalian fodrin stained with antifodrin or non-immune IgG respectively; (c) and (d), cytoplasmic gels from unfertilized sea urchin eggs stained with antifodrin or non-immune IgG respectively, an immunoreactive polypeptide with an apparent molecular weight of 240 Kd is specifically stained by the antifodrin; (e) proteins of the cytoplasmic gels from unfertilized sea urchin eggs visualized by amido black staining; (f) and (g), mitotic sea urchin egg extracts stained with antifodrin or non-immune IgG respectively. (This sample also contains an immunoreactive polypeptide with an apparent molecular weight of 240 Kd that is selectively stained by the antifodrin); (h) proteins of the mitotic sea urchin eggs visualized by amido black staining (actin is removed during the detergent extraction). (Reprinted, with permission, from Schatten *et al.*, 1986*b*).

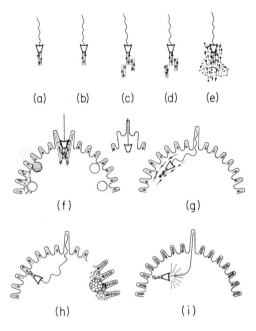

Fig. 9.5. Mechanisms accounting for sperm incorporation in sea urchins. The polarity of actin in the sperm acrosomal process (a,b) and in the microfilaments in the fertlization cone (c,d) preclude a simple actomyosin sliding model since they are in the configurations of (b) and (d). If sliding were to occur, then the sperm might be pushed off the egg surface, rather than pulled into it. Instead, an oligomeric actin-spectrin gel, depicted in (e), might assemble around the entering sperm, effectively anchoring it to the egg surface so that it is not detached during the explosive elevation of the fertilization envelope. The sperm might undergo a lateral displacement along the egg cortex mediated by microfilaments (f,g) and later it is moved into the egg cytoplasm by the growth of microtubules emanating from the sperm centrosome (h,i).

and the distribution of this spectrin are not precisely clear. It might well serve in a capacity similar to its role at the red cell membrane, interaction with actin at the plasma membrane providing gelated cortex.

5   ABSENCE OF THE REQUIREMENT FOR MICROFILAMENT ASSEMBLY DURING MAMMALIAN SPERM INCORPORAT-TION

During mammalian fertilization, microfilaments may not be active during sperm incorporation and may only be required for events different from those requiring microfilaments during sea urchin fertilization. Careful studies of the surface events during sperm incorporation have questioned whether microvilli on the mammalian oocyte actively engulf the successful

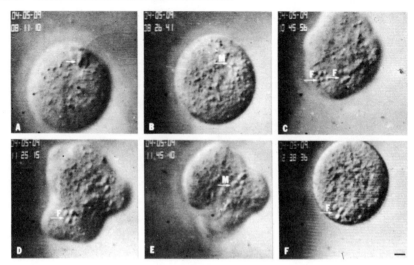

Fig. 9.6. Time lapse video microscopy of mouse fertilization *in vitro*. The incorporation of the sperm head (A) occurs in the presence of 2.6 μM latrunculin, though the incorporation cone does not enlarge normally at four hours post-insemination (B). The male pronucleus develops (B: M). Since the formation of the second polar body is inhibited (B,C), two maternal pronuclei develop, corresponding to a female pronucleus and the second polar body nucleus. The egg becomes unusually active in the pronucleate stage (D–F) but the pronuclei are not moved from the egg cortex and further development is arrested. M: male pronucleus. F: maternal pronuclei. Bar, 10 μm. (Reprinted, with permission, from Schatten *et al*. 1986*a*).

sperm (Shalgi, Phillips, and Kraicer, 1978), as appears to be the case in most lower organisms. However, it is curious that the oocyte surface has specialised regions for sperm-oocyte fusion and for second polar body formation and that the region for sperm-egg fusion is characterised by the presence of microvilli (Nicosia, Wolf, and Inoue, 1977). Microfilament activity in oocytes has been implicated during the formation of the second polar body and the incorporation cone in mammals, and during pronuclear apposition at fertilization (Longo 1978; Shalgi *et al*. 1978; Maro *et al*. 1984; Battaglia and Gaddum-Rose 1985; see Section IV). Neither cytochalasins (Longo 1978; Maro *et al*. 1984) nor latrunculin (Schatten *et al*. 1986*a*) prevent sperm incorporation in mice, indicating that sperm entry may occur independent of both sperm and egg microfilaments in mammals.

The effects of the new microfilament inhibitor latrunculin on mouse fertilization is shown in Figure 9.6. The second polar body does not form after sperm incorporation (Fig. 9.6A), though meiosis is completed as assessed by the appearance of the two maternal pronuclei (F: Fig. 9.6C) corresponding to a female pronucleus and a second polar body nucleus. Sperm incorporation (Fig. 9.6A–C) and pronuclear formation (Fig. 9.6B–C) appears to occur normally, but the sperm and egg nuclei are unable to move into apposition at the egg center (Fig. 9.6C–D). Pseudocleavage is noted at about eight hours post-insemination (Fig. 9.6D–F). In addition, cortical microfilaments are detected in the pseudocleaving mouse eggs with rhodamine-phalloidin. This observation is similar to that shown with cytochalasin B (Wassarman, Ukena, Josefowicz, and Karnovsky 1977), which has also been shown to induce pseudocleavage, but in that case in preovulatory oocytes. Bundles of cortical actin are found along the egg cortex and in the pseudocleavage furrow, but the pronuclei, unlike the meiotic chromosomes, are unable to induce a regional accumulation of microfilaments.

The effects of latrunculin on mouse fertilization differ considerably from its effects on sea urchin fertilization. When the meiotic spindle of ovulated unfertilized mouse oocytes is disrupted with colcemid, the chromosomes disperse along the egg cortex (Schatten, Simerly, and Schatten 1985; Longo and Chen 1985; Maro, Johnson, Webb, and Flach, 1986). Latrunculin will prevent this scattering of the meiotic chromosomes, suggesting that microfilaments are active during this dispersion, and indeed regional accumulations of cortical actin are found adjacent to each chromosome as shown by Maro *et al.* (1986).

Sperm incorporation during mouse fertilization *in vitro* is unaffected by treatment of either the sperm or the oocyte with latrunculin, suggesting that microfilament activity is not required in either gamete during this initial phase of mammalian fertilization, as it is for sperm incorporation in sea urchins. However latrunculin will prevent the apposition of the pronuclei during mouse fertilization, as does cytochalasin (Maro, Johnson, Pickering, and Flach 1984). This finding indicates that, also unlike events during sea urchin fertilization, pronuclear apposition in this mammal requires microfilament function.

6   CHANGES IN THE EGG CORTEX FOLLOWING FERTILIZATION

Microfilaments are also active after fertilization in the restructuring of the egg cortex. In mammalian oocytes this process will involve the formation of the second polar body at the region of the oocyte devoid of microvilli. In addition most eggs undergo bursts of elongation of the egg microvilli after fertilization (Eddy and Shapiro 1976). Though the function of this microvillar elongation in not yet established, it is clear that actin assembly at the egg cortex results in this elongation (Burgess and Schroeder 1977; Begg, Rebhun, and Hyatt 1982) and that inhibition of this event is lethal to the egg (Schatten and Schatten 1980, 1981; Schatten *et al.* 1986a).

A demonstration of the post-fertilisation changes in the egg surface is shown in Figure 9.3, in which sea urchin eggs are examined at thirty minutes post-insemination with and without the addition of latrunculin. Latrunculin at concentrations above 260 nM prevents the normal elongation of the microvilli after fertilization and the formation of the cleavage furrow at first division. Elongate microvilli at thirty minutes after fertilization are depicted in Figures 9.3A and 9.3B. The aberrant cell surface in eggs incubated in 2.6 µM latrunculin added at five minutes post-insemination is shown in Figures 9.3C and 9.3D; the surface is highly convoluted with stubby projections, perhaps the remnants of the short microvilli of the unfertilised egg. At 75 minutes post-insemination, when control eggs have undergone cleavage (Fig. 9.3E and 9.3F), 2.6 µM latrunculin-treated eggs have irregular surfaces with smooth patches that lack microvilli (Fig. 9.3G and 9.3H).

7   FERTILIZATION CONE VS. INCORPORATION CONE

Sperm incorporation typically results in the regional eruption of the egg surface around the successful sperm. The resulting structure is referred to as the 'fertilization cone' in invertebrate systems and the 'incorporation cone' in mammals. Though the two terms could be thought of as roughly synonymous because they are both microfilament-containing swellings near or at the site of sperm-egg fusion, they probably have different roles and are triggered to form by different means. As noted earlier, invertebrate sperm approach the egg in a perpendicular fashion and one initially penetrates vertically through the fusion site. The fertilization cone, perhaps initiated by the regional increase in calcium ion concentration and elevated pH that the sperm introduces, forms precisely around the sperm-egg fusion site and probably is crucial for the very process of sperm incorporation. In contrast, the incorporation cone in mammals seems to be the result, not the cause, of sperm incorporation. The mammalian sperm fuses laterally or tangentially with the oocyte plasma membrane. There may not be one small region of membrane fusion; rather the sperm and oocyte may fuse membranes along their whole contact regions: an analogy might be that of a zipper which starts at the equatorial region of the sperm head. The formation of the incorporation cone, which is sensitive to microfilament inhibitors (Shalgi *et al.* 1978), is not required for sperm incorporation. It might even be the result of a regional interaction between the sperm chromatin and the oocyte cortex. As will be described later, maternal meiotic chromosomes induce regional accumulation of oocyte cortical actin (Longo and Chen 1985; Maro *et al.* 1986, Schatten *et al.* 1986a), and perhaps the sperm chromatin has a similar effect. If so it would result in the formation a microfilament-containing surface protrusion, the mammalian incorporation cone.

8   DIFFERENCES IN SPERM INCORPORATION MECHANISMS AND APPROACHES

The usual description of sperm incorporation for most invertebrates and lower vertebrates involves actin polymerization in the sperm to form the acrosome

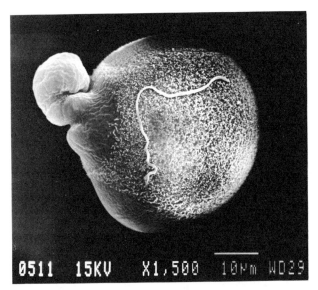

Fig. 9.7. Scanning electron microscopy of sperm incorporation during mouse fertilization. Sperm incorporation in mice occurs with the sperm positioned tangential to the egg surface. The incorporation cone may form at a region distal from the site of sperm entry, and is denoted by its absence of microvilli.

process and in the egg to form initially the fertilization cone and later the global elongation of microvilli. In this description, microfilament assembly in both the sperm and the egg is essential for sperm incorporation, and though myosin may be present at the fusion site, other actin-binding proteins including spectrin, may also be responsible for sperm incorporation. The sperm binds to the egg surface oriented perpendicular to the surface and enters vertically through the fertilization cone region.

Sperm incorporation in mammals occurs in a very different fashion. The presence of assembled actin in the acrosome-reacted sperm is still controversial, and it appears that sperm incorporation can occur in the absence of microfilament assembly in both gametes. The sperm binds and fuses with the oocyte while oriented tangentially, and fusion might even occur all along the sperm's entire length, rather than though a small region. The sperm might be thought of as slowly sinking into the egg laterally rather than as actively drawn in through a small aperture. The incorporation cone forms after sperm incorporation and probably is not required for sperm incorporation (Fig. 9.7).

III   Pronuclear migrations and union.

1   PATTERNS OF PRONUCLEAR MIGRATIONS

The classic depiction of the migrations of the sperm and egg nuclei leading to syngamy derives from the earliest studies nearly a century ago. It was recognised that the sperm introduces a focal point onto which 'the sperm aster' assembles. We now know that the sperm aster is composed of radially arrayed microtubules that extend from the centrosome, a structure of critical importance discussed in Section VI.

As is the case with sperm incorporation, the movements and responsible machinery leading to pronuclear union differ between mammals and most other organisms. The organisation of the sperm aster and role of microtubules will first be considered in invertebrate and lower vertebrate systems, and then the state of knowledge during mammalian fertilization will be addressed.

2   THE SPERM ASTER

The microtubules of the sperm aster are responsible for first moving the male pronucleus (incorporated sperm nucleus) from the inner face of the egg cortex into the egg cytoplasm. Later as the sperm aster increases in size some of the microtubule ends will contact the female pronucleus. Because the sperm may enter at almost any region on the egg surface, and because the female pronucleus is usually eccentrically positioned, there is elegance in this arrangement of the sperm aster (Fig. 9.8). Regardless of the relative positions of the male and female pronucleus, as the microtubules of the sperm aster increase in size they will, sooner or later, contact the female pronuclear surface. At that moment, the female pronucleus, by a still elusive mechanism requiring microtubule disassembly, rapidly translocates along these microtubules to their source, the spot where the male pronucleus resides.

Once the male and female pronuclei are adjacent, the remaining sperm astral microtubules continue to assemble, moving the adjacent pronuclei to the centre of the egg cytoplasm. Typically these microtubules disassemble prior to pronuclear fusion.

That microtubules are required for the pronuclear migrations in systems like sea urchins can be demonstrated by the use of specific assembly and disassembly inhibitors. As shown in Table 9.2, microtubule assembly inhibitors such as colcemid, griseofulvin and nocodazole will prevent the pronuclear migrations, but will not inhibit sperm incorporation. In direct contrast are the effects of microfilament inhibitors like cytochalasins or latrunculin, which will permit the pronuclear migrations if added after sperm incorporation, but which will inhibit sperm incorporation if added prior to insemination.

These results lead to the conclusion that microtubules assembling on the sperm centrosome first push the male pronucleus into the egg cytoplasm, probably as a result of the force of the assembly, though the possibility exists that the individual microtubules act like miniature oars, moving the aster in

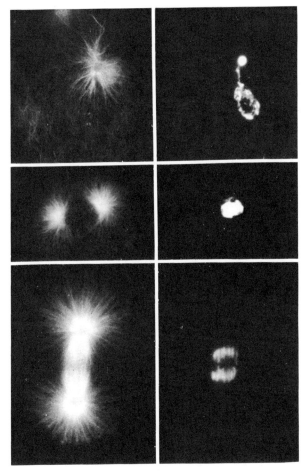

Fig. 9.8. Microtubules during sea urchin fertilization. Microtubules comprising the sperm aster are shown (top panel). The monastral structure develops bipolarity by the 'streak' stage (second panel) and during first division (bottom panel).

the direction with the longest microtubules. Once the microtubule ends contact the female pronuclear surface, a rapid migration, perhaps analogous to the movement of the chromosomes during anaphase, occurs. This motion requires microtubule depolymerisation, but the motive force is not yet known. Later the remaining microtubules, probably again by assembly, move the adjacent pronuclei into intimate contact and to the egg center.

3   CYTOPLASMIC MICROTUBULES DURING MAMMALIAN FERTILIZATION

The pattern of microtubules found during mammalian fertilization differs

Table 9.2
Motility during mouse and sea urchin fertilization

	Sea Urchin	Mouse
Sperm motility	MTs	MTs
Sperm acrosome reaction	MFs	none
Sperm incorporation	Egg MFs	none
Completion of meiosis	N.A.	Egg MTs
Second polar body formation	N.A.	Egg MTs
Pronuclear formation	none	MTs
Pronuclear migrations	MTs	MTs and MFs
First spindle	Fusiform, with asters	Barrel, anastral
Centrosome source	Paternal	Maternal

Abbreviations:
MTs: Microtubule activity; MFs: Microfilament activity; N.A: Not applicable.

considerably from that described above and raises a number of questions concerning the inheritance of microtubule organising centres. These implications will be considered in the section on the centrosome, Section VI.

Mammalian oocytes, like most eggs, are fertilized as oocytes. Sperm incorporation and the final stages of meiotic maturation occur together. In the cases of most mammals, the oocyte is fertilized while arrested at second meiotic metaphase. As seen in Figure 9.9 then, the unfertilized oocyte already has a microtubule array: the second meiotic spindle.

Unexpectedly, after incorporation the sperm does not play any major role in organising microtubules. As shown in Figure 9.9B, over a dozen cytoplasmic foci derived from the oocyte serve as the centres of small asters. During pronuclear development, these microtubule-containing asters increase in size as the cytoplasm becomes filled with a microtubule matrix. The pronuclei are embedded within this matrix, and by a process involving both assembly and disassembly the male and female pronuclei are moved into apposition at the cell centre.

Time-lapse studies demonstrate that the pronuclei move in a seemingly random jostling fashion to reach the egg centre, and they do not follow particular paths at specific times as in other fertilization cases. However, it must be remembered that the fertilization process in mammals, which concludes when the parental genomes intermix at first mitosis (Fig. 9.10), can take as long as a day, whereas in other systems, like sea urchins, syngamy is achieved within fifteen minutes! This rate places severe constraints on the activity of the egg cytoskeleton.

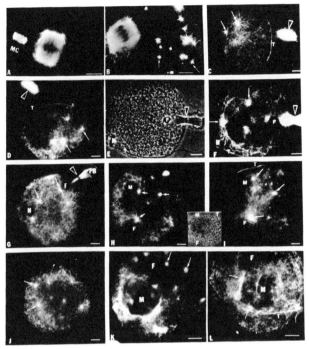

Fig. 9.9. Microtubules during sperm incorporation and in pronucleate eggs; A, unfertilized ooctye; the meiotic spindle is usually the sole microtubule-containing structure in the unfertilized oocyte; it is barrel-shaped and anastral, with broad meiotic poles, and anchored parallel to the oocyte cortex. Inset: meiotic chromosomes (MC) detected with DNA fluorescence. B, Cytoplasmic asters in unfertilized oocyte; several hours after ovulation, about a dozen cytoplasmic asters (arrows) assemble throughout the cytoplasm. C, D, Sperm incorporation; at sperm incorporation, the microtubules of the axoneme (T), the meiotic midbody (triangle), and the cytoplasmic asters (arrows) are apparent. E-F, Early pronucleate eggs; microtubules are found in the incorporated axoneme (T), in the midbody of the rotated meiotic spindle (triangle), and ramifying throughout the cytoplasm as a latticework extending from asters (arrows), some of which are in association with each pronucleus; the asters are not organized by the base of the incorporated sperm axoneme (T); six hours post-ovulation. G-J, Pronucleate eggs; as the male and female pronuclei form, the cytoplasmic asters enlarge, and a pair associate with the pronuclei (arrows); twelve hours post-ovulation. K-L, Late pronucleate eggs; as the pronuclei are moved together to the egg centre, a dense array of microtubules forms; this array has focal sites with the pronuclei embedded within its centre; eighteen hours post-ovulation; *Insets* M, male pronucleus; F, female pronucleus; PB, polar body nucleus; MC, meiotic chromosomes; T, sperm axoneme; arrows, cytoplasmic asters; triangle, meiotic midbody. Bars: 10 μm. (Reprinted, with permission, from Schatten *et al.*, 1986a).

Though the pattern of microtubule configurations during mammalian fertilization is atypical, studies with microtubule inhibitors demonstrate that their activity is required to achieve pronuclear union. As is shown in

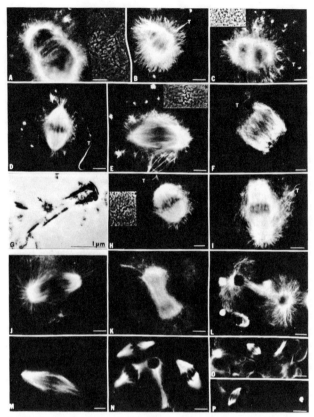

Fig. 9.10. Mitosis and early development, first division to blastocyst. A, At the end of first interphase the cytoplasmic microtubules disassemble from the interior and are replaced by sheaths of microtubules surrounding the adjacent, but still separate, pronuclei; Sixteen hours post-ovulation. *Inset*, chromosomes condensing within the adjacent pronuclei. B-C, prophase; the paternal and maternal chromosome sets are still separated by the perinuclear microtubules within a monaster. C, a spindle begins to emerge as the parental chromosomes meet; the sperm axoneme is apparent. D-G, metaphase; the metaphase spindle is typically barrel-shaped and anastral, with relatively broad mitotic poles; eighteen hours post-ovulation. G, sperm axoneme and centriole complex in a metaphase egg; though numerous parallel microtubules are found in the spindle region, microtubules are not observed near the incorporated sperm axoneme, with its centriole (arrow) and implantation fossa. H, I, anaphase; the spindle lengthens, and sparse microtubules extend from the broad poles towards the cell surface. J, telophase; interzonal microtubules develop, and a few microtubules extend from the wide poles towards the cell surface. K, cleavage; the interzonal microtubules bundle into a midbody. L, second interphase; the daughter nuclei are positioned at the blastomere cell centres within monasters extending from the nuclear surfaces; A midbody persists, and the second polar body remains attached at the left. M, second mitosis; at metaphase the spindle still has broad mitotic poles; twenty-six hours post-ovulation. N, third mitosis; fusiform spindles with well focused mitotic poles are observed at third division; thirty-two hours post-ovulation. O, morula; sixty-four hours post-ovulation. P, blastocyst; eighty hours post-ovulation; typical fusiform mitotic spindles are detetcted at fourth (O) and fifth (P) divisions. Bars: 10 µm, except G. (Reprinted, with permission, from Schatten *et al.*, 1985a).

Table 9.2, inhibitors of both microtubule assembly and disassembly arrest fertilisation, and if pronuclear development occurs, the pronuclei cannot be moved to the egg centre.

4 MICROFILAMENT ACTIVITY DURING THE PRONUCLEAR APPOSITION OF MAMMALIAN FERTILIZATION

Microfilaments are active during mouse fertilization at an unusual site. Although not required for sperm incorporation (Table 9.2), pronuclear apposition is inhibited by microfilament inhibitors. As shown by Maro *et al.* (1984) with cytochalasin D and by Schatten *et al.* (1986a) with latrunculin, microfilament inhibitors will prevent pronuclear apposition but not pronuclear development. In Figure 9.11, microfilaments in mouse eggs during fertilization and first cleavage are detected with rhodamine-phalloidin. In addition to cortical microfilaments, a perinuclear array of microfilaments is found diffusely throughout the cytoplasm. This perinuclear microfilament array appears required, in addition to the cytoplasmic microtubule matrix, for the motions leading to pronuclear apposition during mammalian fertilization.

IV Pronuclear development and nuclear architecture

1 NUCLEAR ARCHITECTURAL ALTERATIONS DURING FERTILIZATION

Fertilization and the onset of development requires several dramatic changes in nuclear organisation (reviewed by Krohne and Benavente 1986). The architecture of the nuclear surface (reviewed by Berezney 1979; Franke *et al.* 1981; Gerace and Blobel 1982; Maul 1982) involves the nuclear lamins, typically three proteins subjacent to the inner nuclear membrane (Fawcett 1966; Gerace *et al.* 1978), and nuclear peripheral proteins, referred to as 'P1' (Chaly, Bladon, Setterfield, Little, Kaplan, and Brown 1984) and 'Perichromin' (McKeon, Tuffanelli, Fukuyama, and Kirschner 1983), which reside between the chromatin and the nuclear lamins. During mitosis in somatic cells, the lamins dissociate from the nuclear envelope at prophase and reappear with the reconstituting envelope at telophase (Gerace and Blobel 1980). Unlike the behaviour of the lamins at mitosis, the peripheral antigens separate from the nuclear periphery and ensheath the condensing chromosomes before breakdown of the nuclear envelope and dissolution of the lamins during mitosis (Chaly *et al.* 1984; McKeon *et al.* 1983). During spermatogenesis, the nuclear lamins are lost or vastly reduced (Stick and Schwarz 1982; Moss, Burham, and Bellvé 1984; Benavente and Krohne 1985; Maul, French, and Bechtol, 1986) whereas during oogenesis the

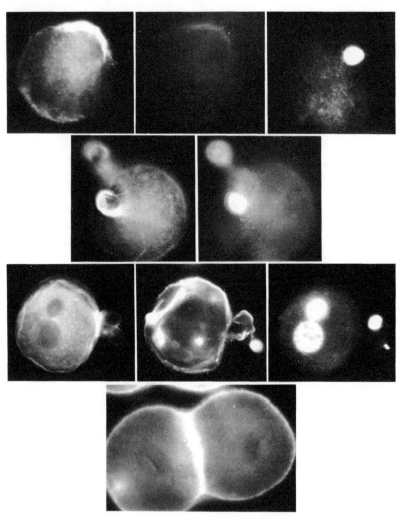

Fig. 9.11. Microfilaments in mouse oocytes during fertilization; microfilaments are found by the forming second polar body (top panel), in the incorporation cone (next panel), as a perinuclear matrix (second panel from bottom) and in the contractile ring during first division (bottom figure).

lamina may comprise only a single lamin (Maul and Avdalovic 1980; Stick and Krohne 1982; Stick and Schwarz 1983; Maul, Battaglia, Newmeyer, and Ohlsson-Wilhem 1984; Maul *et al* 1986b), which differs from somatic lamins (Krohne, Debus, Osborn, Weber, and Franke 1984; Stick and Hausen 1985; Benavente and Krohne 1985).

Because nuclear architecture is modified during gametogenesis, and because fertilization presents the special demand of uniting the parental genomes, the behaviour and appearance of nuclear lamins and other peripheral nuclear antigens is of considerable interest. In addition, alterations in nuclear structure, which might be maintained following determinative cell divisions, has been proposed to play an important role in cell differentiation, particularly during embryogenesis (Blobel 1985).

Changes in nuclear lamins during fertilization have been studied in the amphibian *Xenopus* (Benavente *et al*. 1985; Stick and Hausen 1985), in sea urchins and in mice (Schatten, Maul, Schatten, Chaly, Balczon, Simerly, and Brown 1985b). These systems represent extremes in fertilization mechanisms. For example, the sea urchin egg is spawned as a mature egg with a female pronucleus, and pronuclear fusion or syngamy occurs shortly after sperm incorporation. The ovulated mouse oocyte is arrested at second meiotic metaphase and, because pronuclear fusion does not occur, fertilization is only completed at first mitosis when the parental chromosomes align at metaphase.

2   BEHAVIOUR OF NUCLEAR LAMINS AND PERIPHERAL NUCLEAR ANTIGEN DURING MOUSE FERTILIZATION

Fertilization involves the new appearance of nuclear lamins and rearrangements of the peripheral nuclear antigens. Detectable lamins are vastly reduced in sperm, and the formation of the male pronucleus in both systems, involving dramatic biochemical (reviewed in Longo 1973; Poccia 1982; Zirkin, Soucek, and Chang 1982) and ultrastructural rearrangements (Longo and Anderson 1968; Anderson, Hoppe, Whitten, and Lee 1975), is coupled with the appearance of lamins associated with the male pronuclear envelope. In mouse oocytes, lamins appear on the female pronucleus as it develops after the completion of meiosis. The polar body nucleus has a reduced lamin complement, as judged by fluorescence intensity, perhaps as a result of its restricted access to the cytoplasmic pool of lamins; this circumstance suggests a possible pathway leading to its ultimate degeneration.

Fertilization in the mouse is only formally completed at first division when the parental chromosomes intermix. At prophase, the lamins dissociate from the pronuclei, and the peripheral nuclear antigens ensheath each chromosome, in a pattern typical for somatic cells (Gerace and Blobel 1980; Chaly *et al*. 1984, McKeon *et al*. 1984). After telophase, when diploid nuclei first form, the lamins and peripheral antigens redistribute to the nuclear surface.

In the unfertilized mouse oocyte, peripheral antigens ensheath each meiotic chromosome (MC: Fig. 9.12A) and the lamins are not detected (Fig. 9.12B; see Table 9.3 for results with other tested antibodies). DNA fluorescence of the chromosomes using the Hoechst dye 33258 is shown in Figure 9.12C.

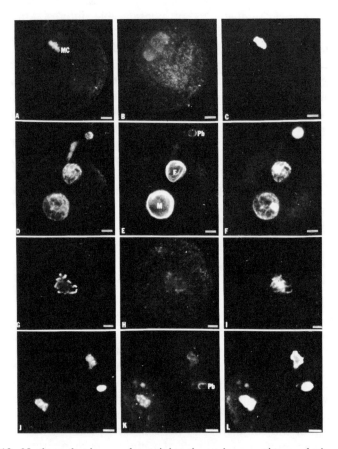

Fig. 9.12. Nuclear lamins and peripheral nuclear antigens during mouse fertilization and early development. A, Unfertilized oocyte; the P1 peripheral antigens ensheath the surface of each meiotic chromosome (MC); B, Lamin staining is lost in the ovulated oocyte, which is arrested at second meiotic metaphase; C, Hoechst DNA fluorescence; D, Pronucleate eggs; E, The peripheral antigens are associate with the rims of the male (M) and female (F) pronuclei and with the polar body nucleus (Pb); the lamins reassociate with the nuclear surface, and characteristically the polar body nucleus stains only weakly; F, Hoechst DNA fluorescence; G, mitotic eggs; at prophase, P1 against the peripheral antigens is redistributed from the pronuclear surfaces to cover each chromosome; H, the lamins dissociate from the mitotic chromosomes; I, Hoechst DNA fluorescence; J, cleavage; as the daughter nuclei reform after first division, the peripheral antigens dissociate from the decondensing chromosomes and reassociate with the nuclear periphery; K, the lamins associate with the reformed nuclear envelope; L, Hoechst DNA fluorescence. Bars: 10 μm. (Reprinted, with permission, from Schatten et al., 1985b).

During fertilization, the developing male and female pronuclei acquire lamins (lamins AC: Fig. 9.12E) and the peripheral antigens redistribute to the nuclear periphery (Fig. 9.12D). The polar body nucleus (Pb: Fig. 9.12E) is only dimly labelled with lamin antibody.

At mitosis the peripheral antigens condense around each chromosome (Fig. 9.12G), as the lamins dissociate from the chromosome mass (Fig. 9.12H). Following first division, the peripheral antigens again redistribute to the periphery of each nucleus (Fig. 9.12J) and the lamins reappear on the reconstituted blastomere nuclei (Fig. 9.12K). Table 9.3 summarises these localizations.

Table 9.3

Distribution of nuclear lamins and peripheral nuclear antigens in mouse and sea urchin eggs during fertilization and embryogenesis

| | Lamins | | | Peripherals |
| | B | A & C | | P1 |
	p-ab	m-ab	p-ab	m-ab
Mouse				
Sperm	+/−	§	−/*	−
Oocytes; GV	+	+	+	+
Meiotic chr.; unfert.	−	−	−	+
Male pronucleus	+	+	+	+
Female pronucleus	+	+	+	+
Polar body nucleus⁻	−	−	−	+
Mitotic chromosomes	−	−	−	+
Blastomere nuclei	+	+	+	+
Morula nuclei	+	−	−	+
Blastocyst nuclei	+	−	−	+
Adult somatic cells (3T3)	+	+	+	+
Sea Urchin				
Sperm	§	§	§	+
Female pronucleus; unfert.	+	−	+	+
Male pronucleus; fert.	+	−	+	+
Female pronucleus; fert.	+	−	+	+
Mitotic chromosomes; first	¶	−	¶	+
Mitotic chromosomes; morula	¶	−	¶	+
Gastrula nuclei	−	+	−	+
Pluteus nuclei	−	+	−	+
Adult cells (coelomocytes)	+	−	+	+

m-ab: monoclonal antibody
p-ab: polyclonal antibody
GV: germinal vesicles
Chr.: chromosones
*: apparent only after extraction with DNAase and 2 M NaCl
§: lamins are localized only at the acrosomal and centriolar fossae in sperm
¶: at chromosome peripheries

3   LAMINS DURING SEA URCHIN FERTILIZATION AND MITOSIS

The sea urchin egg is spawned with a mature female pronucleus, which already has associated lamins. The sperm nucleus, after incorporation into the egg, quickly expands with the concomitant uptake of lamins and peripheral antigens, probably of maternal origin. True pronuclear fusion occurs, and the lamins and peripheral antigens retain their association with the peripheries of the male and female pronuclei and merge to form the zygote nucleus. In contrast to division in somatic cells (Gerace and Blobel 1980) and in the mouse zygotes, the lamins are retained in or around the chromosomes during the first few mitoses in sea urchins. This retention of lamins around the chromosomes could be essential for the swift nuclear envelope reconstitution during the rapid cell cycles in a pattern not dissimilar to that observed in *Drosophila* embryos, where the lamins have been shown to remain near the mitotic spindle (Fuchs *et al*. 1983). Whether the lamins remain attached to the sea urchin mitotic chromosomes or are present together with the membrane vesicles (reviewed in Paweletz 1981) at or near the chromosomes remains to be established.

Lamins are reduced in sea urchin sperm to only the acrosomal (triangles: Fig. 9.13A) and centriolar fossae (arrows: Fig. 9.13A), whereas the peripheral antigens are present around the sea urchin sperm nucleus (Fig. 9.13C). Mouse sperm bind lamin antibodies regionally and only sparsely and apparently do not contain the peripheral antigens (Table 9.3).

The female pronucleus of the unfertilised sea urchin egg is spawned with lamins and peripheral antigens (Table 9.3). After sperm incorporation, the decondensing male pronucleus displays both sets of nuclear structural proteins (lamin B: Fig. 9.13E; peripheral antigens: Fig. 9.13F). At syngamy or pronuclear fusion (Fig. 9.13E and F), the lamins and peripheral antigens remain at the nuclear surface and coalesce to form the zygote nucleus.

The lamins behave in an unusual fashion during mitosis in sea urchins. Unlike mouse zygotes (Fig. 9.11) and somatic cells (Gerace and Blobel 1980), where lamins are never observed on chromosomes, nuclear lamins delineated each of the sea urchin mitotic chromosomes (Fig. 9.12G); frequently the centrosomes are also detected. The peripheral antigens are found to ensheath each chromosome (Fig. 9.12H).

4   SWITCH IN LAMINS EXPRESSED OR EXPOSED DURING EMBRYOGENESIS

During embryogenesis, specific lamins may be replaced. Lamins A and C, closely related proteins (Gerace and Blobel 1982, Kaufman, Gibson, and Shaper 1983), are apparently absent in mouse morulae and blastocysts but reappear later in somatic cells. In sea urchin embryos both lamins A and C

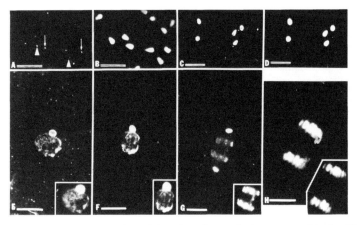

Fig. 9.13. Nuclear lamins and peripheral antigens during sea urchin fertilization and early development. In sperm, lamin antibodies label only the acrosomal (triangles) and centriolar (arrows) fossae, whereas the entire sperm nuclei are outlined with P1, the peripheral antibody (C). B and D, DNA fluorescence; during sea urchin fertilization, pronuclei bind lamin (E) and peripheral (F) antibodies; at syngamy these nuclear structural proteins are found along both pronuclear surfaces. G, at mitosis, the lamins dissociate from the nuclear surface and are found on each chromosome; the centrosomes are frequently detected; the peripheral antigens are redistributed to delineate each chromosome (H). Insets (bottom right of E, F, G, H), Hoechst DNA fluorescence. Bars: 10 μm. (Reprinted, with permission, from Schatten *et al.*, 1985*b*).

and lamin B, as detected with polyclonal antibodies, are lost after the blastula stage, although a different lamin AC epitope emerges as recognised by a monoclonal antibody. The differential disappearance of the lamins during embryogenesis occurs after the first divisions in the mouse and only at the blastula stage in sea urchins (Schatten *et al.* 1985*b*).

Mouse and sea urchin embryogeneses display an unexpected appearance and disappearance of lamin epitopes (Table 9.3). In the mouse, staining of lamins A and C with either monoclonal or polyclonal antibody has become unrecognizable at the morula and blastocyst stages, though lamin B staining is retained. Sea urchin embryogenesis displays a similar phenomenon, with the loss of lamin recognition by human autoimmune antibodies in blastula, gastrula, and plutei and the new recognition of a monoclonal lamin antibody at these stages, which does not bind to egg or morula nuclei (Table 9.3).

The nuclear lamins lost during spermatogenesis are restored at fertilization probably from maternal sources though the new exposure of paternal proteins cannot yet be excluded. The peripheral antigens associate with the surface of chromosomes during meiosis and mitosis, and with the periphery of pronuclei and nuclei during interphase; sea urchin sperm nuclei also have a coating of the peripheral antigens though it appears to be absent in mature mouse sperm nuclei. In the mammalian system, the nuclear lamins behave

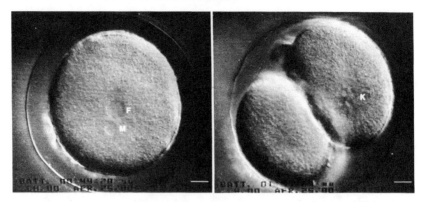

Fig. 9.14. Dithiothreitol prevents syngamy and karyomere fusion during sea urchin fertilization.

during mitosis as observed in somatic cells: they undergo dissolution at late prophase and reassemble at telophase. In contrast, nuclear lamins are retained on the chromosomes during mitosis in sea urchin eggs. During embryogenesis, specific lamins are differentially expressed or exposed in both systems studied. Changes in the architecture of the nuclei which participate in fertilization and embryogenesis may prove crucial for later events leading to development and differentiation.

V   Syngamy and the completion of the fertilization process

Fertilization is only formally completed when the parental genomes merge. This step can occur during interphase when the male and female pronuclear envelopes fuse to form a complete diploid zygote nucleus, as is the case in sea urchins. An alternative case is found in mammals where pronuclear fusion does not occur. Instead the male and female pronuclei remain intact during the first cell cycle and are held in close apposition at the egg centre. At first mitosis, the chromosomes condense within each pronucleus and the genomes only intermix at mitotic metaphase when the chromosomes align along the mitotic spindle equator.

Although it is possible to speculate on the different respective temporal demands on these systems requiring pronuclear fusion in the one case and permitting a less stringent situation in the other, these proposals do not appear to explain fully rather fundamental differences in fertilization mechanisms.

To address this question experimentally, it is now possible to arrest pronuclear fusion, without interfering with the other activities of fertilization, including the cytoskeletal rearrangements and expected motility. In Figure 9.14, frames from a time-lapse video sequence of a sea urchin egg incubated in 2.5 mM dithiothreitol are presented. Remarkably, pronuclear fusion is arrested though the other expected events during fertilization are observed. Of particular importance is the finding that these eggs progress normally through mitosis and cell division, leading to the conclusion that pronuclear fusion during interphase may not even be required in those cases in which it is typically observed. It is also interesting that, following cell division, the chromosomes decondense to form small individual nuclei, karyomeres, which cannot fuse. This finding implies that the processes during the fusion of the pronuclei at fertilization are mirrored each cell cycle when the nuclear envelopes ensheathing each chromosome unite to reconstitute the interphase nucleus.

VI   Centrosomes during fertilization

1   THE IMPORTANCE OF THE CENTROSOME

Centrosomes specify the configurations of microtubules, which in turn direct mitosis (Mazia 1984), the orientation of cellular movements (Albrecht-Buehler 1985), the organisation of the interphase cytoskeleton (Brinkley, Cox, Pepper, Wible, Brenner, and Pardue 1981), nuclear migrations at fertilisation (Schatten 1984), and a variety of other intracellular processes, including the maintenance of cell shape and structure (reviewed by Wheatley 1982 and McIntosh 1983). While the molecular composition of centrosomes is not understood, classical cytologists such as Boveri (1901; reviewed by Wilson 1925 and by Mazia 1984) recognized their critical importance. Laser oblation experiments by Berns and coworkers (Berns, Rattner, Brenner, and Meredith 1977; Peterson and Berns 1978; Koonce, Cloney, and Berns 1984) have demonstrated the importance of the centrosomal region in directing mitotic spindle formation and cellular migrations. Ultrastructural analyses have resolved clouds of osmiophilic material surrounding centrioles termed 'pericentriolar material' (PCM) or 'microtubule organising centres' (MTOCs) in a variety of small cells (Gould and Borisy 1977; Rieder and Borisy 1982; Paweletz, Mazia and Finze 1984). Indeed similar osmiophilic material is observed at expected organising centres in plant cells (Pickett-Heaps 1969; Wick, Seagull, Osborn, Weber, and Gunning 1981; Bajer and Mole-Bajer 1982). Recently autoimmune antibodies have been shown to be reliable markers for centrosomal detection (Calarco-Gillam, Siebert, Hubble, Mitchison, and Kirschner, 1983) providing a new avenue for their exploration and characterisation.

2   THE DOGMA OF THE CENTROSOME DURING FERTILIZATION

During fertilization, centrosomes are thought to be paternally inherited contributed along with the incorporated sperm centriole. Indeed Boveri's theory of fertilization postulated that the centrosome was the 'fertilizing element' that, once imported into the egg by the sperm, established the future mitotic poles (reviewed by Wilson 1925). This scheme mandates extranuclear contributions by both parents. Because centrosomes establish the precise configurations of assembling microtubules, they define mitotic axes, unequal cell divisions, and cytoskeletal patterns for development, differentiation and direction. During the cell cycle centrosomes must reproduce and subsequently separate so that each sibling cell receives a full complement. Our investigation provides experimental evidence demonstrating that centrosomes are indeed flexible structures, as recently proposed by Mazia (1984), which seemingly predict directly the observed microtubule configurations. They reproduce during interphase and aggregate and separate during mitosis. Sea urchins and probably most animals obey Boveri's rules and the centrioles and centrosomes are paternally inherited. Surprisingly mouse centrosomes appear to be of maternal origin.

3   DETECTION OF THE CENTROSOME AT FERTILIZATION

The behaviour of centrosomes during sea urchin and mouse fertilization and cell division has recently been reported by Schatten, Schatten, Mazia, Balczon, and Simerly (1986c). Centrosomes are detected at the bases of sea urchin sperm heads but not in the unfertilised egg. After sperm incorporation, the centrosome is found in the egg initially as an accumulation at the posterior face of the male pronucleus. At this stage the sperm aster is a radially symmetrical array of microtubules. The centrosome appears split after the migration of the female pronucleus, and is found at the junction between the two pronuclei when the sperm astral microtubules extend as an asymmetric crescent from the pronuclei. Following syngamy, the centrosome spreads further into an arc over the zygote nucleus as the sperm aster widens (Fig. 9.15). The incorporated sperm axoneme is also apparent in Figure 9.15 as is a punctate tubulin-containing site probably corresponding to the separating centrioles. At the streak stage two discrete centrosomes are observed and their fluoroescence intensity has increased considerably; the microtubules emanate from these regions and extend from the nuclear surface.

During first division, the centrosomes are initially compact but later flatten and enlarge. At prophase, the centrosomes have moved into the cytoplasm from the chromosome mass and the microtubules appear as two asters. At metaphase (Fig. 9.15), the centrosomes remain spherical and now the spindle microtubules can be distinguished from those in the asters. During anaphase, the centrosomes begin to flatten (Fig. 9.15) as the microtubules at the centres of the asters begin to disassemble. The spindle poles widen and the asters continue to elongate. At telophase the centrosomes enlarge into ellipses with centrosomal antigen concentrated in

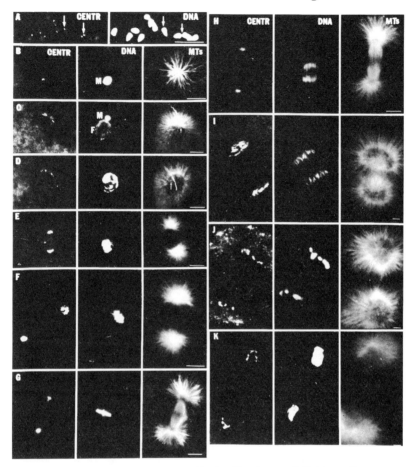

Fig 9.15. Centrosomes during sea urchin fertilization and division. Centrosomes are found at the base of spermheads (arrows: A) but not in unfertilized eggs. After sperm incorporation (B) they appear as a spot (CENTR: left panels) on the male pronucleus (DNA: middle panels) at the centre of the microtubules comprising the sperm aster (MTs: right panels). Following the pronuclear migrations (C) and during pronuclear fusion (D) the centrosomes spread into crescents from which microtubules are organized. Two centrosomes are observed at the streak stage (E) when the bipolar microtubule array extends from the nucleus. At prophase (F) the centrosomes condense and are at the centre a pair of asters. At metaphase (G) they remain as compact spheres from which the astral and spindle microtubules emanate. They flatten at anaphase (H) while the microtubules at the astral peripheries elongate and those at the astral centres disassemble. During telophase (I) the centrosomes expand in the direction of the next mitotic plane and there is a corresponding loss of microtubules at the astral interiors. The centrosomes aggregate on the poleward surfaces of the decondensing karyomeres (J) and reconstituting nuclei (K) during cleavage. G and H are triple stained for centrosomes, microtubules and DNA. Others are double stained for centrosomes and DNA with an antitubulin image at the same stage. M: male pronucleus; F: female pronucleus; Arrows: centrioles in C and D. Bars: 10 μm. (Reprinted with permission from Schatten *et al.* 1986*c*).

more or less compact masses (Fig. 9.15). The interiors of the now prominent asters are free of microtubules with a corresponding increase in the size of the centrosomes. The axes of the centrosomal ellipses are perpendicular to the first mitotic axis and parallel with the second; frequently the centrosomes appear oriented perpendicular to each other in the third dimension. At cleavage the centrosomes condense around the decondensing karyomeres forming crescents along the poleward faces of the reconstituted nuclei. The microtubules correspondingly are organised into partial monasters at cleavage.

4   THE CENTROSOME DURING MAMMALIAN FERTILIZATION

The configurations of microtubules observed during mouse fertilization (Schatten *et al.*, 1985*a*) indicate an unusual organisation. To explore further the behaviour and inheritance of centrosomes in this mammal, Schatten *et al.* (1986*c*) have traced centrosomal antigen throughout the entire course of fertilization and first division. Centrosomal material is found at the spindle poles in unfertilized oocytes (Fig. 9.16) as reported by Calarco-Gillam *et al.* (1983) and as numerous punctate concentrations scattered throughout the cytoplasm (Fig. 9.16; Maro, Howlett, and Webb 1985).

At sperm incorporation when the meiotic spindle rotates, centrosomal material is found throughout the egg cytoplasm (Schatten *et al.* 1986*c*). As the pronuclei develop, centrosomal foci and asters begin to associate with the surfaces of both pronuclei. Later numerous foci are found adjacent to the apposed pronuclei, which are embedded within an array of microtubules (Fig. 9.16).

Towards the end of first interphase, the number of detectable foci increases and they all migrate towards the pronuclear surfaces with several between the pronuclei. At the completion of first interphase, at the first signs of chromosome condensation within the intact pronuclear envelopes, the centrosomal antigen aggregates centrally, forming bright foci that circumscribe each pronucleus.

The centrosomal foci move as two broad clusters to opposing cytoplasmic regions at prophase (Fig. 9.17) as an array of microtubules extends from the centrosomes towards the chromosomes. The chromosomes align between the centrosomes at prometaphase as the mitotic spindle becomes apparent. At metaphase (Fig. 9.17) the centrosomes condense and the spindle is well defined. During anaphase and telophase (Fig. 9.17) the centrosomes remain closely associated as a band or plate composed of several foci from which microtubules extend. The arrangements of the microtubules at the various stages conforms well to the shapes of the centrosomes. As cleavage starts the centrosomes decondense and multiple foci are observed. After cleavage they are found as crescents associated with the poleward faces of the blastomere nuclei. The interzonal microtubules are prominent and typically a partial monaster of microtubules extends from the nuclei.

5 IMPLICATIONS OF CENTROSOMAL INHERITANCE AND BEHAVIOUR

The organisation and arrangements of centrosomes during fertilization, the first cell cycle, and mitosis in sea urchins and mice, shown diagrammatically in Figure 9.18, solves some essential problems in cell biology but raises a number of questions. The appearance of centrosomes in sea urchin sperm but not sea urchin eggs and in mouse eggs but not sperm predicts centrosomal retention during oogenesis in this mammal rather than the typical pattern of centrosome fidelity during spermatogenesis. Typically the mitotic centrosomes appear organized perpendicular to one another in the third dimension. Although this organisation does not affect the next mitotic axis, it may have important consequences during the following division. This shifting in centrosomal axes may prove critical to the embryo's ability to organise future division axes and unequal cleavage planes that may generate pattern.

Comparisons of these fertilization systems provide insights into the interactions between centrosomes and nuclei or chromosomes. Nuclei attract and associate with centrosomes. This behaviour is observed in the mouse after sperm incorporation when both pronuclei acquire centrosomes. The mature sea urchin sperm has a tightly affixed centriole and centrosomes, which after incorporation spread around the decondensing male pronucleus but always remain associated with it. In the absence of centrosomes in the unfertilized sea urchin egg, the female pronucleus can only associate with them after contact with the male pronucleus and from that point on is always found in association with centrosomes. It is of interest that centrosomes always reside between the tightly apposed male and female pronuclei. After first cleavage in both systems each blastomere nucleus is associated with centrosomes.

Chromosomes appear to repel centrosomes. Centrosomes remain associated with the nuclear regions until the breakdown of the nuclear envelopes, when they are displaced into the cytoplasm. In sea urchin eggs the centrosomal particles remain tightly packed around the centrioles whereas in mouse eggs they are only loosely associated. This repulsion of centrosomes from chromosomes may explain the requirement for kinetochores to anchor the opposing microtubule ends. Microtubules organized by centrosomes as asters at interphase may be organized into spindles at meiosis or mitosis by the bundling together of opposing microtubule ends. The stringent requirement for microtubule assembly in meiotic or mitotic cytoplasm and the regional influence of chromosomes in promoting localised microtubule assembly from centrosomes (Karsenti, Newport,

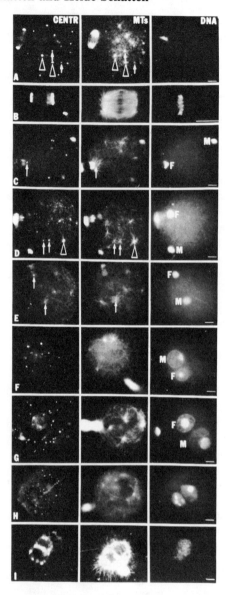

Fig. 9.16. Centrosomes during mouse fertilization. Centrosomes (CENTR: left panels) are found as cytoplasmic foci (A) and at the meiotic spindle poles (A,B) in unfertilized oocytes. Microtubules (MTs: middle panels) extend from the centrosomal material forming the meiotic spindle and cytoplasmic asters; each focus organizes an aster (arrows) with brighter ones associated with larger asters (triangles). Centrosomes are not detected in mouse sperm or with the entering sperm during incorporation (C,D). They associate with the developing pronuclei (C-G) as microtubules fill the cytoplasm. The foci aggregate (H) and condense (I)

Hubble, and Kirschner 1984) explain the loss of cytoplasmic microtubules and the appearance of spindles. The absence of functional kinetochores during interphase is interpretable since centrosomes will interact with the nuclear surface. The loss of kinetochores during mammalian spermatogenesis (Brenner and Brinkley 1982) is particularly noteworthy since had the entering sperm functional kinetochores, like the second meiotic chromosomes, it too might induce the formation of a meiotic spindle and the ejection of some of its chromatin.

Centrosomes mirror chromatin during the cell cycle. At interphase they are dispersed and duplicate. At prophase both the chromatin and the centrosomes condense and loose their associations with the nuclear envelope. At metaphase chromosomes and centrosomes are in their most compact state. During anaphase and telophase both separate, but in different directions; the chromosomes move to the centrosomes while the centrosomes flatten into plates with their axes predicting the next mitotic planes. As the cells enter the next interphase, both the chromosomes and centrosomes decondense and again resume their association with the reconstituted nuclear envelope. Phosphorylation of both nuclear lamins (Gerace and Blobel 1980) and centrosomes (Vandré, Davis, Rao, and Borisy 1984) at prophase and dephosphorylation of both at telophase, and the association of a cyclic AMP-dependent protein kinase with centrosomes (Nigg, Schäfer, Hilz, and Eppenberger 1985), may provide clues to the modifications necessary for these interconversions from interphase structure during mitosis.

VII   Cytoskeletal-karyoskeletal interactions

1   INTERDEPENDENCE OR INDEPENDENCE OF THE CYTO-SKELETAL AND NUCLEAR ARCHITECTURAL CHANGES

This chapter has concentrated on the changes in the cytoskeleton and in nuclear architecture necessary for fertilization. Though many of these events occur sequentially after the onset of fertilization, the mechanisms regulating their initiation, duration, and conclusion are not completely known. In this section, the interdependence or independence of these structural events will be considered.

Though the nuclear and cytoskeletal changes during fertilization have been considered separately, in actuality these structural events occur

---

around the apposed pronuclei at the completion of first interphase as the cytoplasmic microtubules disassemble leaving perinclear sheaths. A-E are triple labelled for centrosomes, microtublues and DNA. F-I are double labelled for centrosomes and DNA with antitubulin images at the same stage. MC: meiotoic chromosomes; M: male pronucleus; F: female pronucleus; Arrows: Centrosomal foci and small asters; Triangles: Corresponding bright centrosomal foci and larger asters. Bars: 10 μm. (Reprinted with permission from Schatten *et al.* 1986c).

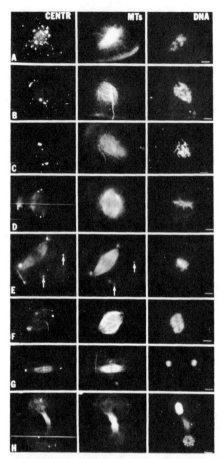

Fig. 9.17. Centrosomes during first division in mouse eggs. Centrosomes (CENTR: left panels) move as two clusters into the cytoplasm at prophase (A and B), as an irregular mass of microtubules (MTs: middle panels) forms around the aligning mitotic chromosomes (DNA: right panels) At prometaphase (C) the centrosomes appear as broad clusters on opposing sides of the chromosome mass as a barrel shaped anastral spindle becomes apparent. At metaphase the centrosomes aggregate into either loose irregular bands (D) or more tightly focused sites (E). Centrosomal foci not associated with spindle poles organize microtubules (arrows: E). During anaphase (F) the centrosomes continue their separation. At cleavage (G and H) the centrosomes are found along the poleward surfaces of the blastomere nuclei and the midbody becomes apparent. All images are triple labelled for centrosomes, microtubules and DNA. Bar: 10 μm. (Reprinted with permission from Schatten et al. 1986c).

synchronously. For example male pronuclear formation, and at times female pronuclear formation, usually occur as microtubules assemble in the cytoplasm and as cortical microfilaments polymerise to form elongate microvilli. Although methods are not yet available to inhibit nuclear

A

B

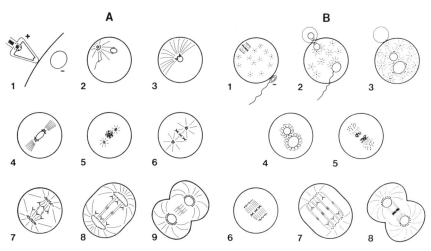

Fig. 9.18. Centrosomes during fertilization and cell division. A. Centrosomes in sea urchin eggs. The unfertilized egg lacks centrosomes, and they are introduced along with the sperm centriole during incorporation (A1). As the microtubules of the sperm aster assemble, the centrosomes spread around the male pronucleus (A2). Following the migration of the female pronucleus, they reside at the junction between the pronuclei (A3) and separate around the time for syngamy. The centrosomes have increased in intensity and are found at opposing poles of the zygote nucleus at the streak stage (A4). At prophase when the nuclear envelope has disintegrated they are displaced into the cytoplasm and nucleate the formation of the bipolar mitotic apparatus (A5). They enlarge by metaphase but retain their spherical configurations (A6) and at anaphase they begin to flatten and spread with axes perpendicular to the mitotic axis (A7). At telophase the centrosomes have expanded into hemispheres as the astral microtubules disassemble within the asters and continue to elongate at the astral peripheries. Centrosomes are found on the poleward faces of the blastomere nuclei in cleaving eggs (A8). B. Centrosomes in mouse eggs. Mouse sperm lack centrosomes and the unfertilized oocyte has sixteen cytoplasmic aggregates of centrosomal antigen as well as centrosomal bands at the meiotic spindle poles (B1). Each centrosomal focus organizes an aster and following sperm incorporation some foci along with their asters begin to associate with the developing male and female pronuclei (B2). When the pronuclei are closely apposed at the egg centre several foci are found in contact with the pronuclei and typically a pair reside between the adjacent pronuclei (B3). Towards the later half of the first cell cycle the number of foci increases. At the end of interphase all the foci condense on the pronuclear surfaces and sheaths of microtubules circumscribe the adjacent pronuclei (B4). At prophase the centrosomes detach from the nuclear regions appearing as two broad clusters (B5) which aggregate into irregular bands at metaphase (B6); the first mitotic spindle is typically barrel-shaped, anastral and organized in the absence of centrioles. At anaphase and telophase the centrosomes widen somewhat (B7) and at cleavage the centrosomes appear on the poleward nuclear faces (B8). Triangles: centrosomal foci; Lines: microtubules. (Reprinted with permission from Schatten *et al.* 1986c).

structural alterations, there are several means to prevent or promote cytoskeletal formation. The effects of cytoskeletal perturbations on the expected nuclear structure and the consequences of the proximity of chromosomes or nuclei on cytoskeletal organisation will be considered.

2   MICROTUBULES, MICROFILAMENTS AND PRONUCLEAR DEVELOPMENT AND ACTIVATION: INDEPENDENCE IN INVERTEBRATES

The question as to whether pronuclear development and activation is dependent on particular cytoskeletal activities has been addressed in several experiments. The results of Schatten, Schatten, Petzelt, and Mazia (1982a), using the microtubule inhibitor griseofulvin, demonstrates that, in the absence of microtubule assembly, the male pronucleus develops on schedule after insemination in sea urchins. Furthermore that work shows that both pronuclei synthesise DNA, as determined with autoradiography with ^{3}H-thymidine. Even though the pronuclei are unable to unite and fuse because of the inhibition of sperm astral microtubule assembly, the zygote undergoes the expected burst in DNA synthesis at times close to normal for the subsequent cell cycles.

In a reciprocal experiment, Schatten, Schatten, Bestor, and Balczon (1982b) using the microtubule assembly promoting drug taxol demonstrated that an increased concentration of microtubules does not appear to affect the rate of male pronuclear development either. Pronuclear fusion was arrested in that study due to the unusually stable microtubules of the sperm aster.

When microfilament inhibitors such as cytochalasin (Schatten and Schatten 1981) or latrunculin (Schatten *et al.* 1986a) are added immediately after insemination in sea urchins, the male pronucleus develops on schedule. Typically, though, the egg perishes before the completion of the first cell cycle, probably because of the effect of these microfilament inhibitors on the bursts in microvillar elongation.

From these experiments, it appears that in sea urchins pronuclear development is not affected by drugs affecting the rate of microtubule or microfilament assembly. It is of interest that the bioelectric events during fertilization are also unaffected by these drugs (Hülser and Schatten 1982). Together these studies led to the conclusion that, whereas many events are initiated at insemination, their progressions through the cell cycle are regulated somewhat independently of one another.

3   REQUIREMENTS OF MICROTUBULE ASSEMBLY FOR PRO-NUCLEAR DEVELOPMENT DURING MOUSE FERTILIZ-ATION

Colcemid disrupts the microtubules in unfertilized oocytes and prevents microtubule formation after insemination (Schatten *et al.* 1985a). The

Fig. 9. 19. Microtubule Inhibitors prevent nuclear lamina acquisition during mouse fertilization. In the presence of colcemid, the peripheral nuclear antigen appear (A) but nuclear lamins (B) do not appear on the condensed chromatin (C) found at twelve hours post-insemination.

chromosomes of unfertilized oocytes treated with colcemid disperse along the oocyte cortex and P_1, but not lamin, antibodies delineate their periphery.

Oocytes fertilized *in vitro* in the presence of colcemid permit sperm incorporation (Fig. 9.19A). The sperm nucleus is outlined with the P_1 antigen (Fig. 9.19B). However neither the scattered meiotic chromosomes nor the incorporated sperm nucleus acquires the expected nuclear lamins (Fig. 9.19C), and they appear to remain in a condensed state, as judged by DNA fluorescence microscopy at twelve hours post-insemination (Fig. 9.19A).

When colcemid is added at four hours after insemination, before the full development of the cytoplasmic microtubule array and the appearance of the pronuclei, and cultured to twelve hours post-insemination, the pronuclei develop normally and acquire the nuclear lamins and P_1 peripheral antigens.

4   NEITHER MICROFILAMENT INHIBITORS NOR MICRO-TUBULE PROMOTER AFFECT PRONUCLEAR DEVELOPMENT DURING MOUSE FERTILIZATION

Neither taxol, which stabilizes microtubules (Schatten *et al.* 1982*b*) nor the microfilament inhibitors cytochalasin (Maro *et al.*, 1984) or latrunculin (Spector *et al.*, 1983) prevent pronuclear development and lamin acquisition; both treatments prevent the cytoplasmic migrations leading to pronuclear apposition.

Taxol stabilizes the microtubules of the meiotic spindle and induces a proliferation of cytoplasmic microtubules. Because the extrusion of the nucleus of the second polar body does not occur properly, this nucleus and the male pronucleus acquire nuclear lamins and decondense normally. The pronuclei are unable to move to the egg centre and remain at the cortex.

Cytochalasin and latrunculin prevent the formation of the second polar body, and the three nuclei develop normally (Fig. 9.6) with associated nuclear lamins. The pronuclei are unable to move to the egg centre (Fig. 9.6).

5   IMPLICATIONS OF MICROTUBULE ASSEMBLY FOR MAMMALIAN PRONUCLEAR DEVELOPMENT

During mouse fertilization, microtubule assembly appears to be a pre-requisite in the cascade of events leading to pronuclear development, nuclear lamina acquisition and DNA synthesis. It does not appear to be required for the acquisition of the peripheral P_1 antigens. This dependency is not found during sea urchin fertilization, and nuclear development appears to proceed spontaneously. This situation may be similar to that in *Xenopus* eggs in which microinjected lambda DNA forms *de novo* into mini-pronuclei (Forbes *et al.* 1983).

Because the mouse egg is fertilized as an oocyte, the requirement for microtubule activity in the meiotic spindle may be coupled with the associated karyoskeletal events. Indeed lamin distribution in pronucleate mouse eggs is unaffected by colcemid application.

Sea urchin eggs are unusual in that they are fertilized as mature pronucleate eggs rather than as oocytes arrested at a meiotic stage. In addition the lamins behave in an unexpected manner at mitosis remaining associated with the chromosome arms. Nevertheless, during interphase the acquisition of lamins on the developing male pronucleus is affected by the cytoskeletal inhibitors tested here.

The hypothesis that the association of lamins is regulated during the cell cycle (Lohka and Maller 1985; Miake-Lye and Kirschner 1985; Burke and Gerace 1986) is supported in both systems studied. In the mouse the pronuclei are found with lamins at interphase, but not while the oocyte is in the final stages of meiotic maturation. In the sea urchin, fertilized at interphase, lamins associate with the incorporated sperm nucleus. However in refertilization experiments performed during mitosis, lamins are not found on the incorporated sperm nucleus until after division. This finding might imply that the lamins associated with the chromosomes are remnants from the interphase nucleus.

In sea urchins, microtubules are implicated in the redistribution of lamins at mitosis (Schatten *et al.* 1986a), though not during fertilization of the pronucleate egg. These findings suggest the hypothesis that microtubules are involved in changing distributions of lamins during meiosis or mitosis, but not during the nuclear reorganisations occurring in interphase.

Microtubule inhibitors have different effects in these two systems. In mice, they inhibit the meiotic-interphase transition and the transition from first mitosis to second interphase (unpublished data). In sea urchins they prevent neither the progression from first interphase to mitosis nor that from mitosis to second interphase. In addition they do not prevent DNA synthesis cycles (Schatten *et al.* 1982a). However, the timing of mitosis is slowed by microtubule inhibitors (Sluder and Begg, 1985).

The sequence and timing of the staged events during fertilization also

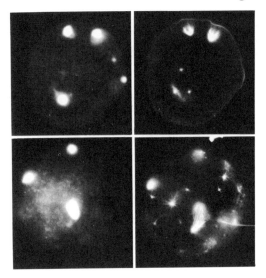

Fig. 9.20. Microtubule recovery after colcemid on unfertilized mouse oocyte. Miniature meiotic spindles form around the maternal chromosomes (upper pair) and around the chromosomes and the sperm head (lower pair) after recovery from colcemid.

differ between mice and sea urchins. In sea urchins, the first cell cycle is as short as an hour, perhaps a result of a time constraint to achieve syngamy. Egg activation is synchronous with fertilization, and cortical microfilament assembly followed by cytoplasmic microtubule formation occurs sequentially. In mice, activation is not tightly coupled with sperm incorporation (Howlett and Boulton 1985) and the cytoskeletal formation appears more complicated. Neither microtubules nor microfilaments seems to be involved in sperm incorporation, whereas both seem necessary to achieve pronuclear union. In this context, it is interesting that microfilament inhibitors prevent pronuclear union in mice (Maro *et al*. 1984; Schatten *et al*. 1986a), but do not prevent pronuclear formation.

In the mouse, cytoskeletal architecture is shaped by nuclear and chromosome position. Longo and Chen (1985) and Maro *et al*. (1986) demonstrate that the meiotic chromosomes induce regional accumulations of cortical actin, and miniature meiotic spindles assemble when chromosomes are adjacent to centrosomal particles (Fig. 9.20; Maro *et al*. 1986).

In conclusion, nuclear organisation during mouse, but not sea urchin, fertilization is dependent on egg cytoskeletal activity as well as under cell cycle controls. Lamins appear on pronuclei in interphase cytoplasm, but not on sperm nuclei in meiotic or mitotic eggs. Progression through meiotic-interphase or mitotic-interphase transitions required microtubules assembly in mice but not in sea urchins.

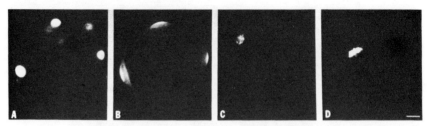

Fig. 9.21. Latrunculin inhibits colcemid-induced chromosome dispersion and blocks the cortical accumulations of actin adjacent to the dispersed meiotic chromosomes. The meiotic chromosomes of unfertilized mouse oocytes treated with 50 μM colcemid disperse along the egg cortex (A: Hoechst DNA fluorescence) and the dispersed chromosomes induce regional accumulations of cortical actin (B: Rhodaminyl-phalloidin microfilament fluorescence). This dispersion is prevented by 2.6 μM latrunculin (C: Hoechst DNA fluorescence). Latrunculin alone does not affect chromosome distribution (D: Hoechst DNA fluorescence). All cells processed at 14 hours post-insemination. Bar, 10 μm. ×550. (Reprinted, with permission, from Schatten *et al.* 1986*a*).

6   CHROMOSOMES INDUCE THE REGIONAL ACCUMULATION OF CORTICAL MICROFILAMENTS IN MOUSE OOCYTES

The microfilament inhibitors cytochalasin (Maro *et al.* 1984) and latrunculin (Schatten *et al.* 1986*a*) have remarkable effects on unfertilized mouse oocytes. When the microtubules of the meiotic spindle in the unfertilized oocyte are disrupted with microtubule inhibitors, the meiotic chromosomes are dispersed along the oocyte cortex (Schatten *et al.* 1985*b*). Recently Longo and Chen (1985) and Maro, Johnson, Webb, and Flach, (1986) have noted regional accumulations of cortical actin adjacent to each dispersed chromosome mass.

The regional accumulation of cortical actin (Fig. 9.21B) occurs adjacent to each mass of dispersed chromosomes (Fig. 9.21A) in colcemid-treated unfertilized oocytes. Latrunculin interferes with this dispersion of the chromosomes (Fig. 9.21C) and cortical microfilaments are restricted to the region that would have formed the second polar body constriction. This experiment indicates that microtubules normally hold the meiotic chromosomes together in the unfertilized oocyte and that there is a counterforce of cortical microfilaments pulling the chromosomes apart.

The meiotic chromosomes of unfertilized mouse oocytes scatter along the egg cortex when the spindle is disrupted with 50 μM colcemid and cortical actin accumulates regionally near the scattered chromosomes. 2.6 μM latrunculin inhibits chromosome dispersion induced by colcemid, similar to the effects of cytochalasin recently reported by Maro *et al.* (1986). 2.6 μM latrunculin alone does not affect chromosome distribution.

An interesting facet of this observation is its temporal nature. It appears that, once the chromosomes decondense into nuclei with annulated nuclear

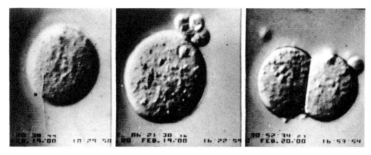

Fig. 9. 22. Time-lapse video of recovery from colcemid in a mouse oocyte. Though multiple second polar bodies are apparent, development occurs normally through first cleavage.

envelopes, they no longer induce this regional assembly of cortical actin. Perhaps the exposed chromatin itself or a peripheral nuclear protein like P_1 can trigger microfilament assembly if located near the cortex. However, perhaps the nuclear envelope or lamins interfere with this induction.

This possible scenario could explain the formation of the incorporation cone. Perhaps the sperm chromatin can induce a cortical microfilament assembly over a small region. This assembly would be reflected in the formation of a surface bleb until such time as the male pronucleus developed with a lamin-containing nuclear envelope. At that time the cone might be expected to be resorbed, as is observed during mammalian fertilization.

VIII   Future Directions

Future investigations of the mechanisms of fertilization hold bright promise for generating both a fuller understanding of this central process in reproduction and for improving our knowledge about the manner in which a cell can manipulate itself to move organelles and change its interior structure.

Some of the cell biological questions that can fruitfully be addressed now include the questions of internal regulatory systems. There appears to be a cascade of events leading to the activation of the unfertilised egg's metabolism (reviewed by Epel 1980). The role of changes in the concentration and distribution of cytoplasmic calcium ions can now be explored within living eggs with digital image processing with fluorescence microscopy (Poenie, Alderton, Tsien, and Steinhardt 1985). Phosphatidyl inositol and related compounds are likely to play a crucial role in initiating this sequence (Whittaker and Baker 1983; Turner, Sheetz, and Jaffe 1985). Other likely regulatory systems include cell cycle dependent fluctuations in cyclic nucleotide concentrations or in the level of phosphorylation (Gerace and Blobel 1980; Vandre *et al*. 1984; Nigg *et al*. 1985).

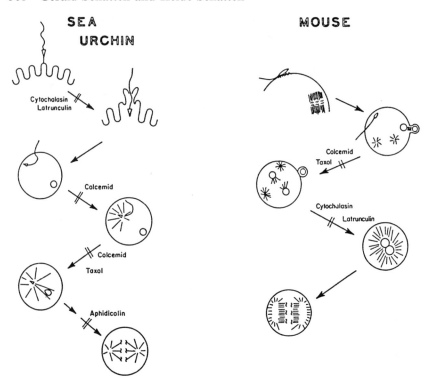

Fig. 9.23. Motility and centrosomes during mouse and sea urchin fertilization. Microfilament inhibitors block sperm incorporation in sea urchins and pronuclear apposition during mouse fertilization. Microtubule inhibitors prevent pronuclear union in both systems. The mouse centrosome is maternally inherited and organizes a plant-like spindle at first division, unlike the typical pattern in animals in which centrosomes derive from the sperm.

At the completion of fertilization, the egg must be left primed to initiate development and embryogenesis. It already appears that nuclear lamins (Benavente *et al.* 1985, Schatten *et al.* 1985b; Stick and Hausen 1985) change from the time of fertilization to early embryogenesis in a few species. This developmental switch might be indicative of a host of events required for the onset of the molecular processes in the embryo. RNA processing requires small nuclear RNA and associated proteins (snRNPs) which are found in the egg and embryonic cytoplasm until transcription begins actively, when they are then located in the nucleus (Nash *et al.* 1986). Nuclear matrix proteins, which might even provide the structural substrate onto which RNA processing occurs (Verheijen *et al.* 1986), only appear after fertilization when the embryo has initiated gene activity. Finally the mode by which exogenous genes are inserted into the genome during transgenic fertilization experiments is not fully clear. Further studies on nuclear structure and on the possibility of employing cytoskeletal events to assist in the production of transgenic eggs might prove useful.

Several questions remain about the respective parental contributions during fertilization. The elegant studies of Surani *et al.* (1986) and McGrath and Soltor (1985) demonstrate that both parental genomes are required for normal mammalian development. However they may not be necessary simultnaeously (Surani *et al.* 1986). The question of whether the mammalian centrosome is maternally inherited can be posed in parthenogenesis and polyspermy studies in which the paternal contribution is absent in the former case and multiplied in the latter. The time of centriole appearance during mammalian fertilization is not known, and the importance of this intricate structure is still open to question. In summary, a more complete characterisation of the pathways and sequence leading to the onset and progression of the fertilization process may provide new avenues for either assisting or interfering with this crucial and central step in reproduction.

References

Afzelius, B. A. (1976). A human syndrome caused by immotile cilia. *Science, N.Y.* **193**, 317–319.

Albrecht-Buehler, G. (1985). Is cytoplasm intelligent too? In *Gene Expression in Muscle* (ed. R. C. Strohman, R. C. and S. Wolf) pp. 1–21. Plenum, NY.

Allen, R. (1981). Cell motility. *J. Cell Biol.* **91**, 149s–155s.

Anderson, E. (1974). Comparative aspects of the ultrastructure of the female gamete. *Int. Rev. Cytol. Suppl.* **4**, 1–70.

—— Hoppe, P. C., Whitten, W. K., and Lee, G. S. (1975). *In vitro* fertilization and early embryogenesis; a cytological analysis. *J. Ultrastruct. Res.* **50**, 231–252.

Austin, C. R. (1968). *Ultrastructure of fertilization.* Holt, Rinehard and Winston, New York.

Baccetti, B., Burrini, A. G., Pallini, V., and Renieri, T. (1981). Human dynein and sperm pathology. *J. Cell Biol.* **88**, 102–107.

Bajer, A. S., and Mole-Bajer, J. (1982). Asters, poles and transport properties in spindlelike microtubule arrays. *Cold Spring Harb. Symp. quant. Biol.* **46**, 263–283.

Balakier, H., and Tarkowski, A. (1976). Diploid parthenogenetic mouse embryos produced by heat-shock and cytochalasin B. *J. Embryol. exp. Morph.* **35**, 25–39.

Balczon, R., and Schatten, G. (1983). Microtubule-containing detergent-extracted cytoskeletons in sea urchin eggs from fertilization through cell division: antitubulin immunofluorescence microscopy. *Cell Motil.* **3**, 213–226.

Banzhaf, W. C., Warren, R. H., and McClay, D. R. (1980). Cortical reorganization following fertilization of sea urchin eggs: sensitivity to cytochasasin B. *Devl. Biol.* **80**, 506–515.

Barak, L., Yocum, R., and Webb, W. (1981). *In vivo* staining of cytoskeletal actin by autointernalization of nontoxic concentrations of nitrobenzoxadiazaole-phallacidin. *J. Cell Biol.* **89**, 368–372.

Battaglia, D. E., and Gaddum-Rose, P. (1985). Distribution of F-actin in the rat egg and its sensitivity to cytochalasin B during fertilization. *J. exp. Zool.* **237**, 97–105.

Bavister, B. (1981). Analysis of culture media for *in vitro* fertilization and criteria for success. In *Fertilization and Embryonic Development in Vitro.* (ed. L. Mastroianni and J. D. Biggers) pp. 41–60. Plenum Press, New York.

Bavister, B. D., Leibfried, M. L., and Lieberman, G. (1983). Development of preimplantation embryos of the golden hamster in a defined medium. *Biol. Reprod.* **28**, 235–247.

Begg, D. A., Rebhun, L. I., and Hyatt, H. (1982). Structural organization of actin in the sea urchin egg cortex: microvillar elongation in the absence of actin filament bundle formation. *J. Cell Biol.* **93**, 24–32.

Benavente, R. and Krohne, G. (1985). Change of the karyoskeleton during spermatogenesis of *Xenopus*: expression of lamin L_{IV}, a nuclear lamina protein specific for the male germ line. *Proc. natn. Acad. Sci. U.S.A.* **82**, 6176–6180.

—— —— and Franke, W. (1985). Cell type-specific expression of nuclear lamina proteins during development in *Xenopus*. *Cell.* **41**, 177–190.

Berezney, R. (1979). Dynamic properties of the nuclear matrix. In *The Cell Nucleus* (ed. H. Busch). **7**, 413–456.

Berns, M. W., Rattner, J. B., Brenner, S., and Meredith, S. (1977). The role of the centriolar region in animal cell mitosis. *J. Cell Biol.* **72**, 351–367.

Bestor, T. H. and Schatten, G. (1981). Anti-tubulin immunofluorescence microscopy of microtubules present during the pronuclear movements of sea urchin fertilization. *Devl. Biol.* **88**, 80–91.

—— —— (1982). Configurations of microtubules in artifically activated eggs of the sea urchin *Lytechinus variegatus*. *Exp. Cell Res.* **141**, 171–178.

Blobel, G. G. (1985). Gene gating: a hypothesis. *Proc. natn. Acad. Sci. U.S.A.* **82**, 8527–8529.

Borisy, G. G., and Olmsted, J. B. (1972). Nucleated assembly of microtubules in porcine brain extracts. *Science N.Y.* **177**, 1196–1197.

Boveri, Th. (1904). *Zellen-Studien IV. Ueber die Natur der Centrosomen.* Fischer, Jena.

Branton, D., Cohen, C., and Tyler, J. (1981). Interaction of cytoskeletal proteins on the human erythrocyte membrane. *Cell.* **24**, 24–32.

Brenner, S. L. and Brinkley, B. R. (1982). Tubulin assembly sites and the organization of microtubule arrays in mammalian cells. *Cold Spring Harb. Symp. quant. Biol.* **46**, 241–254.

Brinkley, B. R., Cox, S. M., Pepper, D. A., Wible, L., Brenner, S. C., and Pardue, R. L. (1981). Tubulin assembly sites and the organization of cytoplasmic microtubules in cultured mammalian cells. *J. Cell Biol.* **90**, 554–562.

Brinkley, B., Fistel, R., Marcum, J., and Pardue, R. (1980). Microtubules in cultured cells: indirect immunofluorescent staining with tubulin antibody. *Int. Rev. Cytol.* **63**, 59–95.

Brinkley, B. R., Fuller, G. M., and Highfield, D. P. (1975). Cytoplasmic microtubules in normal and transformed cells in culture: analysis by tubulin antibody immunofluorescence. *Proc. natn. Acad. Sci. U.S.A.* **72**, 4981–4985.

Bryan, J. and Kane, R. E. (1978). Separation and interaction of the major components of sea urchin actin gel. *J. mol. Biol.* **125**, 207–244.

Burgess, D. R. and Schroeder, T. E. (1977). Polarized bundles of actin filaments within microvilli of fertilized sea urchin eggs. *J. Cell Biol.* **74**, 1032–1037.

Burke, B. and Gerace, L. (1986). A cell free system to study reassembly of the nuclear envelope at the end of mitosis. *Cell* **44**, 639–652.

Burridge, K., Kelly, T., and Mangeat, P. (1982). Nonerythrocyte spectrins: actin-membrane attachment proteins occurring in many cell types. *J. Cell Biol.* **95**, 478–486.

Calarco-Gillam, P. D., Siebert, M. C., Hubble, R., Mitchison, T., and Kirschner, M. (1983). Centrosome development in early mouse embryos as defined by an autoantibody against pericentriolar material. *Cell* **35**, 621–629.

Chaly, N., Bladon, T., Setterfield, G., Little, J. E., Kaplan, J. G., and Brown, D. L. (1984). Changes in distribution of nuclear matrix antigens during the mitotic cell cycle. *J. Cell Biol.* **99** 661–671.

Chandler, D. E. and Heuser, J. (1981). Postfertilization growth of microvilli in the sea urchin egg: new views from eggs that have been quick-frozen, freeze-fractured and deeply etched. *Devl. Biol.* **82**, 393–400.

Cheney, R., Hirokawa, N., Levine, J., and Willard, M. (1983). Intracellular movement of fodrin. *Cell Motil.* **3**, 649–655.

Clarke, G. N., Clarke, F., and Wilson, S. (1982). Actin in human spermatazoa. *Biol. Reprod.* **26**, 319–327.

—— and Yanagimachi, R. (1978): Actin in mammalian sperm heads. *J. exp. Zool.* **205**, 125–132.

Clarke, M. and Spudich, J. (1977). Nonmuscle contractile proteins: the role of actin and myosin in cell motility and shape determination. *A. Rev. Biochem.* **46**, 797–822.

Cline, C., Schatten, H., Balczon, R., and Schatten, G. (1983). Actin-mediated surface motility during sea urchin fertilization and division. *Cell Motil.* **3**, 513–524.

—— and Schatten, G. (1986). Microfilaments during sea urchin fertilization: fluorescence detection with rhodaminyl phalloidin. *Gamete Res.*, **14**, 277–291.

Coffe, G., Foucalt, G., Soyer, M. O., DeBilly, F., and Pudles, J. (1982). State of actin during the cycle of cohesiveness of the cytoplasm in parthogenetically activated sea urchin egg. *Expl. Cell Res.* **142**, 365–372.

—— Rola, F. H., Soyer, M. O., and Pudles, J. (1982). Parthogenetic activation of sea urchin egg induces a cyclical variation of the cytoplasmic resistance to hexylene glycol-triton X-100 treatment. *Expl. Cell Res.* **131**, 63–72.

Donovan, M. J., Mayhew, P. L., and Bellvé, A. R. (1984). Expression of nuclear lamin proteins during fertilization and early embryogenesis in the mouse. *J. Cell Biol.* **99**, 127a.

Dustin, P. (1978). Microtubules. *Springer-Verlag, Berlin.*

Eager, D., Johnson, M. H., and Thurley, K. W. (1975). Ultrastructural studies on the surface membrane of the mouse egg. *J. Cell Sci.* **22**, 345–353.

Eddy, E., and Shapiro, B. (1976). Changes in the topography of the sea urchin egg after fertilization. *J. Cell Biol.* **71**, 35–48.

Epel, D. (1977). The program of fertilization. *Sci. Am.* **237**, 128–138.

—— (1978). Mechanisms of activation of sperm and egg during fertilization of sea urchin gametes. In *Current topics in developmental biology*, **Vol. 12**, (ed. A. A. Moscona and A. Monroy) pp. 186–246. Academic Press, New York.

—— (1980). Ionic triggers in the fertilization of sea urchin eggs. *Ann. N.Y. Acad. Sci.* **339**, 74–85.

Ettensohn, C. A. (1985). Mechanisms of epithelial invagination. *Quart. Rev. Biol.* **60**, 289–325.

Fawcett, D. W. (1966). On the occurrence of a fibrous lamina on the inner aspect of the nuclear envelope in certain cells of vertebrates. *Am. J. Anat.*, **119**, 129–146.

Fishkind, D. J., Bonner, E. M., and Begg, D. A. (1985). Isolation and characterization of sea urchin spectrin: a novel Ca^{++}-sensitive actin crosslinker. *J. Cell Biol.* **101**, 285a.

Flaherty, S. P., Breed, W. G., and Sarafis, V. (1983): Localization of actin in the sperm head of the Plains mouse *Pseudomys australis*. *J. exp. Zool.* **225**, 497–500.

Forbes, D. J., Kirschner, M. W., and Newport, J. W. (1983). Spontaneous formation of nucleus-like structures around bacteriophage DNA microinjected into *Xenopus* eggs. *Cell* **34**, 13–23.

Franke, W. W., Scheer, U., Krohne, G., and Jarasch, E.-D. (1981). The nuclear envelope and the architecture of the nuclear periphery. *J. Cell Biol.* **91**, 39s–50s.

—— Schmid, E., Wellsteed, J., Grund, C., Gigi, O., and Geiger, B. (1983). Change of cytokeratin organization during the cell cycle. *J. Cell Biol.* **97**, 1255–1260.

Fuchs, J.-P., Giloh, H., Kuo, C. H., Saumweber, H., and Sedat, J. (1983). Nuclear structure: determination of the fate of the nuclear envelope in *Drosophila* during mitosis using monoclonal antibodies. *J. Cell Sci.* **64**, 331–349.

Gates, A. H. (1971). Maximizing yield and developmental uniformation in eggs. In *Methods in Mammalian Embryology* (ed. J. C. Daniels, Jr.) pp. 64–75. W. H. Freeman and Co., San Francisco.

Gerace, G., Blum, A., and Blobel, G. (1978). Immunocytochemical localization of the major polypeptides of the nuclear pore complex-lamina fraction. *J. Cell Biol.* **79**, 546–566.

—— and Blobel, G. (1980). The nuclear envelope lamina is reversibly depolymerized during mitosis. *Cell* **19**, 277–287.

—— —— (1982). Nuclear lamina and the structural organization of the nuclear envelope. *Cold Spring Harb. Symp. quant. Biol.* **46**, 967–978.

Gibbons, I. R. (1981). Cilia and flagella of eukaryotes. *J. Cell Biol.* **91**, 107S–124S.

Glenney, J., Jr. and Glenney, P. (1983). Fodrin is the general spectrin-like protein found in most cells whereas spectrin and the TW protein have a restricted distribution. *Cell.* **34**, 503–512.

Goldman, R., Goldman, A. E., and Yang, H.-Y. (1985). Alterations in intermediate filaments and associate proteins during mitosis. *J. Cell Biol.* **99**, 332a.

Goodman, S. R., Zagon, I. S., and Kulikowski, R. R. (1981). Identification of a spectrin-like protein in nonerythroid cells. *Proc. natn. Acad. Sci. U.S.A.* **78**, 7570–7574.

Gould, R. R. and Borisy, G. G. (1977). The pericentriolar material in Chinese hamster ovary cells nucleates microtubule formation. *J. Cell Biol.* **73**, 601–615.

Gould-Somero, M., Holland, L., and Paul, M. (1977). Cytochalasin B inhibits sperm penetration into eggs of *Urechis caupo. Devl. Biol.* **58**, 11–22.

Halenda, R. M., Primakoff, P., and Myles, D. G. (1984). Evidence for the presence of actin in guinea pig spermatogenic cells and its loss prior to formation of mature sperm. *J. Cell Biol.* **99**, 394a.

Hirokawa, N., Cheney, R., and Willard, M. (1983). Location of a protein of the fodrin-spectrin TW260/240 family in the mouse intestinal brush border. *Cell.* **32**, 953–965.

Howlett, S. K., and Bolton, V. N. (1985). Sequence and regulation of morphological and molecular events during the first cell cycle of mouse embryogenesis. *J. Embryol. exp. Morph.* **87**, 175–206.

Hulser, D. and Schatten, G. (1982). Bioelectric responses at fertilization: Separation of the events associated with insemination from those due to the cortical reaction in the sea urchin, *Lytechinus variegatus. Gamete Res.* **5**, 363–377.

Inoue, S. (1981). Cell division and the mitotic spindle. *J. Cell Biol.* **91**, 1315–1328.

Jaffe, L. A. (1976). Fast block to polyspermy in sea urchin eggs is electrically mediated. *Nature, Lond.* **261**, 68–71.

Kane, R. (1980). Induction of either contractile or structural actin-based gels in sea urchin egg cytoplasmic extract. *J. Cell Biol.* **86**, 803–809.

Karsenti, E., Newport, J., Hubble, R., and Kirschner, M. (1984). Interconversion of metaphase and interphase microtubule arrays, as studied by the injection of centrosomes and nuclei into *Xenopus* eggs. *J. Cell Biol.* **98**, 1730–1745.

Kashman, Y., Groweiss, A., and Schmueli, U. (1980). Chemical structure of latrunculin. *Tetrahed lett.* **21**, 3629–3636.

Kaufmann, S. H., Gibson, W., and Shaper, J. H. (1983). Characterization of the major polypeptides in the rat liver nuclear envelope. *J. biol. Chem.*, **258**, 2710–2719.

Keller, C., Gundersen, G., and Shapiro, B. M. (1980). Altered *in vitro* phosphorylation of specific proteins accompanies fertilization of *Strongylocentrotus purpuratus* eggs. *Devl. Biol.* **74**, 86–100.

Kirschner, M. W. (1980). Implications of treadmilling for the stability and polarity of actin and tubulin polymers *in vivo. J. Cell Biol.* **86**, 330–334.

Koonce, M. P., Cloney, R. A., and Berns, M. W. (1984). Laser irradiation of centrosomes in newt eosinophils: Evidence of centriole role in motility. *J. Cell Biol.* **98**, 1999–2010.

Krohne, G., and Benavente, R. (1986). The nuclear lamins: a multigene family of proteins in evolution and differentiation. *Exp. Cell Res.* **162**, 1–10.

—— Debus, E., Osborn, M., Weber, K., and Franke, W. W. (1984). A monoclonal antibody against nuclear lamina protein reveals cell type specificity in *Xenopus laevis. Exp. Cell Res.* **150**, 47–59.

Lazarides, E. (1980). Intermediate filaments as mechanical integrators of cellular space. *Nature, Lond.* **283**, 249–256.

Levine, J., and Willard, M. (1981). Fodrin: axonally transported polypeptides associated with the internal periphery of many cells. *J. Cell Biol.* **90**, 631–643.

—— —— (1983). Redistribution of fodrin (a component of the cortical cytoplasm) accompanying capping of cell surface molecules. *Proc. natn. Acad. Sci. U.S.A.* **80**, 191–195.

Loeb, J. (1913). *Artifical parthenogenesis and fertilization.* University of Chicago Press, Chicago.

Lohka, M. and Maller, J. (1985). Induction of nuclear envelope breakdown, chromosome condensation and spindle formation in cell-free extracts. *J. Cell Biol.* **101**, 518–523.

Longo, F. and Anderson, E. (1968). The fine structure of pronuclear development and fusion in the sea urchin, *Arbacia punctulata. J. Cell Biol.* **39**, 339–363.

Longo, F. J. (1973). Fertilization: a comparative ultrastructural review. *Biol. Reprod.* **9**, 149–215.

—— (1978). Effects of cytochalasin B on sperm-egg interactions. *Devl. Biol.* **67**, 249–265.

—— (1980). Organization of microfilaments in sea urchin (*Arbacia punctulata*) eggs at fertilization: effects of cytochalasin B. *Devl. Biol.* **74**, 422–431.

—— and Chen, D. -Y. (1985). Development of cortical polarity in mouse eggs: Involvement of the meiotic apparatus. *Dev. Biol.* **107**, 382–394.

Mabuchi, I. (1974). A myosin-like protein in the cortical layer of cleaving starfish eggs. *J. Biochem.* **76**, 47–55.

—— Spudich, J. A. (1980). Purification and properties of soluble actin from sea urchin eggs. *J. Biochem.* **87**, 785–802.

Maro, B., Johnson, M., Pickering, S., and Flach, G. (1984). Changes in actin distribution during fertilization of the mouse egg. *J. Embry. exp. Morph.* **81**, 211–237.

—— Johnson, M. H., Webb, M., and Flach, G. (1986). Mechanism of polar body formation in mouse oocytes: an interaction between the chromosomes, the cytoskeleton and the plasma membrane. *J. Embryol. exp. Morph.* **92**, 11–32.

—— Howlett, S. K., and Webb, M. (1985). Non-spindle MTOCs in metaphase II-arrested mouse oocytes. *J. Cell Biol.* **101**, 1665–1672.

Maul, G. G. (1982) *The Nuclear Envelope and the Nuclear Matrix.* Alan Liss, New York.

—— and Avdalovic, N. (1980). Nuclear envelope proteins from *Spisula solidissima* germinal vesicles. *Exp. Cell Res.* **130**, 229–240.

—— French, B. T. and Bechtol, K. B. (1986*a*). Identification and redistribution of lamins during nuclear differentiation in mouse spermatogenesis. *Devl. Biol.* **115**, 68–77.

—— Battaglia, F. A., Newmeyer, D. D., and Ohlsson-Wilhem, B. M. (1984). The major 67000 molecular weight protein of the clam oocyte nuclear envelope, clamin, is lamin-like. *J. Cell Sci.* **67**, 69–85.

Maul, G., and Schatten, G. (1986*b*). Nuclear lamins during gametogenesis, fertilization and embryogenesis. In *Nuclear Architecture.* (ed. W. W. Franke) Springer-Verlag. in press.

—— —— French, B., Lee, W., Pinkus, T., Jimenez, S., and Carrera, A. (1987). Nuclear lamin B localization in oocytes from sea urchins, mice and clams using an autoimmune serum. *Devl. Biol.* in press.

Mazia, D. (1984). Centrosomes and the mitotic poles. *Exp. Cell Res.* **153**, 1–15.

—— Schatten, G., and Sale, W. (1975). Adhesion of cells to surfaces coated with polylysine. *J. Cell Biol.* **66**, 198–200.

McIntosh, J. R. (1983). The centrosome as an organizer of the cytoskeleton. *Modern Cell Biol.* **2**, 115–142.

McGrath, J. and Soltor, D. (1985). Inability of mouse blastomere nuclei transferred to enucleated zygotes to support development. *Science, N.Y.* **226**, 1317–1319.

McKeon, F. D., Tuffanelli, D. L., Fukuyama, K., and Kirschner, M. W. (1983). Autoimmune response directed against conserved determinants of nuclear envelope proteins in a patient with linear scleroderma. *Proc. natn. Acad. Sci. U.S.A.* **80**, 4374–4378.

—— —— Kobayashi, S., and Kirschner, M. W. (1984). The redistribution of a conserved nuclear envelope protein during the cell cycle suggests a pathway for chromosome condensation. *Cell*, **36**, 83–92.

Miake-Lye, R. and Kirschner, M. W. (1985). Induction of early mitotic events in a cell free system. *Cell* **41**, 165–176.

Morton, R. W. and Nishioka, D. (1983). Effects of cytochalasin B on the cortex of the unfertilized sea urchin egg. *Cell Biol Int. Rep.* **7**, 835.

Moss, S. B., Burnham, B. L., and Bellvé, A. R. (1984). Differential occurrence of lamin proteins during mouse spermatogenesis. *J. Cell Biol.*, **99**, 126a.

Nash, M., Angerer, L., Angerer, R., Schatten, H., Schatten, G., and Marzluff, W. (1986). Small nuclear ribonucleoproteins during sea urchin fertilization. *J. Cell. Biol. in press.*

Nicosia, S. V., Wolf, D. P., and Inoue, M. (1977). Cortical granule distribution and cell surface characteristics in mouse eggs. *Dev. Biol.* **57**, 56–74.

Nigg, E. A., Schäfer, G., Hilz, H., and Eppenberger, H. M. (1985). Cyclic-AMP-dependent protein kinase type II is associated with the Golgi complex and with centrosomes. *Cell* **41**, 1039–1051.

Nishioka, D., Balczon, R., and Schatten, G. (1984). Relationship between DNA synthesis and mitotic events in fertilized sea urchin eggs. *Cell Biol. Int. Rep.* **8**, 337–346.

Otto, J., Kane, R., and Bryan, J. (1980). Redistribution of actin and fascin in sea urchin eggs after fertilization. *Cell Motil.* **1**, 31–40.

Paweletz, N. (1981). Membranes in the mitotic apparatus: mini-review. *Cell Biol. Int. Rep.* **5**, 323–336.

—— and Mazia, D. (1979). Fine structure of the mitotic cycle of unfertilized sea urchin eggs activated by ammoniacal sea water. *Eur. J. Cell Biol.* **20**, 37–44.

—— —— and Finze, E.-M. (1984). The centrosome cycle in the mitotic cycle of sea urchin eggs. *Exp. Cell Res.* **152**, 47–65.

Petzelt, C. (1972). Ca²⁺-activated ATPase during the cell cycle of the sea urchin *Strongylocentrotus purpuratus. Exp. Cell Res.* **70**, 333–339.

—— and Hafner, M. (1986). Visualization of the Ca⁺⁺-transport system of the mitotic apparatus of sea urchin eggs with a monoclonal antibody. *Proc. natn. Acad. Sci. U.S.A.* **83**, 1719–1722.

Pickett-Heaps, J. D. (1969). The evolution of the mitotic spindle: an attempt at comparative ultrastructural cytology in dividing plants cells. *Cytobios* **3**, 257–280.

Poccia, D., Greenough, T., Green, G. R., Nash, E., Erickson, J., and Gibbs, M. (1984). Remodeling of sperm chromatin following fertlization: nucleosome repeat length and histone variant transitions in the absence of DNA synthesis. *Dev. Biol.* **104**, 274–286.

—— (1982). Biochemical aspects of sperm nucleus activation by egg cytoplasm. *J. Wash. Acad. Sci.* **72**, 24–33.

Poenie, M., Alderton, J., Tsien, R., and Steinhardt, R. (1985). Changes in free calcium levels with stages of the cell division cycle. *Nature, Lond.* **315**, 47–49.

Pollard, T. (1981). Cytoplasmic contractile proteins. *J. Cell Biol.* **91**, 156s–165s.

Pratt, M. M. (1980). The identification of a dynein ATPase in unfertilized sea urchin eggs. *Devl. Biol.* **74**, 364–378.

—— Otter, T., and Salmon, E. D. (1980). Dynein-like Mg²⁺-ATPase in mitotic spindles isolated from sea urchin embryos (*Strongylocentrotus droebacheinsis*). *J. Cell Biol.* **86**, 738–745.

Rafferty, K. A. (1970). *Methods in Experimental Embryology of the Mouse.* Johns Hopkins Press.

Reparsky, E. A., Granger, B. L., and Lazarides, E. (1982). Widespread occurrence of avian spectrin in nonerythroid cells. *Cell* **29**, 821–833.

Rieder, C. L. and Borisy, G. G. (1982). The centrosome cycle in PtK₂ cells: Asymmetric distribution and structural changes in the pericentriolar material. *Biol. Cell.* **44**, 117–132.

Rothschild, V. (1956). *Fertilization.* Methuen, London.

Sale, W. S. and Gibbons, I. R. (1979). Study of the mechanism of vanadate inhibition of the dynein cross-bridge cycle in sea urchin sperm flagella. *J. Cell Biol.* **82**, 291–298.

Sanger, J. and Sanger, J. (1975). Polymerization of sperm actin in the presence of cytochalasin-B. *J. exp. Zool.* **193**, 441–447.

Satir, P. (1974). How cilia move. *Sci. Am.* **231**, 44–52.

Schackman, R. W., Eddy, E. M., and Shapiro, B. M. (1978). The acrosome reaction of *Strongylocentrotus purpuratus* sperm. *Devl. Biol.* **65**, 483–495.

Schatten, G. (1981). Sperm incorporation, the pronuclear migrations and their relations to the establishment of the first embryonic axis: time lapse video microscopy of the movements during fertilization of the sea urchin *Lytechinus variegatus. Devl. Biol.* **86**, 426–437.

—— (1984). The supramolecular organization of the egg cytoskeleton. *Subcell. Biochem.* **10**, 357–451.

—— Bestor, T., Schatten, H., Balczon, R., and Henson, J. (1985). Intracellular pH shift initiates microtubule-mediated motility during sea urchin fertilization. *Eur. J. cell Biol.* **36**, 116–127.

—— and Hülser, D. (1984). Timing the early events during sea urchin fertilization. *Devl. Biol.* **100**, 244–248.

—— Maul, G., Schatten, H., Chaly, N., Balczon, R., Simerly, C., and Brown, D. L. (1985*b*). Nuclear lamins and peripheral nuclear antigen distribution during fertilization and embryogenesis in mice and sea urchins. *Proc. natn. Acad. Sci. U.S.A.* **82**, 4727–4731.

—— and Mazia, D. (1976). The penetration of the spermatozoon through the sea urchin egg surface at fertilization: observations from the outside on whole eggs and from the inside on isolated surfaces. *Expl. Cell Res.* **98**, 325–337.

—— and Schatten, H. (1981). Effects of motility inhibitors during sea urchin fertilization: microfilament inhibitors prevent sperm incorporation and the restructuring of the fertilized egg cortex, whereas microtubule inhibitors prevent the pronuclear migrations. *Expl. Cell Res.* **135**, 311–330.

—— —— Bestor, T., and Balczon, R. (1982). Taxol inhibits the nuclear movements during fertilization and induces asters in unfertilized sea urchin eggs. *J. Cell Biol.* **94**, 455–465.

—— —— Spector, I., Cline, C., Simerly, C., Paweletz, N., and Petzelt, C. (1986*a*). Latrunculin inhibits the microfilament-mediated processes during fertilization, cytokinesis and early development. *Expl. Cell Res.*, in press.

—— Simerly, C., and Schatten, H. (1985). Microtubule configurations during fertilization, mitosis and early development in the mouse and the requirement for egg microtubule-mediated motility during mammalian fertilization. *Proc. natn. Acad. Sci. U.S.A.* **82**, 4152–4156.

Schatten, H., Cheney, R., Balczon, R., Willard, M., Cline, C., Simerly, C. and Schatten, G. (1986*b*). Localization of fodrin during fertilization and early development in mice and sea urchins. *Devl. Biol.* **118**, 457–466.

—— and Schatten G. (1980). Surface activity at the egg plasma membrane during sperm incorporation and its cytochalasin B sensitivity. *Devl. Biol.* **78**, 435–449.

—— —— Mazia, D., Balczon, R., and Simerly, C. (1986*c*). Behaviour of centrosomes during fertilization and cell division in mice and sea urchins. *Proc. natn. Acad. Sci. U.S.A.* **83**, 105–109.

—— —— Petzelt, C., and Mazia, D. (1982*a*). Effects of griseofulvin on fertilization and early development in sea urchins: independence of DNA synthesis, chromosome condensation and cytokinesis cycles from microtubule-mediated events. *Eur. J. cell Biol.* **27**, 74–89.

Schnapp, B. J., Vale, R. D., Sheetz, M. P., and Reese, T. S. (1985). Single microtubules from squid axoplasm support bidirectional movement of organelles. *Cell* **40**, 455–62.

Scholey, J. M., Porter, M. E., Grissom, P. M., and McIntosh, J. R. (1985). Identification of kinesin in sea urchin eggs, and evidence of its localization in the mitotic spindle. *Nature, Lond.* **318**, 483–486.

Schroeder, T. E. (1981). Interrelations between the cell surface and the cytoskeleton in cleaving sea urchin eggs. In *Cytoskeletal elements and plasma membrane organization* (ed. G. Poste and G. L. Nicolson) pp. 170–216. Elsevier/North Holland, Amsterdam.

Schroeder, T. E. (1978). Microvilli on sea urchin eggs: a second burst of elongation. *Devl. Biol.* **64**, 342–346.

—— (1979). Surface area changes at fertilization: resorption of the mosaic membrane. *Devl. Biol.* **70**, 306–326.

Shalgi, R., Phillips, D. M., and Kraicer, P. F. (1978). Observations on the incorporation cone in the rat. *Gamete Res.* **1**, 27–37.

Shapiro, B. M. and Eddy, E. M. (1980). When sperm meets egg: biochemical mechanisms of gamete interactions. *Int. Rev. Cytol.* **66**, 257–302.

Sluder, G. and Begg, D. (1985). Experimental separation of pronuclei in fertilized sea urchin eggs. *J. Cell Biol.* **100**, 897–903.

Snyder, J. A. and McIntosh, J. R. (1976). Biochemistry and physiology of microtubules. *A. Rev. Biochem.* **45**, 699–745.

Sobel, J. S. (1983). Localization of myosin in the preimplantation mouse embryo. *Devl Biol.* **95**, 227–231.

Sobel, J. S. and Alliegro, M. A. (1985). Changes in the distribution of a spectrin-like protein during development of the preimplantation mouse embryo. *J. Cell Biol.* **100**, 333–336.

—— Opas, M., and Kalnins, V. I. (1985). Spectrin and cell contacts in the early mouse embryo. *J. Cell Biol.* **101**, 1766a.

Spector, I., Shochet, N., Kashman, Y., and Groweiss, A. (1983). Latrunculins: novel marine toxins that disrupt microfilament organization in cultured cells. *Science, N.Y.* **183**, 493–495.

Spudich, A. and Spudich, J. A. (1979). Actin in triton-treated cortical preparations of unfertilized and fertilized sea urchin eggs. *J. Cell Biol.* **82**, 212–226.

Steinhardt, R. A. and Epel, D. (1974). Activation of sea urchin eggs by a calcium ionophore. *Proc. natn. Acad. Sci. U.S.A.* **71**, 1915–1919.

—— Zucker, R., and Schatten, G. (1977). Intracellular calcium release at fertilization in the sea urchin egg. *Devl. Biol.* **58**, 185–196.

Stick, R. and Hausen, P. (1985). Changes in the nuclear lamina composition during early development of *Xeonpus laevis*. *Cell* **41**, 191–200.

—— and Krohne, G. (1982). Immunological localization of the major architectural protein association with the nuclear envelope of the *Xenopus laevis* oocyte. *Expl. Cell Res.* **138**, 319–330.

—— and Schwarz, H. (1982). The disappearance of the nuclear lamina during spermatogenesis: an electron microscopic and immunofluorescence study. *Cell Diff.* **11**, 235–243.

—— —— (1983). Disappearance and reformation of the nuclear lamina structure during specific stages of meiosis in oocytes. *Cell* **33**, 949–958.

Stossel, T. (1984). Contribution of actin to the structure of the cytoplasmic matrix. *J. Cell Biol.* **99**, 15s–21s.

Surani, M. A. H., Barton, S. C., and Norris, M. L. (1986). Nuclear transplantation in the mouse: heritable differences between parental genomes after activation of the embryonic genome. *Cell* **45**, 127–136.

Szöllösi, D., Calarco, P., and Donahue, R. P. (1972). *J. Cell Sci.* **11**, 521–541.

Tilney, L. (1976). The polymerization of actin. II. How non-filamentous actin becomes nonrandomly distributed in sperm: evidence for the associated of this actin with membranes. *J. Cell Biol.* **69**, 51–72.

—— Hatano, S., Ishikawa, H., and Mooseker, M. (1973). The polymerization of actin: its role in the generation of the acrosomal process of certain echinoderm sperm. *J. Cell Biol.* **59**, 109–126.

Tilney, L. G. and Inoue, I. (1981). The kinetics of elongation of the acrosomal process in Thyone sperm and its relation to the assembly of actin. *J. Cell Biol.* **91**, 298a.

—— and Jaffe, L. A. (1980). Actin, microvilli and the fertilization cone of sea urchin eggs. *J. Cell Biol.* **87**, 771–782.

—— and Tilney, M. S. (1984). Observations on how actin filaments become organized in cells. *J. Cell Biol.* **94**, 76s–82s.

Trinkaus, J. (1976). On the mechanism of metazoan cell movements. In *The Cell Surface in Animal Embryogenesis and Development* (ed. G. Poste and G. L. Nicolson) pp. 225–329. Elsevier/North Holland, Amsterdam.

Turner, P., Sheetz, M. P., and Jaffe, J. A. (1984). Fertilization increases polyphosphoinositide content of sea urchin eggs. *Nature, Lond.* **310**, 414–415.

Vacquier, V. D. (1981). Dynamic changes of the egg cortex. *Devl. Biol.* **84**, 1–26.

Vandré, D. D., Davis, F. M., Rao, P. N., and Borisy, G. G. (1984). Phosphoproteins are components of the mitotic microtubule organizing centers. *Proc. natn. Acad. Sci. U.S.A.* **81**, 4439–4443.

378   Gerald Schatten and Heide Schatten

Verheijen, R., Kuijpers, H., Vooijs, P., Venrooij, W. V., and Ramaekers, F. (1986). Distribution of the 70 K RNA–associated protein during interphase and mitosis. Correlation with other U RNP particles and proteins of the nuclear matrix. *J. cell Sci* **86**, 173–190.

Virtanen, I., Bradley, R., Paasivuo, R., and Lehto, V.-P. (1984). Distinct cytoskeletal domains revealed in sperm cells. *J. Cell Biol.* **99**, 1083–1091.

Wang, Y. and Taylor, D. L. (1979). Distribution of fluorescently labeled actin in living
sea urchin eggs during early development. *J. Cell Biol.* **81**, 672–681.

Wassarman, P. M, Ukena, T., Josefowicz, W., and Karnovsky, M. (1977). Asymmetrical distribution of microvilli in cytochalasin B-induced pseudocleavage of mouse oocytes. *Nature, Lond.* **265**, 742–744.

Wassarman, P. M. and Fujiwara, K. (1978). Immunofluorescence anti-tubulin staining of spindles during meiotic maturation of mouse oocytes. *J. Cell Sci.* **29**, 171–188.

Wheatley, D. N. (1982). *The Centriole: a central enigma of cell biology*. Elsevier, Amsterdam.

Whitaker, M. J. and Baker, P. F. (1983). Calcium ions and cortical exocytosis. *Proc. R. Soc., Lond.* **218B**, 397–413.

Whittingham. D. G. (1968). Fertilization of mouse eggs *in vitro*. *Nature, Lond.* **220**, 592–593.

Wick, S. M., Seagull, R. W., Osborn, M., Weber, K., and Gunning, B. E. S. (1981). Immunofluorescence microscopy of organized microtubules in structurally stabilized meristematic plant cells. *J. Cell Biol.* **89**, 685–690.

Wieland, T. (1977). Modification of actins by phallotoxins. *Naturwissenschaften.* **64**, 303–348.

—— and Faulstich, H. (1978). Amatoxins, phallotoxins, phallolysin, and antamanide: the biologically active components of poisonous Amanita mushroons. *CRC Crit. Rev. Biochem.* **5**, 185–260.

Wilson, E. B. (1925). *The cell in development and heredity*. Macmillan, New York.

—— and Leaming, E. (1895). *An atlas of fertilization and karyokinesis*. Macmillan, New York.

—— and Mathews, A. P. (1895). Maturation, fertilization and polarity in the echinoderm egg: new light on the "quadrille of the centers". *J. Morphol.* **10**, 319–342.

Yanagimachi, R. and Noda, Y. D. (1976). Ultrastructural changes in the hamster sperm head during fertilization. *J. Ultrastruct. Res.* **31**, 465–485.

Zimmerman, A. M. and Zimmerman, S. (1967). Action of colcemid in sea urchin eggs. *J. Cell Biol.* **34**, 483–488.

Zirkin, B. R., Soucek, D. A., and Chang, T. S. K. (1982). Sperm nuclear packing and regulation during spermatogenesis and fertilization. *Johns Hopk. Med. J.* **151**, 102–112.

10 Neural transplants and the repair of neuro-endocrine and reproductive deficiencies

H. M. CHARLTON

I General introduction

In this short review the term CNS deficiencies will be confined to cover alterations in the activity of aminergic, cholinergic, or peptidergic neurones which result in aberrant physiological function. The discussion, therefore, not only includes abnormalities in behaviour resulting from pathological changes within the CNS, but also perturbations of peripheral physiology which normally depend directly or indirectly upon peptide hormones secreted by neuroendocrine cells.

Deficiencies in CNS function occur naturally in the normal process of aging or as genetic mutations, and they can also be elicited by environmental factors, such as infection or the ingestion of harmful substances. The laboratory scientist can elicit experimental depletions by the surgical removal of tissue or the interruption of neural connections, and also by reasonably specific electrolytic or chemical lesions. The capacity of CNS grafts to reverse any resulting neurological disorders then becomes amenable to investigation. Although papers reporting transplantation of tissue into the CNS have been published for at least 100 years, in the period 1980–1983 as many papers have been published as in the whole preceding century (Bjorklund and Stenevi 1985). Within the confines of the present review the research on dopaminergic and cholinergic function has concentrated upon the experimentally manipulated rat (see Bjorklund and Stenevi 1984), whilst the ability of CNS grafts to reverse deficiencies in peptide neurohormone production has been largely confined to the antidiuretic hormone (ADH) deficient Brattleboro rat and the gonadotrophin releasing hormone (GnRH) deficient hypogonadal (*hpg*) mouse (Charlton 1986).

II   Technical problems of grafting

In order for a graft to function it must obviously survive and release appropriate neurotransmitters or hormones such that target tissues are stimulated. There are several constraints upon graft survival one of which is the age of the donor tissue, and in the majority of experimental situations, even when donor tissue is grafted between animals of an inbred strain (syngeneic grafts), fetal tissue survives, whereas postnatal tissue is destroyed (Bjorklund and Stenevi 1984; Charlton, Jones, Whitworth, Gibson, Kokoris, and Silverman 1987).

The second major constraint is the relative immunological status of host and donor. It has been known for many years that the CNS is a privileged site with regard to graft survival but the degree of privilege appears to depend upon the extent of genetic disparity between donor and host as well as other factors (Mason, Charlton, Jones, Lavy, Puklavec, and Simmonds 1986).

The use of appropriately aged donor tissue in syngeneic grafts should, theoretically, ensure maximal graft survival, but in addition graft tissue will die unless adequate vascular integration with the host is accomplished.

Until relatively recently solid pieces of CNS tissue containing aminergic or peptidergic cells have been used as transplants. The revascularisation of such blocks of tissue raises problems, as does their capacity to remain *in situ*. Indeed the site of transplantation itself is of primary importance, for in most cases the reinnervation of the host tissue depends upon the graft being placed in close proximity to the cell bodies which have been deprived of their normal input (Dunnett, Bjorklund, Gage, and Stenevi 1985). This problem has to some extent been overcome by aspirating cavities above the denervated region and placing solid tissue grafts into a prepared vascular bed using a delayed surgical procedure (Stenevi, Kromer, Gage, and Bjorklund 1985). Direct intraparenchymal grafting with solid grafts is also possible but unless the region to be supplied lies deep within the CNS the loss of grafted tissue by backtracking along the tract of the cannula used to deposit the graft may be a serious problem (Das 1985).

The ventricles of the brain represent internal cavities which by judicial graft placement can be used to innervate regions of the CNS (Freed 1985).

During the first few hours or days after solid tissue grafting the surgical disruption may result in the donor tissue lying outside the blood brain barrier and, therefore, exposed to the host immune system for differing periods of time. In syngeneic grafting experiments this does not represent a problem as long as enough tissue remains at the selected site to provide adequate reinnervation. However, in allografts where host and donor may differ at major and/or minor histocompatibility loci the operative procedures may well result in sufficient host sensitization for grafts to be destroyed. In cross species xenografts this problem is even more acute (Mason *et al*. 1986). However, the use of immunosuppressant drugs may aid in the survival of allografts and xenografts (Inoue, Kohsaka, Yoshida, Ohtani, Toya, and Tsukada 1985; Brundin, Nilsson, Gage, and Bjorklund 1985).

Many of the problems associated with intraparenchymal grafts of solid tissue can be overcome by the use of isolated cell suspensions where the grafted tissue is dissociated using enzymatic or metal chelating techniques and the subsequent cell suspension implanted into the brain (Gage, Dunnett, Brundin, Isacson, and Bjorklund 1984). There are many theoretical advantages of this technique, for example it should be possible to purify the cell suspensions by removal of glial and endothelial elements, and to 'seed' several regions of the CNS in situations where cholinergic or dopaminergic depletion is more widespread.

As the objective of all the experiments described in this review is the replacement of a missing chemical, it is also theoretically possible that other tissues producing the same transmitter/peptide could be used in transplants. Indeed Freed, Morihisa, Spoor, Hoffer, Olson, Seiger, and Wyatt (1981) and Olson, Stromberg, Herrera-Marschitz, Ungerstedt, and, Ebendal (1985) have utilized adrenal medullary tissue grafted to the lateral ventricle to correct unilateral dopamine depletion caused by lesions in the nigrostriatal pathway (see below).

III Non-endocrine experimental models

In order to gauge the efficiency of CNS grafts to alleviate transmitter/peptide deficiencies experimental models are needed in which it is possible to evaluate success or failure. Obviously some readily observable external phenomenon such as altered behaviour allows for further correlation between anatomical, biochemical and electrophysiological findings.

1 DOPAMINERGIC SYSTEMS

Unilateral destruction of the dopaminergic fibres ascending from the substantia nigra to the striatum can be accomplished by injecting 6-hydroxy-dopamine (6-OHDA) into the substantia nigra itself.

The operation results in spontaneous turning towards the lesioned side for several days or weeks, followed by apparently normal locomotor behaviour. However, in lesioned animals this ipsilateral rotation can be reinitiated by the injection of drugs which can only release dopamine on the side opposite to the lesion. On the other hand, if a dopamine agonist such as apomorphine is injected, then because dopamine depletion results in post-synaptic supersensitivity on the lesioned side, there will be a greater stimulation of the striatum on this side, resulting in contralateral rotation. Using these drug-induced behavioural tests it has been possible first to identify lesioned animals with a suspected massive depletion of dopamine and then, after grafting, to observe a gradual reduction in the rotational behaviour induced by apomorphine.

Transplantation of dopamine neurones as solid tissue grafts into surgically formed cavities, as solid tissue intraventricular grafts, and as cell suspensions have all been effective in counteracting the behavioural effects of the initial 6-OHDA lesion (see Dunnett *et al.* 1985).

Grafts placed in the substantia nigra are ineffective and it seems that many of the effects of grafts placed locally in the striatum can be explained by passive diffusion of dopamine from the grafted tissue: extensive growth of fibres is only seen when grafts are placed in regions which would normally receive an intrinsic dopamine innervation. Olson *et al.* (1985) have also reported that the intensity of reinnervation of the striatum from intraventricular substantia nigra grafts may be even greater than that seen in the normal striatum. Using adrenal medullary cells as a source of dopamine they have demonstrated that catecholamines are only released from the grafts within the ventricle. These grafts are successful in reducing apomorphine induced rotational behaviour and it must, therefore, be assumed that passive diffusion into the surrounding brain tissue can occur and/or that transportation through the CSF may also provide a route whereby the catecholamines reach the striatum.

This behavioural model has been used to investigate the ability of cross species grafts to survive. Bjorklund, Stenevi, Dunnett, and Gage (1982) reported the survival of mouse dopaminergic cells in the rat brain, and more recently Brundin *et al.* (1985) have demonstrated that the survival of cell suspensions of mouse substantia nigra are greatly enhanced when the immunosuppressant drug Cyclosporin A is given daily post-operation.

Finally, transplantation of human adrenal medullary tissue into the striatum has been attempted using the patients' own medullary cells in a case of Parkinson's disease. The outcome of the operations was equivocal (Backlund, Granberg, Hamberger, Sedvall, Seiger, and Olson 1985).

2   CHOLINERGIC SYSTEMS

Cholinergic input to the cerebral cortex has, for many years, been implicated in learning and memory function (Deutsch 1971). The site of origin of the cell bodies providing this innervation is mainly from the ventral forebrain nucleus basalis magnocellularis (NBM). Lesions of this nucleus result in a massive reduction in choline acetyl transferase activity in the cortex. Ibotenic acid, one of the so-called excitotoxins, has the capacity of destroying cell bodies at the site of injection into the brain, but apparently spares nerve fibres of passage. Thus it is possible to destroy a large part of the NBM using such techniques with a resultant massive depletion in the cholinergic input to the cortex. Fine, Dunnett, Bjorklund, Clark, and Iversen (1985) have demonstrated an extensive reinnervation of the cortex in lesioned animals from grafted cell suspension derived from the fetal ventral forebrain. Their cell suspensions contained numerous cholinergic neurones and, when placed directly into the cortex, fibres grew out up to 3 mm from the grafts and apparently were organized in an appropriate laminar pattern.

Dunnett, Taniolo, Fine, Ryan, Bjorklund, and Iversen (1985) subsequently demonstrated that behavioural deficits in the lesioned animals were ameliorated by the ventral forebrain grafts and the results were consistent with a hypothesis linking the basal forebrain – neocortical cholinergic system to certain memory processes.

3   AGING

In the laboratory rat, as in the human, aging is associated with a decline in, for example, sensorimotor co-ordination and learning and memory. Motor co-ordination skills and spatial learning have been improved using neural grafts (Gage, Dunnett, Stenevi, and Bjorklund 1983; Gage, Bjorklund, Stenevi, Dunnett, and Kelly 1984).

In this brief and selective analysis of non-reproductive aspects of neural transplantation several points of neurobiological interest emerge. Although it has not been possible to alleviate the physiological consequences of the dopaminergic lesions by grafts into the substantia nigra, and therefore axons do not, in this case, seem capable of long migration, nevertheless when grafts are placed near the site of end terminal depletions there does seem to be greater specific local branching and outgrowth of fibres with apparent normal topographic innervation. This suggests that the adult brain maintains a capacity to attract and organize incoming axons despite the fact that the original timing of such mechanisms lay in the period of development of the CNS.

IV   Grafting in the CNS and endocrine function

The reproductive biologist in 1987 is well aware of the role of the CNS in controlling anterior pituitary function, and the foundations of this knowledge were based in large part upon experiments involving tissue transplantation (Harris 1972).

The intimate attachment of the anterior pituitary gland to the infundibulum and its enclosure in the sella turcica does not, of course, allow us to define the gland as part of the CNS. Nevertheless a brief resumé of the effects of hypophysectomy and pituitary gland transplantation is not out of place in the present review. Removal of the pituitary results in a cessation of body growth, a failure in thyroid and adrenal function, and an atrophy of the gonads with the cessation of oestrous cycles in the female and spermatogenesis in the male. Grafting the removed pituitaries beneath the kidney capsule, whilst resulting in survival of the gland, does not reverse any of the physiological sequelae of hypophysectomy outlined above. However re-grafting to the sella turcica results in a marked recovery of function. Moving the site of re-grafting into the cranial cavity away from the sella turcica, whilst favouring survival of the pituitary tissue, again failed to reverse the effects of hypophysectomy. From these simple, but elegant, experiments the neurohumoral hypothesis of the control of anterior pituitary functions was established (see Harris 1972).

In the early 1960s the site of production of the hypothalamic releasing factors was unknown and a further phase of CNS transplants was undertaken in experiments where the capacity of pituitary tissue grafted into various CNS regions to reverse the effects of hypophysectomy was investigated. Knigge (1962) demonstrated that reproductive function was stimulated in hypophysectomized rats by pituitary grafts in the tubero-infundibular region of the third ventricle. This led on to the concept of a hypophysiotropic area in the hypothalamus within which pituitary transplants elicited normal physiological responses in hypophysectomized rats (Halasz, Pupp, and Uhlarik 1962; Halasz, Pupp, Uhlarik, and Tima 1963, 1965; Flament-Durand 1965). Recently Kelsey, Hymer, and Page (1981) returned to this technique and they claimed that pituitary transplants can also suppress the action of the pituitary of the recipient host.

The above experiments were readily analyzed because measurable parameters were available to assess graft function, for example the reinstatement of oestrous cycles, growth of the gonads, and indeed body growth itself. There was also a good back-up of histological methods to assess the structure of the grafted pituitary at the termination of experiments. This ability to detect graft function, and methods for assessing biochemical activity and structural integrity were also central to the experimental models of the dopaminergic and cholinergic systems.

The effects of neural grafts upon sexual behaviour have been reported by Arendash and Gorski (1982), Arendash (1983) and Luine, Renner, Frankfurt, and Azmitia (1984). The problems of aging and reproductive senescence have received attention (Rogers, Hoffman, Zornetzer, and Vale 1984) and developmental maturation of the hypothalamus has also been the subject of investigation (Stenevi, Bjorklund, Kromer, Paden,

Gerlach, McEwen, and Silverman 1980; Paden, Silverman, Stenevi, and McEwen 1984; Paden, Gerlach, and McEwen 1985).

In all of the above cases, apart from the phenomena associated with aging, direct experimental perturbations of normal functions were an intergral part of the experimental procedures. However nature also performs experiments in which gene mutation results in alterations in physiological function and two such endocrine models are the ADH deficient Brattleboro rat and the GnRH deficient hypogonadal (*hpg*) mouse.

V   CNS grafts and the restoration of physiological function in the Brattleboro rat

The original discovery of the Brattleboro rat and the physiological consequences of the mutation have been reported by Valtin (1976, 1982). The precursor molecule containing ADH is synthesized and packaged into granules in magnocellular cells of the supra-optic and paraventricular nuclei whose axons project to the posterior lobe of the pituitary. This molecule consists of ADH, its associated neurophysin and a glycoprotein. In the Brattleboro mutant there is a single G residue missing in the glycoprotein coding region of the gene which results in aberrant production of the entire molecule (Schmale and Richter 1984). Although there is some debate as to whether or not this inability to synthesize and secrete ADH is absolute (see Charlton 1986), nevertheless the extreme diabetes insipidus found in the mutant is directly attributable to a massive failure in the production of hypothalamic vasopressin. The Brattleboro rat represents an ideal model in which to test the ability of neural grafts to survive and function. The brain is largely devoid of immunoreactive ADH and therefore grafted cells can be readily identified in the host, and the measurement of water intake and urine output should, theoretically, provide an ideal indicator of physiological recovery.

In the first series of experiments to be reported, Gash, Sladek and Sladek (1980) described the outgrowth of normal ADH axons to the mutant median eminence with a small alleviation of the diabetes insipidus. Since that time several other authors have reported graft survival and apparently normal topographic outgrowth of ADH containing axons, but the early physiological data has not been confirmed (Charlton 1986).

VI   CNS grafts and the restoration of reproductive activity in the *hpg* mouse

1   BACKGROUND

The classical experiments leading up to the formulation of the neurohumoral control of the anterior pituitary gland have been mentioned in Section IV. Normal post-natal reproductive development depends upon the production of at least one hypothalamic releasing factor, the gonadotrophin releasing hormone (GnRH), thought to control both pituitary luteinizing hormone (LH) and follicle stimulating hormone (FSH) synthesis and secretion. As the gonads respond to LH and FSH during the pubertal and adult periods subtle feedback signals from the testes or ovaries are transduced by the CNS and/or anterior pituitary more precisely to regulate spermatogenesis and ovarian follicle turnover. In seasonal breeding, for example, external environmental factors can impose a higher level of control resulting in what has been called 'annual puberty' (Lincoln and Short 1980).

The hypogonadal mouse was disovered in a breeding colony at the Medical Research Council Radiobiology Unit, Harwell, UK. There is a failure in post-natal gonadal growth in both sexes and this is associated with a severe deficiency in hypothalamic GnRH with a consequent depletion in pituitary LH and FSH synthesis and secretion (Cattanach, Iddon, Charlton, Chiappa, and Fink 1977).

Recombinant DNA techniques have demonstrated that there is probably only one copy of the GnRH gene in mammals and that the gene consists of 4 exons which when transcribed and translated produce a peptide containing the decapeptide sequence of GnRH and an additional 56 amino acid sequence, the gonadotrophin releasing hormone associated peptide or GAP (Adelman, Mason, Hayflick, and Seeburg 1986).

Structural analysis of the GnRH gene in hypogonadal mice has shown that there is a deletion removing 2 exons coding for a large part of the GAP moiety (Mason, Hayflick, Zoeller, Young, Phillips, and Seeburg 1986). However this severely truncated gene still contains the codons for GnRH. We have never so far been able to identify immunostainable GnRH in the brains of *hpg* mice (Lyon, Cattanach, and Charlton 1981; Charlton 1986) but using techniques of *in situ*-hybridization Mason *et al.* (1986) have evidence that an aberrant messenger RNA may indeed be formed in the GnRH neurons of *hpg* mice. Presumably this cannot be translated effectively.

Thus we have a mutant in which the effects of the mutation are readily visible phenotypically as early as 2 weeks post-natally, in which pituitary and serum gonadotrophic hormone levels are drastically reduced, and in which the brain is devoid of immunostainable GnRH. This means that any normal tissue grafted into the brain will be easily identified and the biological consequences of such grafts can be assessed at the level of the pituitary and gonads, providing a multitude of parameters by which success or failure can be measured.

2   THE GRAFTING TECHNIQUE

The majority of GnRH neurons are found in the preoptic area (POA) with axons passing laterally and caudally to be concentrated in the midline at the median eminence and proximal pituitary stalk (Merchenthaler, Gorcs, Setalo, Petrusz, and Flerko 1984). Two pieces of late fetal/day 1 neonatal POA tissue were dissected out and placed in a small volume of saline and the tissue taken up with a minimum amount of saline into a modified hypodermic needle with a tight-fitting stainless steel plunger. The needle was then placed in the CNS of anaesthetized mice under stereotaxic control and the POA tissue expelled. In the majority of experiments to be described the site of transplantation was deep within the third ventricle (Krieger, Perlow, Gibson, Davies, Zimmerman, Ferin, and Charlton 1982; Young, Detta, Clayton, Jones, and Charlton 1985).

3   GRAFT STRUCTURE AND DEVELOPMENT

Using immunocytochemical methods Silverman, Zimmerman, Gibson, Perlow, Charlton, Kokoris, and Krieger (1985) demonstrated that GnRH innervation of the median eminence in the region of the pituitary portal vessels could be detected as early as 10 days post-operation. Indeed physiological evidence of early stimulation of the pituitary gonadal axis was reported by Young *et al.* (1985) in experiments where testis weight was significantly increased in 4 out of 10 *hpg* male mice 10 days after POA grafting to the third ventricle. In the study of Silverman *et al.* (1985) the number of immunoreactive GnRH cells found in a graft ranged from 3 to 140 and the branching of axons in the region of the median eminence continued to increase with time. Ultrastructurally the GnRH neurons in the grafts showed evidence of active secretion and the axons in the median eminence were closely associated with specialized ependymal cells or tanycytes (Silverman, Zimmerman, Kokoris, and Gibson (1986).

In all animals which showed a positive stimulation of the gonads, GnRH neurons were present in third ventricle grafts and over 90 per cent of fibres were seen to innervate the median eminence in a normal pattern (Silverman *et al.* 1985; Charlton, Parry, and Jones 1985; Charlton 1986). This suggests that even in the adult the median eminence seems to maintain mechanisms which can attract GnRH axons in a specific fashion (Fig. 10.1). This is in keeping with the data reported in the previous sections on dopaminergic and cholinergic systems.

Grafts between normal and *hpg* mice from our own breeding colony survived and stimulated pituitary and gonadal function for up to 1 year (Charlton *et al.* 1985).

The relative immunological privilege of the brain as a site for tissue transplantation has also allowed us to graft tissue between strains of mice which actually differ in the expression of a neuronal cell surface marker. The glycoprotein Thy-1 is present in two allotypic forms in mice, Thy 1.1 and Thy 1.2. These molecules differ by only one amino acid and although most mice express Thy 1.2 on cell surfaces, the AKR strain expresses Thy 1.1

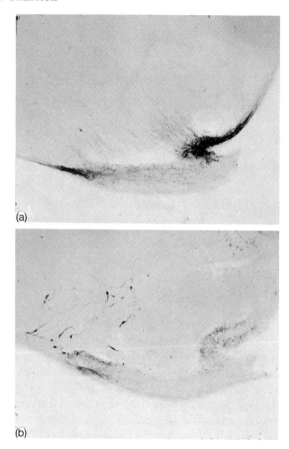

Fig. 10.1. Sagittal sections through the brain of (a) a normal female mouse showing the concentration of GnRH nerve terminals in the median eminence rostrally (on the left) and around the base of the pituitary stalk caudally (on the right). In the *hpg* mouse plus normal POA graft (b) the immunopositive fibres leaving the graft can be seen to innervate exactly the same two regions of the hypothalamus indicating that even in the adult the median eminence retains its capacity specifically to direct the growth of GnRH axons.

Therefore using a monoclonal antibody against Thy 1.1 it is possible to stain the entire grafted tissue in the mutant brain rather than just the GnRH cell bodies and axons (Charlton, Barclay, and Williams 1983). Although Thy 1.1 positive neuronal tissue leaves such grafts superiorly as well as inferiorly, GnRH staining is confined to the inferior region and particularly in the region of the median eminence, thus again confirming the specificity of GnRH containing axonal outgrowth.

It also happens that the rat expresses the Thy 1.1 molecules on neuronal cell surfaces which means that utilizing the monoclonal antibody against Thy

1.1, we can visualize the presence and survival of fetal rat POA material grafted to mice. Intraventricular grafts of such material fail to survive (Mason *et al.* 1986).

At the very outset of this review the problem of the age of grafted tissue was raised, and the effects of age upon survival and function of grafted POA tissue in *hpg* mice has been investigated. Post-natal day 1 (P1) tissue and embryonic day 16–18 (E16–18) both have a 70 per cent success rate with regard to stimulation of gonadal function. Physiologically successful grafts were found in only 20 per cent of animals given post-natal day 5 POA tissue: however the pituitary and testes in these 20 per cent were stimulated as in the E16–18 and P1 groups. Post-natal day 10 tissue failed to survive and function. Thus the age of grafted tissue is critical in the restoration of physiological function to *hpg* mice, but there is no ready answer as to why older tissue fails to survive (Charlton *et al.* 1987).

Thus far we have considered the effect upon survival of the site of transplantation, the immunological relationship of the donor and host, and the age of transplanted tissue. In the section covering dopaminergic grafts it was mentioned that Freed *et al.* (1981) used adrenal medullary tissue as a graft source rather than fetal substantia nigra. Kokoris *et al.* (1984) have reported that GnRH neurones contained in fetal accessory olfactory bulb tissue grafted to the third ventricle of *hpg* mice reversed the mutant hypogonadism, even though these neurones do not normally project to the median eminence.

4   THE PHYSIOLOGICAL ACTIONS OF PREOPTIC AREA (POA) GRAFTS IN *HPG* MICE

Having considered the structure and development of POA grafts an assessment of their physiological function is possible at several levels. The *hpg* pituitary contains only a fraction of the normal amount of LH and FSH (Cattanach *et al.* 1977) and the number of receptors capable of binding GnRH is also drastically reduced (Young, Speight, Charlton, and Clayton 1983). The number of immunostainable gonadotroph cells in the pituitary is also reduced and the number of secretory granules in these cells is subnormal (McDowell, Morris, and Charlton 1982). Preoptic area grafts reverse all of these deficiencies, with pituitary LH and FSH content approaching normal in both sexes (Krieger *et al.* 1982; Gibson, Krieger, Charlton, Zimmerman, Silverman, and Perlow 1984; Young *et al.* 1985). Pituitary GnRH receptor numbers are normalized after POA grafting (Young *et al.* 1985) and the numbers, size and granule content of gonadotrophs also approached normal levels (Morris and Charlton 1983).

At the gonadal level testis growth was increased as early as 10 days post operation (Young *et al.* 1985) and within 14 days some of the grafted female *hpg* mice had open vaginas with stimulated uteri and ovaries (Charlton *et al.*

1985). Although the testes in *hpg* males with POA grafts never reached normal adult proportions, nevertheless all stages of spermatogenesis were present (Krieger *et al.* 1982). The stimulated testes in the vast majority of cases were producing androgens for export as demonstrated by seminal vesicle growth, but in one or two cases seminal vesicle weight did not correlate positively with testis size (Young *et al.* 1985; Silverman *et al.* 1985). This phenomenon may be explained by fewer GnRH fibres reaching the median eminence, although this was not immediately obvious, or perhaps the pattern of a release into the portal vessels was abnormal (this is at present under investigation).

In the grafted *hpg* mice the POA area has been removed from its normal anatomical site and therefore must have been deprived of a great deal of afferent information controlling GnRH release. Will such grafts be susceptible to gonadal steroid feedback? In a previous series of experiments in which the *hpg* GnRH deficiency was corrected by hormone injections, testosterone implants which caused a massive depletion in pituitary LH and FSH in normal male mice could not prevent the GnRH induced increase in FSH synthesis by the mutant pituitary, thus suggesting that a major site of negative feedback of testosterone may lie above the level of the pituitary (Charlton, Halpin, Iddon, Rosie, Levy, McDowell, Megson, Morris, Bramwell, Speight, Ward, Broadhead, Davey-Smith, and Fink 1983). Similar testosterone pellets failed to prevent LH and FSH synthesis in *hpg* males with POA implants, implying an action of testosterone on neural centres regulating either the amplitude or frequency of GnRH secretion, rather than directly on the GnRH neurone itself (Charlton, Jones, Ward, Detta, and Clayton 1987).

Although ovarian activity is dramatically stimulated by POA grafts in *hpg* females none of the grafted animals has shown any evidence of spontaneous ovulation. Indeed the vaginal smear patterns indicate that they all exhibited prolonged periods of constant oestrus interspersed with a few days of leucocytic filtration. Thus in females it would appear that the pattern of GnRH release from the grafts may not be normal, with the result that spontaneous surges of GnRH followed by pituitary LH release and ovulation do not take place. Nevertheless the fact that enough LH and FSH was readily secreted by the pituitaries to stimulate folliculogenesis and steroidogenesis indicated that GnRH had reached the pituitary gland. In very long term grafts the ovary became depleted of follicles but contained a mass of interstitial tissue, probably formed by incorporation of the thecal cells of progressive waves of degenerating follicles (Charlton 1986).

With the stimulation of full spermatogenesis in the males, have the POA grafts repaired the reproductive system to the extent that these males can mate and sire offspring? The *hpg* mutation presents a problem with respect to male mating behaviour in that the mutant male CNS has not been subjected to neonatal androgens. Thus when adult, although treatment with testosterone stimulated full spermatogenesis, no mating behaviour ensued. However if the *hpg* males were given androgen treatment in the early neonatal period, and then testosterone pellets later, the result was fully fertile, sexually active animals (Ward and Charlton 1987). At the present moment we have not grafted POA tissue to *hpg* males pre-treated with androgens on day 1 but it is extremely probable that animals manipulated in this manner would indeed be capable of mating behaviour, but we cannot yet give an absolute answer to the question posed at the beginning of this paragraph.

Normal sexual behaviour in the female rodent is not totally dependent upon perinatal steroid influences, and Ward and Charlton (1981) have shown that *hpg* females treated with oestrogen and progesterone as adults exhibit full female sexual behaviour. Thus although there is no evidence to date for spontaneous ovulation in *hpg* females given POA grafts, nevertheless the stimulation of the uterus and the long periods of vaginal oestrus already mentioned indicate that there may be significant amounts of ovarian steroids in their circulation. It was therefore not surprising that many of these females mated when caged with normal stud males. However a most surprising observation was that many of the animals which mated became pregnant. Apparently then the POA grafts had restored full reproductive capacity to *hpg* females suffering GnRH deficiency (Gibson *et al.* 1984; Charlton *et al.* 1985).

The fact that ovulation occurred argues that this must have been preceded by a surge of LH, and this surge of LH was, in all probability, preceded by a surge of GnRH. In this experimental situation the only likely source of GnRH in the *hpg* females was in the graft. It therefore appears that the mating stimulus had evoked a reflex release of GnRH from the nerve terminals leaving the graft.

What is the neuroanatomical substrate for such reflex release? It is, of course, possible that despite the fact that we have no control over the position of GnRH cell bodies in the graft, nevertheless host axons from an ascending pathway transducing the mating stimulus may specifically seek out such cells or their processes. The possible presence of host axons in the graft is mentioned in the discussion of the paper by Silverman, Zimmerman, Kokoris, and Gibson (1986). If ascending spinal pathways can specifically control GnRH neurones in the grafts, why apparently is there not also an input from centres controlling cyclic release of GnRH? Perhaps a simpler explanation of the reflex release of LH lies in the fact that, in mice, mating is associated with an induced surge of prolactin which is essential for corpus luteum maintenance. The GnRH fibres crossing the median eminence from the graft to the portal vessels must interpolate themselves between many other nerve endings. There is very little evidence for synaptic contacts in this region of the CNS, indeed the accepted picture is that there may be a great degree of axo-axonal contact such that the activity of one set of axons may cause more widespread effects on others in the same region (Ajika 1980). It is therefore possible that the mating stimulus evoking prolactin release also caused GnRH release from donor GnRH nerve terminals in the same region.

Whatever the mechanism evoking the reflex ovulation, in the case of the *hpg* female the genetic deficiency in GnRH production has been fully repaired by CNS grafts. Females, hitherto completely infertile, have mated, ovulated, maintained pregnancies and reared litters within 2 months of transplantation.

In general, control of neuronal function can be exercised in the region of the cell body, at points along axons and dendrites and also at nerve terminals. The fact that both the pituitary and gonads of male and female *hpg* mice are stimulated by neural grafts containing GnRH neurones and that a similar stimulation by synthetic GnRH depends upon multiple daily injections, might be construed as an argument that the grafted neurones may be releasing GnRH in a pulsatile manner. Wherever the pulse generator controlling GnRH in normal animals may reside, it is perfectly possible that its transduction into episodic GnRH secretion into the portal vessels could be activated in whole, or in part, at the level of nerve terminals in the median eminence.

In normal female rodents the daily pulsatile secretion of GnRH and LH is augmented in each oestrous cycle to produce the ovulatory surge of GnRH and LH. This spontaneous surge appears to be absent in POA grafted female mice which suggests that the effects of environmental factors such as day length and olfactory stimuli may influence the GnRH neurone at sites other than nerve terminals (see Fig. 10.2). It is also possible that the positive feedback effects of oestrogens bringing about the LH surge are also mediated by pathways impinging on the GnRH neurone which have been interrupted in the grafting process. This could provide a further explanation for the failure of spontaneous LH surges in the grafted females.

VII   General conclusions

There can be no doubt that neural grafts have proved of therapeutic value in the animal experiments reviewed above. The ability to deliver peptides or amines locally into specific areas of the brain from grafted cells may well have advantages over microinjections of similar compounds via cannulae. The fact that grafted cells may have altered informational input, especially if purified cell suspensions are used, may allow us to answer questions about control mechanisms involved in hormonal output. A point of particular interest with regard to reproduction would be to see if a seasonal breeding animal with a graft of GnRH neurones would still be able to transduce alterations in day length or whether the graft would maintain pituitary activity throughout the year.

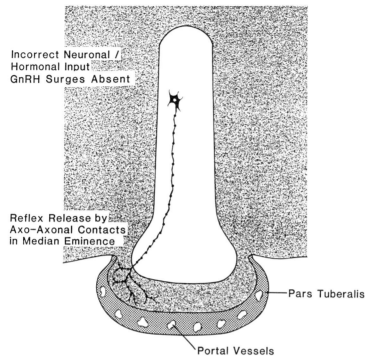

Incorrect Neuronal /
Hormonal Input
GnRH Surges Absent

Reflex Release by
Axo-Axonal Contacts
in Median Eminence

Pars Tuberalis

Portal Vessels

Fig. 10.2. Schematic diagram of a coronal section demonstrating the relationship of a grafted GnRH neurone at the level of the median eminence. As the grafted tissue is removed from its normal anatomical site it may be that the cell body is deprived of much of its normal input. The axon specifically innervates the median eminence where it branches profusely to terminate on portal blood vessels passing to the pituitary. This region of the hypothalamus is intimately related to the pars tuberalis which in rats consists mainly of gonadotroph cells (these points are discussed further in the text).

The interactions between host and donor tissue, in particular the fact that the adult brain seems to maintain the capacity to influence the direction of outgrowth of specific fibres, represents fertile ground for further investigations.

References

Adelman, J. P., Mason, A. J., Hayflick, J. S., and Seeburg, P. H. (1986). Isolation of the gene and hypothalamic cDNA for the common precursor of GnRH and prolactin release-inhibiting factor in human and rat. *Proc. natn. Acad. Sci. U.S.A.* **83**, 179–183.

Ajika, K. (1980). Relationship between catecholaminergic neurons and hypothalamic hormone containing neurons in the hypothalamus. In *Frontiers in neuroendocrinology*, Vol. 6 (ed. W. F. Ganong and L. Martini) pp. 1–32. Raven Press, New York.

Arendash, G. W. (1983). Brain tissue transplantation: a new tool for exploring the sexual differentiation of the brain. *IBRO News* **11**, 7–12.

—— and Gorski, R. A. (1982). Enhancement of sexual behaviour in female rats by neonatal transplantation of brain tissue from males. *Science, N.Y.* **217**, 1276–1278.

Backlund, E. O., Granberg, P. O., Hamberger, B., Sedvall, G., Seiger, A., and Olson, L. (1985). Transplantation of adrenal medullary tissue to striatum in Parkinsonism. In *Neural grafting in the mammalian CNS* (ed. A. Bjorklund and U. Stenevi) pp. 551–558. Elsevier, Amsterdam, New York, Oxford.

Bjorklund, A. and Stenevi, U. (1984). Intracerebral neural implants: neuronal replacement and reconstruction of damaged circuitries. *Rev. Neurosci.* **7**, 279–308.

—— —— (1985). Intracerebral neural grafting: A historical perspective. In *Neural grafting in the mammalian CNS* (ed. A. Bjorklund and U. Stenevi) pp. 3–14. Elsevier, Amsterdam, New York, Oxford.

—— —— Dunnett, S. B., and Gage, F. H. (1982). Cross species neural grafting in the rat model of Parkinson's disease. *Nature, Lond.* **298**, 652–654.

Brundin, P., Nilsson, O. G., Gage, F. H., and Bjorklund, A. (1985). Cyclosporin A increases survival of cross species intrastriatal grafts of embryonic dopamine-containing neurones. *Expl. Brain Res.* **60**, 204–208.

Cattanach, B. M., Iddon, C. A., Charlton, H. M., Chiappa, S. A., and Fink, G. (1977). Gonadotrophin-releasing hormone deficiency in a mutant mouse with hypogonadism. *Nature, Lond.* **269**, 338–340.

Charlton, H. M. (1986). The use of neural transplants to study neuroendocrine mechanisms. In *Frontiers in neuroendocrinology*, Vol. 9 (ed. W. F. Ganong and L. Martini) pp. 77–99. Raven Press, New York.

—— Barclay, A. N., and Williams, A. F. (1983). Detection of neuronal tissue from brain grafts with anti-Thy 1.1 antibody. *Nature, Lond.* **305**, 825–827.

—— Halpin, D. M. G., Iddon, C., Rosie, R., Levy, G., McDowell, I. F. W., Megson, A., Morris, J. F., Bramwell, A., Speight, A., Ward, B. J., Broadhead, J. Davey-Smith, G., and Fink, G. (1983). The effects of daily administration of single and multiple injections of gonadotrophin-releasing hormone on pituitary and gonadal function in the hypogonadal (*hpg*) mouse. *Endocrinology* **113**, 535–544.

—— Jones, A. J., Ward, B. J., Detta, A., and Clayton, R. N. (1987). The effects of castration or testosterone upon pituitary function in hypogonadal (*hpg*) mice bearing normal fetal preoptic area (POA) grafts. *Neuroendocrinology*, in press.

—— —— Whitworth, D., Gibson, M. J., Kokoris, G., and Silverman, A. J. (1987). The effects of age of intracerebroventricular grafts of normal preoptic area tissue upon pituitary and gonadal function in hypogonadal (*hpg*) mice. *Neurosci*, in press.

—— Parry, D. M., and Jones, A. (1985). Thy 1.1 and GnRH fibre output and physiological function in female *hpg* mice given pre-optic area grafts from fetal AKR mice. In *Neural grafting in the mammalian CNS* (ed. A. Bjorklund and U. Stenevi) pp. 637–641. Elsevier, Amsterdam, New York, Oxford.

Das, G. D. (1985). Intraparenchymal transplantation. In *Neural grafting in the mammalian CNS* (ed. A. Bjorklund and U. Stenevi) pp. 23–30. Elsevier, Amsterdam, New York, Oxford.

Deutsch, J. A. (1971). The cholinergic synapse and the site of memory. *Science, N.Y.* **174**, 788.

Dunnett, S. B., Bjorklund, A., Gage, F. H., and Stenevi, U. (1985). Transplantation of mesencephalic dopamine neurons to the striatum of rats. In *Neural grafting in the mammalian CNS* (ed. A. Bjorklund and U. Stenevi) pp. 451–470. Elsevier, Amsterdam, New York, Oxford.

—— Toniolo, G., Fine, A., Ryan, C. N., Bjorklund, A., and Iversen, S. D. (1985). Transplantation of embryonic ventral forebrain neurons to the neocortex of rats with lesions of nucleus basalis magnocellularis-II. Sensorimotor and learning impairments. *Neurosci.* **16**, 787–797.

Fine, A., Dunnett, S. B., Bjorklund, A., Clarke, D., and Iversen, S. D. (1985). Transplantation of embryonic ventral forebrain neurons to the neocortex of rats with lesions of nucleus basalis magnocellularis-I. Biochemical and anatomical observations. *Neurosci.* **16**, 769–786.

Flament-Durand, J. (1965). Observations on pituitary transplants into the hypothalamus of the rat. *Endocrinology* **77**, 446–454.

Freed, W. J., Morihisa, J. M., Spoor, H. E., Hoffer, B. J., Olson, L., Seiger, A., and Wyatt, R. J. (1981). Transplanted adrenal chromaffin cells in rat brain reduce lesion-induced rotational behaviour. *Nature, Lond.* **292**, 351–352.

—— (1985). Transplantation of tissue to the cerebral ventricles: methodological details and rate of graft survival. In *Neural grafting in the mammalian CNS* (ed. A. Bjorklund and U. Stenevi) pp. 31–40. Elsevier, Amsterdam, New York, Oxford.

Gage, F. H., Bjorklund, A., Stenevi, U., Dunnett, S. B. and Kelly, P. A. T. (1984). Intrahippocampal septal grafts ameliorate learning impairment in aged rats. *Science, N.Y.* **225**, 535–536.

—— Dunnett, S. B., Brundin, P., Isacson, P., and Bjorklund, A. (1984). Intracerebral grafting of embryonic neural cells into the adult host brain: an overview of the cell suspension method and its application. *Dev. Neurosci.* **6**, 137–151.

—— —— Stenevi, U., and Bjorklund, A. (1983). Aged rats: recovery of motor impairments by intrastriatal nigral grafts. *Science, N.Y.* **221**, 966–969.

Gash, D. M., Sladek, J. R., and Sladek, C. D. (1980). Functional development of grafted peptidergic neurons. *Science, N.Y.* **210**, 1367–1369.

Gibson, M. J., Krieger, D. T., Charlton, H. M., Zimmerman, E. A., Silverman, A. J., and Perlow, M. J. (1984). Mating and pregnancy can occur in genetically hypogonadal mice with preoptic area brain grafts. *Science, N.Y.* **225**, 949–951.

Halasz, B., Pupp, L., and Uhlarik, S. (1962). Hypophysiotrophic area in the hypothalamus. *J. Endocr.* **25**, 147–159.

—— —— —— and Tima, L. (1963). Growth of hypophysectomised rats bearing pituitary transplants in the hypothalamus. *Acta physiol. Acad. Sci. Hung.* **23**, 287–292.

—— —— —— —— (1965). Further studies on the hormone secretion of the anterior pituitary transplanted into the hypophysiotrophic area of the rat hypothalamus. *Endocrinology* **77**, 343–353.

Harris, G. W. (1972). Humors and hormones. *J. Endocr.* **53**, i–xxiii.

Inoue, H., Kohsaka, S., Yoshida, K., Ohtani, M., Toya, S., and Tsukada, Y. (1985). Cyclosporin A enhances the survivability of mouse cerebral cortex grafted into the third ventricle of the rat brain. *Neurosci. Lett.* **54**, 85–90.

Kelsey, R. C., Hymer, W. C., and Page, R. (1981). Pituitary cell transplants to the cerebral ventricle suppress the pituitary of the recipient host. *Neuroendocrinology* **33**, 312–349.

Knigge, K. M. (1962). Gonadotrophic action of neonatal pituitary glands implanted in the rat brain. *Am. J. Physiol* **202**, 387–391.

Kokoris, G. J., Perlow, M. J., Gibson, M. J., Zimmerman, E. A., Silverman, A. J., Charlton, H. M., and Krieger, D. T. (1984). Effects of graft origin and site of transplantation on reversal of congenital hypogonadism by GnRH transplants. *Soc. Neurosci. Abstr.* **10**, 373.

Krieger, D. T., Perlow, M. J., Gibson, M. J., Davies, T. F., Zimmerman, E. A., Ferin, M., and Charlton, H. M. (1982). Brain grafts reverse hypogonadism of gonadotrophin releasing hormone deficiency. *Nature, Lond.* **298**, 468–471.

Lincoln, G. A. and Short, R. V. (1980). Seasonal breeding: Nature's contraceptive. *Recent Prog. Horm. Res.* **36**, 1–43.

Luine, V. N., Renner, K. J., Frankfurt, M., and Azmitia, E. C. (1984). Facilitated sexual behaviour reversed and serotonin restored by raphe nucleus transplanted into denervated hypothalamus. *Science, N.Y.* **226**, 1436–1439.

Lyon, M. F., Cattanach, B. M., and Charlton, H. M. (1981). Genes affecting sex differentiation in mammals. In *Mechanisms of sex differentiation in animals and man* (ed. C. R. Austin and R. G. Edwards) pp. 367–374. Academic Press, London.

Mason, A. J., Hayflick, J. S., Zoeller, T., Young, W. S., Philips, H. S., and Seeburg, P. H. (1986). A deletion truncating the GnRH gene is responsible for hypogonadism in the *hpg* mouse. *Science, N.Y.* **234**, 1366–1371.

Mason, D. W., Charlton, H. M., Jones, A. J., Lavy, C. B. D., Puklavec, M., and Simmonds, S. J. (1986). The fate of allogeneic and xenogeneic neural tissue transplanted into the third ventricle of rodents. *Neurosci.* in press.

McDowell, I. F. W., Morris, J. F., and Charlton, H. M. (1982). Characterization of the pituitary gonadotroph cells of hypogonadal (*hpg*) male mice: comparison with normal mice. *J. Endocr.* **95**, 321–330.

Merchenthaler, I., Gorcs, T., Setalo, G., Petrusz, P., and Flerko, B., (1984). GnRH neurones and pathways in the rat brain. *Cell Tissue Res.* **237**, 15–29.

Morris, J. F. and Charlton, H. M. (1983). Responses of gonadotrophs in female hypogonadal (*hpg*) mice to intracerebroventricular implantation of preoptic area tissue from normal mice. *J. Anat.* **137**, 803.

Olson, L., Stromberg, I., Herrera-Marschitz, M., Ungerstedt, V., and Ebendal, T. (1985). Adrenal medullary tissue grafted to the dopamine-denervated rat striatum: histochemical and functional effects of additions of nerve growth factor. In *Neural grafting in the mammalian CNS* (ed. A. Bjorklund and U. Stenevi) pp. 505–518. Elsevier, Amsterdam, New York, Oxford.

Paden, C. M., Gerlach, J. L., and McEwen, B. S. (1985). Estrogen and progesterone receptors appear in transplanted fetal hypothalamus preoptic area independently of the steroid environment. *J. Neurosci.* **5**, 2374–2387.

—— Silverman, A. J., Stenevi, U., and McEwen, B. S. (1984). The use of fetal hypothalamic transplants in developmental neuroendocrinology. In *Neural transplants, development and function* (ed. J. R. Sladek and D. M. Gash) pp. 283–304. Plenum Press, New York.

Rogers, J., Hoffman, G. F., Zornetzer, S. F., and Vale, W. W. (1984). Hypothalamic grafts and neuroendocrine cascade theories of aging – immunocytochemical viability of preoptic hypothalamic transplants from fetal to reproductively senescent female rats. In *Neural transplants, development and function* (ed. J. R. Sladek and D. M. Gash) pp. 205–222.

Schmale, H. and Richter, D. (1984). Single base deletion in the vasopressin gene is the cause of diabetes insipidus in Brattleboro rats. *Nature, Lond.* **308**, 705–709.

Silverman, A. J., Zimmerman, E. A., Gibson, M. J., Perlow, M. J., Charlton, H. M., Kokoris, G. J., and Krieger, D. T. (1985). Implantation of normal fetal preoptic area into hypogonadal (*hpg*) mutant mice: temporal relationships of the growth of GnRH neurons and the development of the pituitary/testicular axis. *Neurosci.* **16**, 69–84.

—— —— Kokoris, G., and Gibson, M. J. (1986). Ultrastructure of gonadotrophin releasing hormone (GnRH) neuronal structures derived from normal fetal preoptic area and transplanted into hypogonadal mutant (*hpg*) mice. *Neurosci.* in press.

Stenevi, U., Bjorklund, A., Kromer, L. F., Paden, C. M., Gerlach, J. L., McEwen, B. S., and Silverman, A. J. (1980). Differentiation of embryonic hypothalamic transplants outlined on the choroidal pia in brains of adult rats. *Cell. Tissue Res.* **205**, 217–228.

—— Kromer, L. F., Gage, F. H., and Bjorklund, A. (1985). Solid neural grafts in intracerebral transplantation cavities. In *Neural grafting in the mammalian CNS* (ed. A. Bjorklund and U. Stenevi). pp. 41–50. Elsevier, Amsterdam, New, York, Oxford.

Valtin, H. (1976). Hereditary hypothalamic diabetes insipidus. *J. Pathol.* **83**, 633–636.

—— (1982). The discovery of the Brattleboro rat, recommended nomenclature and the question of proper controls. *Ann. N.Y. Acad. Sci.* **394**, 1–9.

Ward, B. J. and Charlton, H. M. (1981). Female sexual behaviour in the GnRH deficient hypogonadal (*hpg*) mouse. *Physiol. Behav.* **27**, 1107–1109.

—— —— (1986). Male sexual behaviour in the GnRH deficient hypogonadal mouse *Physiol. Behav.* in press.

Young, L. S., Detta, A., Clayton, R. N., Jones, A., and Charlton, H. M. (1985). Pituitary and gonadal function in hypogonadotrophic hypogonadal (*hpg*) mice bearing hypothalamic implants. *J. Reprod. Fert.* **74**, 247–255.

—— Speight, A., Charlton, H. M., and Clayton, R. N. (1983). Pituitary gonadotrophin-releasing hormone receptor regulation in the hypogonadotropic hypogonadal (*hpg*) mouse. *Endocrinology* **113**, 55–61.

11   Growth factors in embryogenesis

D. J. HILL, A. J. STRAIN AND
R. D. G. MILNER

I   Introduction

The growth of the embryo and early fetus, while dependent on a prodigious capacity for cell replication, does not proceed simply by an orderly doubling of cell number. It involves a subtle interaction of cell replication and differentiation together with the condensation of cells into functional organs. At least four separate elements can be identified: 1) the ability of cells to proliferate rapidly; 2) the ordered migration of cells into definable tissues and organ rudiments; 3) the ability of a sub-population of cells to differentiate and synthesize macromolecules for the benefit of the entire embryo; and 4) the temporal integration of each of these processes. While individual aspects of embryonic development have been studied in great detail less is known about the molecules responsible for intercellular communication that may drive and coordinate the whole.

To enable a highly ordered developmental sequence to proceed successfully, cells must be able to communicate. Recent studies have shed light on some of the factors thought to be important. Superficially, these can be divided into: 1) cell adhesion molecules which are cell-surface anchored glycoproteins involved in cell-cell recognition and adhesion during pattern formation and organogenesis; 2) cell-substrate adhesion molecules, required for cell migration and the interaction with extracellular matrix necessary for acquiring cell shape; and 3) peptide growth factors which are diffusable molecules which stimulate cell proliferation and/or differentiation. While much of our understanding, particularly of the biological activity of peptide growth factors, has come from study of their actions in late fetal and post-natal life, recent work suggests that they may also be involved in early embryogenesis. The expression of various 'growth factors' may therefore provide a link between certain classes of genes involved in development and the changing morphology of the embryo.

The proliferative and migratory behaviour of early mammalian embryonic tissue can be illustrated by its anatomy. After four initial mitoses the fertilized ovum has become a morula. The inner cell mass that subsequently arises from the structure will eventually become the embryo and extra-embryonic endoderm. The surrounding peripheral cell layer will form the trophoblast (Hamilton, Boyd, and Mossman 1962). During subsequent blastocyst formation the trophoblast forms an inner cytotrophoblast and an outer syncytiotrophoblast with invasive properties allowing erosion of the endometrium. The trophoblastic tissue will give rise to the outer embryonic membrane, the chorion, and the placenta. The inner mass of cells of the blastula will give rise to the embryo and contribute to the amnion and yolk sac.

In man the first appearance of pattern formation in the blastula is the separation of a layer of primary endoderm from the pluripotential ectodermal stem cells, at the start of gastrulation. This will eventually give rise to the embryonic gut. Fifteen days after conception, an area of ectoderm at the caudal edge of the embryonic disk gives rise to the primitive streak. The lateral movement of cells from the primitive streak to lie between ectoderm and endoderm gives rise to the third germ layer, the mesoderm. The extra-embryonic mesoderm forms the amnion, and together with visceral endoderm, the secondary yolk sac. The ectoderm gives rise to the notochord, and from cells lateral to this the neural tube. Intra-embryonic mesoderm on either side of the notochord becomes thickened to form longitudinal masses, the paraxial mesoderm, which by day 21 has begun to segment into somites. The limb buds appear at approximately 32 days.

The importance of tissue pattern formation and temporally-related cell differentiation during embryogenesis is well illustrated by studies of the development of the embryonic chick wing bud. The tissue rudiments that will give rise to the wrist and forearm of the chick are initially of similar length, yet the growth of the forearm is much more prolific than that of the wrist. Despite the different growth rates of rudiments within the same limb the two chick wings differ by no more than 2% in length at hatching (Lewis 1977). Clearly the cell proliferation needed for wing formation is extremely selective and different elements of the limb have different positional values with regard to mitogenesis. This information may be imparted in the developing vertebrate limb by a specialized area of undifferentiated mesodermal cells at the tip called the progress zone (Summerbell, Lewis, and Wolpert 1973). Positional information along the proximo-distal axis of the chick limb bud is determined by the length of time a cell has spent in the progress zone, counted in cell cycles (Wolpert, Tickle, and Sampford 1979). Consequently, skeletal elements such as cartilage are laid down in a proximo-distal sequence, the primordia of proximal structures differentiating from undifferentiated mesenchyme before distal structures, while the progress zone remains actively proliferating. The maintenance of the progress zone in an undifferentiated state depends on the presence of overlaying pseudostratified columnar epithelium called the apical ectodermal ridge (Saunders and Reuss 1974). Removal of the ridge results in limbs with predictable terminal deficiencies, while transplantation of a progress zone from a young embryo on to a well formed limb of an older embryo results in the formation of a double-wing (Wolpert, Lewis, and Summerbell 1975). This clearly demonstrates the importance of germ layer interaction during embryogenesis.

The pattern of the developing limb is also controlled in the antero-posterior plane by a polarizing region at the posterior margin of the wing bud. When tissue from this region was transplanted to the anterior border a limb with mirror-image symmetry along its long-axis was formed (Smith, Tickle, and Wolpert 1978). The polarizing region appeared to release a diffusable tissue morphogen which determines digit formation, the tissue nearest to the region encountering the greatest concentration and giving rise to digit 4, while that farthest away formed digit 2 (Summerbell, Dhouailly, and Maden 1984; Summerbell and Maden 1985). The actions of the morphogen could be reproduced *in vivo* by the application of a vitamin A derivative, retinoic acid (Summerbell 1983). The polarizing region of the developing chick wing therefore determines differentiated limb symmetry by the release of an unknown morphogen. Since the limb is simultaneously increasing in length a complex interaction of mitogenic and differentiating signals must be occurring, which differs for each aspect of the limb bud.

The evidence summarized above suggests that signalling mechanisms such as those in the developing chick limb could be provided by both cell-surface anchored proteins and freely diffusable growth factors. In the following sections we shall review recent findings relating to the biological actions of these molecules and the control of their expression. This field is one of the most rapidly moving aspects of biological research. The conclusions and views of the authors are based on information available in July 1986.

II   Cell adhesion molecules

Cell adhesion molecules (CAMs) are high molecular weight (125–250 kd) plasma membrane glycoproteins that were first recognized to be important regulatory proteins involved in cell signalling by the ability of specific antibodies to disrupt cell-cell adhesion *in vitro*. In addition to mediating adhesion they are thought to serve an important role in cell-cell recognition which occurs during morphogenesis. It is this property which distinguishes them from other specialized membrane structures which bind cells together (Obrink 1986). Although they do not directly regulate cell proliferation they can be considered as paracrine factors since their expression on the surface of cells has a profound influence on local cellular interactions. Three major CAMs have been isolated and sufficiently purified to allow partial characterization. N-CAM (neural cell adhesion molecule) and L-CAM (liver cell adhesion molecule) are described as primary CAMs because they are expressed during early stages of embryogenesis on cells from more than one germ layer. Ng-CAM (neuron-glia cell adhesion molecule) is a secondary CAM found later in embryogenesis on neuro-ectodermal derivatives.

In addition, a number of related molecules from different species have been described and the reader is referred to several excellent recent reviews for a more detailed consideration (Rutishauser 1984; Edelman 1985; Obrink 1986). There is good evidence that the overall structure and especially the binding domain of CAMs has remained highly conserved throughout evolution. N-CAM and L-CAM are thought to bind by a homophilic self-recognizing mechanism, that is one CAM binds to an identical CAM on an opposing cell (Hoffman and Edelman 1983; Ogou, Yoshida-Noro, and Takeichi 1983). For both molecules the carboxy-terminal portion is attached to the membrane and the binding domain is thought to reside in the projecting amino-terminal region (Cunningham, Hoffman, Rutishauser, Hemperley, and Edelman 1983). In contrast, neuron-glia cell binding appears to operate by a heterophilic mechanism, involving binding between unlike molecules since Ng-CAM has only been detected on the surface of the neuronal cell (Grumet, Hoffman, and Edelman, 1984).

Studies on the character and localization of CAMs in vertebrate embryogenesis have been restricted largely to the chick and mouse. Using immunolocalization techniques, L-CAM has been demonstrated at the 4 cell stage in the mouse embryo (Shirayoshi, Okada, and Takeichi, 1983). N-CAM appears later but, at least in the chick, both L- and N-CAM are distributed uniformly prior to gastrulation (Edelman, Gallin, Delouvee, Cunningham, and Thiery 1983). During gastrulation and early neurogenesis, N-CAM and L-CAM become segregated into different tissues (Edelman *et al.* 1983). N-CAM becomes increasingly abundant in the neural primordium and is also found in many mesodermal tissues (Thiery, Duband, Rutishauser, and Edelman 1982) whereas L-CAM becomes restricted to non-neural ectoderm, to mesodermal derivatives such as the urogenital system and to the endoderm (Thiery, Delouvee, Gallin, Cunningham, and Edelman 1984). Ng-CAM appears later in embryogenesis on cells derived from a derivative of a single germ layer, the neuroectoderm, and is seen mainly on early post-mitotic differentiating neurons (Thiery, Delouvee, Grumet, and Edelman 1985).

The most detailed studies on the function of CAMs have investigated the role of N-CAM in neurogenesis. Much of the current information has been derived from antibody neutralization experiments both *in vitro* with dorsal root ganglion explants and *in vivo* from studies on the development of the chick neural retina. For example, one important aspect of nerve process formation shown to be highly dependent on N-CAM function is neurite fasciculation, involving the formation of large bundles of neurites into nerve tracts. This process can be mimicked *in vitro* by following the outgrowth of neurites, normally consisting of several hundred thick bundles, from dorsal root ganglion explants. In the presence of antibodies to N-CAM the rate of growth and numbers of neurites remain the same, but the diameter of the fascicles is markedly reduced (Rutishauser, Gall, and Edelman 1978). In addition, neurite fibres tend to split apart and associate with other fascicles giving the appearance of a less ordered process. Neurite fascicles probably form *in vivo* by growth of one fibre along another, and Rutishauser and Edelman (1980) have shown, in elegant studies using dorsal root ganglion explants, that anti N-CAM antibody disrupts this process. Early neurite outgrowth in response to a transient nerve growth factor gradient forms the basis for asymmetric growth of subsequent neurites leading to fasciculation. In the presence of antibody most of the neurite growth is uniformly distributed around the ganglion.

The function of CAMs can be regulated by several mechanisms. While N-CAM is qualitatively present on most neural tissue derivatives, developmentally-associated changes in expression have been described. The most striking of these occurs during migration of neural crest cells which are responsible for forming much of the peripheral nervous system. N-CAM disappears from the surface of neural crest cells as they prepare to migrate (Thiery *et al.* 1982). When they reach their destination, prior to formation of ganglia, N-CAM is once again detected on the cell surface. It is not known whether N-CAM is masked during migration or whether it is physically removed and resynthesized. Decreased intercellular adhesion and increased cell motility associated with loss of membrane N-CAM have also been reported *in vitro* using mouse cerebellar cell-lines (Greenberg, Brackenbury, and Edelman 1984), and with primary retinal cells transformed by Rous sarcoma virus (Brackenbury, Greenberg, and Edelman 1984). Another major modification which could regulate N-CAM function is in the degree of glycosylation, particularly that of sialic acid. A comparison of N-CAM from embryonic and adult tissues has revealed a three fold reduction in the amount of sialic acid present, the so-called E-A conversion (Rothbard, Brackenbury, Cunningham, and Edelman 1982), leading to widely differing binding efficiencies (Hoffman and Edelman 1983). Further studies have shown that this heterogeneity also occurs in different regions of embryonic chick and rodent brain (Chuong and Edelman 1984). That this change in N-CAM glycosylation may be crucial to normal neural development is suggested by the finding that normal E-A conversion apparently fails to occur in the brain of mutant *Staggerer* mice (Edelman and Chuong 1982). In the homozygous animal, faults in synapse formation lead to a poorly developed cerebellum resulting in an ataxic animal with a shortened life-span. One possibility, as yet unproven, is that one of the consequences of the genetic defect is the lack of an enzyme responsible for mediating the E-A conversion.

It is apparent from the above discussion that CAMs are of fundamental importance in controlling normal embryogenesis and tissue morphogenesis. However, the relationship between the expression of CAMs by cells and the control of cellular proliferation is not understood. With the recent characterization of several families of peptide growth factors which regulate cell growth and differentiation, this relationship can now be investigated experimentally both *in vitro* and *in vivo*.

III   The nature of peptide growth factors

1   INTRODUCTION

The peptide growth factors represent an ubiquitous class of intercellular messengers which promote extensive cell proliferation and differentiation. Typically they are protein or glycoprotein molecules less than 30 kd molecular weight which may serve both as circulating hormones and/or locally-acting paracrine or autocrine factors. Unlike the classical endocrine hormones they are released from a wide range of separate and anatomically distinct cell types, and are frequently found in association with carrier proteins which extend their biological half-life and may regulate their biological actions.

Peptide growth factors interact with receptors located on the cell membrane. The receptors are usually glycoproteins which communicate with secondary intracellular messenger systems through conformational changes which involve autophosphorylation reactions. While the second messenger systems utilized by individual peptide growth factors may differ, and can involve changes in intracellular free calcium, cyclic AMP, or the phosphatidyl inositol cycle, the gross cellular responses are remarkably similar and are termed the positive pleiotypic response. This includes a stimulation of glucose and amino acid transport leading to the synthesis of RNA and protein. For most peptide growth factors, with the possible exception of nerve growth factor, these events are accompanied by a mitogenic response with a half-maximal concentration of peptide between 1 and 10 nM *in vitro*.

2   PHYSICO-CHEMICAL PROPERTIES AND BIOLOGICAL ACTIONS

i   *Somatomedins/insulin-like growth factors*

The somatomedins/insulin-like growth factors (SM/IGF) are single chain polypeptides with a molecular weight of approximately 7.5 kd which share structural and functional similarities with insulin (Phillips and Vassilopoulou-Sellin 1980). Mature SM-C/IGF I is a 70 amino acid peptide encoded by a gene on human chromosome 12, and IGF II a 67 amino acid peptide encoded by a gene mapped to chromosome 11 in close proximity to the insulin gene (Tricoli, Rall, Scott, Bell, and Shows 1984). Both IGFs show approximately 45 per cent amino acid sequence homology with insulin. However, the IGFs have retained the connecting C peptide which is cleaved in the maturation step from pro-insulin to insulin, and have in addition a D terminal extension (Blundell, Bedarkar, Rinderknecht and Humbel 1978).

The IGFs are potent mitogens for cells derived from all primitive germ layers, and in postnatal life may be especially important for the progression of longitudinal skeletal growth through a stimulation of chondrocyte proliferation and maturation in the epiphyseal growth plate. In this capacity they mediate the growth-promoting actions of pituitary growth hormone. IGF I/SM-C is a more potent growth-promoting peptide than IGF II *in vivo* in the hypophysectomized rat (Schoenle, Zapf, Hauri, Steiner, and Froesch 1985), and is under tighter growth hormone control in man, circulating levels being severely reduced in hypopituitarism and elevated during acromegaly (Zapf, Rinderknecht, Humbel, and Froesch 1978). The structural similarities between IGFs and insulin result in the IGFs demonstrating weak insulin-like metabolic actions *in vitro*, as seen by the promotion of glucose uptake and oxidation by isolated adipocytes, while supraphysiological concentrations of insulin can be directly mitogenic. While the major source of circulating IGFs in postnatal life was previously considered to be the liver there is growing evidence that IGFs are ubiquitous tissue mitogens synthesized in most body tissues in response to growth hormone (D'Ercole, Stiles, and Underwood 1984). In man after birth circulating levels of IGF II are 4 to 5 times greater than those of IGF I/SM-C (Zapf *et al.* 1978).

In addition to its role in metabolic homeostasis, insulin itself is a peptide growth factor. Supra-physiological concentrations of insulin are generally required to promote cell replication in isolated connective tissues during postnatal life. However, the sensitivity of fetal and embryonic tissues to insulin may differ.

ii   Epidermal growth factor

Epidermal growth factor (EGF) is a single chain polypeptide of 53 amino acids and molecular weight 6 kd. It is synthesized as a high molecular weight membrane-bound precursor with a characteristic cystine-rich peptide sequence of 40–50 amino-acids repeated nine times. The terminal copy of this repeat, when cleaved from the precursor, forms the EGF peptide itself (Scott, Urdea, Quiroga, Sanchez-Pescador, Fong, Selby, Rutler, and Bell 1983). Mouse EGF was first isolated from male submaxillary glands and the equivalent human peptide, urogastrone, from urine (Cohen 1962; Gregory 1975). In addition to being present in salivary glands, EGF was shown to have a widespread tissue presence (Kasselberg, Orth, Gray, and Stahlman 1985). It has been localized in the brain, pancreas, Brunner's glands of the duodenum, jejunum, thyroid, mammary gland and kidney during postnatal life. EGF is mitogenic *in vitro* for many cell types: for example, connective tissues, glial cells and ovarian granulosa cells (Carpenter and Cohen 1979). When administered *in vivo* EGF enhanced the proliferation of certain basal epithelial cell layers giving rise to skin thickening and corneal hyperplasia (Gospodarowicz 1981). An associated increase in the rate of collagen synthesis and keratinization caused premature eruption of the teeth and opening of eyelids in neonatal mice. Consequently EGF has shown considerable potential as a wound-healing agent (Buckley, Davidson, Kamerath, Wolt, and Woodward 1985). EGF also influenced gastrointestinal function and in addition to inducing cell proliferation throughout the length of the digestive tract exerted a strong inhibitory action on acid and pepsin secretion by the gastric mucosa, and increased amylase release from cultured mouse pancreatic acini (Bower, Camble, Gregory, Gerring, and Willshire 1975; Logsdon and Williams 1983).

iii   Platelet-derived growth factor

Platelet-derived growth factor (PDGF) was first identified as a peptide stored in the α granule of platelets and released during blood clotting. Human PDGF is a highly basic glycoprotein (pI 10.2) separated into two biologically equipotent molecules, PDGF I which contains 7 per cent carbohydrate and has a molecular weight of 31 kd, and PDGF II with molecular weight 28 kd and 4 per cent carbohydrate (Deuel and Huang 1984). Both forms of PDGF comprise an A-chain of 15–17 kd and a B-chain of 14 kd linked by di-sulphide bridges. The B chain of PDGF shows extensive sequence homology with the viral p28 protein encoded by the v-*sis* oncogene (Waterfield, Scrace, Whittle, Stroobant, Johnsson, Wasteson, Westermark, Heldin, Huang, and Deuel 1983), and the cellular *sis* gene was subsequently shown to encode the precursor peptide for the PDGF B-chain (Johnsson, Heldin, Wasteson, Westermark, Deuel, Huang, Seeburg, Gray, Ullrich, Scrace, Stroobant, and Waterfield 1984).

PDGF stimulates proliferation in many cell types derived from mesoderm including fibroblasts, smooth muscle cells and glial cells (Deuel and Huang 1984). Since PDGF is released from platelets at sites of vascular injury and tissue damage it is thought to act as an important wound healing agent. This is supported by its action as a potent chemotactic protein for inflammatory cells such as neutrophils, in which superoxide synthesis is increased, and monocytes (Deuel, Senior, Huang, and Griffin 1982). Similarly PDGF attracts smooth muscle cells and fibroblasts to the wound site where high local concentrations of the peptide will induce proliferation and aid tissue repair.

iv   Other growth factors

Fibroblast growth factor (FGF) has been isolated from bovine pituitary and brain in both an acidic (pI 4.5) and basic (pI 9.6) form, each of approximately 16–17 kd molecular weight (Gospodarowicz, Cheng, Lui, Baird, and Bohlent 1984; Esch, Baird, Ling, Veno, Hill, Denoroy, Klepper, Gospodarowicz, Bohlen and Guillemin 1985). Several molecular species of basic FGF exist which share a common amino terminal sequence. Acidic FGF was recently shown to share sequence homology with human interleukin I (Thomas, Rios-Candelore, Gimenez-Gallego, Di Salvo, Bennett, Rodkey, and Fitzpatrick 1985). FGF molecules are highly potent mitogens for vascular endothelial cells *in vitro* and *in vivo* (Gospodarowicz, Bialecki, and Thakral 1979), and for other tissues derived from mesenchyme or neuroectoderm such as chondrocytes, fibroblasts, myoblasts, glial cells, adrenal cortex and ovarian granulosa cells (Gospodarowicz 1984). The occurrence of basic FGF in adrenal cortex, corpus luteum, retina, placenta and kidney suggests an ubiquitous distribution.

The transforming growth factors (TGFs) are biologically active polypeptides, first isolated as tumour cell products, that confer a reversible, transformed, phenotype on non-transformed cells (Roberts, Frolik, Anzano, and Sporn 1983). Two classes of TGF exist consisting of TGF-α which has a molecular weight of approximately 5.7 kd, has amino acid sequence homology with EGF and competes for the EGF receptor; and TGF-β which is a distinct homodimeric molecule of 25 kd molecular weight. Neither TGF-α nor TGF-β have the capacity alone to induce anchorage-independent growth in soft agar for most cell lines, but TGF-β exhibits a strong colony-forming activity in the presence of other peptide growth factors including TGF-α, EGF and IGFs (Massague, Kelly, and Mottola 1985). TGF-β is a bi-functional growth factor since it may either stimulate or inhibit cell proliferation depending on the cellular substratum, the embryological origin of the cell type, and whether the cell type is virally transformed (Roberts, Anzano, Wakefield, Roche, Stern, and Sporn 1985). TGF-β may either directly or indirectly induce transmodulation effects on the affinities of receptors to other peptide growth factors as was demonstrated for the EGF receptor (Assoian 1985; Massague 1985). While TGF-β is abundant in platelets, and is likely to play a role in tissue repair, it has also been isolated from kidney, placenta and fibroblasts. Localization of mRNA for TGF-β using a cDNA probe demonstrated its expression in placenta and peripheral blood lymphocytes and a widespread presence in human tumour cell lines (Derynck, Jarrett, Chen, Eaton, Bell, Assoian, Roberts, Sporn, and Goeddel 1985). The ability of TGF-β to modulate the mitogenic actions of other peptide growth factors renders this molecule of great potential interest in embryogenesis.

Nerve growth factor (NGF) is the only recognized peptide growth factor which has no mitogenic activity, at least during postnatal development. Rather, NGF maintains the survival of certain populations of neurons, the sympathetic and peripheral sensory neurones, and has chemotactic properties which may aid innervation of tissues during regeneration (Levi-Montalcini 1976). NGF also regulates the expression of catecholamine biosynthetic enzymes such as tyrosine hydroxylase, and of substance P and somatostatin (Black 1982). The peptide was first purified from male mouse salivary glands as βNGF, a dimer of two identical monomeric forms with a molecular weight 13 kd (Darling, Petrides, and Beguin 1983).

NGF could act as a retrograde intracellular messenger between the peripheral sensory target tissues and the innervating neurons. It is internalized by sympathetic and sensory nerve terminals following binding to a specific receptor population and is transported back to the neuronal cell bodies in membranous vesicles. However this may simply represent a clearance of NGF by the cell while the biological information is relayed by second messenger systems following receptor activation (Thoenen, Korsching, Heumann, and Acheson 1985). High levels of NGF in sympathetic ganglia therefore derive from retrograde transport, the sites of synthesis most- probably being the innervated tissues. In addition to the salivary glands, mRNA for NGF was found in the heart and vas deferens. NGF was found to be secreted by isolated mouse heart cells (Furukawa, Furukawa, Satoyoshi, and Hyashi 1984) and rat iris where it was also localized immunohistochemically (Barth, Korsching, and Thoenen 1984; Rush 1984). An identical peptide was purified from bovine seminal fluid (Harper, Glanville, and Thoenen 1982).

In the central nervous system adrenergic neurons are unresponsive to NGF but cholinergic neurons leading from the basal forebrain nuclei to the hippocampus and cortex manifest retrograde transport, and respond to NGF with a synthesis of choline acetyltransferase in the newborn rat (Seiler and Schwab 1984). The chemotactic properties of NGF have been demonstrated both *in vitro* and *in vivo*. When exposed to a local concentration gradient of NGF *in vitro*, growth cones of chick sensory neurites rapidly turned and grew towards the source of NGF (Gunderson and Barrett 1980).

3   GROWTH FACTOR RECEPTORS

The biological cross-reactivity between IGFs and insulin is explained not only by amino acid sequence homology between the peptides but also by similarity between their respective tissue receptor glycoproteins. Two classes of IGF receptor exist. The type I IGF receptor shows close homology with the insulin receptor and consists of two α subunits with approximate molecular weight 135 kd, linked by disulphide bonds in a tetrameric structure to two β subunits of 95 kd (Jacobs and Cuatrecasas 1982). The α subunit of the type I receptor is 8 kd smaller than that of the insulin receptor. On both the insulin and type I IGF receptor the peptide recognition site is located on the extracellular portion of the α subunit. The type I IGF receptor has a higher affinity for IGF I than IGF II and binds insulin with low affinity. Conversely, the insulin receptor binds insulin with higher affinity than IGF I or II. A second class of IGF receptor exists, called type II, which has a single chain glycoprotein of molecular weight 250 kd (Massague and Czech 1982). The type II receptor binds IGF II with greater affinity than IGF I but does not bind insulin. However it has been questioned whether the type II receptor mediates a mitogenic signal (Mottola and Czech 1984). The receptors for both EGF and PDGF are single chain glycoproteins of approximately 180 kd (Carpenter and Cohen 1979; Frackleton, Tremble, and Williams 1984).

Both the insulin and the type I IGF receptor are activated by peptide binding, leading to phosphorylation of tyrosine residues within the intracellular region of the β subunit (Kasuga, Karlsson, and Kahn 1982; Sasaki, Rees-Jones, Zick, Nissley, and Rechler 1985). Tyrosine phosphorylation is catalized by a tyrosine-specific protein kinase within the β subunit structure. Similar tyrosine-specific autophosphorylation follows occupation of the EGF and PDGF receptor (Frackleton, Tremble, and Williams 1984; Ullrich, Coussens, Hayflick, Dull, Gray, Tam, Lee, Yarden, Lieberman, Schlessinger, Downward, Mayes, Whittle, Waterfield, and Seeburg 1984). The peptide-receptor complexes are subsequently internalized and the peptide growth factors degraded by lysozymes (Stoscheek and Carpenter 1984).

4 BINDING PROTEINS

The IGFs circulate in plasma bound to specific carrier proteins and no free hormone is detectable. Most IGFs are associated in postnatal life with a growth hormone-dependent binding protein of 150 kd molecular weight which may be constructed as a hexamer of smaller units, each of 24–28 kd (Wilkins and D'Ercole 1985). An additional and separate binding protein of 40–60 kd molecular weight also exists in the circulation which is not growth hormone dependent, and is present in most extra-vascular fluids including amniotic fluid (D'Ercole, Drop, and Kortleve 1985). This form of binding protein is invariably released by tissues *in vitro* in parallel with IGF peptides and may be intimately associated with the regulation of IGF actions (Hill, Crace, Nissley, Morrell, Holder, and Milner 1985). The 150 kd binding protein is limited to the circulation and does not appear until approximately 27 weeks gestation in the human fetus (D'Ercole, Wilson, and Underwood 1980). Both forms of binding protein greatly extend the biological half life of IGF peptides compared to other free peptide growth factors.

EGF extracted from whole tissues also exists in association with a high molecular weight binding protein which was identified as a serine protease (Lundgren, Ronne, Rask, and Peterson 1984). Although PDGF is not normally present in human plasma, a binding protein is present and was identified as an alpha$_2$-macroglobulin (Raines, Bowen-Pope, and Ross 1984). This may serve as a clearance mechanism to prevent PDGF from entering the systemic circulation following its release from platelets at the site of injury.

5 INTERACTIONS BETWEEN PEPTIDE GROWTH FACTORS DURING CELL PROLIFERATION

The apparently overlapping biological activity of several peptide growth factors during the initiation of cell replication has been partly resolved by studies which determine the precise points of action of each growth factor within the cell cycle. The experimental model which has yielded most information has been the growth of the Balb/c-3T3 cell *in vitro* (O'Keefe and Pledger 1983).

It was noted that PDGF alone, at concentrations thought to reproduce tissue levels following local platelet lysis, was unable to initiate optimal proliferation of Balb/c-3T3 fibroblasts in density-arrested cultures, and that the addition of plasma was required for cells to enter S phase. Similarly plasma alone did not promote cell division, but transient exposure of cells to PDGF followed by incubation with plasma-containing medium induced full mitogenic activity. PDGF induced a state of 'competence' which rendered quiescent cells capable of traversing the cell cycle from G_0 to G_1 in response to other agents present in plasma. For this cell line FGF fulfilled a similar role to PDGF as a 'competence' factor. The 'progression' of cells through G_1 in response to plasma could be reproduced in a defined culture medium containing only physiological concentrations of IGF I/SM-C or MSA, and EGF. Supra-physiological concentrations of insulin could substitute for the presence of IGFs (Van Wyk, Underwood, D'Ercole, Clemmons, Pledger, Wharton, and Loef 1981). The temporal order of exposure to 'progression' factors was important for subsequent entry into DNA synthesis, EGF being necessary for the first six hours and IGF's for the remainder of the 12 hour G_1 period required for growth arrested cells to enter S phase following the induction of competence. Thus, growth-arrested Balb/c-3T3 cells could be induced to traverse the cell cycle into S phase in defined medium containing only PDGF, EGF and IGFs which were required in a strict order of exposure. Cells could complete S, G_2 and M phases without further growth factor stimulation.

The ability of Balb/c-3T3 cells to traverse the cell cycle in defined growth factor-supplemented medium was dependent on the availability of amino acids. In medium deficient in amino acids cells became growth arrested at a point mid-way through G_1 coincident with the time from which exposure to IGFs was required (Stiles, Isberg, Pledger, Antoniades, and Scher 1979). This suggests that an important feature of IGF action in G_1 involves nutrient transport into the cell. Analogous studies with human neonatal fibroblasts revealed a much later amino acid restriction point 2-3 hours before the G_1/S boundary (Cooper and Wharton 1985).

An important difference between the growth regulation of early passage human fibroblasts and that of established rodent cell lines is that the former release substantial amounts of IGF I/SM-C, thus partly obviating the need for exogenous peptides during replication (Clemmons, Underwood, and Van Wyk 1981). However the differences in cell cycle regulation between human fibroblasts and other cell types may be more profound since exposure of human cells to a single mitogen, either PDGF or EGF, induced growth-arrested cells to progress to DNA synthesis. Additionally the division between 'competence' and 'progression' may be less well defined since either PDGF or EGF must be optimally present for up to 24 hours (Westermark and Heldin 1985). The additive nature of these two peptides during short pulse exposure suggests that they may share common pathways of action.

While the mechanism of interaction of peptide growth factors during cell replication has not been fully resolved these molecules induce separate cellular functions which act synergistically during DNA synthesis. Similar mechanisms most probably control embryonic development.

IV Growth factors in early development

1 INTRODUCTION

The presence and actions of growth factors during late fetal development has received intensive study. Much less is known about the role of growth factors as autocrine or paracrine regulators during embryonic development, and especially prior to implantation.

Pre-implantation growth and cleavage can be achieved *in vitro* in defined culture media without the requirement for exogenous growth factors or other macromolecules. This suggests that the early embryo is self-sufficient with regard to the expression of intercellular messengers. Further growth *in vitro*, in which the embryo mimics post-implantation development, can be achieved and two-cell mouse embryos have been maintained until the development of an early-somite embryo with a beating heart (Hsu and Basker 1974), and even further to an embryo with limb buds, primordia of lung, liver and a dorsal pancreas (Chen and Hsu 1982). The sequence of development and differential events *in vitro* reproduce those seen *in utero* but occur more slowly. The development of a mouse blastocyst, isolated on day 3.5 of gestation, into an embryo with somites morphologically equivalent to an animal *in utero* of 9.5 to 10 days gestational age requires the presence of fetal calf serum and human cord serum over a ten day period, the calf serum being necessary for the first four days and the cord serum for the next 6 days (Hsu 1980). These studies demonstrate that once the time of implantation has past, supplementation of culture medium with several specific, fetally-derived macromolecules is necessary for growth and differentiation *in vitro*. Hsu (1980) has identified, but not fully characterized, at least three growth and differentiation peptides that are necessary.

As described earlier in this chapter peptide growth factors act in synergy during the cell cycle. It is therefore important to consider which growth factors are present during early human development. This becomes particularly important when considering paracrine or autocrine growth mechanisms where a local cocktail of growth factors could transiently exist within the microenvironment of a particular tissue as suggested by Hsu (1980). Of the known growth factors IGFs, EGF, PDGF and TGF-β are recognized to be present and almost certainly active during embryogenesis.

2 TERATOCARCINOMAS

Much of the available information concerning mechanisms of growth in the early post-implantation embryo has arisen from the study of mouse embryonal carcinoma cells. These are obtained from teratocarcinoma tumours experimentally derived from the primitive ectoderm of the early embryo. The undifferentiated, pluripotential stem cells show many morphological and biological features in common with the primitive ectoderm of 6 day post-implantation mouse embryos, including a common series of cell surface antigens (Artzt, Dubois, Bennet, Condamine, Babinet, and Jacob 1973; Martin 1975, 1980). Moreover, certain embryonal carcinoma cells participate in normal growth following injection into the pre-implantation embryo and contribute to differentiated adult tissues (Brinster 1974; Papaioannou, McBurney, Gardner, and Evans 1975; Mintz and Cronmiller 1978). In prolonged culture at high cell density many teratocarcinoma cells will spontaneously differentiate to yield derivatives of all three germ layers (Graham 1977). This process can be aided *in vitro* by various experimental manipulations. For example, if the PC13 embryonal carcinoma line is treated with retinoic acid, the stem cells will differentiate into PC13 END cells, which phenotypically and morphologically resemble the extra-embryonic mesoderm found in the amnion and yolk sac of the post-implantation embryo (Rayner and Graham 1982). Similarly the F9 teratocarcinoma line, which normally undergoes very limited spontaneous differentiation *in vitro*, will rapidly differentiate following exposure to retinoic acid to yield predominantly parietal endoderm-like cells (Strickland and Mahodavi 1978). Prolonged exposure (44 days) of F9 cells to retinoic acid yields an additional cell type (Dif 5) with properties of both parietal and visceral endoderm which may be unable to complete full differentiation to either tissue (Nagarajan, Jetten, and Anderson 1983). Hence the use of retinoic acid as a differentiating agent provides a range of cell types with embryonic characteristics which can be assayed for the presence and action of peptide growth factors.

Embryonal carcinoma cells proliferate readily in minimal tissue culture media with a low requirement for exogenous macromolecules (Nagarajan and Anderson 1982; Heath and Deller 1983). A complete absence of serum components, however, leads to cell death. The PC13 line expressed both type I and type II IGF receptors, but not insulin receptors (Heath and Deller 1983; Heath and Shi 1986). In contrast, the F9 cell line contained insulin receptors which most probably mediated the growth-promoting actions of physiological concentrations of insulin (Nagarajan and Anderson 1982). Unlike the PC13 line, F9 cells grow rapidly in response to both insulin and rat IGF II (Nagarajan, Nissley, Rechler, and Anderson 1982). However, while specific cellular binding of IGF II occurs it has not been confirmed whether this represents the presence of type I or type II receptors, or both.

In summary, both cell lines have receptor populations which could mediate the anabolic actions of growth factors such as IGFs and insulin. By analogy it seems likely that similar populations of receptors may exist on the equivalent stem cells *in vivo*. The F9 line exhibits growth stimulation in response to exogenous insulin-like peptides, but the PC13 line does not. These observations have two possible interpretations. The inability of the PC13 cell line to respond to IGFs may suggest that the cells already grow at an optimal rate in response to autocrine factors. Alternatively PC13 cells may be growth factor-independent in their proliferation.

Cell differentiation in response to retinoic acid has revealed dramatic changes in the expression of peptide growth factors and their receptors in teratocarcinoma cell lines. The growth of both Dif 5 and PC13 END cells *in vitro*, in contrast to their undifferentiated counterparts, is largely dependent on exogenous growth factors in fetal calf serum and is of finite duration (Nagarajan *et al.* 1983; Heath and Rees 1985). The endoderm-like Dif 5 cells exhibited density-dependent growth inhibition at high cell density, as do normal fibroblast cell lines, although this was less apparent in PC13 END cells. Again this suggests that either autocrine growth factors may be important in the mesodermal cell type, or that the cells are simply less dependent on growth factor proliferation. Both Dif 5 primitive endoderm cells and differentiated PC13 END cells were subsequently found to release radioimmunoassayable and biologically active IGF II together with a 60 kd molecular weight IGF binding protein (Nagarajan, Anderson, Nissley, Rechler, and Jetten 1985; Heath and Shi 1986). A more extensive survey revealed that three undifferentiated embryonal carcinoma stem cell lines, F9, Nulli and PCC4, produced little or no rat IGF II immunoreactivity, while differentiated parietal endoderm-like cells (PYS-2), visceral endoderm-like cells (PSA-5E) and Dif 5 cells all released substantial amounts (30–50 ng per 10^6 cells per 24 hr) (Nagarajan *et al.* 1985). The differentiation of teratocarcinoma cell lines is therefore generally accompanied by activation of the gene for IGF II.

Interestingly, the expression of IGF II by PC13 END cells was associated with a 60 per cent reduction in the apparent binding capacity by the cells of IGF II (Heath and Shi 1986). Whether this reflects a real decrease in receptor number or simply occupation of receptor sites by endogenous growth factors, as observed in several malignant cell types, has not yet been determined.

The differentiation of PC13 cells also leads to changes in the expression of other growth factor receptors. PC13 END cells, but not their undifferentiated precursors, were induced to grow by both insulin and EGF and this was probably associated with the appearance of insulin and EGF receptors on PC13 END cells following differentiation (Rees, Adamson, and Graham 1979; Heath, Bell and Rees 1981).

While differentiated teratocarcinoma cell types express IGF II, separate peptide growth factors may be expressed by the undifferentiated precursor cells. A single chain peptide of molecular weight approximately 17.5 kd has been isolated from serum-free medium conditioned by PC13 cells, and referred to as embryonal carcinoma-derived growth factor (Heath and Isacke 1984). In addition to the possible autocrine stimulation of PC13 cell growth, this factor was found to promote the growth *in vitro* of PC13 END cells and extra-embryonic yolk sac mesoderm cells, but not the growth of either visceral or parietal endoderm (Heath and Rees 1985). It is therefore possible that there is cross-regulation of embryonal stem cell growth between differentiated and undifferentiated tissues. The Dif 5 and PC13 END cells release IGF II which may contribute to the growth of their undifferentiated precursors. Conversely there is evidence that the embryonal carcinoma-derived growth factor can regulate the growth of its differentiated progeny. Studies with a spontaneously-differentiating embryonal carcinoma cell line, OC15S1, showed that exogenous embryonal carcinoma-derived growth factor inhibited cell differentiation and maintained rapidly growing cell populations (Heath and Rees 1985). This effect was over-ridden by retinoic acid.

3   GROWTH FACTORS IN NORMAL EMBRYOLOGY

i   IGFs and insulin in the embryo and early fetus

While observations from teratocarcinoma cell lines possibly reveal growth factor interactions in the early post-implantation period, it is essential that parallel studies are undertaken to confirm analogous findings during normal development. The presence of IGFs and insulin in the human and mammalian fetus during the third trimester or at term, and their relationship to fetal size and development have recently been extensively reviewed (Underwood and D'Ercole 1984; Gluckman and Liggins 1984; Hill and Milner 1985*a*) and will not be described in depth in this chapter. Rather, recent data concerning the expression and action of insulin-like molecules during the first and early second trimester will be described. Much of the work reviewed has concerned the role of the IGFs in the human fetus.

Studies with experimental animals have established that hormones such as insulin, growth hormone and IGF I do not cross the placenta from mother to fetus (Goodner and Freinkel 1961; Laron, Mannheimer, and Guttman 1966; Underwood and D'Ercole 1984). In addition, the lack of an umbilical arterial/venous difference in circulating IGF levels at term suggests that the same is true in man (Franklin, Pepperell, Rennie, and Cameron 1979). Fetal IGFs are thought to be predominantly synthesized within fetal tissues. Insulin and both IGF I and IGF II are detectable in the human fetal circulation from 12 weeks gestation (Obershain, Adam, King, Teramo, Raivio, Raiha, and Schwartz 1970; Ashton, Zapf, Einshenk, and McKenzie 1985) which is also about the time that immunoreactive IGF I/SM-C is detected in tissue extracts (D'Ercole, Hill, Strain and Underwood 1986). Since circulating levels of IGF peptides probably reflect synthesis at multiple tissue sites, it appears likely that both IGF I/SM-C and IGF II are distributed within human fetal tissues from at least late in the first trimester.

Insulin and pro-insulin have been identified in the fetal rat yolk sac by both RNA/DNA cross-hybridization using specific DNA probes, and by radioimmunoassay (Muglia and Locker 1984). It has also been reported that both insulin and insulin receptors are present in the fertilized chick embryo as early as day 2 of embryogenesis, that is 2–3 days prior to the differentiation of the endocrine pancreas (De Pablo, Roth, Hernandez, and Pruss 1982; Hendricks, De Pablo, and Roth 1984). Immuno-precipitation of conditioned culture media from extra-embryonal mouse tissues 9.5 days post-conception showed that the amnion and extra-embryonic visceral yolk sac were the principle sites of IGF II and IGF binding protein release *in vitro* (Heath and Shi 1986). Both these tissues contain mesodermal elements derived from primitive ectoderm. Isolated cells prepared enzymically from limb bud mesenchyme taken from 11 day gestation mouse embryos released immunoassayable IGF I/SM-C as did explants of mouse embryo liver (D'Ercole, Applewhite, and Underwood 1980).

We recently localized IGF peptides by immunocytochemistry in tissue sections from human fetuses of 14 to 18 weeks gestation (Han, Hill, D'Ercole, Strain, Askin, Lauder, and Underwood 1986), although it was not possible to distinguish between IGF I/SM-C and IGF II. However we were able to locate IGF peptides in the liver parenchyma, the dermis, striated and cardiac muscle, the fetal medullary zone of the adrenal, and in epithelial cell layers in the distal convoluted tubules of the kidney, the alveoli of the lung, and the intestine. We were also able to extract immunoreactive IGF I/SM-C from a wide variety of human fetal tissues of 9 to 19 weeks gestation (D'Ercole *et al.* 1986). In contrast, although transcripts of mRNA for IGF II have been identified at high levels comparable to those seen in certain malignant tumours such as Wilms' tumour, in most human fetal tissues mRNA transcripts for IGF I/SM-C were reported to be absent (Scott, Cowell, Robertson, Priestley, Wadey, Hopkins, Pritchard, Bell, Rall, Graham, and Knott 1985).

The reasons for this discrepancy are unclear but one possibility that should be considered is that additional forms of IGF peptides may be present in the fetus, either as separate gene products or as molecular variants produced by differential mRNA processing. Such variants have recently been identified as minor molecular species in adults (Zumstein, Luthi, and Humbel 1985) and could be invoked to account for an immunological cross-reactivity with anti-sera for IGF I/SM-C as well as poor hybridization of mRNA species to some IGF I/SM-C DNA probes. Evidence for a specific fetal IGF was presented by Sara, Hall, Rodeck, and Wetterberg (1981) who found that human fetal brain cell membranes used in a radioligand assay detected considerably more IGF in cord blood than could be measured in the same sample by a radioimmunoassay detecting a mixture of IGF I/SM-C and IGF II. Thorsson and Hintz (1985) have suggested, however, that this observation was partly due to interference by IGF binding proteins tightly associated with the membrane preparation in the radioligand assay. The existence of additional forms of IGF peptides with biological activity during early development remain speculative at present.

ii *IGF release* in vitro

Many primary or early passage fetal cells or tissues maintained in culture have the capacity to release IGF peptides. This strongly suggests that they do this *in vivo* also. Human tissues taken between 12 and 20 weeks gestation and maintained *in vitro* as either myoblasts, dermal fibroblasts or cartilage explants, released immunoreactive IGF I/SM-C, myoblasts releasing more IGF per unit cell protein than did fibroblasts or cartilage (Hill, Crace, and Milner 1985). The IGF I/SM-C present in cell-conditioned culture medium could be separated by gel filtration at acid pH into free IGF I/SM-C of molecular weight 7.6 kd together with an IGF binding protein of approximately 40–60 kd. Similarly, immunoreactive IGF I/SM-C could be extracted from human fetal pancreas in second trimester, and from culture medium conditioned by isolated human fetal hepatocytes (Hill, Frazer, Swenne, Wirdnam, and Milner 1987; Strain, Hill, Swenne, and Milner 1987). IGF peptides were localized by immunocytochemistry to the pancreatic β-cells and to a sub-population of hepatocytes. The established line of WI38 human embryonic lung fibroblasts also releases IGF I/SM-C (Atkinson and Bala 1981).

Little information exists as to whether cultured human fetal tissues release IGF II, although by extrapolation from studies with fetal animal tissues taken in late gestation they might be expected to do so. Fetal rat fibroblasts released IGF II almost exclusively during culture, while postnatal rat fibroblasts released predominantly IGF I/SM-C (Adams, Nissley, Greenstein, Yang, and Rechler 1983; Adams, Nissley, Handwerger, and Rechler 1983). This developmental change mirrored that seen in circulating IGF peptides. However, fibroblasts are not representative of all fetal rat tissues in late gestation. Fetal rat myoblasts released approximately equimolar amounts of IGF I/SM-C and IGF II, while isolated fetal and neonatal rat islets of Langerhans released IGF I/SM-C exclusively (Hill, Crace, Nissley, Morrell, Holder, and Milner 1985; Romanus, Rabinovitch, and Rechler 1985; Swenne, Hill, Strain, and Milner 1987a). Clearly the distribution of IGF peptide synthesis differs between tissues, and possibly with the stage of development.

Early experiments suggested that the placenta was not a source of IGF peptides, little IGF I/SM-C being released from explants of fetal mouse placenta (D'Ercole *et al.* 1980). More recently human placental explants, and fibroblasts isolated from placenta, were found to release immunoassay-able IGF I/SM-C (Fant, Munro, and Moses 1986) while mRNA prepared from first trimester and term human placenta directed the translation of immunoprecipitable IGF I/SM-C-like proteins in a cell-free reticulocyte lysate system *in vitro* (Mills, Underwood, and D'Ercole 1985). While it seems likely that the placenta is a source of IGF peptides as early as the first trimester there is no evidence that extra-fetal tissue IGFs contribute to the growth of the fetus as a whole, the release of peptide from placental cells *in vitro* being no greater per cell than that from other fetal cell types (Fant *et al.* 1985; Hill, Crace, and Milner 1985). Placental IGFs may act predominantly, as do other fetally-derived IGFs, as paracrine growth factors. Placenta is rich in IGF receptor sites (Daughaday, Mariz, and Trevedi 1981). Importantly, a paracrine action of IGF peptides on placental growth will influence fetal size indirectly through an increased placental capacity for nutrient and oxygen transfer.

iii Receptors for IGFs and insulin

Insulin receptors appear in the placenta early in the first trimester but are predominantly on the maternal surface of the syncytiotrophoblast (Posner 1974; Whitsett, Johnson, and Hawkins 1979). The human placenta contains both type I and type II IGF receptors (Daughaday *et al.* 1981; Bhaumick, Bala, and Hollenberg 1981). Studies with experimental animals generally showed an increase in IGF binding per unit weight of cell membrane with gestational age (D'Ercole, Foushee, and Underwood 1976).

Numerous first trimester human fetal tissues including lung, liver, brain, heart and kidney possess specific receptors for IGF peptides and for insulin (Potau, Ruidor, and Ballabriga 1981; Sara, Hall, Misaki, Fryklund, Christenson, and Wetterberg 1983). At the earliest ages studied fetal brain preferentially bound IGF II rather than IGF I/SM-C. Before the end of the first trimester a low affinity IGF I/SM-C receptor appeared but had been replaced at 17 weeks gestation by a high affinity IGF I/SM-C receptor. By 25 weeks gestation the IGF binding pattern of fetal brain resembled that seen in adult tissue. Early brain development, prior to 18 weeks, includes a period of rapid cell replication predominantly due to neuroblast formation (Dobbing and Sands 1973). Thereafter glial cell formation is thought to predominate. The appearance of a high affinity IGF I receptor in whole brain during glial cell formation may be related to the finding that cultured rat astroglial cells possess both type I and type II IGF receptors, and respond to exogenous IGFs by increasing DNA synthesis (Han, D'Ercole, Lauder, and Svoboda 1985).

In contrast to brain, human fetal liver contained predominantly a type II IGF receptor which did not alter in characteristics with gestational age (Sara *et al.* 1983). Hepatic insulin receptors increased in number per unit weight of tissue with increasing age until 25 weeks gestation, after which only the binding affinity increased in late gestation (Neufeld, Scott, and Kaplan 1980). Both the insulin and IGF receptor populations may undergo significant changes in the human fetus near term. Monocytes for human cord blood had five times the insulin binding capacity of adult cells and a greater IGF I/SM-C binding capacity (Thorsson and Hintz 1977; Rosenfeld, Thorsson, and Hintz 1979). Similarly cord blood erythrocytes bound greater amounts of insulin than adult cells (Herzberg, Boughter, Carlisle, Ahmad, and Hill 1980). It is possible, therefore, that fetal tissues in late gestation may have a high sensitivity to insulin, not only as an anabolic hormone but as a mitogen.

iv   IGFs, insulin and tissue growth

To assess the physiological relevance of IGF peptides in the early human fetus it is necessary, in addition to demonstrating specific tissue receptors, to show biological responses in fetal tissues *in vitro*. Both IGF I/SM-C and rat IGF II promoted DNA synthesis and the uptake of the non-metabolizable amino acid, α-aminoisobutyric acid, by isolated human fetal myoblasts and dermal fibroblasts (Conover, Rosenfeld, and Hintz 1986; Hill, Crace, Strain, and Milner 1986a). The mean half-maximal effective concentrations (ED_{50}) for the two IGF peptides were similar for promotion of amino acid uptake into both cell types, and ranged from 0.9 nM to 1.9 nM. The mean ED_{50} values for DNA synthesis ranged from 2.0 nM to 4.3 nM. We have recently obtained similar results with human IGF II (D. J. Hill, unpublished observations). Thus, human fetal myoblasts and fibroblasts did not show a differential sensitivity to IGF II compared with IGF I/SM-C. The finding that IGF II was at least as effective as IGF I/SM-C may be physiologically significant since IGF II is a much weaker growth promoting agent than IGF I/SM-C *in vivo* in the adult rat (Schoenle *et al.* 1985).

   The ability of IGFs to promote chondrocyte proliferation and differentiation within the epiphyseal growth plates makes them of particular interest in relation to longitudinal skeletal growth *in utero*. Ashton and Otremski (1986) reported that rat IGF II stimulated DNA synthesis in cartilage explants taken from human fetuses of 15–18 weeks, but that tissue from younger or older fetuses was unresponsive, as was post-natal cartilage. However their previous studies had shown that fetal cartilage outside this developmental period, and postnatal cartilage, increased both DNA synthesis and mucopolysaccharide production in response to partially purified IGF I/SM-C and bioassayable plasma somatomedin activity (Ashton and Phizackerly 1981; Ashton and Spencer 1983). Thus, although human fetal cartilage responded to IGF I/SM-C during the late first and early second trimester, a response to IGF II was limited to a 4 week period of development only. By contrast no such limited developmental response was found with other non-cartilaginous connective tissues (Hill *et al.* 1986*a*). While some human fetal tissues may possess a heightened sensitivity to IGF II the ED_{50} values for DNA synthesis by human fetal cartilage in response to partially pure IGF I/SM-C (Ashton and Spencer 1983), or by human fetal fibroblasts in response to the pure peptide (Hill *et al.* 1986*a*), were no greater than the corresponding values seen with postnatal tissues (Conover, Dollar, Rosenfeld, and Hintz 1985).

Insulin was shown to stimulate DNA synthesis directly in fetal rat myoblasts isolated in late gestation (Crace, Hill, and Milner 1985). However, insulin in physiological concentrations did not increase DNA synthesis in second trimester human fetal connective tissues (Hill *et al.* 1986*a*), although insulin was a potent stimulator of amino acid uptake by isolated human fetal tissues with mean ED_{50} values below 2 nM. This contrasts sharply with the poor ability of insulin to increase amino acid uptake by newborn human foreskin fibroblasts with an ED_{50} of greater than 17 nM, and by the L6 rat myoblast cell line with an ED_{50} of 14 nM (Kaplowitz, D'Ercole, Underwood, and Van Wyk 1984; Florini, Ewton, Falen, and Van Wyk 1985). Insulin was reported to induce cytoplasmic enlargement in fetal cell types independently of any action on proliferation (Cheek, Brayton, and Scott 1974; Zetterberg, Engstrom, and Dafgard 1984). This is consistent with our findings that insulin was a potent anabolic agent but not a mitogen for isolated human fetal connective tissues, thereby giving rise to tissue hypertrophy but not to hyperplasia at physiological concentrations.

When the information available on circulating and tissue levels of IGF peptides, tissue receptors and tissue responses to the peptides is assessed for the human fetus during the first and second trimester, it is clear that IGFs are potentially important during early development. However, there is no consistent evidence for enhanced IGF release or increased tissue sensitivity of musculo-skeletal tissues to IGF peptides *in utero* compared to postnatal life that could account for rapid fetal growth. While human fetal connective tissues are extremely sensitive to the metabolic actions of insulin in the second trimester, insulin does not apparently act as a mitogen. A more useful understanding may come from consideration of IGF action not in isolation, but in synergy with other peptide growth factors.

v Epidermal growth factor

During mouse embryogenesis, receptors for EGF are detectable as early as the appearance of the first differentiated cell type, primary trophoecto-derm, identified in outgrowths of whole blastocysts which had been cultured for 3 days (Adamson and Meek 1984). The trophoblastic cells form a major proportion of the placenta during later fetal development and probably account for the high specific binding of EGF to placental plasma membrane preparations at term (O'Keefe, Hollenberg, and Cuatrecasas 1974). Placental EGF receptors, which increase in number per unit tissue weight with fetal age (Lei and Guyda 1984), are predominantly located in the microvillus plasma membranes exposed to the maternal circulation and in the basolateral plasma membranes in close proximity to the fetal circulation (Rao *et al.* 1985). Receptors for EGF are also present on the human chorion, amnion and decidua, the chorion demonstrating the greatest binding (Roa, Carman, Ghegini, and Schultz 1984). The physiological significance of EGF binding to placenta and fetal membranes is not understood, but exogenous EGF introduced into the fetal sheep was shown to stimulate placental lactogen release into the maternal circulation (Thorburn, Waters, Young, Dolling, Buntine, and Hopkins 1981).

Specific binding to EGF by whole embryonic tissues was first measurable on day 11 of mouse gestation and the number of receptors per unit tissue weight increased in all tissues until term (Adamson and Meek 1984). However the receptor affinity declined with time in the lean body mass and placenta while remaining constant in liver and brain. Clearly developmental changes are occurring in receptor presence and behaviour within individual tissues. EGF-like peptides were detected in various fetal mouse tissues by radioimmunoassay, although higher levels were apparently measured by radioreceptor assay (Nexo, Hollenberg, Figueroa, and Pratt 1980). The authors proposed that a fetal form of EGF may exist which exhibited only partial immunological cross-reactivity. Matrisian, Pathak, and Magun (1982) isolated an EGF-like peptide from the rat fetus with larger molecular size than mouse EGF. This was almost certainly a complex of the EGF analogue, TGF-α and another molecule, possibly a binding protein, since the material displaced iodinated EGF from membrane receptors yet also induced a transformed phenotype in cells grown in soft agar. It is not yet known whether EGF or TGF-α is the dominant molecular species in the rodent embryo and early fetus.

EGF has been localized by immunocytochemistry in several human fetal tissues during early second trimester (Kasselberg, Orth, Gray, and Stahlman 1985). EGF was present in the pituitary where EGF-positive cells also stained for FSH, LH or TSH, but not for growth hormone, prolactin or ACTH. EGF-positive cells were noted in the stomach exclusively at the base of gastric and pyloric glands and in the duodenum in the Brunner's glands. The trachea contained EGF-positive cells arranged in clusters within the pseudostratified columnar epithelium. EGF was also localized in the placenta and salivary glands. Additionally we have identified EGF-containing cells by immunocytochemistry in the tubular epithelium of the distal convoluted tubule of the human fetal kidney (D. J. Hill, unpublished observations).

The biological actions of exogenous EGF in the fetus are numerous. Many concern maturational events and will be dealt with later. Of those actions dealing with fetal growth, in only one report was EGF infused directly into the fetal compartment. Administration of mouse EGF to the fetal lamb for up to two weeks in the third trimester caused gross hypertrophy of the skin epithelium and led to an increase in the weight of the adrenal, thyroid, liver and kidney (Thorburn *et al.* 1981). Total body weight and length were unaltered. Exogenous EGF promoted DNA synthesis in isolated amnion, limb bud and lung taken from 15-day mouse embryos (Adamson *et al.* 1981). We found that mouse EGF promoted tritiated thymidine uptake by late gestation fetal rat costal cartilage explants (Hill, Holder, Seid, Preece, Tomlinson, and Milner 1983), and by human fetal fibroblasts grown from late first and early second trimester skin (Hill, Strain, Elstow, Swenne, and Milner 1986). Fetal rat hepatocytes showed an increased DNA synthesis in response to EGF, although this may have been mediated by a release of IGF II (Richman, Benedict, Florini, and Toly 1985). This is in line with several other reports which suggest that IGF release or synthesis in fetal or postnatal tissues may be activated *in vitro* by EGF (Atkinson, Bala, and Hollenberg 1984; Atkinson, Hayden, Bala, and Hollenberg 1984*b*; Clemmons 1984).

The interaction between EGF and other peptide growth factors during the proliferation of fetal tissues *in vitro* was clearly demonstrated by Kaplowitz, D'Ercole, and Underwood (1982). Undifferentiated fetal mouse cells were removed from mesenchymal limb buds and cultured in the presence of peptide growth factors either alone or in combination. When grown in monolayer no single growth factor promoted cell proliferation in a basal medium. However the presence of a unique fetal mouse liver-derived growth factor increased cell number in the presence of EGF and insulin. In high density cultures EGF, FGF, rat IGF II and IGF I/SM-C each induced an increase in cell number alone but were more effective in the presence of the liver factor. Experiments with combinations of growth factors showed that IGF I/SM-C, IGF II and insulin all fulfilled a similar role during the growth of undifferentiated mesenchymal cells, and were not additive at maximal concentrations. At a later stage of fetal development we found that IGF I/SM-C and EGF were synergistic during DNA synthesis in fetal rat cartilage explants (Hill *et al.* 1983). The accumulated evidence suggests that EGF, like the IGFs and insulin, is present and biologically active throughout prenatal life.

One further recent observation adds intriguing new insight into the role of EGF or EGF-like peptides during development. A gene, designated *Notch*, which appears to regulate early neural development in the fruit-fly *Drosophila melanogaster* has been sequenced and found to encode a protein with numerous cystine-rich 40–50 amino-acid repeats characteristic of EGF, and therefore to be homologous to the EGF precursor molecule which contains nine such sequences (Wharton, Johansen, Xu, and Artavanis-Tsakonas 1985). This protein is thought to be membrane bound but in addition may have an extracellular EGF-like domain. Interestingly, several other mammalian cell proteins, both membrane bound, such as the low density lipoprotein receptor, and secreted proteins, including the clotting factors urokinase and plasminogen-activator, also share a common EGF-like domain (Gunzler, Steffens, Otting, Klim, Frankus, and Flohe 1982; Ny, Elgh, and Lund 1984; Russell, Schneider, Yamamoto, Laskey, Brown, and Goldstein 1984). While the role of the EGF-like domain in these divergent proteins is a mystery, evidence from studies with *Drosophila* suggest that the *Notch* gene product may be responsible for the transmission of developmental signals between adjacent cells (Wharton *et al.* 1985). This finding raises the possibility for the first time of a link between membrane-anchored regulatory proteins such as the CAMs and diffusable growth factors. However, it must also be considered that the occurrence of EGF-like peptide domains in several molecules may indicate a structural rather than a functional role.

vi   Other peptide growth factors

While there is no information concerning peptide growth factor expression in human embryonic tissues during the first month of life, the work of Goustin, Betsholtz, Pfeifer-Ohlsson, Persson, Rydnert, Bywater, Holmgren, Heldin, Westermark, and Ohlsson (1985) has demonstrated the expression of the gene coding for the B chain of PDGF, which is closely related to the cellular oncogene *sis*, in human placenta from as early as 21 days post-conception. By *in situ* hybridization of c-*sis* mRNA to a c-*sis* DNA probe in histological sections, expression was localized to just one cell type, the trophoblast. Explants of first trimester placenta released radioreceptor-assayable PDGF-activity, while cultured trophoblast cell lines were rich in high-affinity PDGF receptors, and responded to exogenous PDGF with an increased expression of the proto-oncogene c-*myc*, accompanied by DNA synthesis. The early trophoblast therefore appears to provide an example of an autocrine growth mechanism whereby the expression of PDGF by a discrete cell type may directly activate the proliferation of the same cell. Preliminary evidence indicated that the c-*sis* gene is also expressed in whole human embryos during the first trimester (Goustin *et al.* 1985).

It is not known whether PDGF is present in human tissues other than platelets beyond the first trimester, but cultured rat aortic smooth muscle cells isolated two weeks following birth expressed a PDGF-like molecule which was not released from equivalent adult cells (Seifert, Schwartz, and Bowen-Pope 1984). Vascular smooth muscle cells also released IGF peptides *in vitro* (Clemmons and Van Wyk 1985), and a synergistic interaction was seen between exogenous PDGF and either exogenous or endogenous IGFs during DNA synthesis in immature porcine smooth muscle cells (Clemmons 1985). Indeed, IGF release by vascular smooth muscle may be enhanced by PDGF as well as other growth factors (Pfeifle and Ditschuneit 1984). Human umbilical endothelial cells expressed the c-*sis* gene during tissue culture (Collins, Ginsberg, Boss, Orkin, and Pober 1985) and released PDGF into conditioned culture medium (Di Corleto and Bowen-Pope 1983). Hence paracrine and autocrine interactions between smooth muscle and endothelial-derived peptide growth factors may play a central role in the rapid angiogenesis necessary during embryonic and fetal development.

The known biological properties of TGF-β suggest that this peptide may contribute to a rapid embryonic growth rate. The presence of TGF-β in fresh fetal mouse tissues was reported by Proper, Bjornson, and Moses (1982) and in term human placenta by Frolik, Dart, Meyers, Smith, and Sporn (1983). We recently noted that TGF-β-like bioactivity was present in eight separate fetal rat tissues removed on day 21 of gestation and subjected to acetic acid extraction (Hill, Strain, and Milner 1986). The TGF-β-like activity was most abundant in muscle, lung, liver and kidney, and least in skin and brain. Tissue levels were considerably higher during the final five days *in utero* than in the adult, the decline occurring a few days after birth. Fetal rat tissue-derived TGF-β had a molecular weight of 25–40 kd following acidic chromatography.

Lawrence, Pircher, Krycere-Martinerie, and Jullien (1984) demonstrated that normal chicken, mouse and human embryo fibroblasts released a TGF-β-like bioactivity in basal culture medium, but that this activity was present in a latent form. The TGF-β was activated by chromatography at acid pH whereupon the chick embryo fibroblast-derived material eluted with molecular sizes of approximately 500 kd, 125 kd and 20 kd. The smallest peak is similar to pure TGF-β (molecular weight 25 kd) obtained from platelets but it is not yet understood whether the larger molecular forms represent the presence of binding proteins or precursor polypeptides. The physiological activation mechanism of TGF-β is unknown. We confirmed that conditioned culture medium from human fetal fibroblasts contained TGF-β-like activity following acidic chromatography, but obtained bioactivity with molecular size of 12–14 kd (Elstow, Hill, Strain, Swenne, Crace, and Milner 1985). Centrella and Canalis (1985) demonstrated that isolated fetal rat calvaria released TGF-β-like activity *in vitro* of molecular weight 25–50 kd.

As outlined previously, TGF-β acts as a bifunctional growth regulator for several cell types *in vitro* and can influence the manner in which cells respond to other peptide growth factors, either positively or negatively. The reversible, transformed phenotype conferred on non-transformed fibroblastic indicator cells, both in soft agar and in monolayer culture, is morphologically reproduced during the growth of early passage human fetal fibroblasts in monolayer (Elstow *et al.* 1985). These cells did not arrest their growth at high cell density but formed overlapping layers. While this occurred rapidly in the presence of fetal calf serum it was also apparent over a considerably longer period in a basal culture medium without serum. The ability of human fetal fibroblasts to proliferate in serum-free conditions may be explained by the observation that they release peptide growth factors such as IGF I/SM-C (Hill, Crace, and Milner 1985).

When human fetal fibroblasts were exposed to exogenous platelet-derived TGF-β a bifunctional response was observed, involving either an inhibition of DNA synthesis resulting in growth arrest, or an increased DNA synthesis accompanied by cell proliferation (Hill, Strain, Elstow, Swenne, and Milner 1986). The stimulatory growth response to TGF-β was only apparent for cells derived from fetuses of less than 50 g body weight, while a growth inhibitory response was only seen with cells from fetuses of 100 g body weight or greater. The bifunctional response was not due to variation in cell density during culture, or to passage number, or change in the duration of the cell cycle. For both positive and negative growth responses maximal effects were apparent at very low concentrations of TGF-β (approximately 40 pM). Additionally, when human fetal fibroblasts were exposed to other peptide growth factors such as IGF I/SM-C, EGF or PDGF, the resulting increase in DNA synthesis was either potentiated or inhibited by exposure to TGF-β, according to the weight of the donor fetus. A developmental change may therefore occur within human fetal connective tissues in their response to endogenous TGF-β around 13–16 weeks gestation. The physiological significance of these observations is reinforced by data which indicated there is a change from an accelerating to a slowly diminishing rate of human fetal growth, as measured by crown-rump and crown-heel length, at this gestational age range (Birbeck 1981).

We have also recently shown that TGF-β inhibited DNA synthesis in hepatocytes from human fetuses of 35 g to 240 g weight at extremely low concentrations (0.7 pM), while non-parenchymal cells in the same cultures remained resistant to doses of TGF-β over 100 fold higher (Strain, Hill, and Milner 1986). Further indications of the biological role of TGF-β came with the finding that the exogenous peptide stimulated the production of prostaglandin E_2 and bone resorption in fetal rat calvaria (Tashjian, Voelkel, Lazzaro, Singer, Roberts, Derynck, Winkler and Levine 1985). The ability of TGF-β to modify the mitogenic responses of tissues to other peptide growth factors may be contributory to the existence of a rapid growth rate *in utero* despite there being similar or lower circulating levels of peptides such as EGF and the IGFs compared to those present in childhood and adult life.

The role of NGF as a survival and differentiating factor for a number of cell types derived from the neural crest, including sympathetic and sensory neurones, is well established (Thoenen and Barde, 1980). However there have been no convincing reports that NGF has a mitogenic role for neuronal tissue during embryonal development. Recently, NGF was shown to elicit a mitogenic response in cultured adrenal chromaffin cells isolated from rats one week following birth (Lillien and Claude 1985). These are endocrine cells derived from the neural crest which are closely related to sympathetic neurones, yet continue to proliferate for a short period following birth. In response to NGF these cells exhibit neural traits (Doupe, Landis, and Patterson 1985), and following extended exposure will become morphologically neuronal. Thus chromaffin cells appear to have retained elements of the pleuripotential nature of sympatho-adrenal precursor cells, and the observations suggest that NGF may act as a mitogen for other neural crest-derived tissues during early development. There is limited information on the presence of NGF in the embryo and fetus. By analogy with postnatal life NGF might be expected to be synthesized within tissues innervated by the sympathetic nervous system, and the release of nerve growth-stimulating factors has been established for chick embryo heart cells (Norrgren, Ebendal, and Wikstrom 1984). NGF is present in the human placenta (Goldstein, Reynolds, and Perez-Polo 1978) and a venous/arterial gradient exists in human umbilical cord serum, suggesting a possible placental secretion (Walker, Tarris, Weichsel, Scott, and Fisher 1981).

It is clear from the above discussion that peptide growth factors appear very early in post-implantation development. Their potent growth-promoting and functional actions on isolated embryonic and fetal tissues suggest an important contribution to development *in utero*.

4   CONTROL OF GROWTH FACTOR RELEASE

The development of the fetus is under multifactorial control in which the inherent drive to body growth determined by the fetal genome is tempered by environmental factors such as maternal size, an adequate maternal nutrition and the sufficiency of the placenta in supplying both nutrients and oxygen to the fetal compartment (see Milner 1987). Since fetal development may involve the expression of peptide growth factors for signalling the control of mitogenesis and differentiation within and between tissues, the regulation of growth factor release may thereby be dependent on nutritional sufficiency and gaseous exchange. However there is evidence that factors such as insulin and placental hormones can also influence growth factor expression. At present almost all available evidence concerns the control of IGF release.

In postnatal life peptide growth factor synthesis is strongly influenced by endocrine hormones. The major growth factor for longitudinal skeletal growth, IGF I/SM-C, is largely dependent on the presence of pituitary growth hormone for its synthesis in tissues and appearance in the circulation (D'Ercole, Stiles, and Underwood 1984). Furthermore, both insulin and thyroid hormones are necessary for an adequate expression of IGF I/SM-C release (Miller, Schalch, and Draznin 1981; Marek, Schullerova, Schreiberova, and Limanova 1981). However the hormonal control of IGF I/SM-C synthesis is overridden by adequate nutritional supply, such that in human or experimental animal malnutrition circulating levels of the growth factor are substantially reduced (Phillips and Young 1976; Clemmons, Klibanski, Underwood, McArthur, Ridgway, Beitins, and Van Wyk 1981). Other peptide growth factors such as EGF and NGF are dependent on the presence of thyroid hormones and androgens for the maintenance of tissue levels in postnatal life (Walker, Weichsel, Hoath, Poland, and Fisher 1981), while the influence of nutrition on the synthesis of these molecules is not known.

Animal experimentation has demonstrated that pituitary growth hormone is not necessary for either fetal growth or IGF expression in the circulation until after birth (Brinsmead and Liggins 1979; Hill, Davidson, and Milner 1979; Parkes and Hill 1985). Similarly, in the human anencephalic fetus body weight can fall within the normal corrected weight range for gestational age (Grumbach and Kaplan 1973), and circulating levels of IGF I/SM-C and IGF II are not altered in anencephalic fetuses during the second trimester compared to normal fetuses (Ashton *et al.* 1985).

The role of insulin as a growth factor during the latter half of fetal development and its relationship to the IGFs have recently been extensively reviewed (Hill and Milner 1985*b*; Gluckman 1986) and will not be described in detail here. While extremes of human birth weight can be associated with abnormality of insulin secretion or action, as in the small infant with leprechaunism which is associated with extreme insulin resistance, or in the large-for-gestational-age hyperinsulinaemic infant of the poorly controlled diabetic mother (Pederson 1977; Schilling, Rechler, Grunfeld, and Rosenberg 1979), much of the variation can be accounted for by adiposity. This does not become apparent until after weeks 24–28 of gestation when the fetal pancreas becomes acutely responsive to glucose as an insulin secretogogue, and adipocytes gain insulin receptors (Milner 1974).

Fetal hypoinsulinaemia, subsequent to induced hypoglycaemia, is invariably associated with depressed circulating IGF values and fetal growth retardation (De Prins, Hill, Fekete, Robsen, Fieller, Van Assche, and Milner 1984). Under these conditions it is unclear whether lowered circulating IGF values are due primarily to reduced nutritional availability or a decreased insulin-dependent nutritional incorporation by the tissues. This was partly answered by pancreatectomy *in utero* in the fetal sheep (Fowden and Comline 1984). Following a three week period of hypoinsulinaemia slightly elevated levels of glucose and fructose were found in the circulation of the pancreatectomized fetus compared to controls. Despite adequate nutrition, fetal body weight showed a 30 per cent reduction for gestational age and circulating levels of IGF I/SM-C were severely reduced (Gluckman, Fowden, Butler, and Comline 1985). In the absence of the anabolic actions of insulin the release of IGF peptides from isolated fetal tissues may be largely independent of physiological variations in nutrition.

If, as appears likely, peptide growth factors act as paracrine or autocrine factors in early development, their synthesis and release from individual tissues may be largely controlled by the permissive interactions of nutrient availability and insulin-dependent nutrient uptake.

Animal models do not readily exist for the control of growth factor expression *in vivo* during the first half of pregnancy, and conclusions must argely be drawn from tissue culture experiments. The lack of dependency of fetal IGF levels on pituitary growth hormone seen *in vivo* has been reproduced *in vitro*. Growth hormone failed to promote either DNA synthesis or IGF I/SM-C release from human fetal fibroblasts or myoblasts isolated in the late first and early second trimester (Hill, Crace, and Milner 1985), or from human fetal pancreas explants (Swenne, Hill, Strain, and Milner 1987*b*). While the distribution of growth hormone receptors on human fetal tissues is unknown such receptors were absent from the liver of the fetal sheep and rat (Maes, De Hertogh, Watrin-Granger, and Ketelslegers 1983; Gluckman, Butler, and Elliot 1983). It was therefore surprising that growth hormone increased both the labelling index of human fetal hepatocytes, isolated in first and second trimester, and the release of IGF I/SM-C from these cells (Strain *et al*. 1987). Maximal biological action occurred between 50 and 250 ng/ml which is within the physiological range for the human fetus (Grumbach and Kaplan 1973). It is clearly unwise to extrapolate freely from fetal animal studies when assessing the control of IGF synthesis in the human fetus.

Evidence from studies with fetal animal tissues has suggested that placental lactogens may exert numerous anabolic effects *in utero*. Isolated diaphragm muscle from fetal rats showed increased amino acid uptake following exposure to ovine placental lactogen (oPL), but not growth hormone (Freemark and Handwerger 1983). oPL promoted glycogen synthesis and inhibited glycogenolysis in isolated fetal rat hepatocytes (Freemark and Handwerger 1984, 1985). oPL, but not growth hormone, increased the release of IGF II from fetal rat fibroblasts *in vitro* (Adams, Nissley, Handwerger, and Greenstein 1983). Recently Freemark and Handwerger (1986) demonstrated the presence of a specific oPL receptor on fetal sheep liver plasma membranes. While these findings suggest a growth promoting action in the fetus, the majority of studies have used oPL, which is known to be a more potent somatotrophic hormone than human placental lactogen (hPL) when injected into the hypophysectomized rat (Brinsmead, Waters, and Thorburn 1980). We therefore undertook a series of experiments to investigate the possible anabolic actions of hPL on human fetal tissues during late first and second trimester.

We found that hPL promoted DNA synthesis and/or cell replication in human fetal myoblasts, fibroblasts, hepatocytes and pancreas explants isolated in the first half of gestation (Hill, Crace, and Milner 1985; Strain *et al*. 1987; Swenne *et al*. 1987*b*). The half-maximal concentrations of hPL necessary to promote DNA synthesis in fetal connective tissues were between 2 and 8 nM (Hill, Crace, Strain, and Milner 1986). This encompasses the range of hPL values found in the human fetal circulation at this stage of gestation (D. J. Hill and M. Freemark, unpublished observations). The mitogenic actions of hPL were accompanied, for each tissue so far examined, by a parallel release of immunoassayable IGF I/SM-C. The ability of hPL to induce DNA replication in fetal fibroblasts was prevented in the presence of excess antibody to IGF I/SM-C, suggesting that the mitogenic actions of the placental hormone were largely mediated by a paracrine release of IGF peptides (Hill, Crace, Strain and Milner 1986). However a more rapid action of hPL on human fetal tissues, that of amino acid transport, was not impeded by antibody to IGF I/SM-C, suggesting that at least some actions of hPL may be direct. hPL and IGF I/SM-C were not additive at maximal concentrations during DNA synthesis in human fetal fibroblasts, but were additive during amino acid transport suggesting separate pathways. The influence of hPL on peptide hormone release *in utero* may not be limited to the IGFs since exposure to hPL also increased insulin release from human fetal pancreas explants (Swenne *et al.* 1987*b*).

Tissue culture experimentation has therefore shown that hPL could potentially function *in vivo* as a fetal growth factor. However, the ability of infants lacking the hPL genes to achieve normal birth weight for gestational age (Wurzel, Parks, Herd, and Nielson 1982) suggests that hPL is not essential for human fetal development. There may, however, be more than one molecular species of placental lactogen, as occurs in the rat (Robertson, Owens, McCoshen, and Friesen 1984).

5   GROWTH FACTORS AND DIFFERENTIATION

Classical embryological dogma states that the proliferation and differentiation of tissues should be mutually exclusive. Such views have arisen from observations that undifferentiated stem cells do not undertake differentiated cell functions but provide a pool of proliferative cells which may selectively expand when and where they are required during embryogenesis, and later during tissue remodelling or wound repair. Conversely differentiated, metabolically active cells such as hepatocytes, myotubes and neurones do not normally proliferate. A simplistic approach to growth factor action during embryogenesis would suggest a role largely restricted to the proliferation of progenitor cell populations. This does not appear to be so since molecules such as FGF and the IGFs may influence both proliferation and differentiation in the same target tissue, and may have divergent actions during specific maturational events.

One of the best studied models of tissue maturation has been the apparently spontaneous differentiation of proliferative myoblasts into post-mitotic, polynuclear, fused myotubes. While this process largely takes place *in utero* during human and bovine development, an extensive population of myoblasts still exists in the rat at birth. Subsequent large-scale muscle differentiation occurs 7 to 10 days after birth (Dubowitz 1967). Following enzymatic dispersal of fetal or neonatal rat myoblasts, differentiation will occur *in vitro* once cell monolayers approach confluence, and can be followed experimentally by changes in cell morphology, the appearance of acetyl choline receptors, or the acquisition of myotube-specific enzymes such as creatine phosphokinase and myokinase (Ewton and Florini 1981; Hill, Crace, Fowler, Holder, and Milner 1984). Additionally myoblast cell lines exist which will fully differentiate *in vitro* (Yaffe's L6 rat line) or differentiate without myotube fusion (BC3H-1 mouse line).

Physiological concentrations of IGF I/SM-C and IGF II, and supraphysiological amounts of insulin, promoted cell replication in myoblasts from the fetal rat (Hill, Crace, Nissley, Morrell, Holder, and Milner 1985a), human fetus (Hill, Crace, Strain, and Milner 1986), chick (Schmid, Steiner, and Froesch 1983a), and the L6 cell line (Ewton and Florini 1980). The insulin-like group of peptides also enhanced the rate of differentiation of the L6 line, cells leaving the proliferative cycle following two or three divisions in response to the growth factors. This outcome was not simply due to increased cell density since IGF-induced differentiation was also apparent in cultures growth-arrested by exposure to cytosine arabinoside (Ewton and Florini 1981). The potency of IGF I/SM-C as a differentiating agent was ten times greater than that of insulin and more than 100 times greater than IGF II (Florini, Ewton, Falen, and Van Wyk 1985). IGF-induced differentiation required the presence of extracellular calcium (Shainberg, Yagil, and Yaffe 1971) and involved the activation of intracellular polyamines (Ewton, Erwin, Pegg, and Florini 1984) and a rapid fall in the expression of the proto-oncogene c-*myc* (Sejerson, Sumegi, and Ringertz 1985). While undifferentiated BC3H-1 myocytes possess both type I and type II IGF receptors they gain insulin receptors only following differentiation (De Vroede, Romanus, Standaerdt, Pollett, Nissley, and Rechler 1984; Standaert, Schimmel, and Pollet 1984).

Primary and established myoblast cultures also proliferate in response to FGF. However FGF inhibited myoblast differentiation and potentiated the progenitor cell population (Gospodarowicz, Weseman, Moran, and Lindstrom 1976; Linkhardt, Clegg, and Hauscha 1981). Separate peptide growth factors, which in short-term experiments are mitogenic for isolated myoblasts, therefore give rise to very different outcomes. The presence of FGF favours the maintenance of a stem cell population and may well suppress the differentiating actions of IGFs and insulin. This may be especially relevant in early development when fetal myoblasts release IGF peptides *in vitro* (Hill, Crace, Nissley, Morrell, Holder, and Milner 1985a; Hill, Crace, and Milner 1985b). Incubation with IGFs also induced differentiated function in cultured fetal rat calvaria cells (Canalis 1980;

Schmid, Steiner and Froesch 1983*b*), chondrocytes (Guenther, Guenther, Froesch, and Fleisch 1982) and erythroid precursor cells (Kurtz, Jelkmann, and Bauer 1982).

In late gestation treatment of the fetal lamb or rabbit with EGF induced lung epithelial maturation and surfactant production (Catterton, Escobedo, Sexson, Gray, Sundell and Stahlman 1979; Sundell, Gray, Serenius, Escobedo, and Stahlman 1980). However in some embryonic tissues exposure to EGF may prevent ordered differentiation. One example concerns the influence of EGF on the formation of the secondary palate in the fetal mouse. The secondary palate first becomes morphologically distinct on day 12 of gestation and consists of bilateral extensions of the maxillary processes. The palatal processes change their direction of growth between days 14 and 15 and fuse medially above the tongue, separating the nasal and oral cavities. To enable fusion to progress programmed cell death must occur within the medial epithelial lamina, allowing the mesodermal tissues of the two processes to become confluent (Hassell and Pratt 1977). This does not depend on the presence of the underlying mesenchyme. By day 17 of gestation the basal palatal epithelium has differentiated into pseudo-stratified ciliated columnar cells, while the ventral epithelium has formed the stratified squamous epithelium of the mouth.

Tissue culture studies showed that EGF inhibited palatal fusion by preventing the degeneration of the medial edge palatal epithelium (Grove and Pratt 1984). This was achieved by the hypertrophy and keratinization of medial edge cells. EGF added early in the fusion process initiated epithelial cell proliferation in both the oral and nasal regions. If EGF was added after the medial edge cells had ceased proliferation, prior to fusion, it still prevented cell death but was no longer mitogenic. During the final stages of differentiation on day 14 the addition of EGF no longer stimulated cell proliferation or prevented fusion (Pratt, Figueroa, Nexo, and Hollenberg 1978).

The ability of EGF to inhibit cell death in the fusion zone of the palate was dependent on the presence of underlying mesenchyme, as was its mitogenic action on oral and nasal presumptive epithelia in early differentiation (Tyler and Pratt 1980). This may be partly due to the synthesis of a specific EGF-dependent basement membrane between the epithelial and mesenchymal components (Silver, Murray, and Pratt 1984). The release of glycosaminoglycans by the palate in response to EGF was greatest during medial edge fusion and subsequently declined. The mesenchymal-epithelial interactions may also involve a release of EGF-dependent IGF activity from the mesenchymal tissues since this was observed for cultured mouse palate (Atkinson, Bala, and Hollenberg 1984). IGF release was apparent both before and after palate fusion, but not during epithelial cell disruption. The mesenchymal cells of the human embryonic palate proliferated in response to EGF when grown in monolayer, and were growth-inhibited in response to high levels of glucocorticoids (Yoneda and Pratt 1981*a*, *b*). The developing palatal mesenchyme contains glucocorticoid receptors and glucocorticoids may physiologically induce key enzymes and modulate EGF receptor levels during ordered palatal fusion (Pratt, Kim, and Grove 1984). Hence a precise quantitative and temporal interaction between glucocorticoids, EGF, and epithelial and mesenchymal tissues may give rise to the normally different-iated palate. The injection of pregnant mice with cortisone induced fetal cleft palate which was potentiated by further injection with EGF (Bedrick and Ladda 1978). The experimental model of the developing palate presents the best available example of how a temporal presence or responsiveness to growth factors may regulate normal embryogenesis, and how abnormality of growth factor presence may be linked with teratogenesis.

Of the peptide growth factors so far characterized NGF has the best documented action on tissue differentiation. NGF is essential for the survival of isolated sympathetic neurones and for the outgrowth of neuronal processes during tissue culture (Mains and Patterson 1973). Treatment of newborn mice with NGF increased the size of the sympathetic ganglia and mitotic activity in undifferentiated sympathetic precursor cells (Levi-Mon-talcini 1976), while injection of NGF antiserum into fetal rats curtailed the development of sensory nerve cells and sympathetic precursor cells. Recently evidence of co-operation between NGF and insulin-like peptides has emerged. Both IGF II and insulin stimulated neurite formation in sensory and sympathetic neurones and enhanced the survival of embryonic neurons (Recio-Pinto and Ishii 1984). Additionally both IGF II and insulin enhanced levels of mRNA for tubulin in a human neuroblastoma cell line, an important element in axonal microfilament structure (Mill, Chao, and Ishii 1985). An interaction may therefore occur between insulin-like peptides and NGF during neurite outgrowth, and may be partly explained by a requirement for insulin or IGF II for the maintenance of NGF receptors on cultured human neuroblastoma cells (Recio-Pinto, Lang, and Ishii 1984). An added dimension to neuronal development became apparent with the finding that IGF I/SM-C was a potent growth stimulator for oligodendro-cytes isolated from the newborn rat (McMorris, Smith, De Salvo, and Furlanetto 1986). Oligodendrocytes play a critical role in nervous function by the synthesis and maintenance of myelin.

6   ONCOGENE EXPRESSION AND GROWTH FACTOR ACTION

Oncogenes are normal components of the genome which, through an inappropriate degree of expression, translocation, or point mutation, are no longer under precise developmental control. This leads to the expression of structurally or quantitatively aberrant oncogene-encoded proteins and poorly controlled cell growth. Central to this hypothesis was the finding that the viral oncogenes introduced into cells by retroviruses, and known since

1970 to encode proteins which served as cellular tumour markers, are closely homologous to normal cellular genes (c-*onc*) (Bishop 1983). During the evolution of retroviruses, these normal cellular genes have become integrated into the viral genome. Following subsequent retroviral infection of cells these genes, now designated v-*oncs*, are expressed in aberrant forms leading to neoplastic growth of the cell. While the majority of human tumours are not thought to be viral in origin, a series of c-*oncs*, of which approximately 30 are recognized, may be selectively expressed in cancer cells encoding inappropriate amounts or altered forms of tumour-related proteins, perhaps in response to carcinogens.

Oncogenic proteins have been shown to possess biological activities intrinsically associated with the mechanisms of cell proliferation including many aspects of peptide growth factor presence and action (Table 11.1). For instance the v-*ras* gene product is the p21 protein which exhibits GTP-binding activity. This is related to, but does not represent, the GTP binding protein involved in the activation of adenylate cyclase (Beckner, Hattori, and Shih 1985). Many oncogenic proteins, such as the p60 protein of v-*src*, encode tyrosine protein kinases associated with the cell membrane. These have been shown to bear homology with the tyrosine kinases located within the receptor molecules for peptide growth factors, such as the insulin receptor (Ullrich, Bell, Chen, Herrera, Petruzzelli, Dull, Gray, Coussens, Lias, Tsubokaura, Mason, Seeburg, Grunfeld, Rosen, and Ramacharidran 1985). Other oncogenes, such as v-*mos*, encode cytoplasmic protein kinases that may represent second messenger systems for peptide growth factors, or, like v-*myc* and v-*fos*, nuclear proteins which bind DNA and may directly modulate the expression of other genes. The first indications that oncogenes could also encode whole peptide growth factors or their membrane receptors came with the observations that the v-*sis* gene of the simian sarcoma virus encoded a protein, p28, which was structurally similar to the B-chain of PDGF (Waterfield, Scrace, Whittle, Stroobant, Johnsson, Wasteson, Westermark, Heldin, Huang, and Deuel 1983; Doolittle, Hunkapillar, Hood, Devare, Robbins, Aaronson and Antoniades 1983), and that the v-*erb*-B gene of the chicken erythroblastosis virus encoded a homologue of the EGF receptor (Downward, Yarden, Mayes, Scrace, Totty, Stockwell, Ullrich, Schlessinger, and Waterfield 1984). It seems likely that each stage of peptide growth factor action on cell proliferation can be reproduced in an uncontrolled manner by aberrant oncogene expression during cancer. It is reasonable to consider that normally-regulated cellular oncogenes perform similar functions during ordered growth and development, including embryogenesis.

The expression of several c-*oncs* has been specifically associated with the mitogenic actions of particular peptide growth factors *in vitro*. PDGF and FGF were found to increase levels of c-*myc* or c-*fos* mRNA in lymphocytes and fibroblasts (Kelly, Cochran, Stiles, and Pledger 1983; Armelin,

Table 11.1
Viral oncogenes which code for peptides related to growth factors, their receptors, or their intracellular messengers

Viral oncogene	Protein Product		Associated Tumori- genecity
	Biochemical Function	Cellular Location	
sis	PDGF B-chain	Extracellular	Sarcoma
mos	Tyrosine kinase sharing homology with pro-EGF	Membrane	"
erb-B	Portion of EGF receptor	Membrane	Erythroleukemia, sarcoma
neu	Related to EGF receptor	"	Neuro- and glio- blastoma
fms	Related to macrophage-specific growth factor receptor	"	Sarcoma
met	Tyrosine kinase, some similarity to insulin receptor kinase	Membrane	Similar gene locus on chromosome 7 to cystic fibrosis
ros			
yes	Tyrosine kinase	"	
src	Tyrosine kinase, some similarity to EGF receptor kinase	"	Sarcoma
abl	Tyrosine kinase	"	Human leukemia
ras	GTP binding protein	Membrane	Sarcoma, erythro- leukemia
fos	DNA binding protein	Nuclear	Carcinoma, sarcoma, myelocytoma, Myoblastic leukemia
myc	" " "	"	
myb	" " "	"	
erb A	Glucocorticoid receptor	"	

Armelin, Kelly, Stewart, Leder, Cochran, and Stiles 1984; Muller, Bravo, Burckhardt, and Curran 1984). The effects were both rapid and transient, maximal levels of mRNA being detectable 1–3 hours following exposure to growth factor. From these studies it would appear that c-*myc* expression may be strongly repressed in cells arrested in G_0 phase, but expressed as part of a 'competence' procedure enabling cells to enter G_1. However the reverse may also be true since although little mRNA was detectable in growth-arrested lung fibroblasts the c-*myc* gene was transcribed at a high rate (Blanchard, Piechaczyk, Dani, Chambard, Franchi, Pouyssegur, and Jeanteur 1985). The dramatic increase in mRNA detectable at the G_0/G_1 boundary was due to post-transcriptional changes in the quantity of mRNA with no absolute change in the rate of gene transcription, and control may be at the level of mRNA degradation. Changes in c-*myc* and c-*fos* mRNA were specific responses to the 'competence-inducing' growth factors, and levels were only slightly elevated or unaltered following exposure to EGF or insulin (Muller *et al.* 1984; Blanchard *et al.* 1985).

One of the most common oncogene products found to be expressed in human tumours is the p21 protein of the *ras* genes. A specific point mutation is responsible for converting the cellular proto-oncogene product into a peptide capable of inducing cellular transformation. However micro-injection of high concentrations of the non-transforming, cellular p21 protein into NIH 3T3 cells produced a transformed phenotype, while injections of antibodies against *ras* proteins prevented cells entering S phase of the cell cycle (Mulcahy, Smith, and Stacey 1985). Exposure of cell membranes from NRK fibroblasts to EGF stimulated GTP binding activity and phosphorylation of the p21 proteins (Kamuta and Feramisco 1984). Similar results were obtained with insulin but not with PDGF. Therefore, while changes in c-*myc* and c-*fos* function accompany cell 'competence', and are influenced by PDGF and FGF, the expression of *ras* proteins appears to be important during the G_1 phase for cell 'progression' and is influenced by 'progression' factors such as EGF and the insulin-like molecules.

7   ONCOGENE EXPRESSION DURING EMBRYOLOGY

The biochemical nature of the proteins encoded by oncogenes has led to considerable speculation that the cellular oncogenes contribute to the rapid development occurring in embryogenesis. The expression of the c-*sis* gene and its peptide product, the PDGF B chain, during early development of the placental trophoblast has already been described. This, so far, is the only recognized example of an oncogene-encoded growth factor expressed in human embryogenesis. However, there is now considerable evidence, the majority coming from animal experimentation, to suggest cellular oncogene involvement in growth factor receptor expression and in intracellular regulation of growth in early life.

A survey of oncogene expression in whole mouse embryos from day 7 of gestation until term revealed the presence of detectable mRNA from five cellular oncogenes, these being c-*myc*, c-*erb*, c-Ha-*ras*, c-*src* and c-*sis* (Slamon and Cline 1984). The quantity of mRNA for c-*ras* and c-*myc* remained stable throughout gestation although a transient increase in the latter occurred on day 17. The expression of c-*erb* and c-*src* increased steadily throughout gestation until approximately day 15 before declining, with levels of c-*src* falling to barely detectable levels by term. While c-*sis* was expressed throughout gestation mRNA levels were greatest at day 8 suggesting that PDGF may be present in the mouse embryo from early in development. Whole mouse embryos were also found to contain transcripts of the c-*mos* gene (Propst and Vande Woude 1985).

Subsequent studies have identified oncogene expression within individual fetal tissues. We undertook a detailed analysis of oncogene expression in the human fetus in the early second trimester. At this time only two c-*onc* genes, c-Ha-*ras* and c-N-*ras*, were found to be preferentially expressed in fetal compared to adult tissues (Mellersh, Strain, and Hill 1987). Expression of the *ras* genes was noted in each of twelve separate tissues, the greatest number of transcripts being found in brain and lung. Since the p21 protein product of the *ras* genes is a GTP binding protein thought to act as an intracellular messenger during the mitogenic response of cells to peptide growth factors, its widespread expression during early development may simply reflect a high overall tissue growth rate. However the expression of other cellular oncogenes has been shown to reside in specific embryonal or fetal tissues.

Gene transcripts for c-*fos* were up to one hundred times greater in term human amnion, chorion and placenta than in human fetal fibroblasts (Muller, Tremblay, Adamson, and Verma 1983). The expression of c-*fos* was also detected in amnion from the fetal mouse at 13.5 days gestation and additionally in parietal endoderm (Mason, Murphy, and Hogan 1985). Transfection and expression of exogenous c-*fos* genes into F9 teratocarcinoma cells induced cellular differentiation suggesting that the c-*fos* gene may be active during differentiation (Muller and Wagner 1984). This concept is supported by a rise in c-*fos* expression which accompanies NGF-induced neuronal differentiation in PC12 pheochromocytoma cells (Kruijer, Schubert, and Verma 1985). However, a note of caution is necessary since the induction of differentiation of F9 teratocarcinoma cells into parietal-endoderm-like cells by retinoic acid was not accompanied by any transient rise in c-*fos* RNA expression (Mason *et al.* 1985). Additionally, c-*fos* may be expressed in terminally differentiated cells at different times within the cell cycle to that seen in proliferating tissues (Muller, Curran, Muller, and Guilbert 1985). This suggests that the *fos* proteins may serve separate roles in different cell types. The protein product of the c-*src* gene, p60, was detected by immunocytochemistry in the neural region of the embryonal chick retina (Sorge, Levy, and Maness 1984) and in cultured neurones from the brains of 14 day gestation rat embryos (Brugge, Cotton, Queral, Barrett, Nonner, and Keane 1985). The *src* gene appears to be preferentially active in embryonic neural tissues such as eye and brain in both the chick and human embryo (Cotton and Brugge 1983; Jacobs and Ruebsamen 1983). The greater expression during mid- and late gestation rather than during early embryogenesis in chick brain suggests that the *src* gene may be more closely associated with neural differentiation rather than proliferation.

Particular attention has been paid to the expression of the c-*myc* genes during early development since *myc* expression accompanies PDGF and FGF-dependent cell proliferation in several tissue types (Huang and Huang 1985). The *myc* group of genes comprises c-*myc*, N-*myc* and L-*myc*. While c-*myc* is expressed in many rapidly dividing cells and elevated expression is characteristic of many classes of tumour the N-*myc* gene is restricted to a group of tumours with neural characteristics including human neuroblastoma, and retinoblastoma (Kohl, Gee, and Alt 1984). The three *myc* genes were found to be preferentially expressed in the fetal and neonatal rat brain with no detectable expression in the adult (Zimmerman, Yancopoulos, Collum, Smith, Kohl, Denis, Nau, Witte, Toran-Allerand, Gee, Minna, and Alt 1986). However *myc* expression was not limited to neural tissues since mRNA from each gene was also identified in kidney, intestine, lung, and heart during early life. Low levels of N-*myc* mRNA could be demonstrated in other fetal tissues such as liver, adrenal and spleen. To ascertain whether the N-*myc* expression found in fetal mouse liver could be accounted for by the presence of a differentiating haemopoietic cell population the authors examined gene expression in cell lines representing different stages of B-cell differentiation. A stage-specific expression of N-*myc* was seen.

An expression of N-*myc* in the fetal mouse was also reported by Jakobovits, Schwab, Bishop, and Martin (1985), being strongest between days 7 and 12 of gestation. These authors also investigated N-*myc* expression in a variety of mouse and human teratocarcinoma cell lines. Three tumorigenic cell lines, shown to have common properties with, or to be derived from, the inner cell mass of the 4.5 day mouse embryo (PSA-1, F9 and embryonic stem cells), were found to express N-*myc* abundantly. Additionally N-*myc* mRNA was located in a human teratocarcinoma, whereas a normal mouse embryo fibroblast line and a teratocarcinoma-derived endoderm cell line, PYS, which is not tumorigenic, showed no N-*myc* expression. Confirmation that differentiated endodermal cells contained substantially fewer N-*myc* transcripts than their teratocarcinoma cell progenitors was obtained with the spontaneously differentiating PSA-1 cell line and F9 cells induced to differentiate with retinoic acid. Whether the expression of *myc* genes is directly related to the demonstrated autocrine release of peptide growth factors from teratocarcinoma cells is not known.

A convincing demonstration of c-*myc* activation in human embryonic tissues was provided by a series of three papers by Pfeifer-Ohlsson, Goustin, Rydnert, Wahlstrom, Byersing, Stehelin and Ohlsson (1984), Goustin, Betsholtz, Pfeifer-Ohlsson, Persson, Rydnert, Bywater, Holmgren, Heldin, Westermark, and Ohlsson (1985), and Pfeifer-Ohlsson, Rydnert, Goustin, Larsson, Betsholtz and Ohlsson (1985). Using *in situ* mRNA/cDNA hybridization techniques to demonstrate mRNA transcripts directly in histological sections, a co-expression of the proto-oncogenes c-*sis* and c-*myc* was found in the trophoblastic cells of the first trimester human placenta, particularly the highly proliferative and invasive cytotrophoblastic shell. This suggests that a highly localized autocrine growth stimulation may occur by the production of PDGF and the subsequent induction of trophoblast proliferation involving the expression of c-*myc*. Cultured trophoblastic cells responded to exogenous PDGF with an increase in c-*myc* mRNA transcription and DNA synthesis. The use of *in situ* hybridization enabled these investigators to extend their studies to the human embryo during the first trimester. Transcripts for c-*myc* were identified in particular in several epithelial cell layers including the basal and luminal layers of the intestine, and in the skin with particularly strong hybridization apparent in the mouth. This may correlate with the extensive remodelling of embryonic facial skin which occurs 7 to 8 weeks post-conception in the human embryo. Cells positive for *myc* transcripts were also found in the brain and in epithelial elements of the kidney and lung. These findings suggest that c-*myc* expression is not simply a marker of cell proliferation, since rapidly dividing human embryonic mesenchyme and cartilage cells of the limb buds showed no *in situ* hybridization. Since c-*myc* mRNA levels rise transiently during cell proliferation *in vitro* (Blanchard *et al.* 1985) the finding of high transcript levels in precise cellular structures may indicate a degree of cell synchronization during tissue remodelling.

It seems likely that not only c-*myc* expression, but that of many other cellular oncogenes, may be both anatomically and temporally dependent during embryogenesis. Such expression is under tight developmental control, and from what is already understood of the function of oncogene-encoded proteins most probably reflects multiple aspects of peptide growth factor action during growth and differentiation.

V   Concluding remarks

The information reviewed in this chapter presents considerable evidence that several of the best-characterized peptide growth factors are present, and most probably biologically active, during the growth and differentiation of the embryo and early fetus. Peptide growth factors are detectable in embryonic tissues as early in development as present techniques will allow measurement, and certainly coincident with the appearance of the first differentiated cell types. However, most evidence is so far largely circumstantial since, although there are occasional reports of growth factors administered to the fetus in late gestation, there have been no studies in which peptide growth factors have been introduced, or the actions of endogenous growth factors blocked, in the early pre- or post-implantation embryo. Only by such approaches can the theoretical implications of growth factor presence be directly linked to the morphological observation of pattern formation.

The discovery of CAMs has done much to improve our understanding of the regulatory systems governing cell migration during germ layer formation. However, the tissues must proliferate and differentiate coincident with the physical movement of positional modelling, or in some instances are required to enter programmed cell death. Certain aspects of tissue pattern formation invite obvious speculative links with growth factor expression and action. For instance in the experimental model of the chick limb bud, positional information for limb development along the proximo-distal axis is determined by a progress zone. The information is specified to each cell by the time spent by that cell within the progress zone, and time may be quantified by the number of cell divisions (Wolpert, Tickle, and Sampford 1979). The number of cell cycles undertaken, and the rate of progression of the zone, may be dependent on peptide growth factor presence within that structure. It is also tempting to speculate that the ability of retinoic acid to promote the differentiation of embryonal cells *in vitro* in association with an expression of IGF II may be related to the abilities of retinoic acid gradients to influence digit formation in the chick limb bud (Summerbell 1983). An added dimension to the consideration of peptide growth factors within spatial development is the observation that these peptides bind avidly to elements of extracellular matrix such as collagen (Smith, Singh, Lillquist, Goon, and Stiles 1982). Hence, growth factors may act as fixed, directional signals during tissue modelling in addition to a role as diffusable growth regulators. The extension of a sophisticated growth factor technology to the study of tissue form in addition to tissue function may unify fragmented aspects of our understanding of embryology, and may additionally elucidate mechanisms of teratogenesis.

Acknowledgements

A. J. Strain is supported by the Yorkshire Cancer Research Campaign (YCRC). Studies performed by the authors and their colleagues were supported financially by the British Diabetic Association, the Yorkshire Cancer Research Campaign, Birthright, the Nuffield Foundation, the Smith-Kline Foundation, the Hawley Trust and the Medical Research Council. We are grateful to Dr B. Hogan for helpful discussion and Miss L. Richardson for preparation of the manuscript.

Glossary

GROWTH FACTORS AND THEIR COMMON ABBREVIATIONS
Cell adhesion molecules (CAMs)
Somatomedins — somatomedin-C/insulin-like growth factor I (SM-C/IGF
  I) — insulin-like growth factor II (IGF II)
Epidermal growth factor (EGF)
Platelet-derived growth factor (PDGF)
Fibroblast growth factor (FGF)
Transforming growth factor α (TGF-α), β (TGF-β)
Nerve growth factor (NGF)
Embryonal carcinoma-derived growth factor (ECDGF)

References

Adamson, E. D., Deller, M. J., and Warshaw, J. B. (1981). Functional EFG receptors are present on mouse embryo tissues. *Nature, Lond.* **291**, 656–659
—— Meek, J. (1984). The ontogeny of epidermal growth factor receptors during mouse development. *Devl. Biol.* **103**, 62–70.
Adams, S. O., Nissley, S. P., Greenstein, L. A., Yang, Y. W. H., and Rechler, M. M. (1983). Synthesis of multiplication-stimulating activity (rat insulin-like growth factor II) by rat embryo fibroblasts. *Endocrinology* **112**, 979–987.
—— —— Handwerger, S., and Rechler, M. M. (1983). Development patterns of insulin-like growth factor I and II synthesis and regulation in rat fibroblasts. *Nature, Lond.* **302**, 150–153.
Armelin, H. A., Armelin, M. C. S., Kelly, K., Stewart, T., Leder, P., Cochran, B. H., and Stiles, C. D. (1984). Functional role for c-*myc* in mitogenic responses to platelet-derived growth factor. *Nature, Lond.* **310**, 655–660.
Artzt, K., Dubois, P., Bennet, D., Condamine, H., Babinet, C., and Jacob, F. (1973). Surface antigens common to mouse cleavage embryos and primitive teratocarcinoma cells in culture. *Proc. natl. Acad. Sci. U.S.A.* **70**, 2988–2992.
Ashton, I. K. and Otremski, I. (1986). *In vitro* effect of multiplication stimulating activity (MSA) on human fetal and postnatal cartilage. *Early Hum. Devl.* **13**, 161–167.
—— and Phizackerley, S. (1981). Human fetal cartilage response to plasma somatomedin activity in relation to gestational age. *Calcif. Tissue Int.* **33**, 205–209.
—— and Spencer, E. M. (1983). Effect of partially purified human somatomedin on human fetal and postnatal cartilage *in vitro*. *Early Hum. Devl.* **8**, 135–140.
—— Zapf, J., Einshenk, I., and McKenzie, I. Z. (1985). Insulin like growth factors (IGF) I and II in human foetal plasma and relationship to gestational age and foetal size during mid pregnancy. *Acta endocr. Copenh.* **110**, 558–563.
Assoian, R. K. (1985). Biphasic effects of type β transforming growth factor on epidermal growth factor receptors in NRK fibroblasts. *J. biol. Chem.* **260**, 9613–9617.
Atkinson, P. R. and Bala, R. M. (1981). Partial characterization of a mitogenic factor with somatomedin-like activity produced by cultured WI38 human fibroblasts. *J. cell. Physiol.* **107**, 317–327.
—— and Hollenberg, M. D. (1984). Somatomedin-like activity from cultured embryo-derived cells: Partial characterization and stimulation of production by epidermal growth factor (urogastrone). *Can. J. Biochem. Cell Biol.* **62**, 1335–1342.
—— Hayden, L. J., Bala, R. M., and Hollenberg, M. D. (1984). Production of

somatomedin-like activity by human adult tumour-derived, transformed and normal cell cultures and by cultured rat hepatocytes: effect of culture conditions and of epidermal growth factor (urogastrone). *Can. J. Biochem. Cell. Biol.* **62**, 1343–1350.

Barth, E. M., Korsching, S., and Thoenen, H. (1984). Regulation of nerve growth factor synthesis and release in organ cultures of rat iris. *J. Cell Biol.* **99**, 839–843.

Beckner, S. K., Hattori, S., and Shih, T. Y., (1985). The *ras* oncogene product p21 is not a regulatory component of adenylate cyclase. *Nature, Lond.* **317**, 71–72.

Bedrick, A. D. and Ladda, R. L. (1978). Epidermal growth factor potentiates cortisone-induced cleft palate in the mouse. *Teratology* **17**, 13–18.

Bhaumick, B., Bala, R. M., and Hollenberg, M. D. (1981). Somatomedin receptor of human placenta: solubilization, photolabelling, partial purification and comparison with insulin receptor. *Proc. natn. Acad. Sci. U.S.A.* **78**, 4279–4283.

Birbeck, J. L. (1981). Fetal growth and endocrinology. In *Clinical Paediatric Endocrinology, 1981* (ed. C. G. D. Brook) pp. 3–34. Blackwell Scientific Publications, Oxford.

Bishop, M. J. (1983). Cellular oncogenes and retroviruses. *A. Rev. Biochem.* **52**, 301–354.

Black, I. B. (1982). Stages of neurotransmitter development in autonomic neurons. *Science, N.Y.* **215**, 1198–1204.

Blanchard, J.-M., Piechaczyk, M., Dani, C., Chambard, J.-C., Franchi, A., Pouyssegur, J., and Jeanteur, P. (1985). c-*myc* gene is transcribed at high rate in G_0-arrested fibroblasts and is post-transcriptionally regulated in response to growth factors. *Nature, Lond.* **317**, 443–445.

Blundell, T. L., Bedarkar, S., Rinderknecht, E., and Humbel, R. E. (1978). Insulin-like growth factors: a model for tertiary structure accounting for immunoreactivity and receptor binding. *Proc. natn. Acad. Sci. U.S.A.* **75**, 180–184.

Bower, J. M., Camble, R., Gregory, H., Gerring, E. L., and Willshire, J. R. (1975). The inhibition of gastric acid secretion by epidermal growth factor. *Experientia* **31**, 825–826.

Brackenbury, R., Greenburg, M. E., and Edelman, G. M. (1984). Phenotypic changes and loss of N-CAM-mediated adhesion in transformed embryonic chicken retinal cells. *J. Cell Biol.* **99**, 1944–1954.

Brinsmead, M. W. and Liggins, G. C. (1979). Serum somatomedin activity after hypophysectomy and during parturition in fetal lambs. *Endocrinology* **105**, 297–305.

—— Waters, M. T., and Thorburn, G. D. (1980). Placental lactogen and fetal growth. In *Endocrinology, 1980* (eds. I. A. Cumming, J. R. Funder and F. A. O. Mendelsohn) pp. 457–460. Australian Academy of Science, Canberra.

Brinster, R. L. (1974). The effect of cells transferred into the mouse blastocyst on subsequent development. *J. exp. Med.* **140**, 1049–1056.

Brugge, J. S., Cotton, P. C., Queral, A. E., Barrett, J. N., Nonner, D., and Keane, R. W. (1985). Neurones express high levels of a structurally modified activated form of pp60 c-src. *Nature, Lond.* **316**, 554–557.

Buckley, A., Davidson, J. M., Kamerath, C. D., Wolt, T. B., and Woodward, S. C. (1985). Sustained release of epidermal growth factor accelerates wound repair. *Proc. natn. Acad. Sci. U.S.A.* **82**, 7340–7344.

Canalis, E. (1980). Effect of insulin like growth factor I on DNA and protein synthesis in cultured rat calvaria. *J. clin. Invest.* **66**, 709–719.

Carpenter, G. and Cohen, S. (1979). Epidermal growth factor. *A. Rev. Biochem.* **48**, 193–216.

Catterton, W. Z., Escobedo, M. B., Sexson, W. R., Gray, M. E., Sundell, H. W., and Stahlman, M. T. (1979). Effect of epidermal growth factor on lung maturation in fetal rabbits. *Pediat. Res.* **13**, 104–108.

Centrella, M. and Canalis, E. (1985). Transforming and non-transforming growth factors are present in medium conditioned by fetal rat calvariae. *Proc. natn. Acad. Sci. U.S.A.* **82**, 7335–7339.

Cheek, D. B., Brayton, J. B., and Scott, R. E. (1974). Overnutrition, overgrowth and hormones (with special reference to the infant born of the diabetic mother). In *Advances in experimental medicine and biology*, (eds. A. F. Roche and F. Falkner). pp. 47–72. Plenum Press, London.

Chen, L. T. and Hsu, Y. C. (1982). Development of mouse embryos *in vitro*: Pre-implantation to the limb bud stage. *Science, N.Y.* **218**, 66–68.

Chuong, C. M. and Edelman, G. M. (1984). Alterations in neural cell adhesion molecules during development of different regions of the nervous system. *J. Neurosci.* **4**, 2354–2368.

Clemmons, D. R. (1984). Multiple hormones stimulate the production of somatomedin by cultured human fibroblasts. *J. clin. Endocr. Metab.* **58**, 850–856.

—— (1985). Exposure to platelet-derived growth factor modulates the porcine aortic smooth muscle cell response to somatomedin-C. *Endocrinology* **117**, 77–83.

—— Klibanski, A., Underwood, L. E., McArthur, J. W., Ridgway, E. C., Beitins, I. Z., and Van Wyk, J. J. (1981). Reduction of plasma immunoreactive somatomedin-C during fasting in humans. *J. clin. Endocr. Metab.* **53**, 1247–1250.

—— Underwood, L. E., and Van Wyk, J. J. (1981). Hormonal control of immunoreactive somatomedin production by cultured human fibroblasts. *J. clin. Invest.* **67**, 10–19.

—— and Van Wyk, J. J. (1985). Evidence for a functional role of endogenously produced somatomedin-like peptides in the stimulation of human fibroblast and porcine smooth muscle cell DNA synthesis. *J. clin. Invest.* **75**, 1914–1918.

Cohen, S. (1962). Isolation of a submaxillary gland protein accelerating incisor eruption and eyelid opening in the newborn animal. *J. biol. Chem.* **237**, 1555–1562.

Collins, T., Ginsberg, D., Boss, J. M., Orkin, S. H., and Pober, J. S. (1985). Cultured human endothelial cells express platelet-derived growth factor B chain: cDNA cloning and structural analysis. *Nature, Lond.* **316**, 748–750.

Conover, C. A., Rosenfeld, R. G., and Hintz, R. L. (1986). Hormonal control of the replication of human fetal fibroblasts: role of somatomedin C/insulin-like growth factor I. *J. cell. Physiol.* **128**, 47–54.

—— Dollar, L. A., Rosenfeld, R. G., and Hintz, R. L. (1985). Somatomedin C-binding and action in fibroblasts from aged and progeric subjects. *J. clin. Endocr. Metab.* **60**, 685–691.

Cooper, J. L. and Wharton, W. (1985). Late G_1 amino acid restriction point in human dermal fibroblasts. *J. cell. Physiol.* **124**, 433–438.

Cotton, P. C. and Brugge, J. S. (1983). Neural tissues express high levels of the cellular *src* gene product pp60c-*src*. *Molec. cell. Biol.* **3**, 1157–1162.

Crace, C. J., Hill, D. J., and Milner, R. D. G. (1985). Mitogenic actions of insulin on fetal and neonatal rat cells *in vitro*. *J. Endocr.* **104**, 63–68.

Cunningham, B. A., Hoffman, S., Rutishauser, U., Hemperley, J. J., and Edelman, G. M. (1983). Molecular topography of the neural cell adhesion molecule N-CAM: surface orientation and location of sialic acid-rich and binding regions. *Proc. natn. Acad. Sci. U.S.A.* **80**, 3116–3120.

Darling, T. L. J., Petrides, P. E., and Beguin, P. (1983). The biosynthesis and processing of proteins in the mouse 7S nerve growth factor complex. *Cold Spring Harb. Symp. quant. Biol.* **47**, 427–434.

Daughaday, W. H., Mariz, I. K., and Trevedi, B. (1981). A preferential binding site for insulin-like growth factor II in human and rat placental membranes. *J. clin. Endocr. Metab.* **53**, 282–288.

De Pablo, F., Roth, J., Hernandez, E., and Pruss, R. M. (1982). Insulin is present in chick eggs and early chick embryos. *Endocrinology* **111**, 1909–1916.

De Prins, F. A., Hill, D. J., Fekete, M., Robsen, D. J., Fieller, N. R. J., Van Assche, F. A., and Milner, R. D. G. (1984). Reduced plasma somatomedin activity and costal cartilage sulfate incorporation activity during experimental growth retardation in the rat. *Pediatr. Res.* **18**, 1100–1104.

D'Ercole, A. J., Applewhite, G. T., and Underwood, L. E. (1980). Evidence that somatomedin is synthesized by multiple tissues in the fetus. *Devl. Biol.* **75**, 315–328.

—— Drop, S. L. S., and Kortleve, D. J. (1985). Somatomedin-C/insulin-like growth factor I-binding proteins in human amniotic fluid and in fetal and postnatal blood: evidence of immunological homology. *J. clin. Endocr. Metab.* **61**, 612–617.

—— Foushee, D. B., and Underwood, L. E. (1976). Somatomedin-C receptor ontogeny and levels in porcine, fetal and human cord serum. *J. clin. Endocr. Metab.* **43**, 1069–1077.

—— Hill, D. J., Strain, A. J., and Underwood, L. E. (1986). Tissue and plasma somatomedin-C/insulin-like growth factor I (SM-C/IGF I) concentrations in the human fetus during the first half of gestation. *Pediatr. Res.* **20**, 253–255.

—— Stiles, A. D., and Underwood, L. E. (1984). Tissue concentrations of somatomedin-C: further evidence for multiple sites of synthesis and para-crine/autocrine mechanisms of action. *Proc. natn. Acad. Sci. U.S.A.* **81**, 935–939.

—— Wilson, D. F., and Underwood, L. E. (1980). Changes in the circulating form of serum somatomedin-C during fetal life. *J. clin. Endocr. Metab.* **51**, 674–676.

Derynck, R., Jarrett, J. A., Chen, E. Y., Eaton, D. H., Bell, J. R., Assoian, R. K., Roberts, A. B., Sporn, M. B., and Goeddel, D. V. (1985). Human transforming growth factor-β complementary DNA sequence and expression in normal and transformed cells. *Nature, Lond.* **316**, 701–705.

Deuel, T. F. and Huang, J. S. (1984). Platelet-derived growth factor. Structure, function, and roles in normal and transformed cells. *J. clin. Invest.* **74**, 669–676.

—— Senior, R. M., Huang, J. S., and Griffin, G. L. (1982). Chemotaxis of monocytes and neutrophils to platelet-derived growth factor. *J. clin. Invest.* **69**, 1046–1049.

De Vroede, M. A., Romanus, J. A., Standaert, M. L., Pollet, R. J., Nissley, S. P., and Rechler, M. M. (1984). Interaction of insulin-like growth factors with a non-fusing mouse muscle cell line: binding, actions, and receptor down-regulation. *Endocrinology* **114**, 1917–1929.

Di Corleto, P. E. and Bowen-Pope, D. F. (1983). Cultured endothelial cells produce a platelet-derived growth factor-like protein. *Proc. natn. Acad. Sci. U.S.A.* **80**, 1919–1923.

Dobbing, J. and Sands, J. (1973). Quantitative growth and development of human brain. *Archs. Dis. Child.* **48**, 757–767.

Doolittle, R. F., Hunkapillar, M. W., Hood, L. E., Devare, S. G., Robbins, K., Aaronson, S. A., and Antonaides, H. N. (1983). Simian sarcoma virus onc gene, v-*sis*, is derived from the gene (or genes) encoding a platelet-derived growth factor. *Science, N.Y.* **221**, 275–277.

Doupe, A. J., Landis, S. C., and Patterson, P. H. (1985). Environmental influences on the development of neural crest derivatives: glucocorticoids, growth factors and chromaffin cell plasticity. *J. Neurosci.* **5**, 2119–2142.

Downward, J., Yarden, Y., Mayes, E., Scrace, G., Totty, N., Stockwell, P., Ullrich, A., Schlessinger, J., and Waterfield, M. D. (1984). Close similarity of epidermal growth factor receptor and v-erb B oncogene protein sequences. *Nature, Lond.* **307**, 521–527.

Dubowitz, V. (1967). In *Exploratory concepts in Muscular Dystrophy* (ed. A. T. Milka) pp. 164–182. Excerpta Medica Found, New York.

Edelman, G. M. (1985). Cell adhesion and the molecular processes of morphogenesis. *A. Rev. Biochem.* **54**, 135–169.

—— and Chuong, C. M. (1982). Embryonic to adult conversion of neural cell adhesion molecules in normal and staggerer mice. *Proc. natn. Acad. Sci. U.S.A.* **79**, 7036–7040.

—— Gallin, W. J., Delouvee, A., Cunningham, B. A., and Thiery, J. P. (1983). Early epochal maps of two different cell adhesion molecules. *Proc. natn. Acad. Sci. U.S.A.* **80**, 4384–4388.

Elstow, S. F., Hill, D. J., Strain, A. J., Swenne, I., Crace, C. J., and Milner, R. D. G. (1985). Production and partial purification of TGF-β-like activity from early passage human foetal fibroblasts. *J. Endocr.* **107**, Suppl. Abstr. 100.

Esch, F., Baird, A., Ling, N., Veno, N., Hill, F., Denoroy, L., Klepper, R., Gospodarowicz, D., Bohlen, P., and Guillemin, R. (1985). Primary structure of bovine pituitary basic fibroblast growth factor (FGF) and comparison with the amino-terminal sequence of bovine brain acidic FGF. *Proc. natn. Acad. Sci. U.S.A.* **82**, 6507–6511.

Ewton, D. Z., Erwin, B. G., Pegg, A. E., and Florini, J. R. (1984). The role of polyamines in somatomedin-stimulated differentiation of L6 myoblasts. *J. cell. Physiol.* **120**, 263–270.

—— and Florini, J. R. (1980). Relative effects of the somatomedins, multiplication-stimulating activity, and growth hormone on myoblasts and myotubes in culture. *Endocrinology* **106**, 577–583.

—— —— (1981). Effects of somatomedins and insulin on myoblast differentiation *in vitro. Devel. Biol.* **56**, 31–39.

Fant, M. E., Munro, H. N., and Moses, A. C. (1986). An autocrine/paracrine role for insulin-like growth factors in the regulation of human placental growth. *J. clin. Endocr. Metab.* **63**, 499–505.

Florini, J. R., Ewton, D. Z., Falen, S. L., and Van Wyk, J. J. (1985). Stimulation of differentiation by epidermal growth factor and somatomedins. In *Somatomedins and other peptide growth factors: Relevance to Paediatrics.* (Eds. R. L. Hintz and L. E. Underwood). pp. 102–107. Report of the 89th Ross Conference on Paediatric Research. Ross Laboratories, Columbus, Ohio.

Fowden, A. L. and Comline, L. S. (1984). The effects of pancreatectomy on the sheep fetus *in utero. J. exp. Physiol.* **69**, 319–330.

Frackleton, A. R., Tremble, P. M., and Williams, L. T. (1984). Evidence for the platelet-derived growth factor-stimulated tyrosine phosphorylation of the platelet-derived growth factor receptor *in vivo J. biol. Chem.* **259**, 7909–7915.

Franklin, R. C., Pepperell, R. J., Rennie, G. C., and Cameron, D. P. (1979). Acid-ethanol-extractable non-suppressible insulin-like activity (NSILA-S) during pregnancy and the puerperium, and in cord serum at term. *J. clin. Endocr. Metab.* **48**, 695–699.

Freemark, M. and Handwerger, S. (1983). Ovine placental lactogen, but not growth hormone, stimulates amino acid transport in fetal rat diaphragm. *Endocrinology* **112**, 402–404.

—— —— (1984). Ovine placental lactogen stimulates glycogen synthesis in fetal rat hepatocytes. *Am. J. Physiol.* **246**, E21–25.

—— —— (1985). Ovine placental lactogen inhibits glucagon-induced glycogenolysis in fetal rat hepatocytes. *Endocrinology* **116**, 1275–1280.

—— —— (1986). The glycogenic effects of placental lactogen and growth hormone in ovine fetal liver are mediated through binding to specific fetal ovine placental lactogen receptors. *Endocrinology* **118**, 613–618.

Frolik, C. A., Dart, L. L., Meyers, C. A., Smiths, D. M., and Sporn, M. B. (1983). Purification and initial characterization of a type-α transforming growth factor from human placenta. *Proc. natn. Acad. Sci. U.S.A.* **80**, 3676–3680.

Furukawa, Y., Furukawa, S., Satoyoshi, E., and Hyashi, K. (1984). Nerve growth factor secreted by mouse heart cells in culture. *J. biol. Chem.* **259**, 1259–1264.

Gluckman, P. D. (1986). The role of pituitary hormones, growth factors and insulin in the regulation of fetal growth. In *Oxford Reviews of Reproductive Biology* vol. 8 (ed. J. R. Clarke) pp. 1–60. Clarendon Press, Oxford.

—— Butler, J. H., and Elliot, T. (1983). The ontogeny of somatotropic binding sites in ovine hepatic membranes. *Endocrinology* **112**, 1607–1612.

—— Fowden, A. L., Butler, J. H., and Comline, L. S. (1985). Insulin-like growth factor-1 and -2 concentrations in the pancreatectomized ovine fetus – Evidence for a regulatory influence by nutritional substrates. In *Abst. 67th Ann. Meeting Endocr. Soc.* Baltimore, MD, p. 16. Abstr. 64.

—— and Liggins, G. C. (1984). Regulation of fetal growth. In *Fetal physiology and medicine* (eds. R. W. Beard and P. W. Nathanielsz). pp 511–557. Marcel Dekker, New York.

Goldstein, L. D., Reynolds, C. P., and Perez-Polo, J. R. (1978). Isolation of human nerve growth factor from placental tissue. *Neurochem. Res.* **3**, 175–183.

Goodner, C. J. and Freinkel, N. (1961). Carbohydrate metabolism in pregnancy. IV. Studies on the permeability of the rat placenta to I^{131} insulin. *Diabetes* **10**, 383–392.

Gospodarowicz, D. (1981). Epidermal and nerve growth factors in mammalian development. *A. Rev. Physiol.* **43**, 251–263.

—— (1984). Biological activity *in vivo* and *in vitro* of pituitary and brain fibroblast growth factor. In *Mediators in cell growth and differentiation* (ed. W. J. Pagel). pp. 109–134. Raven Press, New York.

—— Bialecki, H., and Thakral, T. (1979). The angiogenic activity of the fibroblast and epidermal growth factor. *Exp. Eye Res.* **28**, 501–514.

—— Cheng, J., Lui, G-M., Baird, A., and Bohlent, P. (1984). Isolation of brain fibroblast growth factor by heparin-sepharose affinity chromatography: identity with pituitary fibroblast growth factor. *Proc. natn. Acad. Sci. U.S.A.* **81**, 6963–6967.

—— and Mescher, A. L. (1977). A comparison of the responses of cultured myoblasts and chondrocytes to fibroblast and epidermal growth factors. *J. cell. Physiol.* **93**, 117–128.

—— Weseman, J., Moran, J. S., and Lindstrom, J. (1976). Effect of fibroblast growth factor on the division and fusion of bovine myoblasts. *J. Cell. Biol.* **70**, 395–405.

Goustin, A. S., Betsholtz, C., Pfeifer-Ohlsson, S., Persson, H., Rydnert, J., Bywater, M., Holmgren, G., Heldin, C-H., Westermark, B., and Ohlsson, R. (1985). Co-expression of the *sis* and *myc* proto-oncogenes in developing human placenta suggests autocrine control of trophoblast growth. *Cell* **41**, 301–312.

Graham, C. F. (1977). Teratocarcinoma cells and normal mouse embryogenesis. In *Concepts in mammalian embryogenesis* (ed. M. I. Sherman) pp. 315–394. The MIT Press, Cambridge, MA.

Greenberg, M. E., Brackenbury, R., and Edelman, G. M. (1984). Alteration of neural cell adhesion molecule (N-CAM) expression after neuronal cell transformation by Rous sarcoma virus. *Proc. natn. Acad. Sci. U.S.A.* **81**, 969–973.

Gregory, H. (1975). Isolation and structure of urogastrone and its relationship to epidermal growth factor. *Nature, Lond.* **257**, 325–327.

Grove, R. I. and Pratt, R. M. (1984). Influence of epidermal growth factor and cyclic AMP on growth and differentiation of palatal epithelial cells in culture. *Devl. Biol.* **106**, 427–437.

Grumbach, M. M. and Kaplan, S. L. (1973). Ontogenesis of growth hormone, insulin, prolactin and gonadotropin secretion in the human fetus. In *Foetal and neonatal physiology* (eds. R. S. Comline, K. W. Cross, G. S. Dawes and P. W. Nathanielsz) pp. 462–487. Cambridge University Press, London.

Grumet, M., Hoffman, S., and Edelman, G. M. (1984). Two antigenically related neuronal cell adhesion molecules of different specificities mediate neuron-neuron and neuron-glia adhesion. *Proc. natn. Acad. Sci. U.S.A.* **81**, 267–271.

Guenther, H. L., Guenther, H. E., Froesch, E. R., and Fleisch, H. (1982). Effect of insulin-like growth factor on collagen and glycosaminoglycan synthesis by rabbit articular chondrocytes in culture. *Experimentia* **38**, 979–980.

Gunderson, R. W. and Barrett, J. W. (1980). Characterization of the turning response of dorsal root neurites towards nerve growth factor. *J. Cell Biol.* **87**, 546–554.

Gunzler, W. A., Steffens, G. J., Otting, F., Klim, S. M. A., Frankus, E., and Flohe, L. (1982). The primary structure of high molecular mass urokinase from human urine. *Hoppe-Seyler's Z. Physiol. Chem.* **363**, 1155–1165.

Hamilton, W. J., Boyd, J. D., and Mossman, H. W. (1962). *Human Embryology.* Heffer, Cambridge.

Han, V. K. R., D'Ercole, A. J., Lauder, J. M., and Svoboda, M. E. (1985). Receptor binding of somatomedin/insulin-like growth factors (SM/IGFs) on cultured rat astroglial cells. *Pediat. Res.* **19**, p. 620, Abstr. 100.

—— Hill, D. J., D'Ercole, A. J., Strain, A. J., Askin, F., Lauder, J. M., and Underwood, L. E. (1986). Immuno-localization of somatomedins/insulin-like growth factors in human fetal tissues. *J. Endocr.* **108**, *Suppl.* Abstr. 217.

Harper, G. P., Glanville, R. W., and Thoenen, H. (1982). The purification of nerve growth factor from bovine seminal plasma. *J. biol. Chem.* **257**, 8541–8548.

Hassell, J. R. and Pratt, R. M. (1977). Elevated levels of cAMP alters the effect of epidermal growth factor *in vitro* on programmed cell death in the secondary palatal epithelium. *Exp. Cell Res.* **106**, 55–62.

Heath, J. K., Bell, S., and Rees, A. R. (1981). Appearance of functional insulin receptors during the differentiation of embryonal carcinoma cells. *J. Cell Biol.* **91**, 293–297.

—— and Deller, M. J. (1983). Serum-free culture of PC13 murine embryonal carcinoma cells. *J. cell. Physiol.* **115**, 225–230.

—— and Isacke, C. M. (1984). PC13 Embryonal carcinoma derived growth factor. *EMBO J.* **3**, 2957–2962.

—— and Rees, A. R. (1985). Growth factors in mammalian embryogenesis. In *Growth factors in biology and medicine* (ed. D. Evered). pp. 3–22. Ciba Fdn. Symp. No. 116, Longmans, London.

—— and Shi, W-k. (1986). Developmentally regulated expression of insulin-like growth factors by differentiated murine teratocarcinomas and extra-embryonic mesoderm. *J. Embryol. exp. Morph.* **95**, 193–212.

Hendricks, A., De Pablo, F., and Roth, J. (1984). Early developmental and tissue-specific patterns of insulin binding in chick embryo. *Endocrinology* **115**, 1315–1323.

Herzberg, V. L., Boughter, J. M., Carlisle, S. K., Ahmad, F., and Hill, D. S. (1980). ^{125}I-insulin receptor binding to cord blood erythrocytes of varying gestational age and comparison with adult values. *Pediat. Res.* **14**, 4–7.

Hill, D. J., Crace, C. J., Fowler, L., Holder, A. T., and Milner, R. D. G. (1984). Cultured fetal rat myoblasts release peptide growth factors which are immunologically and biologically similar to somatomedin. *J. cell. Physiol.* **119**, 349–358.

—— —— and Milner, R. D. G. (1985b). Incorporation of [^{3}H] thymidine by isolated human fetal myoblasts and fibroblasts in response to human placental lactogen (HPL): possible mediation of HPL action by release of immunoreactive SM-C. *J. cell. Physiol.* **125**, 337–344.

—— —— Nissley, S. P., Morrell, D., Holder, A. T., and Milner, R. D. G. (1985a). Fetal rat myoblasts release both rat somatomedin-C (SM-C)/insulin-like growth factor I (IGF I) and multiplication-stimulating activity *in vitro*: partial characterization and biological activity of myoblast-derived SM-C/IGF I. *Endocrinology* **117**, 2061–2072.

—— —— Strain, A. J., and Milner, R. D. G. (1986a). Regulation of amino acid uptake and DNA synthesis in isolated human fetal fibroblasts and myoblasts: effect of human placental lactogen, somatomedin-C, multiplication-stimulating activity and insulin. *J. clin. Endocr. Metab.* **62**, 753–760.

—— Davidson, P., and Milner, R. D. G. (1979). Retention of plasma somatomedin activity in the foetal rabbit following decapitation *in utero*. *J. Endocr.* **81**, 93–102.

—— Frazer, A., Swenne, I., Wirdnam, P. K., and Milner, R. D. G. (1987). Somatomedin-C in the human fetal pancreas: cellular localization and release during organ culture. *Diabetes*. In Press.

—— Holder, A. T., Seid, J., Preece, M. A., Tomlinson, S., and Milner, R. D. G. (1983). Increased thymidine incorporation into fetal rat cartilage *in vitro* in the presence of human somatomedin, epidermal growth factor and other growth factors. *J. Endocr.* **96**, 489–497.

—— Milner, R. D. G. (1985a). The role of peptide growth factors and hormones in the control of fetal growth, In *Recent advances in perinatal medicine* (ed. M. L. Chiswick). pp. 79–102. Churchill Livingstone, London.

—— —— (1985b). Insulin as a growth factor. *Pediat. Res.* **19**, 879–886.

—— Strain, A. J., Elstow, S. F., Swenne, I., and Milner, R. D. G. (1986). Bi-functional action of transforming growth factor-β on DNA synthesis in early passage human fetal fibroblasts. *J. cell. Physiol.* **128**, 322–328.

Hill, D. J., Strain, A. J., and Milner, R. D. G. (1986), Presence of transforming growth factor-β-like activity in multiple fetal rat tissues. *Cell Biol. internat. Rep.* **10**, 915–922.

Hoffman, S. and Edelman, G. M. (1983). Kinetics of homophilic binding by embryonic and adult forms of the neural cell adhesion molecule. *Proc. natn. Acad. Sci. U.S.A.* **80**, 5762–5766.

Hsu, Y. C. (1980). Embryo growth and differentiation factors in embryonic sera of animals. *Devl. Biol.* **78**, 465–474.

—— and Basker, J. (1974). Differentiation *in vitro* of normal mouse embryos and mouse embryonal carcinoma. *J. natn. Cancer Inst.* **53**, 177–185.

Huang, J. S. and Huang, S. S. (1985). Role of growth factors in oncogenesis: growth factor-proto-oncogene pathways of mitogenesis. In *Growth factors in biology and medicine* (ed. D. Evered) pp. 46–65. Ciba Fdn. Symp. No. 116, Longmans, London.

Jacobovits, A., Schwab, M., Bishop, J. M., and Martin, G. R. (1985). Expression of N-*myc* in teratocarcinoma stem cells and mouse embryos. *Nature, Lond* **318**, 188–191.

Jacobs, S. and Cuatrecasas, P. (1982). Insulin receptors and insulin receptor antibodies: structure-function relationships. In *Receptors, antibodies and disease* pp. 82–90. Ciba Fdn. Symp. No. 90, Pitman, London.

Jacobs, C. and Ruebsamen, H. (1983). Expression of pp60c-*src* protein kinase in adult and fetal human tissue: high activities in some sarcomas and mammary carcinomas. *Cancer Res.* **43**, 1696–1702.

Johnsson, A., Heldin, C-H., Wasteson, A., Westermark, B., Deuel, T. F., Huang, J. S., Seeburg, P. H., Gray, A., Ullrich, A., Scrace, G., Stroobant, P., and

Waterfield, M. D. (1984). The c-*sis* gene encodes a precursor of the B chain of platelet-derived growth factor. *EMBO J*. **3**, 921–928.

Kamuta, T. and Fermaisco. J. R. (1984). Epidermal growth factor stimulates guanine nucleotide binding activity and phosphorylation of *ras* oncogene products. *Nature, Lond*. **310**, 147–150.

Kaplowitz, P. B., D'Ercole, A. J., and Underwood, L. E. (1982). Stimulation of embryonic mouse limb bud mesenchymal cell growth by peptide growth factors. *J. cell. Physiol*. **112**, 353–359.

—— —— —— and Van Wyk, J. J. (1984). Stimulation by somatomedin-C and of aminoisobutyric acid uptake in human fibroblast: a possible test for cellular responsiveness to somatomedins. *J. clin. Endocr. Metab*. **58**, 176–181.

Kasselberg, A. G., Orth, D. N., Gray, M. E., and Stahlman, M. T. (1985). Immunocytochemical localization of human epidermal growth factor/urogastrone in several human tissues. *J. Histochem. Cytochem*. **33**, 315–322.

Kasuga, M., Karlsson, F. A., and Kahn, C. R. (1982). Insulin stimulates the phosphorylation of the 95,000 dalton subunit of its own receptor. *Science, N.Y.* **215**, 185–187.

Kelly, K., Cochran, B. H., Stiles, C. D., and Leder, P. (1983). Cell-specific regulation of the c-*myc* gene by lymphocyte mitogens and platelet-derived growth factor. *Cell* **35**, 603–610.

Kohl, N. E., Gee, C. E., and Alt, F. W. (1984). Activated expression of N-*myc* gene in human neuroblastomas and related tumours. *Science, N.Y.* **226**, 1335–1336.

Kruijer, W., Schubert, D., and Verma, I. M. (1985). Induction of the proto-oncogene *fos* by nerve growth factor. *Proc. natn. Acad. Sci. U.S.A.* **82**, 7330–7334.

Kurtz, A., Jelkmann, W., and Bauer, C. (1982). A new candidate for the regulation of erythropoiesis. *FEBS Let*. **149**, 105–108.

Laron, Z., Mannheimer, S., and Guttman, S. (1966). Lack of transplacental passage of growth hormone in rabbit. *Experimentia* **22**, 831–832.

Lawrence, D. A., Pircher, R., Krycere-Martinerie, C., and Jullien, P. (1984). Normal embryo fibroblasts release transforming growth factors in a latent form. *J. cell. Physiol*. **121**, 184–188.

Lei, W. H. and Guyda, H. J. (1984). Characterization of epidermal growth factor receptors in human placental cell cultures. *J. clin. Endocr. Metab*. **58**, 344–357.

Levi-Montalcini, R. (1976). The nerve growth factor: its role in growth differentiation and function of the sympathetic adrenergic neuron. *Prog. Brain Res*. **45**, 235–238.

Lewis, J. H. (1977). Growth and determination in the developing limb. In *Vertebrate limb and somite morphogenesis* (eds. D. A. Ede, J. R. Hinchliffe and M. Balls) pp. 215–228. Cambridge University Press, London.

Lillien, L. E. and Claude, P. (1985). Nerve growth factor is a mitogen for cultured chromaffin cells. *Nature, Lond*. **317**, 632–634.

Linkhardt, T. A., Clegg, C. H., and Hauscha, S. D. (1981). Myogenic differentiation in permanent clonal mouse myoblast cell lines: regulation by macromolecular growth factors in the culture medium. *Devl. Biol*. **86**, 19–30.

Logsdon, C. D. and Williams, J. A. (1983). Epidermal growth factor binding and biologic effects on mouse pancreatic acini. *Gastroenterology* **85**, 339–345.

Lundgren, S., Ronne, H., Rask, L., and Peterson, P. A. (1984). Sequence of an epidermal growth factor-binding protein. *J. biol. Chem*. **259**, 7780–7784.

Maes, M., De Hertogh, R., Watrin-Granger, P., and Ketelslegers, J. M. (1983). Ontogeny of liver somatotropic and lactogenic binding sites in male and female rats. *Endocrinology* **13**, 1325–1332.

Mains, R. E. and Patterson, P. H. (1973). Primary cultures of dissociated sympathetic neurones. 1. Establishment of long-term growth in culture and studies on differentiated properties. *J. Cell Biol*. **59**, 329–345.

Marek, J., Schullerova, M., Schveiberova, O., and Limanova, Z. (1981). Effect of thyroid

448   D. J. Hill, A. J. Strain and R. D. G. Milner

function on serum somatomedin activity. *Acta endocr. Copenh.* **96**, 491–497.

Martin, G. R. (1975). Teratocarcinomas as a model system for the study of embryogenesis and neoplasia. *Cell* **5**, 229–243.

—— (1980). Teratocarcinomas and mammalian embryogenesis. *Science, N.Y.* **209**, 768–776.

Mason, I., Murphy, D., and Hogan, L. M. (1985). Expression of c-*fos* in parietal endoderm, amnion and differentiating F9 teratocarcinoma cells. *Differentiation* **30**, 76–81.

Massague, J. (1985). Transforming growth factor-β modulates the high affinity receptors for epidermal growth factor and transforming growth factor α. *J. Cell Biol.* **100**, 1508–1514.

—— and Czech, M. P. (1982). The subunit structures of two distinct receptors for insulin-like growth factors I and II and their relationship to the insulin receptor. *J. biol. Chem.* **257**, 5038–5045.

—— Kelly, B., and Mottola, C. (1985). Stimulation by insulin-like growth factors is required for cellular transformation by type β transforming growth factor. *J. biol. Chem.* **260**, 4551–4554.

Matrisian, L. M., Pathak, M., and Magun, B. E. (1982). Identification of an epidermal growth factor-related transforming growth factor from rat fetuses. *Biochem. Biophys. Res. Commun.* **107**, 761–769.

McMorris, F. A., Smith, T. M., De Salvo, S., and Furlanetto, R. W. (1986). Insulin-like growth factor I/somatomedin C: a potent inducer of oligodendrocyte development. *Proc. natn. Acad. Sci. U.S.A.* **83**, 822–826.

Mellersh, H., Strain, A. J., and Hill, D. J. (1986).Expression of the protooncogenes C-H-*ras* and N-*ras* in early second trimester human fetal tissues. *Biochem. Biophys. Res. Commun.* **141**, 510–516.

Mill, J. F., Chao, M. V., and Ishii, D. N. (1985). Insulin, insulin-like growth factor II, and nerve growth factor effects on tubulin mRNA levels and neurite formation. *Proc. natn. Acad. Sci. U.S.A.* **82**, 7126–7130.

Miller, L. L., Schalch, D. S., and Draznin, B. (1981). Role of the liver in regulating somatomedin activity: effects of streptozotocin diabetes and starvation on the synthesis and release of insulin-like growth factor and its carrier protein by the isolated perfused rat liver. *Endocrinology* **108**, 1265–1271.

Miller, R., Slamon, D. J., Tremblay, J. M., Cline, M. J., and Verma, I. M. (1982). Differential expression of cellular oncogenes during pre- and postnatal development of the mouse. *Nature, Lond.* **299**, 640–645.

Mills, N. C., Underwood, L. E., and D'Ercole, A. J. (1985). Expression of insulin-like growth factors (somatomedins) in the human placenta. In *Abst. 67th Ann. Meeting Endocr. Soc.* Baltimore, MD, p. 268, Abstr. 1072.

Milner, R. D. G. (1974). Growth and development of the endocrine glands – endocrine pancreas. In *Scientific foundation of paediatrics* (Eds. J. A. Davies and W. Dobbing) pp. 507–513. W. Heinemann, London.

—— (1986). Determinants of growth in utero. In *The study of linear growth in third world children*. Proc. 13th Nestle Nutrition Workshop, Petchburi, Thailand. In press.

Mintz, B. and Cronmiller, C. (1978). Normal blood cells of anemic genotype in teratocarcinoma-derived mosaic mice. *Proc. natn. Acad. Sci. U.S.A.* **75**, 6247–6251.

Mottola, C. and Czech, M. P. (1984). The type II insulin-like growth factor receptor does not mediate increased DNA synthesis in H-35 hepatoma cells. *J. biol. Chem.* **259**, 12705–12713.

Muglia, L. and Locker, J. (1984). Extrapancreatic insulin gene expression in the fetal rat. *Proc. natn. Acad. Sci. U.S.A.* **81**, 3635–3639.

Mulcahy, L. S., Smith, M. R., and Stacey, D. W. (1985). Requirement for *ras* proto-oncogene function during serum-stimulated growth of NIH 3T3 cells.

Nature, Lond. **313**, 241–243.

Muller, R., Bravo, R., Burckhardt, J., and Curran T. (1984). Induction of c-*fos* gene and protein by growth factors precedes activation of c-*myc*. *Nature, Lond.* **312**, 716–721.

—— Curran, R., Muller, D., and Guilbert, L. (1985). Inducation of c-*fos* during myelomonocytic differentiation and macrophage proliferation. *Nature, Lond.* **314**, 546–548.

—— Tremblay, J. M., Adamson, E. D., and Verma, I. M. (1983). Tissue and cell type-specific expression of two human c-*onc* genes. *Nature, Lond.* **304**, 454–456.

—— and Wagner, E. F. (1984). Differentiation of F9 teratocarcinoma stem cells after transfer of c-*fos* proto-oncogenes. *Nature, Lond.* **311**, 438–442.

Nagarajan, L. and Anderson, W. B. (1982). Insulin promotes the growth of F9 embryonal carcinoma cells apparently by acting through its own receptor. *Biochem. Biophys. Res. Commun.* **106**, 974–980.

—— —— Nissley, S. P., Rechler, M. M., and Jetten, A. M. (1985). Production of insulin-like growth factor II (MSA) by endoderm-like cells derived from embryonal carcinoma cells: possible mediator of embryonic cell growth. *J. cell. Physiol.* **124**, 199–206.

—— Jetten, A. M., and Anderson, W. B. (1983). A new differentiated cell line (Dif 5) derived by retinoic acid treatment of F9 teratocarcinoma cells capable of extracellular matrix production and growth in the absence of serum. *Expl. Cell Res.* **147**, 315–327.

—— Nissley, S. P., Rechler, M. M., and Anderson, W. B. (1982). Multiplication-stimulating activity stimulates the multiplication of F9 embryonal carcinoma cells. *Endocrinology* **110**, 1231–1237.

Neufeld, N. D., Scott, M., and Kaplan, S. A. (1980). Ontogeny of the mammalian insulin receptor. *Devl. Biol.* **78**, 151–160.

Nexo, E., Hollenberg, M. D., Figueroa, A., and Pratt, R. M. (1980). Detection of epidermal growth factor-urogastrone and its receptor in fetal mouse development. *Proc. natn. Acad. Sci. U.S.A.* **77**, 2782–2785.

Norrgren, G., Ebendel, T., and Wikstrom, H. (1984). Production of nerve growth-stimulating factor(s) from chick embryo heart cells. *Expl. Cell Res.* **152**, 427–435.

Ny, T., Elgh, F., and Lund, B. (1984). The structure of the human tissue-type plasminogen activator gene: correlation of intron and exon structures to functional and structural domains. *Proc. natn. Acad. Sci. U.S.A.* **81**, 5355–5359.

Obershain, S. S., Adam, P. A. J., King, K. C., Teramo, K., Raivio, K. O., Raiha, N., and Schwartz, R. (1970). Human fetal insulin response to sustained maternal hyperglycaemia. *New Engl. J. Med.* **283**, 566–570.

Obrink, B. (1986). Epithelial cell adhesion molecules. *Expl. Cell Res.* **163**, 1-21.

Ogou, S. I., Yoshida-Noro, C., and Takeichi, M. (1983). Calcium-dependent cell-cell adhesion molecules common to hepatocytes and teratocarcinoma stem cells. *J. Cell Biol.* **97**, 944–948.

O'Keefe, E., Hollenberg, M. D., and Cuatrecasas, P. (1974). Epidermal growth factor: characteristics of specific binding in membranes from liver, placenta, and other tissues. *Archs. Biochem. Biophys.* **164**, 518–527.

—— and Pledger, W. J. (1983). A model of cell cycle control: sequential events regulated by growth factors. *Molec. Cell. Endocr.* **31**, 167–186.

Papaioannou, V. E., McBurney, M., Gardner, R. L., and Evans, M. J. (1975). Fate of teratocarcinoma cells injected into mouse blastocysts. *Nature, Lond.* **258**, 70–73.

Parkes, M. J. and Hill, D. J. (1985). Lack of growth hormone-dependent somatomedins on growth retardation in hypophysectomized fetal lambs. *J. Endocr.* **104**, 193–199.

Pederson, J. (1977). *The pregnant diabetic and her newborn.* 2nd edition. Williams and Wilkins, Baltimore, MD.

Pfeifer-Ohlsson, S., Goustin, A. S., Rydnert, J., Wahlstrom, T., Byersing, L., Stehelin, D., and Ohlsson, R. (1984). Spatial and temporal pattern of cellular *myc* oncogene expression in developing human placenta: implications for embryonic cell proliferation. *Cell* **38**, 585–596.

—— Rydnert, J., Goustin, A. S., Larsson, E., Betsholtz, C., and Ohlsson, R. (1985). Cell-type-specific pattern of *myc* proto-oncogene expression in developing human embryos. *Proc. natn. Acad. Sci. U.S.A.* **82**, 5050–5054.

Pfeifle, B. and Ditschuneit, H. (1984). Insulin-like growth factor I (IGF I): regulation by platelet-derived growth factor (PDGF), epidermal growth factor (EGF) and fibroblast growth factor (FGF). In *Abst. 7th Int. Cong. Endocr.*, Quebec, Canada, p. 1139. Anstr. 1758.

Phillips, L. S. and Vassilopoulou-Sellin, R. (1980). Somatomedins. *New Engl. J. Med.* **302**, 371–380.

—— and Young, H. S. (1976). Nutrition and somatomedin I. Effect of fasting and refeeding on serum somatomedin activity and cartilage growth activity in rats. *Endocrinology* **99**, 304–314.

Posner, B. E. (1974). Insulin receptors in human and animal placental tissue. *Diabetes* **23**, 209–217.

Potau, N., Riudor, E., and Ballabriga, A. (1981). Insulin receptors in human brain in relation to gestational age. In *Current views on insulin receptors* (ed. D. Andreain). Serono Symp. 41, pp. Academic Press, London.

Pratt, R. M., Figueroa, A. A., Nexo, E., and Hollenberg, M. D. (1978). Involvement of epidermal growth factor during secondary palatal development. *J. Cell Biol.* **79**, 24a. Abstr. CD128.

—— Kim, C. S., and Grove, R. I. (1984). Role of glucocorticoids and epidermal growth factor in normal and abnormal palatal development. *Curr. Top. Devl. Biol.* **19**, 81–101.

Proper, J. A., Bjornson, C. L., and Moses, H. L. (1982). Mouse embryos contain polypeptide growth factors capable of inducing a reversible neoplastic phenotype in non-transformed cells in culture. *J. cell. Physiol.* **110**, 169–174.

Propst, F. and Vande Woude, G. F. (1985). Expression of c-*mos* proto-oncogene transcripts in mouse tissues. *Nature, Lond.* **315**, 516–518.

Raines, E. W., Bowen-Pope, D. F., and Ross, R. (1984). Plasma binding proteins for platelet-derived growth factor that inhibit its binding to cell-surface receptors. *Proc. natn. Acad. Sci. U.S.A.* **81**, 3424–3428.

Rao, C. V., Carman, F. R., Ghegini, N., and Schultz, G. S. (1984). Binding sites for epidermal growth factor in human fetal membranes. *J. clin. Endocr. Metab.* **58**, 1034–1042.

—— Ramani, N., Chegini, N., Stadie, B. K., Carman, F. R., Woost, P. G., Schultz, E. S., and Cook, C. L. (1985). Topography of human placental receptors for epidermal growth factor. *J. biol. Chem.* **260**, 1705–1710.

Rayner, M. J. and Graham, C. F. (1982). Clonal analysis of the change in growth phenotype during embryonal carcinoma differentiation. *J. Cell Sci.* **58**, 331–334.

Recio-Pinto, E. and Ishii, D. N. (1984). Effects of insulin, insulin-like growth factor-II and nerve growth factor on neurite outgrowth in cultured human neuroblastoma cells. *Brain Res.* **302**, 323–334.

—— Lang, F. F., and Ishii, D. N. (1984). Insulin and insulin-like growth factor II permit nerve growth factor binding and the neurite formation response in cultured human neuroblastoma cells. *Proc. natn. Acad. Sci. U.S.A.* **81**, 2562–2566.

Rees, A. R., Adamson, E. D., and Graham, C. F. (1979). Epidermal growth factor receptors increase during the differentiation of embryonal carcinoma cells. *Nature, Lond.* **281**, 309–311.

Richman, R. A., Benedict, M. R., Florini, J. R., and Toly, B. A. (1985). Hormonal regulation of somatomedin secretion by fetal rat hepatocytes in primary culture.

Endocrinology **116**, 180–188.

Roberts, A. B., Anzano, M. A., Wakefield, L. M., Roche, N. S., Stern, D. F., and Sporn, M. B. (1985). Type β transforming growth factor: a bifunctional regulator of cellular growth. *Proc. natn. Acad. Sci. U.S.A.* **82**, 119–123.

—— Frolik, C. A., Anzano, M. A., and Sporn, M. B. (1983). Transforming growth factors from neoplastic and non-neoplastic tissues. *Fedn. Proc.* **42**, 2621–2626.

Robertson, M. C., Owens, R. E., McCoshen, J. A., and Friesen, H. G. (1984). Ovarian factors inhibit and fetal factors stimulate the secretion of rat placental lactogen. *Endocrinology* **114**, 22–30.

Romanus, J. A., Rabinovitch, A., and Rechler, M. M. (1985). Neonatal islet cell cultures synthesize insulin-like growth factor I. *Diabetes* **34**, 696–702.

Rosenfeld, R., Thorsson, A. V., and Hintz, R. L. (1979). Increased somatomedin receptor sites in newborn circulating mononuclear cells. *J. clin. Endocr. Metab.* **48**, 456–461.

Rothbard, J. B., Brackenbury, R., Cunningham, B. A., and Edelman, G. M. (1982). Differences in the carbohydrate structures of neural cell adhesion molecules from adult and embryonic chicken brains. *J. biol. Chem.* **257**, 11064–11069.

Rush, R. A. (1984). Immunohistochemical localization of endogenous nerve growth factor. *Nature, Lond.* **312**, 364–367.

Russell, D. W., Schneider, W. J., Yamamoto, T., Laskey, K. L., Brown, M. S., and Goldstein, J. L. (1984). Domain map of the LDL receptor: sequence homology with the epidermal growth factor precursor. *Cell* **37**, 577–585.

Rutishauser, U. (1984). Developmental biology of a neural cell adhesion molecule. *Nature, Lond.* **310**, 549–554.

—— and Edelman, G. M. (1980). Effects of fasciculation on the outgrowth of neurites from spinal ganglia in culture. *J. Cell Biol.* **87**, 370–378.

—— Gall, W. E., and Edelman, G. M. (1978). Adhesion among neural cells of the chick embryo. IV. Role of the cell surface molecule CAM in the formation of neurite bundles in cultures of spinal ganglia. *J. Cell Biol.* **79**, 382–393.

Sara, V. R., Hall, K., Misaki, M., Fryklund, L., Christenson, L., and Wetterberg, L. (1983). Ontogenesis of somatomedin and insulin receptors in the human fetus. *J. clin. Invest.* **71**, 1084–1094.

—— Hall, K., Rodeck, C. H., and Wetterberg, L. (1981). Human embryonic somatomedin. *Proc. natn. Acad. Sci. U.S.A.* **78**, 3175–3179.

Sasaki, N., Rees-Jones, R. W., Zick, Y., Nissley, S. P., and Rechler, M. M. (1985). Characterization of insulin-like growth factor I stimulated tyrosine kinase activity associated with the β-subunit of type I insulin-like growth factor receptors of rat liver cells. *J. biol. Chem.* **260**, 9793–9804.

Saunders, J. W. and Reuss, C. (1974). Inductive and axial properties of prospective wing-bud mesoderm in the chick embryo. *Devl. Biol.* **38**, 41–50.

Schilling, E. E., Rechler, M. M., Grunfeld, C., and Rosenberg, A. M. (1979). Primary defect of insulin receptors in skin fibroblasts cultured from an adult with leprechaunism and insulin resistance. *Proc. natn. Acad. Sci. U.S.A.* **76**, 5877–5881.

Schmid, C., Steiner, T., and Froesch, E. R. (1983a). Preferential enhancement of myoblast differentiation by insulin-like growth factors (IGF I and IGF II) in primary cultures of chicken embryonic cells. *FEBS Lett.* **161**, 117–121.

—— —— —— (1983b). Insulin-like growth factors stimulate synthesis of nucleic acids and glycogen in cultured calvaria cells. *Calcif. Tissue Int.* **35**, 578–585.

Schoenle, E., Zapf, J., Hauri, C., Steiner, T., and Froesch, E. R. (1985). Comparison of in vivo effects of insulin-like growth factors I and II and of growth hormone in hypophysectomized rats. *Acta endocr. Copenh.* **108**, 167–174.

Scott, J., Cowell, J., Robertson, M. E., Priestley, L. M., Wadey, R., Hopkins, B.,

Pritchard, J., Bell, G. I., Rall, L. B., Graham, L. F., and Knott, T. J. (1985). Insulin-like growth factor-II gene expression in Wilms' tumour and embryonic tissues. *Nature, Lond.* **317**, 260–262.

—— Urdea, M., Quiroga, M., Sanchez-Pescador, R., Fong, N., Selby, M., Rutler, W. J., and Bell, G. I. (1983). Structure of a mouse sub-maxillary messenger RNA encoding epidermal growth factor and seven related proteins. *Science, N.Y.* **221**, 236–240.

Seifert, R. A., Schwartz, S. M., and Bowen-Pope, D. F. (1984). Developmentally regulated production of platelet-derived growth factory-like molecules. *Nature, Lond*, **311**, 669–671.

Seiler, M. and Schwab, M. E. (1984). Specific retrograde transport of nerve growth factor (NGF) from neocortex to nucleus basalis in the rat. *Brain Res.* **300**, 33–39.

Sejerson, T., Sumegi, J., and Ringertz, N. R. (1985). Density-dependent arrest of DNA replication is accompanied by decreased levels of c-*myc* mRNA in myogenic but not in differentiation-defective myoblasts. *J. cell. Physiol.* **125**, 465–470.

Shainberg, A., Yagil, G., and Yaffe, D. (1971). Alterations of enzymatic activities during muscle differentiation *in vitro*. *Devl. Biol.* **25**, 1–29.

Shirayoshi, Y., Okada, T. S., and Takeichi, M. (1983). The calcium-dependent cell-cell adhesion system regulates inner cell mass formation and cell surface polarization in early mouse development. *Cell* **35**, 631–638.

Silver, M. H., Murray, J. C., and Pratt, R. M. (1984). Epidermal growth factor stimulates type-V collagen synthesis in cultured murine palatal shelves. *Differentiation* **27**, 205–208.

Slamon, D. J. and Cline, M. J. (1984). Expression of cellular oncogenes during embryonic and fetal development of the mouse. *Proc. natn. Acad. Sci. U.S.A.* **81**, 7141–7145.

Smith, J. C., Singh, J. P., Lillquist, J. S., Goon, D. S., and Stiles, C. D. (1982). Growth factors adherent to cell substrate are mitogenically active *in situ*. *Nature, Lond.* **296**, 154–156.

—— Tickle, G., and Wolpert, L. (1978). Attenuation of positional signalling in the chick limb by high doses of gamma-radiation. *Nature, Lond.* **272**, 612–613.

Sorge, L. K., Levy, B. T., and Maness, P. F. (1984). pp60$^{c\text{-}src}$ is developmentally regulated in the neural retina. *Cell* **36**, 249–257.

Standaert, M. L., Schimmel, S. D., and Pollet, R. J. (1984). The development of insulin receptors and responses in the differentiation of non-fusing muscle cell line BC3H-1. *J. biol. Chem.* **259**, 2337–2345.

Stiles, C. D., Isberg, R. R., Pledger, W. J., Antoniades, H. N., and Scher, C. D. (1979). Control of the Balb/c-3T3 cell cycle by nutrients and serum factors: analysis using platelet-derived growth factor and platelet-poor plasma. *J. cell. Physiol.* **99**, 395–406.

Stoscheek, C. M. and Carpenter, G. (1984). Down-regulation of epidermal growth factor receptors: direct demonstration of receptor degradation in human fibroblasts. *J. Cell Biol.* **98**, 1048–1053.

Strain, A. J., Hill, D. J., and Milner, R. D. G. (1986). Divergent action of transforming growth factor β on human foetal liver cells. *Cell Biol. internat. Rep.* **10**, 855–860.

—— —— Swenne, I., and Milner, R. D. G. (1987). Stimulation of DNA synthesis in isolated human foetal hepatocytes by insulin-like growth factor I, placental lactogen and growth hormone. *J. cell. Physiol.* in press.

Strickland, S. and Mahodavi, V. (1978). The induction of differentiation in terato-carcinoma cells by retinoic acid. *Cell* **15**, 393–403.

Summerbell, D. (1979). The zone of polarizing activity: evidence for a role in normal chick wing morphogenesis. *J. Embryol. exp. Morph.* **50**, 217–233.

—— (1983). The effect of local application of retinoic acid to the anterior margin of the developing chick limb. *J. Embryol. exp. Morph.* **78**, 269–258.

—— Dhouailly, D., and Maden, M. (1984). Vitamin A and the control of development. *Prog. clin. biol. Res.* **151**, 439–451.

—— and Maden, M. (1985). Regulation of size and pattern and effect of vitamin A in regulation and development. *Prog. clin. biol. Res.* **163C**, 121–125.

—— Lewis, J. H., and Wolpert, L. (1973). Positional information in chick limb morphogenesis. *Nature, Lond.* **244**, 492–496.

Sundell, H. W., Gray, M. E., Serenius, F. S., Escobedo, M. B., and Stahlman, M. T. (1980). Effects of epidermal growth factor on lung maturation in fetal lambs. *Am. J. Path.* **100**, 707–726.

Swenne, I., Hill, D. J., Strain, A. J., and Milner, R. D. G. (1987*a*). Growth hormone regulation of somatomedin-C/insulin-like growth factor I production and DNA replication in fetal rat islets in tissue culture. *Diabetes.* **36**, 288–294.

—— —— —— —— (1987*b*). Effects of human placental lactogen and growth hormone on the production of insulin and somatomedin-C/insulin-like growth factor I by human fetal pancreas in tissue culture. *J. Endocr.* in press.

Tashjian, A. H., Voelkel, E. F., Lazzaro, M., Singer, F. R., Roberts, A. B., Derynck, R., Winkler, M. E., and Levine, L. (1985). α and β human transforming growth factors stimulate prostaglandin production and bone resorption in cultured mouse calvaria. *Proc. natn. Acad. Sci. U.S.A.* **82**, 4535–4538.

Thiery, J. P., Delouvee, A., Gallin, W. J., Cunningham, B. A., and Edelman, G. M. (1984). Ontogenetic expression of cell adhesion molecules: L-CAM is found in epithelia derived from three primary germ layers. *Devl. Biol.* **102**, 61–78.

—— —— Grumet, M., and Edelman, G. M. (1985). Initial appearance and regional distribution of the neuron-glial cell adhesion molecule (Ng-CAM) in the chick embryo. *J. Cell Biol.* **100**, 442–456.

—— Duband, J. L., Rutishauser, U., and Edelman, G. M. (1982). Cell adhesion molecules in early chicken embryogenesis. *Proc. natn. Acad. Sci. U.S.A.* **79**, 6737–6741.

Thoenen, H. and Barde, Y. A. (1980). Physiology of nerve growth factor., *Physiol. Rev.* **60**, 1284–1335.

Thoenen, H., Korsching, S., Heumann, R., and Acheson, A. (1985). Nerve growth factor. In *Growth factors in biology and medicine.* (ed. D. Evered). pp. 113–128. *Ciba Fdn. Symp.* No. 116, Pitman, London.

Thomas, K. A., Rios-Candelore, M., Gimenez-Gallego, G., Di Salvo, J., Bennett, C., Rodkey, J., and Fitzpatrick, S. (1985). Pure brain-derived acidic fibroblast growth factor is a potent angiogenic vascular endothelial cell mitogen with sequence homology to interleukin I. *Proc. natn. Acad. Sci. U.S.A.* **82**, 6409–6413.

Thorburn, G. C., Waters, M. J., Young, I. R., Dolling, M., Buntine, D., and Hopkins, P. S. (1981). Epidermal growth factor: a critical factor in fetal maturation? In *The Fetus and independent life.* (ed. J. Wheelan). pp. 172–191. *Ciba Fdn. Symp.* No. 86, Pitman, London.

Thorsson, A. V. and Hintz, R. L. (1977). Insulin receptors in the newborn. *New Engl. J. Med.* **297**, 908–912.

—— —— (1985). Interactions of human cord sera with somatomedin receptors of human fetal brain and placental membranes. In *Abst. 67th Ann. Meeting Endocr. Soc.*, Baltimore, MD, p. 49, Abstr. 195.

Tricoli, J. V., Rall, L. B., Scott, J., Bell, G. I., and Shows, T. B. (1984). Localization of insulin-like growth factor genes to human chromosomes 11 and 12. *Nature, Lond.* **310**, 784–786.

Tyler, M. S. and Pratt, R. M. (1980). Effect of epidermal growth factor on secondary palatal epithelium in vitro: tissue isolation and recombination studies. *J. Embryol. exp. Morphol.* **58**, 93–106.

Underwood, L. E. and D'Ercole, A. J. (1984). Insulin and somatomedins/insulin like growth factors in fetal and neonatal development. *Clin. Endocr. Metab.* **13**, 69–89.

Ullrich, A., Bell, J. R., Chen, E. Y., Herrera, R., Petruzzelli, L. M., Dull, T. J., Gray, A., Coussens, L., Lias, Y-C., Tsubokaura, M., Mason, A., Seeburg, P. H., Grunfeld, C., Rosen, O. M., and Ramacharidran, J. (1985). Human insulin receptor and its relationship to the tyrosine kinase family of oncogenes. *Nature, Lond.* **313**, 756–761.

—— Coussens, L., Hayflick, J. S., Dull, T. J., Gray, A., Tam, A. W., Lee, Y., Yarden, Y., Lieberman, T. A., Schlessinger, J., Downard, J., Mayes, E. L., Whittle, N., Waterfield, M. D., and Seeburg, P. H. (1984). Human epidermal growth factor receptor cDNA sequence and aberrant expression of the amplified gene in A431 epidermoid carcinoma cells. *Nature, Lond.* **309**, 418–425.

Van Wyk, J. J., Underwood, L. E., D'Ercole, A. J., Clemmons, D. R., Pledger, W. J., Wharton, W. R., and Loef, E. B. (1981). Role of somatomedin in cellular proliferation. In *Biology of normal human growth* (eds. M. Ritzen, A. Aperia, K. Hall, A. Larsson, A. Zetterberg and R. Zellerstrom). pp. 223–239. Raven Press. New York.

Walker, P., Tarris, R. H., Weichsel, M. E., Scott, S. M., and Fisher, D. A. (1981). Nerve growth factor in umbilical cord serum: Demonstration of a veno-arterial gradient. *J. clin. Endocr. Metab.* **53**, 218–220.

—— Weichsel, M. E., Hoath, S. B., Poland, R. E., and Fisher, D. A. (1981). Effect of thyroxine, testosterone, and corticosterone on nerve growth factor (NGF) and epidermal growth factor (EGF) concentrations in adult female submaxillary gland: dissociation of NGF and EGF responses. *Endocrinology* **109**, 582–587.

Waterfield, M. D., Scrace, T., Whittle, N., Stroobant, P., Johnsson, A., Wasteson, A., Westermark, B., Heldin, C. H., Huang, J. S., and Deuel, T. (1983). Platelet-derived growth factor is structurally related to the putative transforming protein of p28^{sis} of simian sarcoma virus. *Nature, Lond.* **304**, 35–39.

Westermark, B. and Heldin, C-H. (1985). Similar action of platelet-derived growth factor and epidermal growth factor in the prereplicative phase of human fibroblasts suggests a common intracellular pathway. *J. cell. Physiol.* **124**, 43–48.

Wharton, K. A., Johansen, K. M., Xu, T., and Artavanis-Tsakonas, S. (1985). Nucleotide sequence from the neurogenic locus notch implies a gene product that shares homology with proteins containing EGF-like receptors. *Cell* **43**, 567–587.

Whitsett, J. A., Johnson, C. L., and Hawkins, K. (1979). Differences in localization of insulin receptors and adenylate cyclase in the human placenta. *Am. J. Obstet. Gynec.* **133**, 204–207.

Wilkins, J. R. and D'Ercole, A. J. (1985). Affinity-labelled plasma somatomedin-C/insulin like growth factor I binding proteins. *J. clin. Invest.* **75**, 1350–1358.

Wolpert, L., Lewis, J. H., and Summerbell, D. (1975). Morphogenesis of the vertebrate limb. In *Cell Patterning*. pp. 95–130. Ciba Fdn. Symp. No. 29, Associated Scientific Publishers, Amsterdam.

—— Tickle, C., and Sampford, M. (1979). The effect of killing by X-irradiation on pattern formation in the chick limb. *J. Embryol. exp. Morph.* **50**, 175–198.

Wurzel, J., Parks, J., Herd, J., and Nielson, P. (1982). A gene deletion is responsible for absence of human chorionic somatomammotropin. *DNA* **1**, 251–257.

Yoneda, T. and Pratt, R. M. (1981a). Interaction between glucocorticoids and epidermal growth factor in vitro in the growth of palatal mesenchymal cells from the human embryo. *Differentiation* **19**, 194–198.

—— —— (1981b). Mesenchymal cells from the human embryonic palate are highly responsive to epidermal growth factor. *Science, N.Y.* **213**, 563–565.

Zapf, J., Rinderknecht, E., Humbel, R. E., and Froesch, E. R. (1978).

Non-suppressible insulin-like activity (NSILA) from human serum: recent accomplishments and their physiological implications. *Metabolism* **27**, 1803–1829.

Zetterberg, A., Engstrom, W., and Dafgard, E. (1984). The relative effects of different types of growth factors on DNA replication, mitosis, and cellular enlargement. *Cytometry* **5**, 368–376.

Zimmerman, K. A., Yancopoulos, G. D., Collum, R. G., Smith, R. K., Kohl, N. E., Denis, K. A., Nau, M. M., Witte, O. N., Toran-Allerand, D., Gee, C. E., Minna, J. D., and Alt, F. W. (1986). Differential expression of *myc* family genes during murine development. *Nature, Lond.* **319**, 780–783.

Zumstein, P. P., Luthi, C., and Humbel, R. E. (1985). Amino acid sequence of a variant pro-form of insulin-like growth factor II. *Proc. natn. Acad. Sci. U.S.A.* **82**, 3169–3172.

12   Influence of the adrenal axis upon the gonads
GARY P. MOBERG

I   Introduction

II   Adrenal and gonadal interactions

III   Adrenal axis and gonadotrophin secretion
    1   Influence of corticosteroids on gonadotrophin secretion
    2   Direct effect of adrenal axis on the gonadotrophs
        i   *in vivo* studies
        ii   *in vitro* studies
        iii   Influence on steroid feedback
    3   Other adrenal steroids
    4   Direct effect of ACTH
    5   Influence on hypothalamic secretion of GnRH

IV   Influence of the adrenal axis on the testis

V   Effect of the adrenal axis on the ovary

VI   Concluding remarks

I   Introduction

Since Hans Selye's early work there has been a continued interest in how the hypothalamic-pituitary-adrenocortical axis interacts with the hypothalamic-pituitary-gonadal axis. While the hypothalamic-pituitary-gonadal axis (or more simply stated, the gonadal axis) has some influence on the regulation of the adrenocortical system (see Kime, Vinson, Major, and Kilpatrick 1980), most research has been directed at understanding the role that the hormones of the hypothalamic-pituitary-adrenocortical axis (or adrenal axis) play in modulating the regulation of the gonadal axis, and in turn reproduction. This review focuses on the work conducted during the past decade to define the physiological mechanisms that account for the ability of the adrenal axis to modulate regulation of the gonadal axis.

The primary impetus for such studies has been to provide an explanation for how stress disrupts reproduction. Selye (1939) argued that during stress, pituitary regulation favoured the secretion of adrenocorticotrophin (ACTH) at the expense of the gonadotrophins, accounting for the lowered reproductive success that occurs during stress. As our understanding of the neuroendocrinology of stress developed, it became evident that this part of Selye's hypothesis was incorrect – during stress the pituitary production of one pituitary hormone does not shift in favour of another. In fact, depending upon the nature of the stressor, many different patterns of hormone secretion may occur (Moberg 1985a). However, Selye's observation that stress alters gonadotrophin secretion proved correct and interest has shifted toward defining the role that the adrenocortical response to stress has in modulating the hypothalamic-pituitary-gonadal axis.

The influence of stress on reproductive success has been offered as an explanation for several natural phenomena. Disruption of normal reproduction by the adrenal stress response has been proposed as an important factor in regulating the population size of wild animals (Christian, Lloyd, and Davis 1965). The effect of stress on reproduction has provided a working explanation for the influence social hierarchies have on the reproductive success of individuals within a social group (Sapolsky 1985). Also, stress may account for the diminished reproductive success observed in domestic animals maintained under intensive management conditions (Moberg 1976, 1984). Although there are several ways stress can disrupt reproduction, an important mechanism appears to stem from the adrenal cortical response to stress disrupting the normal regulation of the gonadal axis.

Secondly, clinicians have been interested in how various adrenal diseases, such as Cushing's syndrome, diminish fertility in their patients. They have also been concerned about the potential side effects that might result from the therapeutic use of adrenal corticosteroids and their synthetic analogs, a concern that was recently exemplified by a report that asthmatic males requiring continuous treatment with a synthetic corticosteroid (prednisone) had lowered circulating levels of testosterone (Reid, Ibbertson, France, and Pybus 1985). It is such clinical studies that have provided us with the limited amount of human data that exists concerning the influence of the adrenal hormones on the function of the gonadal axis.

A third interest in the influence of the adrenal axis on reproduction has stemmed from our efforts to describe the interaction of various neuroendocrine systems with the gonadal axis. The adrenal axis has been suggested as having a modulating role in such events as puberty (Ramaley 1974) and the timing of the preovulatory release of luteinizing hormone (LH) (Rogers, Schwartz, and Nequin 1974). In fact, there is evidence that the adrenal axis plays a role in regulating the gonadal axis in all vertebrates. Adrenal hormones have been found to influence reproduction in fish (Van den Hurk, Gielen, and Terlou 1984), amphibians (Moore and Zoeller 1985), birds (Deviche 1983), rats (Baldwin and Sawyer 1974), swine (Juniewicz and Johnson 1983), sheep (Matteri, Watson, and Moberg 1984), cattle (Li and Wagner 1983a), horses (Asa and Ginther 1982), monkeys (Moberg, Watson, and Hayashi 1982) and humans (Cunningham, Capteron, and Goldzieher 1975). Such a broad effect across the classes of vertebrates implies a fundamental role for the adrenal in modulating reproduction, a role which developed early in the evolution of vertebrates and must be of critical importance to have persisted throughout. Interestingly, this role of the adrenal axis may have changed during evolution, facilitating reproduction in lower vertebrates, such as fish (Sundararaj, Goswami, and Lamba 1982; Van den Hurk *et al.* 1984), and disrupting reproduction in the higher vertebrates, as in the case of primates. Clarification of the adrenal gland's role in reproduction in vertebrate evolution offers an interesting direction for future studies in comparative endocrinology.

Although the influence of the adrenal axis on reproduction is widespread, the emphasis of this review is on mammals, since the bulk of evidence concerning the adrenal influence on the gonadal axis has been obtained from these species. The consensus of scientific opinion is that the adrenal hormones suppress reproduction. This is consistent with the conventional view that stress is generally harmful to an individual, with reproduction being one of the individual's biological functions that suffers. Certainly, there is ample evidence to support such a conclusion. However, as will be discussed, careful examination of the available data reveals conditions where the adrenal hormones have no effect or, at times, may even facilitate reproductive processes.

II   Adrenal and gonadal interactions

For scientist and layman alike, the pituitary-adrenal axis (Fig. 12.1) has

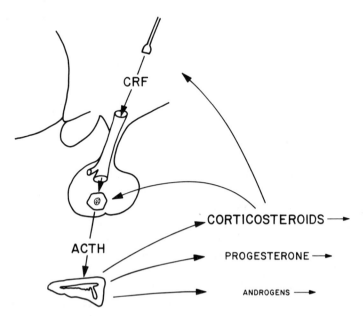

Fig. 12.1. Diagrammatic representation of the hypothalamic-pituitary-adrenocortical axis indicating the major hormones being considered in the review.

been associated with stress. A variety of stressors, especially emotional stressors, will stimulate this neuroendocrine system (reviewed by Moberg 1985*a*). Although the adrenal cortex secretes several categories of steroids, it is the glucocorticoids, such as cortisol or corticosterone, that are generally viewed as the most important adrenal steroids in the stress response. Thus, in this discussion the glucocorticoids, as a group, will subsequently be referred to as corticosteroids.

Because stress is usually considered harmful to the individual, the adrenal corticosteroids are frequently associated with negative effects on an individual's normal physiology. In fact, probably the opposite is true. The adrenal response to a stressor is essential for those adjustments in biological function that allow an individual to cope with the stressor. Life for the adrenalectomized animal, void of corticosteroids, is precarious at best. The corticosteroids are equally essential in the "nonstressed" animal to regulate normal biological functions such as metabolism, lactation, immunity and, most likely, reproduction. Undoubtedly, as in the case of immunity (Munck, Guyre, and Holbrook 1984), the adrenal hormones have one effect on reproduction during nonstressed times and an entirely different function when the adrenal hormones are elevated over a prolonged period as occurs during stress.

While the adrenal hormones affect several aspects of reproduction, the most profound effect in disrupting reproduction appears to result from the adrenal axis's modulation of the gonadal axis (Moberg 1985*b*). The female seems to be especially vulnerable to these effects since she relies on a series of carefully orchestrated endocrine events for successful reproduction, events whose temporal relationship is so critical that any disruption in the sequence will jeopardize reproductive success. As will be subsequently discussed, there are numerous reports of stressors influencing the secretion of gonadotrophins and gonadal steroids in males; unfortunately no attempt has been made to establish whether the stress-related changes in the reproductive hormones are correlated with any loss in fertility.

Logic dictates that if the biological response of the adrenal axis is capable of influencing the regulation of the gonadal axis, then there should be evidence of a correlation between increased circulating adrenal corticosteroids occurring during stress or adrenal disease and a concomitant suppression of the gonadal axis. Such evidence does exist. For example, immobilization of female rats has been reported to result in an increased concentration of circulating corticosteroids and a corresponding drop in plasma LH (Tache, Du Ruisseau, Tache, Selye, and Collu 1976; Du Ruisseau, Tache, Brazeau, and Collu 1979). However, the response of follicle stimulating hormone (FSH) to immobilization was much more variable (Tache *et al.* 1976), sometimes being suppressed by the stress and other times unaffected. A comparable differential effect on the secretion of the two gonadotrophins has been observed during social stress (Bronson

1973). Male mice engaged in fighting for social dominance show an increase in plasma concentrations of corticosterone and a concomitant drop in circulating LH (94 per cent) and FSH (19 per cent). During this period of stress the dominant animals' plasma corticosteroid values return to nonstress (control) levels in 1 to 3 days followed by the return of the subordinate animals' plasma corticosteroid values to normal in 3 to 6 days. However, the plasma concentration of LH for both groups remained depressed even longer. Thus, the correlation between increased circulating corticosteroid concentration and depressed gonadotrophin secretion was observed only during the early phases of the stress. These data illustrate two points about the gonadal axis's response to stress: the control of the two gonadotrophins is not similarly affected; and not all of the changes in gonadotrophin secretion can be attributed to the response of the adrenal axis to stress (Gray, Smith, Damassa, Ehrenkranz, and Davidson 1978; Moberg 1985b).

Two studies using rats with adrenal diseases provide additional support for the argument that adrenal steroids can alter the regulation of gonadotrophin secretion. Harris Rice, Mason, and Bartke (1982) found that transplanting adrenocortical carcinomas into immature male rats resulted in elevated concentrations of plasma corticosterone and progesterone along with a simultaneous reduction in plasma concentrations of testosterone, LH and FSH. Similarly, male rats with ACTH-prolactin secreting tumours had decreased circulating concentrations of LH and FSH, but the effects of these tumours on gonadotrophin secretion could be abolished by adrenalectomy, indicating that the adrenal steroids being secreted in response to the tumour ACTH were responsible for the suppression of gonadotrophin secretion (Weber, Ooms, and Vreeburg 1982). Thus, in rodents a correlation exists between elevated plasma corticosteroid concentrations and suppression of gonadotrophin secretion.

In domestic animals not only has comparable evidence established a correlation between stress induced elevations of corticosteroids and suppressed gonadal axis acitivity, but investigations have extended the findings in the rat by demonstrating both a temporal and quantitative relationship between the adrenal response and suppression of gonadal function. Bulls exposed to the procedures of collecting semen by electro-ejaculation exhibit a temporal relationship between increasing concentrations of circulating corticosteroids and a decline in serum LH and testosterone, with fewer episodic LH and testosterone peaks occurring (Welsh and Johnson 1981). As in rats, restraining cows will suppress the pulsatile secretion of LH in those animals which have a marked (10 to 20 fold) increase in plasma cortisol, while in cows where the plasma cortisol increased only 2 to 4 times above nonstressed levels there was no change in circulating LH (Echternkamp 1984). These data suggest that not only do stress-induced increases in corticosteroids depress LH secretion, but that a marked increase in the secretion of adrenal steroids is required before LH secretion is affected.

As in other species, humans also exhibit reduced gonadal function during stress. Even simple hospitalization can delay the preovulatory release of LH in women (Peyser, Ayalon, Harell, Toaff, and Cordova 1973). Surgery has been found not only to raise corticosteroid concentrations but also to decrease plasma LH and FSH values during postoperative recovery (Charters, O'Dell and Thompson 1969).

III   Adrenal axis and gonadotrophin secretion

1   INFLUENCE OF CORTICOSTEROIDS ON GONADOTROPHIN SECRETION

The correlation between depressed gonadotrophin secretion and elevated concentrations of plasma corticosteroids during stress and adrenal disease indicate that adrenal steroids may have a direct effect on gonadotrophin secretion. To investigate this relationship further the next step has been to administer exogenous adrenal hormones in order to determine if the association between adrenal hormones and gonadotrophin secretion was causal or coincidental. Two experimental strategies have been used: administration of ACTH to stimulate adrenal steroid secretion or the direct administration of natural or synthetic corticosteroids.

The administration of ACTH to bring about adrenal activation has certain advantages. First, ACTH stimulates the adrenal cortex to secrete all of the adrenal steroids that are released during stress, not just the corticosteroids. Secondly, the amount of the corticosteroids secreted in response to a single administration of ACTH will be limited by the biological capacity of the adrenal cortex, eliminating the potential of a pharmacological effect occurring, as is possible when excessive amounts of exogenous corticosteroids are administered. Of course, this approach is predicated on the assumption that ACTH does not have an independent, direct effect on gonadotrophin secretion, an assumption that will be examined later.

Since the primary steroids secreted in response to stress are the corticosteroids, most investigators have chosen to administer directly either the natural corticosteroids (cortisol or corticosterone) or synthetic corticosteroids such as dexamethasone or triamcinolone. A frequent problem with this approach is that an attempt is seldom made to establish plasma concentrations of the corticosteroids that correlate with what the animal experiences under physiological conditions. The problem is further complicated by the use of synthetic corticosteroids, where at best it is difficult to equate their potency with that of the natural steroids.

It should be noted that in the majority of the studies to be discussed, only the secretion of LH has been monitored because it has been easier to measure LH than FSH, and because many investigators have been primarily interested in the influence of the adrenal axis on ovulation, thus focusing their attention on the preovulatory release of LH. Further, it has been generally assumed that any effect of the adrenal axis on FSH secretion would be similar to the effect on LH, an assumption based on the predominant view that only one releasing hormone regulates the secretion of both gonadotrophins. However, this assumption has recently been challenged by data indicating that the adrenal hormones may have separate unrelated effects on LH and FSH release (see below).

In general, it can be concluded from existing data that corticosteroids can block the preovulatory release of LH. However, whether a comparable effect exists for the basal release of LH and FSH is more controversial and seems dependent upon the species, sex and length of corticosteroid treatment. Because of these species differences, the following discussion of this evidence will be divided into three sections: rats, domestic animals, and primates.

Rats – Baldwin and Sawyer (1974) found that dexamethasone administration during the second day of dioestrus and the morning of proestrus could block the preovulatory release of LH. Likewise, the subcutaneous implantation of cortisol was equally effective in blocking the preovulatory release of LH (Baldwin, 1979). Thus, in female rats corticosteroids appear capable of blocking the preovulatory release of LH. With respect to basal secretion of gonadotrophins there is no evidence of the corticosteroids having a comparable effect in female rats. On the other hand, Ringstrom and Schwartz (1984) found in male rats that cortisol treatment for four days prior to combined castration and adrenalectomy prevented the usual postcastration rise in plasma LH levels without any effect on the increase in circulating FSH.

Domestic animals – As in the rat, corticosteroids are capable of blocking the preovulatory release of LH in domestic animals. In cows, administration of ACTH blocked a synchronized, preovulatory release of LH and lowered basal LH (Stoebel and Moberg 1982). When these authors attempted to duplicate the ACTH effects by infusing cortisol, the preovulatory release of LH was again blocked, but no comparable effect was seen on the basal secretion of LH. Similar results were obtained by Li and Wagner (1983a) with $ACTH_{1-24}$ (first 24 amino acids of the ACTH molecule which stimulate corticosteroid secretion). As in the previous study, the $ACTH_{1-24}$ treatment decreased basal LH levels and prevented the preovulatory release of LH. Moreover, the investigators were able to abolish this effect of $ACTH_{1-24}$ by adrenalectomy, supporting the conclusion that the effect of ACTH was being mediated through the release of adrenal steroids. When Li and Wagner infused the adrenalectomized cows with cortisol, the preovulatory release of LH was again blocked, but as in the earlier study of Stoebel and Moberg, cortisol alone had no effect on the basal secretion of LH.

These two studies indicate that in cows activation of the adrenal cortex can block the preovulatory release of LH, but the animals' corticosteroids do not have a significant effect on lowering the basal secretion of LH. However, ACTH treatment did have a slight, but significant effect on the basal secretion of LH, suggesting either ACTH or one of the other adrenal steroids was influencing the basal LH secretion.

In bulls, unlike cows, there have been reports that corticosteroid treatment can lower the basal secretion of LH. Thibier and Rolland (1976) found the administration of dexamethasone effective in decreasing the basal secretion of LH. A subsequent study in bulls demonstrated that exogenous ACTH resulted in a six hour peak in plasma corticosteroids with a concomitant suppression of LH secretion (Johnson, Welsh, and Juniewicz 1982). When the circulating concentrations of corticosteroids returned to baseline, the episodic release of LH recommenced. These results are consistent with the laboratory's earlier work (Welsh and Johnson 1981) correlating stress induced increases in the plasma concentration of corticosteroids with suppressed secretion of LH.

In disagreement with these results is one study which reported that a single injection of ACTH had no effect on the basal secretion of LH or FSH in either bulls or steers, even though the ACTH treatment resulted in a marked increase in circulating cortisol (Barnes, Kazmer, Birrenkott, and Grimes 1983). This discrepancy with previous findings suggests that the length of exposure to elevated corticosteroids is critical for the basal secretion to be affected. We have also found that for three hours following the administration of ACTH there was no significant effect on the basal secretion of LH in castrated rams, although circulating corticosteroid concentrations were significantly elevated (Scott and Moberg unpublished observations). Thus, for circulating corticosteroids to have an influence on the basal LH secretion in the male, plasma concentrations of corticosteroids may need to be elevated for an extended period of time.

Juniewicz and Johnson (1981) also reported that a single injection of ACTH had no effect on the circulating levels of LH in either intact or castrated male pigs. However, prolonged treatment for 12 days with either ACTH or cortisol blocked the preovulatory release of LH in female pigs (Barb, Kraeling, Rampacek, Fonda, and Kiser 1982). In a separate study they found that pretreating prepuberal female pigs for 3 to 10 days with ACTH before castration, prevented the elevation in basal LH and the increased frequency in episodic LH secretion that normally occurs following castration (Fonda, Rampacek, and Kraeling 1984). This effect of ACTH was abolished by adrenalectomy, but in both adrenalectomized and adrenal intact animals, cortisol was still capable of blocking the postcastration increase in basal LH and of lowering the magnitude of the episodic LH peaks. Cortisol treatment, however, had no influence on the frequency of the LH peaks. Thus, at least two of the effects of ACTH, the suppressed basal secretion of LH and the lowered magnitude of episodic LH peaks, may be attributed to cortisol secreted in response to the exogenous ACTH. However, the ability of ACTH to suppress the frequency of episodic LH release must be by some mechanism other than cortisol release.

In summary, the preovulatory release of LH in domestic animals appears to be readily suppressed by exogenous corticosteroids. However, these hormones must be administered for an extended period of time to alter the basal secretion of LH.

Primates – In nonhuman primates the limited evidence available indicates that adrenal hormones can suppress the secretion of gonadotrophins, especially LH. Moberg, Watson and Hayashi (1982) found that the repeated administration of ACTH during the oestrous cycle of rhesus monkeys abolished the normal increase in plasma oestrogen that occurs during the follicular phase of the cycle and blocked the preovulatory surge of LH. Likewise, treatment of the adult male rhesus monkey for five days with either ACTH or cortisol lowered the basal levels of LH (Hayashi and Moberg unpublished observations). Consistent with these findings are the observations in prepuberal rhesus monkeys that the administration of cortisol to either castrated (Dubey and Plant 1985) or both adrenalect-omized and castrated males (Plant and Zorub 1984) lowered the circulating concentrations of plasma LH and FSH.

Women suffering from Cushing's syndrome frequently have subnormal levels of circulating LH and FSH (Boccuzzi, Angeli, Bisbocci, Fonzo, Gaidano, and Ceresa 1975). In fact, a direct correlation appears to exist between the amount of urinary free cortisol and the extent that the circulating LH and FSH was suppressed. The higher the urinary cortisol concentrations the more marked the suppression of gonadotrophin secretion (White, Sanderson, Mashiter, and Joplin 1981). When the patients underwent successful therapy, the circulating gonadotrophins usually returned to normal as the amount of corticosteroids secreted declined.

In normal subjects, the administration of synthetic corticosteroids alters gonadotrophin secretion. Dexamethasone has been found to suppress the basal secretion of both LH and FSH in women (Sowers, Rice, and Blanchard 1979). Women treated with triamcinolone during the first and second day of the menstrual cycle fail to have the normal preovulatory surge of LH and FSH (Cunningham, Goldzieher, de la Pena, and Oliver 1978).

In summary – From the available evidence it appears that the preovulat-ory release of LH is sensitive to the effects of corticosteroids. The effect of the steroids on the basal secretion of LH is less conclusive. If the circulating concentration of corticosteroids is markedly elevated for a period of time, then at least LH and perhaps FSH secretion is altered. Secondly, the effect of ACTH on the basal secretion of LH is different from the effect of the corticosteroids, suggesting that adrenal hormones, other than cortisol (or corticosterone) may have an influence on the basal secretion of the gonadotrophins.

2   DIRECT EFFECT OF THE ADRENAL AXIS ON THE GONADOTROPHS

From the evidence reviewed, it is evident that the hormones of the adrenal axis, especially the corticosteroids, can alter the pattern of gonadotrophin secretion. These adrenal hormones could be acting at the level of the hypothalamus, blocking the secretion of GnRH, or directly on the anterior pituitary cells secreting gonadotrophins, the gonadotrophs, preventing GnRH from stimulating the synthesis and release of gonadotrophins. Evidence exists for a site of action at both levels of gonadotrophin control, with the importance of each site varying between species. Because it has been technically difficult to measure the effect of corticosteroids on the hypothalamic secretion of GnRH, most of the research to date has been directed at describing the direct effect of the corticosteroids on the gonadotrophs.

The prevailing experimental approach has been to determine how the adrenal hormones alter the gonadotrophs' response to exogenous GnRH. The logic of this experimental strategy is that if the corticosteroids have no effect on the gonadotrophs, then the pituitary response to the releasing hormone will not be altered. On the other hand, if the pituitary's response to GnRH is altered by the corticosteroids, then the steroids are directly altering the cellular function of the gonadotrophs. It should be noted that the latter result does not preclude action at both levels of the gonadal axis, hypothalamic secretion of GnRH as well as the gonadotrophs' response to GnRH.

These studies have provided a complex variety of frequently contradictory data, undoubtedly as a result of investigators addressing independently the corticosteroids' effects on five different aspects of the gonadotrophs' role in LH and FSH secretion: basal secretion of the gonadotrophins, synthesis of gonadotrophins, responsiveness of the gonadotrophs to GnRH, the response of gonadotrophs sensitized by oestrogen prior to GnRH administration, and the response of the gonadotrophs to the negative feedback of gonadal steroids. Unfortunately, no research group has systematically studied all these parameters in the same experimental system.

i   In vivo studies

Rats – The earliest approach to this problem was the study of Hagino, Watanabe, and Golzieher (1969) demonstrating that the administration of ACTH could block pregnant mare serum gonadotrophin (PMSG) from inducing ovulation in immature rats. While these authors did not directly measure LH, they were able to provide evidence that the ACTH treatment was preventing the preovulatory release of LH. They were able to override the ACTH block with injections of human chorionic gonadotrophin indicating that the ovary was still responsive to LH activity. In addition, their data indicated that the effect of ACTH was being mediated by the release of corticosteroids since ACTH could not block ovulation in adrenalectomized animals, but dexamethasone could. They also found that dexamethasone treatment prevented electrical stimulation of the medial preoptic area (MPOA) of the hypothalamus from inducing ovulation, indicating that the dexamethasone was preventing gonadotrophs from responding to the GnRH released by MPOA stimulation. Finally, they observed that dexamethasone implants into the MPOA were capable of blocking ovulation, leading them to conclude that the corticosteroids had a second, direct effect on the hypothalamic release of the GnRH. However, this latter data must be interpreted with caution since dexamethasone may have reached the pituitary by way of the hypothalamic-hypophyseal portal system, thus potentially having a direct effect on the gonadotrophs. One disturbing aspect of their work was that they could not duplicate dexamethasone's effect with the rat's natural corticosteroid, corticosterone, raising the question as to whether the animal's own pituitary- adrenal axis could have a comparable influence or if the results they observed were pharmacological.

Two subsequent studies (Baldwin and Sawyer 1974; Baldwin 1979) confirmed that dexamethasone treatment could block the preovulatory surge of LH by preventing gonadotrophs from secreting LH in response to exogenous GnRH. Moreover, this effect could be duplicated by cortisol implants. Baldwin (1979) convincingly argued that the corticosteroids acted on gonadotrophs by preventing oestrogen from sensitizing the cells to GnRH. He found that the synthetic corticosteroid triamcinolone eliminated the enhancing effect of oestrogen on the gonadotrophs' response to GnRH, but was without effect on the response of non-oestrogen-sensitized gonado-trophs to GnRH (Fig. 12.2). From these studies it is clear that in rats the adrenal corticosteroids have the potential of blocking the preovulatory release of LH by acting directly on the gonadotrophs, probably by preventing oestrogen from sensitizing the pituitary cells to GnRH.

With respect to the basal secretion of gonadotrophins, the effect of corticosteroids appears consistent with the thesis that adrenal corticoster-oids can suppress gonadotrophin secretion. Weber, de Greef, Koning and Vreeburg (1983) reported that adult male rats implanted with a tumour secreting both ACTH and prolactin have circulating concentrations of LH and FSH that are suppressed 45 and 70 per cent, respectively. Inasmuch as adrenalectomy abolished the ability of the tumour to suppress gonado-trophin secretion, this effect of the tumour was apparently being mediated by the increased secretion of corticosteroids. It should be noted, however, the

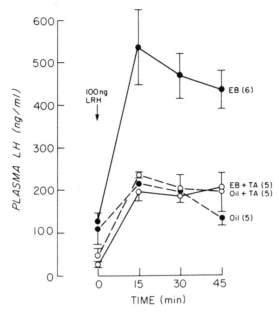

Fig. 12.2. Effect of triamcinolone (TA) on the LH response to 100 ng of GnRH in ovariectomized rats given oestradiol (EB) or oil. Oestradiol or oil was administered at 1000 h and TA at 1800 h on the third day after ovariectomy. The following day GnRH (LRH) was given at 1300 h. (Reproduced with permission from Baldwin, 1979).

corticosteroids' effect appeared to be also at the level of the hypothalamus as the concentration of GnRH in the hypophyseal stalk blood of the tumour-bearing animals dropped to 61 pg/h as opposed to 122 pg/h in the control animals.

In contrast to the above findings one study reported an enhancing effect of corticosteroids on the response of the rat gonadotroph to GnRH (Cohen and Mann 1981). They found that adrenalectomized, ovariectomized rats maintained under constant light and receiving corticosterone replacement therapy had a greater LH response six hours after the administration of GnRH. However, the interpretation of these data is somewhat difficult as experimental conditions were extreme (that is, combined adrenalectomy and ovariectomy) and the animals were maintained under constant light, which in itself disrupts reproduction (Wurtman 1967). Nevertheless, these results are of interest because they are consistent with recent *in vitro* findings that corticosteroids may actually increase gonadotrophin secretion (Suter and Schwartz 1985*a*).

The majority of the *in vivo* data indicate that the primary effect of corticosteroids on the rat gonadotroph is to block the preovulatory release of LH. Moreover, if the circulating steroids are elevated over an extended period of time, they may diminish the basal secretion of gonadotrophins.

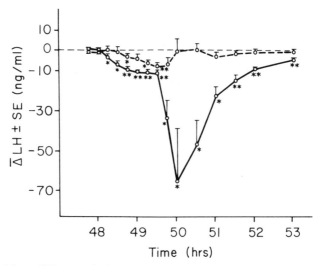

Fig. 12.3. Mean difference in LH levels following GnRH administration in cows treated with either cortisol (broken line) or ACTH (solid line), and control animals. GnRH (100 ng) was administered (i.v.) at 48 and 49.5 h following the initiation of cortisol or ACTH treatment. The mean change in LH represents the average difference (± S.E.M.) in plasma LH between the control and hormone-treated animals at each sample time (n = 4); for example if the mean change in LH was = 0, then no suppression is indicated. *P ≤ 0.05 compared with the zero line in paired t tests. (Data from Matteri and Moberg, 1982).

Domestic Animals – In cows whose oestrous cycles had been synchronized by treatment with progesterone and oestradiol Matteri and Moberg (1982) found that the administration of either ACTH or cortisol can diminish the amount of LH secreted in response to exogenous GnRH. However the response differs between the two hormones (Fig. 12.3). In this study each animal received two injections of GnRH 1.5 hours apart, resulting in two peaks of LH with the second LH response (the primed response) being greater. When ACTH was administered for 48 hours prior to the GnRH challenges, the total amount of LH secreted was markedly diminished. However, cortisol administered during a comparable time period had a less profound effect, reducing the amount of LH secreted in response to the first GnRH challenge but not influencing the primed LH response. These results indicate that bovine gonadotrophs are sensitive to the hormones of the adrenal axis, but cortisol alone cannot totally account for the effect of ACTH on the gonadotrophs. Li and Wagner (1983b) confirmed that ACTH administration can significantly reduce the amount of LH secreted in response to GnRH and furthermore this effect of ACTH can be prevented by adrenalectomy. But if they infused adrenalectomized cows with cortisol they were able only to alter the qualitative response to GnRH; that is the latency to peak response was longer and the peak response was lower. However, the total amount of LH secreted in the cortisol treated cows was comparable to the amount of LH secreted in response to GnRH in the non-treated animals. Taken together, these two studies indicated that in cows ACTH, by stimulating the secretion of adrenal steroids, can reduce the gonadotrophs' response to GnRH. The ACTH effects, however, cannot be entirely accounted for by cortisol.

Chantaraprateep and Thibier (1978, 1979) found that, as in the female, corticosteroids decrease the responsiveness of the male bovine gonadotrophs to GnRH. Dexamethasone administered 5 hours before a GnRH challenge diminished the LH response, although when the experiment was repeated the next day, dexamethasone had no effect on the LH response to the second GnRH challenge, a result similar to cortisol having no effect on the priming response to GnRH in the cow (Matteri and Moberg 1982).

In apparent disagreement with other studies of cattle are the findings by Barnes *et al.* (1983) that a single injection of ACTH to bulls or steers will not reduce the amount of LH secreted in response to a 200 µg GnRH challenge, although the FSH response in the bulls was reduced. The discrepancy between these results and those obtained in other studies may have resulted from the dose of 200 µg of GnRH exceeding the threshold for the corticosteroid's suppressing effect on the gonadotrophs. We have found that a dose effect for GnRH exists in rams where cortisol treatment was not able to reduce the ovine gonadotrophs' response to 100 µg of GnRH (Fuquay and Moberg 1983), but the same cortisol treatment was effective in diminishing the animals' response to 10 µg of GnRH (Kuck and Moberg, unpublished observations). Thus, the effectiveness of corticosteroids to modify gonadotroph function depends upon how much GnRH is administered. Pharmacological doses of GnRH can override the suppressive effect of the corticosteroids on gonadotrophs.

Primates – In humans, exposure to elevated concentrations of corticosteroids may have a biphasic effect on the response to exogenous GnRH, blunting the LH response but having no marked influence on the FSH response. Female patients undergoing treatment with a synthetic corticosteroid, prednisolone, secreted less LH in response to a GnRH challenge than controls (Sakaakura, Takebe, and Nakagawa 1975), but this treatment had no effect on the secretion of FSH (Sakakura, Yoshioka, and Takebe 1978). However, four days of dexamethasone treatment has been found to suppress the baseline concentrations of serum LH and FSH as well as to blunt the response of both gonadotrophins to GnRH (Sowers *et al.* 1979).

Comparable data has been obtained from patients with Cushing's disease who have elevated plasma concentrations of corticosteroids. For example, in one group, five of the six patients had no increased secretion of LH in response to a GnRH challenge and the sixth patient's response was subnormal (Boccuzzi *et al.* 1975). Once again, the elevated plasma levels of corticosteroids had no effect on FSH secretion.

In humans it appears that the corticosteroids have a direct effect on the gonadotrophs, modifying the cells' response to GnRH. However, in these studies the subjects were exposed to the corticosteroids for an extended period of time, Thus, the decreased pituitary responsiveness of these individuals could have resulted from corticosteroids suppressing the secretion of GnRH from the hypothalamus, which in turn would deprive the gonadotrophs of the priming effect of GnRH (Fink, Chiappa, and Aiyer 1976), rendering the cells hyporesponsive to exogenous GnRH. Such an indirect effect of corticosteroids on gonadotrophs would be consistent with data obtained in nonhuman primates where the primary effect of corticosteroids on gonadotrophin secretion appears to result from blocking the hypothalamic release of GnRH, with the steroids having no significant direct effect on the pituitary gonadotrophs.

Sapolosky (1985) administered dexamethasone to male baboons six hours prior to injecting GnRH and found no effect on the amount of LH secreted in response to the subsequent administration of the GnRH, even though the corticosteroid treatment had significantly depressed the gonadal secretion of testosterone. Even a more prolonged exposure to elevated concentrations of corticosteroids seems to have no marked effect on the pituitary's responsiveness to GnRH. For 62 days, Dubey and Plant (1985) injected castrated monkeys daily with cortisol. During this period they observed a correlation between elevated serum cortisol and the suppression of circulating levels of LH and FSH. Furthermore, they found that this inverse relationship was most strongly correlated when the concentrations of serum cortisol were compared with the values of gonadotrophins found in blood samples that were collected two weeks later, suggesting that there is a time lag before the corticosteroids are able to modify gonadotrophin secretion. Moreover, the corticosteroids have an abiding effect on gonadotrophin secretion as serum gonadotrophins did not return to normal until 2–4 weeks after the cessation of corticosteroid treatment. In their study the primary site of action of the corticosteroids also appears to be at the hypothalamus, preventing the release of GnRH. During the period of corticosteroid-induced suppression of gonadotrophin secretion, the infusion of GnRH was still able to stimulate LH and FSH secretion (Fig. 12.4), indicating that the corticosteroids can suppress GnRH secretion without having any direct effect on the gonadotrophs. With the intact adult male monkey, we have obtained data supporting Dubey and Plant's findings (Hayashi and Moberg in press). When we administered either ACTH or cortisol for five days, the basal concentrations of serum LH were lowered, but the amount of LH secreted in response to a GnRH challenge was unaffected by the treatments.

ii   In vitro *studies*

In as much as it has proved difficult in the whole animal to confirm a direct effect of the corticosteroids on gonadotrophs, conducting the experi-

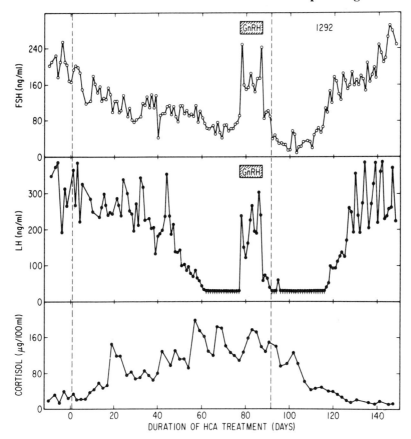

Fig. 12.4. Effect of GnRH infusion (0.1 µg per min for 3 m. every hour) on LH and FSH secretion in castrated male monkeys undergoing hydrocortisone acetate (HCA) treatment (20 mg per kg BW·day). The horizontal bar shows time of GnRH infusion and the vertical dotted lines indicate duration of hydrocortisone treatment. (Reproduced with permission from Dubey and Plant 1985).

ments *in vitro* has offered an attractive alternative, in spite of the disadvantages of using cell cultures. One potential problem is that the cultured gonadotrophs are deprived of the trophic effects of the gonado-trophins and no longer experience the general hormonal milieu of the animal, such as gonadal steroid feedback or the possible permissive effects of other hormones. Nevertheless, *in vitro* studies have provided additional information about the direct effect of the corticosteroids on gonadotrophs. Unfortunately, there are only a few studies and these have been limited to rat and bovine tissue. These results, however, do confirm a direct effect of the corticosteroids on gonadotrophs and reaffirm a species difference in this response.

Basal LH Secretion – Short term exposure of cultured female rat pituitary cells to cortisol for six hours has no effect on the basal secretion of LH (Fujihara and Shiino 1980). However, a more prolonged exposure of cultured cells to cortisol for 48 hours decreases the basal secretion of LH without modifying the cellular contents of LH (Suter and Schwartz 1985*a*). This suggests that the steroid is capable of modifying the secretion of LH in the absence of GnRH without having any influence on the synthesis of LH. Unlike with the cultured female pituitary cells, exposure of cultured male pituitary cells to corticosteroids seems to have no effect on the basal secretion of LH, but does increase the cellular content of LH (Suter and Schwartz 1985*b*).

As in the rat studies, Padmanabhan, Keech, and Convey (1983) saw no effect on the basal secretion of LH by cultured pituitary cells from cows after only 6 hours' exposure to corticosteroids. However, treatment of the pituitary cells with high doses of cortisol (10^{-6} M) for 8 hours suppressed the basal secretion of LH, and if the exposure of the cells to the adrenal steroids was extended to 20 hours, even lower doses of cortisol (10^{-8} M) were capable of depressing basal LH secretion.

These data clearly indicate that corticosteroids have a limited effect on the basal secretion of LH by the gonadotrophs, but only after the cells have experienced prolonged exposure to the steroids. Such an effect may be of limited importance in the normal physiology of the animal and may have an influence only when individuals experience prolonged elevations of corticosteroids as occurs during adrenal disease or corticosteroid therapy.

Basal FSH Secretion – The only data available concerning the basal secretion of FSH was obtained in the recent studies of Suter and Schwartz (1982*a*, *b*). Corticosteroids appear to have a subtle, but positive, effect on FSH secretion. Corticosteroid treatment for 48 hours enhanced the basal secretion of FSH and the cellular content of FSH in cultured female rat pituitary cells, indicating a direct effect on FSH synthesis by a mechanism that is independent of LH synthesis. The female appears far more sensitive to this enhancing effect of corticosteroids than the male since the male gonadotrophs did not show any increase in basal FSH secretion until after 96 hours of exposure, and then only with very high doses of corticosterone.

Response to GnRH – In the two studies by Suter and Schwartz (1985*a*, *b*), corticosteroids had no effect on the ability of the cultured pituitary cells to secrete LH in response to a GnRH challenge. This is consistent with Baldwin's (1979) earlier observation that corticosteroids only affected LH secretion from gonadotrophs that had been sensitized by oestrogen pretreatment, a condition that Suter and Schwartz did not investigate. The latter authors also found that while corticosteroid treatment had no effect on the FSH response to GnRH in the male, the steroids did enhance the amount of FSH secreted in response to GnRH in the female which undoubtedly reflects the marked increase in FSH synthesis they observed in the female pituitary in response to corticosteroid treatment.

An apparent contradiction to these results is found in an earlier study (Fujihara and Shiino 1980) where corticosterone treatment for 6 hours enhanced the amount of LH secreted in response to GnRH. However, these data must be interpreted with caution since only a small amount of LH was secreted by the nontreated, control, cells and may reflect a lack of sensitivity of the system used. It is also possible that this discrepancy in findings between the two studies resulted from the difference in time the cells were exposed to corticosteroids. Fujihara and Shiino examined the GnRH response after only 6 hours of treatment while Suter and Schwartz treated their preparation with corticosterone for 48 hours, suggesting that the nature of the corticosteroid effect in modulating the GnRH response may vary with the length of exposure to the steroids. Consistent with this latter interpretation was the observation of Liptrap and Raeside (1983) that a one hour exposure to corticosteroids enhanced the GnRH response in pigs, while prolonged treatment with GnRH blocked the preovulatory release of LH.

In two independent studies on cultured bovine pituitary cells, cortisol decreased the amount of LH secreted in response to a GnRH challenge (Li and Wagner, 1983b; Padmanabhan et al. 1983) which is consistent with the majority of the in vivo data discussed previously. Furthermore, in the study by Padmanabhan, Keech, and Convey, cortisol treatment affected only the first response to GnRH, having no effect on the second, primed response of the gonadotrophs. This finding is in agreement with the earlier report of Matteri and Moberg (1982) where cortisol diminished the LH secreted in response to the first GnRH challenge, but did not affect the subsequent primed response in the cow (see Fig. 12.3). Whether this difference in corticosteroid response reflects the steroids differentially affecting the release of stored LH, as opposed to synthesis, remains to be investigated, but such an interpretation is in agreement with the data which indicated that corticosteroids seemed to have only a limited effect on the synthesis of LH in the rat.

The mechanisms by which corticosteroids alter gonadotroph function is unknown. Padmanabhan et al. (1983) found that cortisol and dexamethasone had a comparable effect on LH secretion. The corticosteroids decreased the slope of the LH response curve suggesting that the steroid decreased cell responsiveness, although this effect is probably not related to altered binding: at least for rats, there is no evidence that the corticosteroids have any effect on receptor binding of GnRH, (Naess, Cusan, Brekke, Purvis, Trojesen, and Hansson 1981; Tibolt and Childs 1985). This would suggest that the effect of the adrenal steroids must be at some post-binding step in regulation.

Summarizing the in vitro data, it appears that in the bovine pituitary, corticosteroids can diminish the initial response of gonadotrophs to GnRH, but has little effect on subsequent primed responses. In rats, corticosteroids have little if any significant effect on the secretion of LH in response to GnRH, except when the pituitary has been previously sensitized by oestrogen, suggesting that many of the *in vivo* effects of corticosteroids may reflect the effect of the corticosteroids on the secretion of GnRH from the hypothalamus.

iii   Influence on steroid feedback

Another potential mechanism for adrenal hormones to alter gonadotrophin secretion is to modify the effect that gonadal steroids (oestrogen and testosterone) have on gonadotrophs. The gonadal steroids have two distinct effects on the gonadotroph regulation: 1) oestrogen sensitization of the gonadotrophs in females, as part of the physiological events leading to the preovulatory release of gonadotrophins and 2) the inhibition of gonado-trophin secretion by classical negative feedback on the pituitary.

I have already reviewed the evidence that in the rat corticosteroids appear to prevent oestrogen from sensitizing gonadotrophs to GnRH (see Fig. 12.2). From that data Baldwin (1979) argued that the action of the corticosteroids is to block oestrogen sensitization of the pituitary, preventing the preovulatory release of LH. Such a mechanism of action would account for the corticosteroids having a profound effect in blocking the preovulatory release of LH while having only a marginal effect on other aspects of gonadotroph function, such as synthesis and basal secretion.

Whether the influence of corticosteroids on oestrogen sensitization is unique to the rat, or is more widespread in nature, has not been thoroughly investigated. However, two studies raise questions about whether a compar-able effect exists in other species. Since the anoestrous ewe responds to the administration of oestrogen with a marked secretion of LH, we have used this preparation as a model to study the influence of corticosteroids on the pituitary response to oestrogen. We found that pretreatment of anoestrous ewes with large doses of either dexamethasone or cortisol did not prevent oestrogen from initiating an LH release (Moberg, Watson, Stoebel, and Cook 1981). At least in this model, oestrogen's effect on gonadotrophs was not modified by corticosteroids.

Using a different approach to address the same question, we have recently found that in the female monkey whose preovulatory release of LH was blocked by prolonged ACTH treatment (Moberg et al. 1982), it was still possible to induce a release of LH with an injection of oestrogen (Hayashi and Moberg, unpublished). While neither of these studies duplicate the phenom-enon of the oestrogen sensitization reported in the rat by Baldwin, they do indicate a need for additional studies to determine whether the corticosteroid prevention of gonadotroph sensitization by oestrogen is unique to the rat.

There is even less information concerning the influence of corticosteroids on the negative feedback of gonadal steroids on the pituitary. One recent study did examine the effect of combined corticosteroid and testosterone replacement on serum concentrations of LH and FSH in adrenalectomized rats (Vreeburg, DeGreef, Doms, VanWouw, and Weber 1984). It was found that neither corticosteroids nor testosterone alone had a marked effect on levels of serum gonadotrophins, but if the two steroids were administered simultaneously, the circulating concentration of LH dropped from 373 to 9 ng per ml and FSH from 1549 to 364 ng per ml. The authors concluded that corticosteroids may sensitize the gonadotrophs to the negative feedback effects of testosterone.

In spite of the sparsity of evidence supporting a corticosteroid influence on the action of gonadal steroids at the level of the gonadotrophs, there is sufficient data to encourage further study of the question. Such an effect of corticosteroids on gonadal steroid action might explain additional corticosteroid influences on other biological actions of gonadal steroids that are essential for normal reproduction. For example, it has been found that corticosteroid pretreatment of ovariectomized sows (Ford and Christenson 1981), rats (DeCantazaro and Gorzalka 1980), monkeys (Everitt and Herbert 1971) and ewes (Moberg 1985b) will prevent exogenous oestrogen from inducing oestrous behaviour. While the mechanism by which the adrenal steroids block oestrogen from inducing behavioural changes is not known, the intriguing possibility exists that the mechanism may be comparable to the way corticosteroids alter pituitary function in the proestrous rat.

3   OTHER ADRENAL STEROIDS

The adrenal cortex secretes steroids other than the glucocorticoids. Of these, progesterone and testosterone are of special interest for the regulation of the gonadal axis. Both of these steroids, whose primary source is the gonads, are important components of the neuroendocrine axis controlling gonadal function. In the previous discussion on the possible effects of corticosteroids on gonadotrophin secretion, there were several instances where corticosteroid administration did not duplicate the effect that exogenous ACTH had on gonadotrophin secretion. For example, Stoebel and Moberg (1982) observed that ACTH depressed the basal secretion of LH but cortisol did not. Likewise, Li and Wagner (1983a) blunted the LH response to GnRH with ACTH treatments but cortisol only changed the qualitative nature of the response. In the ram, ACTH diminished the LH response to a large GnRH challenge, but cortisol was ineffective in blocking the pituitary response to the same dose of GnRH (Fuquay and Moberg 1983). These findings indicate that in response to exogenous ACTH some steroid, other than the glucocorticoids, may be secreted from the adrenal cortex and has the capability of influencing gonadotrophin secretion. A critical question is whether under physiological conditions the adrenal cortex secretes sufficient amounts of the steroids to influence regulation of gonadotrophin secretion.

Of the two steroids, adrenal progesterone has attracted the most attention, undoubtedly because of the recognized capability of the steroid either to stimulate or inhibit gonadotrophin secretion as well as the possible role that adrenal progesterone may have in synchronizing the preovulatory release of LH (Mann, Korwitz, and Barraclough 1975). However, if adrenal progesterone is going to have a meaningful role in regulating the gonadal axis, then the adrenal gland must be capable of secreting significant amounts of progesterone under physiological conditions.

Administration of ACTH results in an increase in circulating plasma progesterone in a variety of species (Resko 1969; Stoebel and Moberg 1982; De Silva, Kaltenbach, and Dunn 1983; Watson and Munro 1984). While many of these studies used large doses of ACTH, Watson and Munro's work indicates that small doses of ACTH (10 μg) administered intravenously in cows can result in a significant elevation of plasma progesterone. This indicates that the amount of ACTH secreted in response to a stressful stimulus should be adequate to cause a meaningful increase in the adrenal gland's secretion of progesterone. Nevertheless, there is at present no conclusive evidence that adrenal progesterone does have a direct effect on gonadotrophin regulation.

Adrenal testosterone is also a tempting candidate as a non-glucocorticoid adrenal steroid that might alter gonadotrophin secretion, either by its own direct action on the hypothalamic-pituitary axis or indirectly by its conversion to oestrogen. As in the case of adrenal progesterone, there is no evidence of physiological stimuli resulting in sufficient secretion of testosterone from the adrenal cortex to have any effect on the gonadotrophs.

Thus, although adrenal steroids other than the glucocorticoids are attractive candidates for modulating the gonadal axis, conclusive evidence for such a role is lacking.

4   DIRECT EFFECT OF ACTH

It has been the prevailing view that ACTH has no direct effect of its own on the gonadal axis. Any effect that does result from the administration of ACTH has been attributed to adrenal steroids secreted in response to the ACTH. Indeed, there is considerable evidence supporting such a conclusion. For example, while $ACTH_{1-24}$ was effective in lowering concentrations of circulating LH in cows, adrenalectomy abolished the ACTH effect (Li and Wagner 1983a). Likewise, cultured bovine pituitary cells which were responsive to corticosteroids were unaffected by ACTH (Padmanabhan et al. 1983). While these data are consistent with the conclusion that ACTH has no direct effect on the gonadal axis, these findings do not explain the results from a limited number of studies that strongly suggest

ACTH has an effect on the gonadal axis that is distinct from the release of adrenal steroids.

In his early work describing the possible role of the adrenal in controlling the population size of wild animals, Christian reported that ACTH was able to inhibit the reproductive function in both sexes of adrenalectomized, white-footed mice (*Peromyscus leucopus*), even though the animals were being maintained on corticosteroids (Pasley and Christian 1972). Subsequently, Ogle (1977) reported that the administration of ACTH to adrenalectomized rats reduced the anticipated increase in LH following ovariectomy. Furthermore, adrenalectomy alone markedly diminished the expected rise in LH following ovariectomy, presumably because the increased amount of circulating ACTH that occurs following adrenalectomy directly suppressed the secretion of LH. Recently, Vreeburg and colleagues obtained additional evidence for ACTH having a direct effect on LH secretion when they observed that ACTH slightly reduced serum LH, but not serum FSH, in adrenalectomized male rats (Vreeburg *et al.* 1984). Combined, these studies suggest that ACTH has an influence on the regulation of the gonadal axis that does not require adrenal participation.

Earlier it was emphasized in the study of Matteri and Moberg (1982) that exogenous ACTH was far more effective in cows in suppressing the pituitary's response to GnRH than was cortisol alone (see Fig. 12.3), even though the plasma concentrations of corticosteroids were comparable during both treatments. One explanation for the differential effects of ACTH and cortisol is that, in addition to cortisol, the exogenous ACTH had stimulated the secretion of other adrenal steroids, especially progesterone, and that these steroids acted either separately or synergistically with the cortisol to cause the more pronounced suppression of LH secretion observed in the ACTH treated animals. Appealing as this interpretation is, at this time there is a paucity of evidence supporting any significant effect for adrenal progesterone in regulating the gonadal axis. An alternate explanation for these findings is that the ACTH may have had an effect on the gonadal axis that was not adrenal related, an assumption that has now received some experimental support.

Using rams as an experimental model, we examined the importance of the adrenal gland in ACTH's action on the secretion of LH (Fuquay and Moberg 1983). We observed that ACTH depressed the amount of LH secreted in response to GnRH, but infusion of cortisol had no similar effect on the GnRH response. In order to determine if the more potent effect of ACTH was related to adrenal steroids other than cortisol, the ACTH treatment was repeated in rams before and after adrenalectomy. Absence of the adrenal glands had no influence on the ability of ACTH to depress the GnRH response (Fig. 12.5), suggesting that ACTH was acting on gonadotrophs by a nonadrenal mechanism.

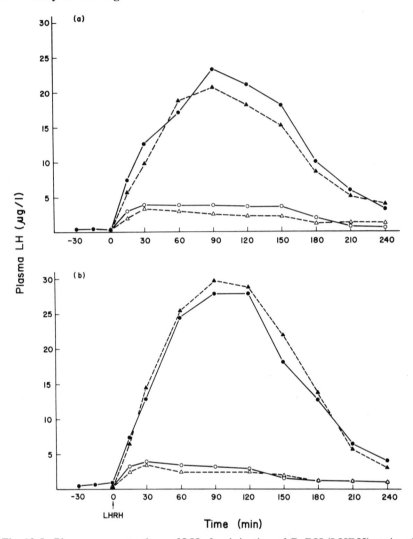

Fig. 12.5. Plasma concentrations of LH after injection of GnRH (LHRH) at time 0 in two rams (a and b) before and after adrenalectomy. Control (●), ACTH treatment (○), adrenalectomized control (▲), adrenalectomized and ACTH treated (△). (Reproduced with permission from Fuquay and Moberg, 1983).

To understand this more completely, we have subsequently examined *in vitro* the effect of ACTH on both LH and FSH release from perifused ovine pituitary tissue (Matteri, Moberg, and Watson 1986). In this system, we found that synthetic ACTH ($ACTH_{1-39}$) has a biphasic effect on the release of LH and FSH, stimulating the basal secretion of the gonadotrophins but diminishing the amount of the gonadotrophins released in response to a

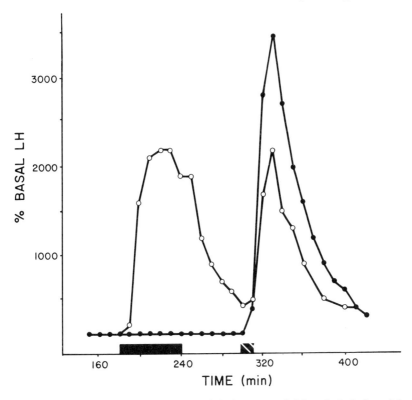

Fig. 12.6. Data from one pituitary gland during control (closed circles) and 5 × 10⁻⁷M ACTH 1–39 (open circles) treatment. The treatment period is delineated by the dark bar and the 10⁻⁸ GnRH challenge is represented by the diagonally-striped bar. (Reproduced with permission from Matteri et al., 1986).

subsequent GnRH challenge (Fig. 12.6). When this response of the pituitary tissue to ACTH was compared to the effect of peptide fragments of ACTH, we found that the peptide containing the last 22 amino acids of ACTH (ACTH$_{18-39}$) stimulated basal secretion of LH and FSH (see Fig. 12.7) without altering the GnRH response (Fig. 12.8). On the other hand, that portion of the ACTH molecule associated with stimulation of the adrenal cortex, that is the first 24 amino acids of ACTH (ACTH$_{1-24}$), had no effect on the basal release of LH and FSH, but this peptide reduced the amount of the gonadotrophins secreted in response to GnRH. These data suggest that each of the biphasic effects of ACTH on the gonadotroph can be attributed to a different portion of the ACTH molecule.

The physiological significance of the effects of ACTH on the release of gonadotrophin is open to speculation since ACTH released in response to exogenous corticotrophin-releasing hormone (CRH) has not been demonstrated to alter gonadotrophin secretion. There is no effect on gonadotrophin

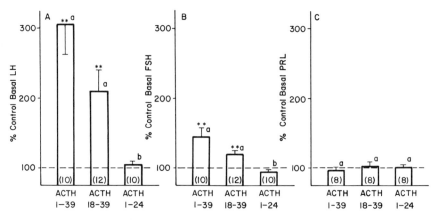

Fig. 12.7. The LH (A), FSH (B) and prolactin (C) released during the ACTH peptide pretreatment period (average percent of secretion in untreated tissue in each pituitary gland ± S.E.M.). Bars bearing different lower case letters are significantly different (P < 0.05). The numbers within parentheses indicate the sample sizes. Significant treatment effects (different from the 100% control level) are also indicated (*, P < 0.05; **, P < 0.01). (Reproduced with permission from Matteri et al., 1986).

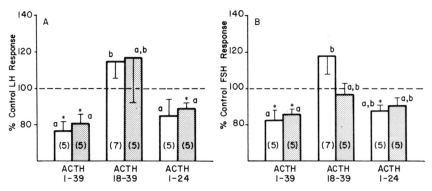

Fig. 12.8. The LH (A) and FSH (B) responses to GnRH following ACTH peptide pretreatment are shown (average percent of secretion in untreated tissue in each pituitary gland ± S.E.M.). The open and stippled bars represent the mean responses to 10^{-8} and 10^{-9} M GnRH, respectively. Bars bearing different lower case letters are significantly different (P < 0.05). The numbers within parentheses show the sample sizes. Significant treatment effects (different from the 100% control level) are also indicated (*, P < 0.05). (Reproduced with permission from Matteri et al., 1986).

secretion by CRH administered peripherally to rats (Rivier and Vale 1984) or applied to cultured rat pituitary cells (Wehrenberg, Baird, Ying, Rivier, Ling, and Guillemin 1984). However, in these studies only one dose of CRH was administered and the levels of ACTH induced *in vitro* were approximately 200 times less than those administered in the perifused pituitary study (Matteri *et al.* 1986). If stress can create high enough levels of ACTH in the pituitary gland, ACTH might possibly influence gonadotrophin release. Acute stress has been found to produce a rapid, transient elevation of LH secretion in the male rat (Krulich, Hefco, Illner, and Read 1974) and in the male rhesus monkey (Hayashi and Moberg unpublished) which might be related to the release of ACTH.

The inhibitory effect of $ACTH_{1-39}$ and $ACTH_{1-24}$ on the gonadotrophin response to subsequent GnRH challenges is surprising as *in vitro* studies have reported no effect by $ACTH_{1-24}$ on GnRH sensitivity (Padmanabhan *et al.* 1983; Li and Wagner 1983*b*). The reason for this discrepancy in findings is unclear. Certainly differences in species, doses and durations of treatment may have contributed to the divergent results as well as possible differences in the *in vitro* systems used. Rat pituitary tissue has been shown to differ physiologically from dispersed cells (Speight and Fink 1981). Additionally, GnRH self-priming occurs in the tissue from castrated ram lambs used in our perifusion system (Moberg, unpublished observations), while dispersed steer pituitary cells do not show this property (Padmanabhan, Kesner, and Convey 1981).

The similar stimulatory effect of $ACTH_{1-39}$ and GnRH on basal LH secretion suggests that ACTH might be acting as a GnRH agonist. To explore this possibility further, we pretreated the tissue with a GnRH antagonist and then administered GnRH or ACTH. The antagonist inhibited the LH and FSH response to both ACTH and GnRH (Fig. 12.9),

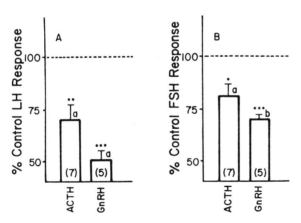

Fig. 12.9. The LH (A) and FSH (B) release due to treatment with 10^{-7} M ACTH or 10^{-10} M GnRH, following GnRH antagonist pretreatment (average percent of secretion in tissue not pretreated with antagonist $\pm$ S.E.M.). Bars bearing different lower case letters are significantly different (P $<$ 0.05). The numbers within parentheses show the sample sizes. Significant treatment effects (different from the 100% control level) are also indicated (*, P $<$ 0.05; **, P $<$ 0.01). (Reproduced with permission from Matteri et al., 1986).

providing preliminary evidence that ACTH may be acting in some manner on the gonadotrophs' GnRH receptors.

Several interesting questions concerning the relationship between ACTH and gonadotrophin secretion have been raised by these data. It is known that $ACTH_{1-24}$ is fully bioactive at the adrenal cortex, but it does not behave as the intact hormone in our *in vitro* system. This may be of importance, as prior *in vivo* pituitary studies reporting no measurable ACTH effects on gonadotrophin secretion have used the 1–24 peptide, rather than the full 1–39 sequence of ACTH. One implication from these data is that various fragments of ACTH, or more important, peptide fragments of the precursor molecule, pro-opiomelanocorticotrophin (POMC) may act on the gonadotroph. Alpha-melanocyte stimulating hormone (which is equivalent to $ACTH_{1-13}$) has been reported to induce gonadotrophin release in man (Reid, Ling, and Yen 1984) and in rat pituitary glands *in vitro* (Miyake and Yen 1981), although the latter finding has been disputed (Khorram, De Palatis, and McCann 1984).

Recently, we have found in our *in vitro* system that other fragments of the POMC molecule can alter LH secretion (Matteri and Moberg 1985). Gamma-endorphin or human beta-endorphin (but not ovine beta-endorphin) elevated basal LH secretion followed by an inhibition of the response of the gonadotroph to a subsequent GnRH challenge. As in the case of ACTH, the stimulatory effect of beta-endorphin on basal LH secretion could be suppressed by pretreatment with a GnRH antagonist but not by pretreating the tissue with the classical opiate antagonist, naloxone.

Combined, these data raise the possibility that various fragments of the POMC molecule may have a direct effect on the pituitary gonadotroph. However, the data are so limited that it is not possible to decide if the effects should be ascribed to a pharmacological response or are indicative of a yet undescribed physiological role. Nevertheless, these results do raise an intriguing and potentially important consideration. Classically, the only portion of the POMC molecule considered as part of the adrenal axis is ACTH, because of its action on the adrenal cortex. Yet during stress, other portions of this precursor molecule are released from the pituitary in equimolar concentrations with ACTH. If some of these peptides eventually prove to have a physiological role in directly modulating the gonadotroph, then our view of what comprises the adrenal axis' influence on the gonadal axis may need revision.

5   INFLUENCE ON HYPOTHALAMIC SECRETION OF GnRH

Not all of the effects of corticosteroids on gonadotrophin secretion can be attributed to a direct action of the steroids on the pituitary gonadotroph. As pointed out earlier, unlike other species, in nonhuman primates there is at present no evidence of any direct effect of corticosteroids on the gonadotroph; instead the primary effect appears to be on the hypothalamic release of GnRH (Dubey and Plant 1985; Sapolsky 1985). In rats, ACTH secreting tumours decreased the concentration of GnRH in the portal blood (Weber, de Greef, de Koning, and Vreeburg 1983). Unfortunately, measurement of GnRH release from the hypothalamus is technically difficult, hampering studies on how corticosteroids may alter the secretion of GnRH from the hypothalamus. While corticosteroids do bind in the hypothalamus (McEwen 1973; Stith, Person, and Dana 1976), it remains to be determined whether the corticosteroids will bind to GnRH secreting cells.

At this time we have no indication of how the corticosteroids regulate GnRH secretion. From indirect evidence it is possible to speculate that one of the mechanisms might be altering the activity of hypothalamic peptidases that degrade GnRH. It has been proposed that such degradation of GnRH by peptidases may contribute to the regulation of GnRH secretion (Advis, Krause, and McKelvy 1982). Coincubation of ovine hypothalamic extracts with corticosteroids increases the activity of those peptidases (Swift and Crighton, 1979). Such a possibility awaits further study.

Recently, another component of the adrenal axis, corticotrophin releasing hormone (CRH), has been found to inhibit the hypothalamic release of LH, but not FSH (Ono, Lumpkin, Samson, McDonald, and McCann 1984; Rivier and Vale 1984). When Rivier and Vale administered CRH into the third ventricle of female rats that previously had been either ovariectomized or both ovariectomized and adrenalectomized, there was a rapid dose related depression of LH secretion. The effect of CRH appeared independent of the remaining components of the adrenal axis since comparable amounts of CRH administered peripherally had no effect on LH secretion. As the CRH was effective in adrenalectomized animals, the authors concluded that the releasing hormone directly inhibited the secretion of GnRH from the hypothalamic neurosecretory cells. Furthermore, recent evidence suggests that the action of CRH on GnRH secretion may be of physiological importance since the central administration of the CRH antagonist (alpha-helical ovine CRH residue $_{9-14}$) prevents the depression of LH secretion which normally occurs during electroshock stress (Rivier, Rivier, and Vale 1986).

It remains to be determined whether the same CRH neurosecretory cells that inhibit GnRH are also the cells that release CRH into the hypothalamic-hypophyseal portal system to control the pituitary cortico-trophs. There are examples of hypothalamic factors (such as dopamine, GnRH and TRH) that, in addition to regulating the pituitary, also appear to be putative neurotransmitters in systems whose function is unrelated to the control of the pituitary (Elde and Hökfelt 1978). In any event these findings suggest the potential of CRH regulating the secretion of LH by a direct effect on the hypothalamic release of GnRH.

IV   Influence of the adrenal axis on the testis

The major influence of the adrenal axis in regulating the gonads is by modulating the secretion of gonadotrophins. However, the corticosteroids also have a direct action on the testis that is independent of gonadotrophin regulation. The primary effect of the corticosteroids is to depress testosterone secretion, although there are instances in some species where the corticosteroid's initial response is to enhance testosterone secretion. Even in these cases, persistant elevation of corticosteroids will suppress the release of testosterone.

In human patients, both stress and adrenal related diseases have been found to reduce testosterone secretion even when serum concentrations of gonadotrophins appear to be unaffected. For example, patients with elevated plasma concentrations of corticosteroids as a result of Cushing's disease or medullary thyroid carcinoma have depressed levels of testosterone. When the hypercortisol condition is reversed by treatment, testosterone secretion returns to normal (Hajjar, Hill, and Samaan 1975; Luton, Thiebolt, Valcke, Mahoudeau, and Bricaire 1977; McKenna, Lorber, Lacroix, and Rabin 1979). In some of these patients testosterone secretion was suppressed even though they had normal levels of gonadotrophins (McKenna *et al.* 1979). Likewise, following surgery patients may have suppressed levels of testosterone even when gonadotrophin levels are normal (Wang, Chan, and Yeung 1978). The explanation for these findings is that corticosteroids released in response to disease or stress act directly at the testis to suppress testosterone release.

Administration of exogenous corticosteroids or ACTH to normal males also appears to have a direct effect on the testis. In several studies ACTH or dexamethasone suppressed plasma levels of testosterone without altering the plasma concentrations of LH or FSH. In addition, dexamethasone suppressed the testicular response to human chorionic gonadotrophin (hCG) stimulation, further supporting the argument for adrenal steroids directly blocking the normal testicular response to gonadotrophins (Beitins, Bayard, Kowarski, and Migeon 1973; Doerr and Pirke 1976; Schaison, Durand, and Mowszowicz 1978).

Animal experiments have provided even more support for a direct adrenal steroid effect on the testis. Unlike female rats (Tache *et al.* 1976), chronic immobilization stress will induce in male rats a drastic fall in plasma testosterone without necessarily having any detectable effect on plasma LH (Charpenet, Tache, Forest, Haour, Saez, Bernier, Duchartie, and Collu 1981). In bulls, an inverse relationship has been found between the plasma concentrations of cortisol and the amount of testosterone secreted in

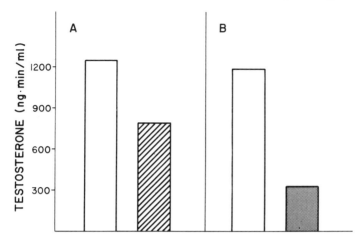

Fig. 12.10. Effect of ACTH (A) or cortisol (B) treatment on GnRH-induced testosterone secretion by male monkeys. Panel A compares the amount of testosterone secreted for 6 hours after the i.v. administration of GnRH (2 μg/kg BW) during control (open bar) and ACTH (striped bar) treatments. ACTH (40 IU) had been administered i.m. three hours before the injection of GnRH. The amount of GnRH-induced testosterone secreted during ACTH treatment was significantly less (P < 0.01) than during the control phase. Panel B compares the amount of GnRH-induced testosterone secreted during control (open bar) and cortisol treatments (stippled bar). Cortisol (100 mg twice daily) was administered for 5 days before the GnRH challenge. The amount of GnRH-induced testosterone secreted during cortisol treatment was significantly less (P < 0.01) than during the control phase.

response to exogenous LH (Welsh, McGraw, and Johnson 1979). Recently, Sapolsky (1985) observed in baboons that although dexamethasone treatment did not alter the ability of the pituitary to release LH in response to GnRH, dexamethasone did suppress the LH-induced testicular secretion of testosterone.

We have been able to confirm Sapolsky's findings in rhesus monkeys with nonsynthetic adrenal hormones (Hayashi and Moberg in press). The administration of ACTH 3 hours before an injection of GnRH resulted in a marked increase in circulating corticosteroids but had no effect on the amount of LH secreted. However, the ACTH treatment did suppress the amount of LH-induced testosterone secreted (Fig. 12.10-A). If the exposure to elevated corticosteroids was prolonged by injecting cortisol for 5 days before GnRH, an even greater suppression of LH-induced testosterone secretion resulted (Fig. 12.10-B). Combined, the data from these studies are consistent with the hypothesis that corticosteroids act directly on the testis to alter the Leydig cells response to LH.

At the Leydig cell, the corticosteroids appear to act in two ways: by altering the function of the membrane LH receptors and by modifying the intracellular synthesis of testosterone. The concomitant administration of either natural corticosteroids or dexamethasone will decrease in a dose dependent manner the hCG-induced testosterone production of primary cultures of testicular cells (Welsh, Bambino, and Hsueh 1982). Dexamethasone, but not corticosterone, will decreased the hCG binding capacity of the cells and both dexamethasone and corticosterone will decrease the number of LH/hCG receptors that FSH can induce in immature hypophysectomized rats (Bambino and Hseuh 1981). The effects of corticosteroids on binding are consistent with the decreased hCG response observed in rats during stress (Charpenet, Tache, Bernier, Ducharme, and Collu 1982).

When primary cultures of testicular cells were concomitantly treated with dexamethasone and hCG, there was a 53 per cent reduction of hCG-stimulated cAMP production (Welsh et al. 1982). In these same cell cultures dexamethasone also reduced the ability of cholera toxin and dibutyryl cAMP to stimulate testosterone production. This is consistent with earlier work which indicated that testicular cells from chronically stressed rats were less responsive to hCG, dibutyryl cAMP and cholera toxin (Charpenet et al. 1981).

Welsh et al. (1982) also found evidence in their cell culture studies that dexamethasone acts on post-cAMP pathways to block 17 α-hydroxylase activity without altering 17–20-lyase enzyme activity. This resulted in a reduction in progesterone production as well as a decrease in hCG-stimulated production of androstenedione and 17 α-hydroxy progesterone.

From the rat data, it can be concluded that elevated corticosteroids can suppress testosterone production as a result of the steroids acting at the membrane to alter LH binding to membrane receptors and by disrupting normal pre- and post-cAMP pathways involved in testosterone synthesis.

In swine different effects have been observed in vivo since corticosteroids appear to have a biphasic effect on testosterone production. Administration of ACTH to boars results in a rapid increase in testosterone production that is independent of any changes in LH secretion (Liptrap and Raeside 1975; Pitzel, Fenske, Holtz, and Konig 1979; Hahmeir, Fenske, Pitzel, Holtz, and Konig 1980; Juniewicz and Johnson 1981). With prolonged ACTH treatment, the production of testosterone is suppressed as in other species (Liptrap and Raeside 1975; Pitzel et al. 1979; Knight, Kattesh, Gwazdauskas, Thomas, and Kornegay 1982). The stimulatory effect of ACTH appears to be mediated by the adrenal cortex since adrenalectomy abolishes the stimulatory effect of ACTH (Pitzel et al. 1979). This increased secretion of testosterone has been attributed to the corticosteroids being converted to testosterone since exogenous cortisol reportedly stimulates testosterone secretion in boars (Liptrap and Raeside 1975). However, Juniewicz and Johnson (1984) recently questioned this interpretation since they could find no increased production of testosterone in boars following the administration of either cortisol or progesterone. Their results reopen the question as to what adrenal factor is responsible for the stimulation of testosterone production. Of equal interest is whether this testicular response to ACTH seen in swine is in some way related to the porcine testes utilizing the $\triangle$–5 pathway for the synthesis of testosterone as opposed to the more classical $\triangle$–4 biosynthetic pathway (Nakajin and Hall 1981; Nakajin, Hall, and Onoda 1981). Nevertheless, while the adrenal axis may have an initial stimulatory effect on testosterone production in swine, prolonged exposure to elevated corticosteroids will still suppress testosterone production, presumably by mechanisms similar to those of other species.

V   Effect of the adrenal axis on the ovary

Adrenal corticosteroids also have the ability to act directly on the ovary to influence the synthesis of ovarian steroids. This has been demonstrated both *in vitro* with cultured granulosa cells and *in vivo* by the administration of exogenous corticosteroids. However, the biological importance of this action is unknown since it remains to be demonstrated whether the amount of adrenal corticosteroids secreted as a biological response to stress or disease is sufficient to influence directly oestrogen synthesis.

Women treated with the synthetic corticosteroid triamcinolone exhibited a variable suppression of plasma oestrogens that was not correlated with comparable changes in gonadotrophin levels (Cunningham *et al.* 1978). This apparent direct effect of the corticosteroids on the ovarian synthesis of oestrogens is consistent with the findings of *in vitro* studies with cultured rat granulosa cells where corticosteroids reduced the synthesis of oestrogen (Hsueh and Erickson 1978).

Corticosteroids apparently bind with receptors in the ovarian cells (Schreiber, Nakamura, and Erickson 1982) to inhibit the FSH induction of steroid synthesizing enzymes and to inhibit the FSH induction of LH receptors in granulosa cells. In cultured rat granulosa cells from hypophysectomized, diethylstilbesterol treated rats, concomitant treatment with dexamethasone inhibited FSH-induced increase in aromatase activity by 50 per cent but did not interfere with pre-existing aromatic activity (Hsueh and Erickson 1978). In addition, dexamethasone decreased the concentration of specific high-affinity LH/hCG-binding sites during FSH treatment without influencing receptor affinity (Schoonmaker and Erickson 1983). Since both of these effects of the corticosteroids were seen only in the presence of very high concentrations of corticosteroids, there is a question as to whether the corticosteroids have a significant impact during normal reproduction.

In contrast to the inhibitory effect the corticosteroids had on oestrogen synthesis *in vitro*, these steroids enhanced FSH-stimulated progesterone synthesis, possibly by increasing FSH stimulated activity of 3-β-hydroxysteroid dehydrogenase and a FSH-independent suppression of the metabolizing 20-α-hydroxysteroid dehydrogenase (Adashi, Jones, and Hsueh 1981). This enhancement of progesterone may explain how dexamethasone in women suppress all steroid secretion except progesterone (Lachilin, Barnett, Hopper, Brink, and Yen 1979) and may also account for the initial increase of progesterone secretion observed in cows following ACTH treatment (da Rosa and Wagner 1981).

From these limited data, the adrenal axis appears to have the potential to act directly on the ovary to modulate ovarian steroid synthesis. However, unlike the testis, there is still reservation about whether physiological amounts of the corticosteroids have any effect in altering normal steroid synthesis and any important influence on reproduction.

VI   Concluding remarks

The adrenal axis has the capacity to modify the regulation of the gonadal axis. This influence apparently exists in all vertebrates and may have played an important role in the evolution of vertebrates. While significant advances have been made in elucidating the role of the adrenal axis in reproduction, several critical questions remain.

The most important influence of the adrenal axis on the regulation of the gonadal axis is by modifying the secretion of gonadotrophins. In mammals, the preovulatory release of LH is especially sensitive to the inhibitory effects of the corticosteroids. Recent evidence indicates that sufficient corticosteroids are released during stress to block the pituitary's response to GnRH (Moberg 1985b). This action of the adrenal steroids provides a plausible mechanism for the effect of stress in blocking ovulation. However the extent of the adrenal responses needed to block ovulation and to disrupt female reproduction has not been determined.

A question regarding the influence that the corticosteroids have on the synthesis and basal secretion of gonadotrophins still remains. The recent studies of Suter and Schwartz (1985a, b) offer data suggesting that the corticosteroids have the potential of influencing both the synthesis and release of gonadotrophins, if the adrenal steroids are elevated for extended periods of time. Few stressors result in such a prolonged elevation in plasma corticosteroids. However, adrenal disease or the clinical use of corticosteroids can expose individuals to such prolonged, high levels of corticosteroids and may have an important influence on the synthesis of gonadotrophins. There is a need to establish what influence corticosteroids have on the synthesis and release of the two gonadotrophins, especially in nonrodent species.

The difference between the effect of ACTH on gonadotrophin secretion and the effect of corticosteroids remains unresolved. The preliminary observation of ACTH having its own direct effect on gonadotrophs offers intriguing possibilities, but awaits additional evidence that this response is of physiological significance. The inability of the corticosteroids to duplicate the effects of ACTH on gonadotrophin secretion also may be attributed to the adrenal progesterone secreted in response to ACTH. However, the importance of adrenal progesterone in regulating gonadotrophin secretion has not been established. Logic dictates that this adrenal steroid should have an important effect if secreted in sufficient quantities, but the necessary proof is lacking.

The influence of adrenal corticosteroids on testosterone secretion offers another potentially important effect on reproduction. Unfortunately, little attempt has been made to correlate the influence of the adrenal axis on testosterone secretion with meaningful changes in male reproduction. The effect of suppressed testosterone secretion on the expression of male behaviour may prove to be the most important effect of the adrenal gland on male reproduction.

Finally, there is a need for expanding our studies on the influence of the adrenal axis on reproduction to include other species, particularly nonmammals. The rat has been the experimental animal of choice in many reproduction studies, but it may be unsuitable for investigating the influence of the adrenal axis on reproduction. The laboratory rat, as we know it, has been selected over the years to reproduce under frequently adverse conditions. Female rats which did not reproduce, for whatever reason, were removed from the breeding colony. The result is that we may have unconsciously selected animals whose reproduction is relatively stress resistant, and thus less responsive to the influence of adrenal steroids. Such a selection might account for the adrenal hormones having a less profound effect in laboratory rats as compared to nonrodent species. Certainly, comparative studies are needed to identify the extent that the adrenal axis influences regulation of the gonadal axis.

References

Adashi, E. Y., Jones, P. B. C., and Hsueh, A. J. W. (1981). Synergistic effect of glucocorticoids on the stimulation of progesterone production by follicle-stimulating hormone in cultured rat granulosa cells. *Endocrinology* **109**, 1888–1894.

Advis, J. P., Krause, J. E., and McKelvy, J. F. (1982). Luteinizing hormone-releasing hormone peptidase activities in discrete hypothalamic regions and anterior pituitary of the rat: apparent regulation during the prepubertal period and first estrous cycle at puberty. *Endocrinology* **110**, 1238–1245.

Asa, C. S. and Ginther, O. J. (1982). Glucocorticoid suppression of oestrus, follicles, LH and ovulation in the mare. *J. Reprod. Fert. Suppl.* **32**, 247–251.

Baldwin, D. M. (1979). The effect of glucocorticoids on estrogen-dependent luteinizing hormone release in the ovariectomized rat and on gonadotrophin secretion in the intact female rat. *Endocrinology* **105**, 120–127.

—— and Sawyer, C. H. (1974). Effects of dexamethasone on LH release and ovulation in the cyclic rat. *Endocrinology* **94**, 1397–1403.

Bambino, T. H. and Hsueh, A. J. W. (1981). Direct inhibitory effect of glucocorticoids upon testicular luteinizing hormone receptor and steroidogenesis *in vivo* and *in vitro*. *Endocrinology* **108**, 2142–2147.

490   Gary P. Moberg

Barb, C. R., Kraeling, R. R., Rampacek, G. B., Fonda, E. S., and Kiser, T. E. (1982). Inhibition of ovulation and LH secretion in the gilt after treatment with ACTH or hydrocortisone. *J. Reprod. Fert.* **64**, 85–92.

Barnes, M. A., Kazmer, G. W., Birrenkott, G. P., and Grimes, L. W. (1983). Induced gonadotrophin release in adrenocorticotropin-treated bulls and steers. *J. Anim. Sci.* **56**, 155–161.

Beitins, I. Z., Bayard, F., Kowarski, A., and Migeon, C. J. (1973). The effect of ACTH administration on plasma testosterone, dihydrotestosterone and serum LH concentrations in normal men. *Steroids* **21**, 553–563.

Boccuzzi, G., Angeli, A., Bisbocci, D., Fonzo, D., Gaidano, G. P., and Ceresa, F. (1975). Effect of synthetic luteinizing hormone releasing hormone (LH-RH) on the release of gonadotrophins in Cushing's disease. *J. Endocr.* **40**, 892–895.

Bronson, F. H. (1973). Establishment of social rank among grouped male mice: relative effects on circulating FSH, LH and corticosterone. *Physiol. Behav.* **10**, 947–951.

Chantaraprateep, P. and Thibier, M. (1978). Effects of dexamethasone on the responses of luteinizing hormone and testosterone to two injections of luteinizing hormone releasing hormone in young prepubertal bulls. *J. Endocr.* **77**, 389–395.

—— —— (1979). Effects of dexamethasone and testosterone propionate on LH response to gonadoliberin (LRH) in young post-pubertal bulls. *Andrologia* **11**, 25–32.

Charpenet, G., Tache, Y., Bernier, M., Ducharme, J. R., and Collu, R. (1982). Stress-induced testicular hyposensitivty to gonadotropin in rats. Role of the pituitary gland. *Biol. Reprod.* **27**, 616–623.

—— —— Forest, M. G., Haour, F., Saez, J. M., Bernier, M., Duchartie, J. R., and Collu, R. (1981). Effects of chronic immobilization stress on rat testicular androgenic function. *Endocrinology* **109**, 1254–1258.

Charters, A. C., O'Dell, W. D., and Thompson, J. C. (1969). Anterior pituitary function during surgical stress and convalescence radioimmunoassay measurement of blood TSH, LH, FSH, and growth hormone. *J. Endocr.* **29**, 63–71.

Christian, J. J., Lloyd, J. A., and Davis, D. E. (1965). The role of endocrines in self-regulation of mammalian populations. *Recent Prog. Horm. Res.* **21**, 501–578.

Cohen, I. R. and Mann, D. R. (1981). Influence on corticosterone on the response to gonadotropin-releasing hormone in rats. *Neuroendocrinology* **32**, 1–6.

Cunningham, G. R., Caperton, E. M., Jr., and Goldzieher, J. W. (1975). Antiovulatory activity of synthetic corticoids. *J. clin. Endocr. Metab.* **40**, 265–267.

—— Goldzieher, J. W., De la Pena, A., and Oliver, M. (1978). The mechanism of ovulation inhibition by triamcinolone acetonide. *J. clin. Endocr. Metab.* **46**, 8–14.

Da Rosa, G. O. and Wagner, W. C. (1981). Adrenal-gonad interactions in cattle. Corpus luteum function in intact and adrenalectomized heifers. *J. Anim. Sci.* **52**, 1098–1105.

De Cantazaro, D. and Gorzalka, B. B. (1980). Effects of dexamethasone, corticosterone, and ACTH on lordosis in ovariectomized and adrenalectomized-ovariectomized rats. *Pharmac. Biochem. Behav.* **12**, 201–206.

De Silva, M., Kaltenbach, C. C., and Dunn, T. G. (1983). Serum cortisol and progesterone after administration of adrenocorticotropin and (or) prolactin to sheep. *J. Anim. Sci.* **57**, 1525–1529.

Deviche, P. (1983). Interactions between adrenal function and reproduction in male birds. In *Avain endocrinology: environmental and ecological perspectives* (eds. S. Mikami *et al.*) pp. 243–254. Springer-Verlag, Berlin.

Doerr, P. and Pirke, K. M. (1976). Cortisol-induced suppression of plasma testosterone in normal adult males. *J. clin. Endocr. Metab.* **43**, 622–629.

Dubey, A. K. and Plant, T. M. (1985). A suppression of gonadotropin secretion by cortisol in castrated male rhesus monkey (*Macaca mulatta*) mediated by the interruption of hypothalamic gonadotropin-releasing hormone release. *Biol. Reprod.* **33**, 423–431.

Du Ruisseau, P., Tache, Y., Brazeau, P., and Collu, R. (1979). Effects of chronic immobilization stress on pituitary hormone secretion, on hypothalamic factor levels, and on pituitary responsiveness to LHRH and TRH in female rats. *Neuroendocrinology* **29**, 90–99.

Echternkamp, S. E. (1984). Relationship between LH and cortisol in acutely stressed beef cows. *Theriogenology* **22**, 305–311.

Elde, R. and Hökfelt, T. (1978). Distribution of hypothalamic hormones and other peptides in the brain. In *Frontiers in neuroendocrinology, Volume 5* (ed. by W. F. Ganong and L. Martini). pp. 1–33. Raven Press, New York.

Erickson, G. F. (1983). Primary cultures of ovarian cells in serum-free medium as models of hormone-dependent differentiation. *Molec. cell Endocr.* **29**, 21–49.

Evain, D., Morera, A. M., and Saez, J. M. (1976). Glucocorticoid receptors in interstitial cells of the rat testis. *J. steroid Biochem.* **7**, 1135–1139.

Everitt, B. J. and Herbert, J. (1971). The effects of dexamethasone and androgens on sexual receptivity of female rhesus monkeys. *J. Endocr.* **51**, 575–588.

Fink, G., Chiappa, S., and Aiyer, M. (1976). Priming effect of luteinizing hormone releasing factor elicited by preoptic stimulation and by intravenous infusion and multiple injections of the synthetic decapeptide. *J. Endocr.* **69**, 359–372.

Fonda, E. S., Rampacek, G. B., and Kraeling, R. R. (1984). The effect of adrenocorticotropin or hydrocortisone on serum luteinizing hormone concentration after adrenalectomy and/or ovariectomy in the prepuberal gilt. *Endocrinology* **114**, 268–273.

Ford, J. J. and Christenson, R. K. (1981). Glucocorticoid inhibition of estrus in ovariectomized pigs: relationship to progesterone action. *Horm. Behav.* **15**, 427–435.

Fujihara, N. and Shiino, M. (1980). The participation of corticosterone in luteinizing hormone releasing hormone (LH-RH) action on luteinizing hormone (LH) releasing from anterior pituitary cells *in vitro*. *Life Sci.* **26**, 777–781.

Fuquay, J. W. and Moberg, G. P. (1983). Influence of the pituitary-adrenal axis on the induced release of luteinizing hormone in rams. *J. Endocr.* **99**, 151–155.

Gray, G. D., Smith, E. R., Damassa, D. A., Ehrenkranz, J. R. L., and Davidson, J. M. (1978). Neuroendocrine mechanisms mediating the suppression of circulating testosterone levels associated with chronic stress in male rats. *Neuroendocrinology* **25**, 247–256.

Hagino, N., Watanabe, M., and Goldzieher, J. W. (1969). Inhibition by adrenocorticotrophin of gonadotrophin-induced ovulation in immature female rats. *Endocrinology* **84**, 308–314.

Hahmeier, W., Fenske, M., Pitzel, L., Holtz, W., and Konig, A. (1980). Corticotropin- and lysine-vasopressin induced changes of plasma corticosteroids and testosterone in the adult male pig. *Acta endocr. Copenh.* **95**, 518–522.

Hajjar, R. A., Hill, L. S., Jr. and Samaan, N. A. (1975). Adrenal mediation of the effect of excess ACTH on testosterone levels in the male: a study of a patient with probable ACTH secreting medullary thyroid carcinoma. *Acta endocr. Copenh.* **80**, 339–343.

Harris Rice, M. E., Mason, J. I., and Bartke, A. (1982). Suppression of pituitary and testicular function in rats by a transplanted adrenocortical carcinoma. *Int. J. Androl.* **5**, 613–618.

Hayshi, K. T. and Moberg, G. P. (1987). Influence of acute stress and the adrenal axis on regulation of LH and testosterone in the male rhesus monkey (*Macaca mulatta*). *Am. J. Primat.* in press.

492   Gary P. Moberg

Hsueh, A. J., Adashi, E. Y., Jones, P. B., and Welsh, T. H., Jr. (1984). Hormonal regulation of the differentiation of cultured ovarian granulosa cells. *Endocr. Rev.* **5**, 76–127.

—— and Erickson, G. F. (1978). Glucocorticoid inhibition of FSH-induced estrogen production in cultured rat granulosa cells. *Steroids* **32**, 639–648.

Johnson, B. H., Welsh, T. H., Jr., and Juniewicz, P. E. (1982). Suppression of luteinizing hormone and testosterone secretion in bulls following adrenocorticotropin hormone treatment. *Biol. Reprod.* **26**, 305–310.

Juniewicz, P. E. and Johnson, B. H. (1981). Influence of adrenal steroids upon testosterone secretion by the boar testis. *Biol. Reprod.* **25**, 725–733.

—— —— (1983). Phenotypic variation in testosterone and luteinizing hormone production among boars: differential response to gonadotropin releasing hormone and adrenocorticotropic hormone. *Biol. Reprod.* **29**, 464–471.

—— —— (1984). Ability of cortisol and progesterone to mediate the stimulatory effect of adrenocorticotropic hormone upon testosterone production by the porcine testis. *Biol. Reprod.* **30**, 134–142.

Khorram, O., De Platis, L. R., and McCann, S. M. (1984). The effect and possible mode of action of melanocyte-stimulating hormone on gonadotropin release in the ovariectomized rat: an *in vivo* and *in vitro* analysis. *Endocrinology* **114**, 227–233.

Kime, D. E., Vinson, G. P., Major, P. W., and Kilpatrick, R. (1980). Adrenal-gonad relationships. In *General, comparative and clinical endocrinology of the adrenal cortex* (ed. I. C. Jones and I. W. Henderson) pp. 183–264. Academic Press, London.

Knight, J. W., Kattesh, H. G., Gwazdauskas, F. C., Thomas, H. R., and Kornegay, E. T. (1982). Peripheral testosterone in boars after administration of hCG, ACTH and testosterone at three ages. *Theriogenology* **17**, 383–392.

Krulich, L., Hefco, E., Illner, P., and Read, C. B. (1974). The effects of acute stress on the secretion of LH, FSH, prolactin and GH in the normal male rat, with comments on their statistical evaluation. *Neuroendocrinology* **16**, 293–311.

Lachelin, G. C. L., Barnett, M., Hopper, B. R., Brink, G., and Yen, S. S. C. (1979). Adrenal function in normal women and women with the polycystic ovary syndrome. *J. clin. Endocr. Metab.* **49**, 892–898.

Li, P. S. and Wagner, W. C. (1983*a*). Effect of hyperadrenal states on luteinizing hormone in cattle. *Biol. Reprod.* **29**, 11–24.

—— —— (1983*b*). *In vivo* and *in vitro* studies on the effect of adrenocorticotropic hormone or cortisol on the pituitary response to gonadotropin releasing hormone. *Biol. Reprod.* **29**, 25–37.

Liptrap, R. M. and Raeside, J. I. (1975). Increase in plasma testosterone concentration after injection of adrenocorticotrophin into the boar. *J. Endocr.* **66**, 123–131.

—— —— (1983). Effect of cortisol on the response to gonadotrophin releasing hormone in the boar. *J. Endocr.* **97**, 75–81.

Luton, J., Thieblot, P., Valcke, J., Mahoudeau, J. A., and Bricaire, H. (1977). Reversible gonadotropin deficiency in male Cushing's disease. *J. clin. Endocr. Metab.* **45**, 488–495.

Mann, D. R., Korowitz, C. D., and Barraclough, C. A. (1975). Adrenal gland involvement in synchronizing the preovulatory release of LH in rats. *Proc. Soc. exp. Biol. Med.* **150**, 115–120.

Matteri, R. L. and Moberg, G. P. (1982). Effect of cortisol or adrenocorticotrophin on release of luteinizing hormone induced by luteinizing hormone releasing hormone in the dairy heifer. *J. Endocr.* **92**, 141–146.

—— —— (1985). The effect of opioid peptides on ovine pituitary gonadotropin secretion *in vitro*. *Peptides* **6**, 957–963.

—— —— and Watson, J. G. (1986). Adrenocorticotropin-induced changes in ovine pituitary gonadotropin secretion *in vitro*. *Endocrinology* **118**, 2091–2096.

—— Watson, J. G., and Moberg, G. P. (1984). Stress or acute adrenocortico-trophin treatment suppresses LHRH-induced LH release in the ram. *J. Reprod. Fert.* **72**, 385–393.

McEwen, B. S. (1973). Glucocorticoid binding sites in rat brain: subcellular and anatomical localizations. *Prog. Brain Res.* **39**, 87–97.

McKenna, T. J., Lorber, D., Lacroix, A., and Rabin, D. (1979). Testicular activity in Cushing's disease. *Acta endocr. Copenh.* **91**, 501–510.

Miyake, A. and Yen, S. S. C. (1981). Direct *in vitro* stimulation of pituitary LH release by alpha melanocyte stimulating hormone. *Life Sci.* **29**, 2637–2640.

Moberg, Gary P. (1976). Effect of environment and management stress on reproduction in dairy cows. *J. Dairy Sci.* **59**, 1618–1624.

—— (1984). Adrenal-pituitary interactions: effects on reproduction. *Proc. X Int. Cong. Anim. Reprod. Art. Insemin.* **4**, 29–36.

—— (1985a). Biological response to stress: key to assessing animal well-being. In *Animal stress*. (ed. G. P. Moberg) pp. 27–49. American Physiological Society, Bethesda.

—— (1985b). Influence of stress on reproduction: measure of well-being. In *Animal stress* (ed. G. P. Moberg) **14**, 245–267.

—— Watson, J. G., and Hayashi, K. T. (1982). Effects of adrenocorticotropin treatment on estrogen, luteinizing hormone and progesterone secretion in the female rhesus monkey. *J. med. Primat.* **11**, 235–241.

—— —— Stoebel, D. P., and Cook, R. (1981). Effect of cortisol and dexametha-sone on the oestrogen-induced release of luteinizing hormone in the anoestrous ewe. *J. Endocr.* **90**, 221–225.

Moore, F. L. and Zoeller, R. T. (1985). Stress-induced inhibition of reproduction: evidence of suppressed secretion of LH-RH in an amphibian. *Gen. comp. Endocr.* **60**, 252–258.

Munck, A., Guyre, P. M., and Holbrook, N. J. (1984). Physiological functions of glucocorticoids in stress and their relationship to pharmacological agents. *Endocr. Rev.* **5**, 25–44.

Naess, O., Cusan, L., Brekke, I., Purvis, K., Torjesen, P., and Hansson, V. (1981). Effects of castration, sex steroids, LHRH and glucocorticoids on LHRH binding in the anterior pituitary of male rats. *Int. J. Androl.* **4**, 685–690.

Nakajin, S., Hall, P. F., and Onoda, M. (1981). Testicular microsomal cytochrome P-450 for C_{21} steroid side chain cleavage. *J. biol. Chem.* **256**, 6134–6139.

—— —— (1981). Side-chain cleavage of C_{21} steroids to C_{19} steroids by testicular microsomal cytochrome P-450:17 -hydroxy C_{21} steroids as obligatory inter-mediates. *J. steroid Biochem.* **14**, 1249–1252.

Ogle, T. F. (1977). Modification of serum luteinizing hormone and prolactin concentrations by corticotropin and adrenalectomy in ovariectomized rats. *Endocrinology* **101**, 494–497.

Ono, N., Lumpkin, M. D., Samson, W. K., McDonald, J. K., and McCann, S. M. (1984). Intrahypothalamic action of corticotropin-releasing factor (CRF) to inhibit growth hormone and LH release in the rat. *Life Sci.* **35**, 1117–1123.

Padmanabhan, V., Keech, C., and Convey, E. M. (1983). Cortisol inhibits and adrenocorticotropin has no effect on luteinizing hormone-releasing hormone-induced release of luteinizing hormone from bovine pituitary cells *in vitro*. *Endocrinology* **112**, 1782–1787.

494   Gary P. Moberg

—— Kesner, J. S., and Convey, E. M. (1981). A priming effect of luteinizing hormone releasing hormone on bovine pituitary cells *in vitro. J. Anim. Sci.* **52**, 1137–1142.

Pasley, J. N. and Christian, J. J. (1972). The effect of ACTH, group caging, and adrenalectomy in *Peromyscus leucopus* with emphasis on suppression of reproductive function. *Proc. Soc. exp. Biol. Med.* **139**, 921–925.

Peyser, M. R., Ayalon, D., Harell, A., Toaff, R., and Cordova, T. (1973). Stress induced delay of ovulation. *Obstet. Gynecol.* **42**, 667–671.

Pitzel, L., Fenske, M., Holtz, W., and Konig, A. (1979). Corticotrophin-induced testicular testosterone release in the Göttingen miniature pig. *Acta endocr. Copenh. Suppl.* **225**, 94.

Plant, T. M. and Zorub, D. S. (1984). A study on the role of the adrenal glands in the imitation of the hiatus in gonadotropin secretion during prepubertal development in the male rhesus monkey (*Macaca mulatta*). *Endocrinology* **114**, 560–565.

Ramaley, J. A. (1974). Adrenal-gonadal interactions at puberty. *Life Sci.* **14**, 1623–1633.

Reid, I. R., Ibbertson, H. K., France, J. T., and Pybus, J. (1985). Plasma testosterone concentrations in asthmatic men treated with glucocorticoids. *Br. med. J.* **291**, 574.

Reid, R. L., Ling, N., and Yen, S. S. C. (1984). Gonadotropin-releasing activity of alpha-melanocyte-stimulating hormone in normal subjects and in subjects with hypothalamic-pituitary dysfunction. *J. clin. Endocr. Metab.* **58**, 773–777.

Resko, J. A. (1969). Endocrine central, adrenal progesterone secretions in the ovariectomized rat. *Science, N.Y.* **164**, 70–71.

Ringstrom, S. J. and Schwartz, N. B. (1984). Examination of prolactin and pituitary-adrenal axis components as intervening variables in the adrenalectomy-induced inhibition of gonadotropin response to castration. *Endocrinology* **114**, 880–887.

Rivier, C., Rivier, J., and Vale, W. (1986). Stress-induced inhibition of reproductive functions: role of endogenous corticotropin-releasing factor. *Science, N.Y.* **231**, 607–609.

—— and Vale, W. (1984). Influence of corticotropin-releasing factor on reproductive functions in the rat. *Endocrinology* **114**, 914–921.

Rogers, C. H., Schwartz, N. B., and Nequin, L. G. (1974). Interaction between the ovarian and adrenocortical regulating systems: occurrence of ovulation. In *Biological rhythms in neuroendocrine activity* (ed. M. Kawakami). pp. 241–252. Jgaku Shoin Ltd., Tokyo.

Saez, J. M., Morera, A. M., Haour, F., and Evain, O. (1977). Effects of *in vivo* administration of dexamethasone, corticotropin and human chorionic gonado-tropin on steroidgenesis and protein and DNA synthesis of testicular interstitial cells in prepuberal rats. *Endocrinology* **101**, 1256–1263.

Sapolsky, R. M. (1985). Stress-induced suppression of testicular function in the wild baboon: role of glucocorticoids. *Endocrinology* **116**, 2273–2278.

Sakaakura, M., Takebe, K., and Nakagawa, S. (1975). Inhibition of luteinizing hormone secretion induced by synthetic LRH by long-term treatment with glucocorticoids in human subjects. *J. clin. Endocr. Metab.* **40**, 774–779.

—— Yoshioka, M., and Takebe, K. (1978). The effect of prednisolone and metyrapone on FSH release induced by the administration of LRH. *Endocrinology, Japan* **25**, 335–339.

Schaison, G., Durand, F., and Mowszowicz, I. (1978). Effect of glucocorticoids on plasma testosterone in men. *Acta endocr. Copenh.* **89**, 126–131.

Schreiber, J. R., Nakamura, K., and Erickson, G. F. (1982). Rat ovary glucocorticoid receptor: identification and characterization. *Steroids* **39**, 569–584.

Selye, H. (1939). The effect of adaption to various damaging agents on the female sex organs in the rat. *Endocrinology* **25**, 651–624.

Shoomaker, J. N. and Erickson, G. F. (1983). Glucocorticoid modulation of follicle-stimulating hormone-mediated granulosa cell differentiation. *Endocrinology* **113**, 1356–1363.

Sowers, J. R., Rice, B. F., and Blanchard, S. (1979). Effect of dexamethasone on luteinizing hormone and follicle stimulating hormone responses to LHRH and to clomiphene in the follicular phase of women with normal menstrual cycles. *Horm. metab. Res.* **11**, 478–480.

Speight, A. and Fink, G. (1981). Comparison of steroid and LH-RH effects on the responsiveness of hemipituitary glands and dispersed pituitary cells. *Molec. cell. Endocr.* **24**, 267–281.

Stith, R. D., Person, R. J., and Dana, R. C. (1976). Uptake and binding of H-hydrocortisone by various pig brain regions. *Brain Res.* **117**, 115–124.

Stoebel, D. P. and Moberg, G. P. (1982). Effect of adrenocorticotropin and cortisol on luteinizing hormone surge and estrous behaviour of cows. *J. Dairy Sci.* **65**, 1016–1024.

Sundararaj, B., Goswami, S., and Lamba, V. (1982). Role of testosterone, estradiol-17 beta, and cortisol during vitellogenin synthesis in the catfish, *Heteropneustes fossilis* (Bloch). *Gen. comp. Endocr.* **48**, 390–397.

Suter, D. E. and Schwartz, N. B. (1985*b*). Effects of glucocorticoids on responsiveness of luteinizing hormone and follicle-stimulating hormone to gonadotropin-releasing hormone by male rat pituitary cells in vitro. *Endocrinology* **117**, 855–859.

—— —— (1985*a*). Effects of glucocorticoids on secretion of luteinizing hormone and follicle-stimulating hormone by female rat pituitary cells *in vitro*. *Endocrinology* **117**, 849–854.

Swift, A. D. and Crighton, D. B. (1979). The effects of certain steroid hormones on the activity of ovine hypothalamic luteinizing hormone-releasing hormone (LH-RH) degrading enzymes. *Febs Letters* **100**, 110–112.

Tache, Y., Du Ruisseau, P., Taché, J., Selye, H., and Collu, R. (1976). Shift in adenohypophyseal activity during chronic intermittent immobilization of rats. *Neuroendocrinology* **22**, 325–336.

Thibier, M. and Rolland, O. (1976). The effect of dexamethasone (DXM) on circulating testosterone (T) and luteinizing hormone (LH) in young post-pubertal bulls. *Theriogenology* **5**, 53–60.

Tibolt, R., and Childs, G. (1985). Cytochemical and cytophysiological studies of gonadotropin-releasing hormone (GnRH) target cells in the male rat pituitary: differential effects of androgens and corticosterone on GnRH binding and gonadotrophin release. *Endocrinology* **117**, 396–404.

Van Den Hurk, R., Gielen, J. Th., and Terlou, M. (1984). Accumulation of glycoprotein gonadotropin in the pituitary of juvenile rainbow trout in response to androgens and C_{21}-steroids, including 11-steroids. *Cell Tissue Res.* **235**, 635–642.

Vreeburg, J. T. M., De Greef, W. J., Doms, M. P., Van Wouw, P., and Weber, R. F. A. (1984). Effects of adrenocorticotropin and corticosterone on the negative feedback action of testosterone in the adult male rat. *Endocrinology* **115**, 997–983.

Wang, C., Chan, V., and Yeung, R. T. T. (1978). Effect of surgical stress on pituitary-testicular function. *Clin. Endocr.* **9**, 255–266.

Watson, E. D. and Munro, C. D. (1984). Adrenal progesterone production in the cow. *Br. vet. J.* **140**, 300–306.

Weber, R. F. A., de Greef, W. J., de Koning, J., and Vreeburg, J. T. M. (1983). LH-RH and dopamine levels in hypophysical stalk plasma and their relationship to plasma gonadotrophins and prolactin levels in male rats bearing a prolactin- and

496   Gary P. Moberg

adrenocorticotrophin-secreting pituitary tumor. *Neuroendocrinology* **36**, 205–210.

—— Ooms, M. P., and Vreeburg, J. T. M. (1982). Effects of a prolactin- and adrenocorticotropin-secreting tumor on gonadotropin levels and accessory sex organ weights in adult male rats: a possible role in the adrenals. *Endocrinology* **111**, 412–417.

Wehrenberg, W. B., Baird, A., Ying, S-Y., Rivier, C., Ling, N., and Guillemin, R. (1984). Multiple stimulation of the adenohypophysis by combinations of hypothalamic releasing factors. *Endocrinology* **114**, 1995–2001.

Welsh, T. H., Jr., Bambino, T. H., and Hsueh, A. J. W. (1982). Mechanism of glucocorticoid-induced suppression of testicular androgen biosynthesis *in vitro*. *Biol. Reprod.* **27**, 1138–1146.

—— and Johnson, B. H. (1981). Stress-induced alterations in secretion of corticosteroids, progesterone, luteinizing hormone, and testosterone in bulls. *Endocrinology* **109**, 185–190.

—— McGraw, R. L., and Johnson, B. H. (1979). Influence of corticosteroids on testosterone production in the bull. *Biol. Reprod.* **21**, 755–763.

White, M. C., Sanderson, J., Mashiter, K., and Joplin, G. F. (1981). Gonadotrophin levels in women with Cushing's Syndrome before and after treatment. *Clin. Endocr.* **14**, 23–29.

Wurtman, R. J. (1967). Effects of light and visual stimuli on endocrine function. In *Neuroendocrinology II*, (ed. by L. Martini and W. F. Ganong). pp. 19–59. Academic Press, London.

Index